AUDITORY BIOCHEMISTRY

AUDITORY BIOCHEMISTRY

Edited By

DENNIS G. DRESCHER

*Laboratory of Bio-otology, Department of Otolaryngology
and Department of Biochemistry
Wayne State University School of Medicine
Detroit, Michigan, U.S.A.*

CHARLES C THOMAS • PUBLISHER
Springfield • Illinois • U.S.A.

Published and Distributed Throughout the World by

CHARLES C THOMAS • PUBLISHER
2600 South First Street
Springfield, Illinois 62717

This book is protected by copyright. No part of it may be reproduced in any manner without written permission from the publisher.

© 1985 by CHARLES C THOMAS • PUBLISHER

ISBN 0-398-05122-4
Library of Congress Catalog Card Number: 85-2588

With THOMAS BOOKS *careful attention is given to all details of manufacturing and design. It is the Publisher's desire to present books that are satisfactory as to their physical qualities and artistic possibilities and appropriate for their particular use.* THOMAS BOOKS *will be true to those laws of quality that assure a good name and good will.*

Printed in the United States of America
SC-R-3

Library of Congress Cataloging in Publication Data
Main entry under title:

Auditory biochemistry.

  Based in part on material presented at a symposium
held Feb. 5-7, 1984 in St. Petersburg Beach, Fla.
during the Midwinter Research Meeting of the Association
for Research in Otolaryngology.
  Includes index.
  1. Auditory pathways—Congresses.   2. Hearing—
Congresses.   3. Biological chemistry—Congresses.
I. Drescher, Dennis G.   II. Association for Research
in Otolaryngology (U.S.). Midwinter Research Meeting
(1984 : Saint Petersburg Beach, Fla.) [DNLM: 1. Auditory
Pathways—congresses.   2. Biochemistry—congresses.
3. Neuroregulators—congresses. WV 272 A9121 1984]
QP460.A93   1985   612'.85   85-2588
ISBN 0-398-05122-4

CONTRIBUTORS

Maxwell Abramson, Department of Otolaryngology, College of Physicians and Surgeons, Columbia University, New York, New York 10032, U.S.A.

Richard Altschuler, Laboratory of Neuro-otolaryngology, National Institute of Neurological and Communicative Disorders and Stroke, National Institutes of Health, Bethesda, Maryland 20205, U.S.A.

Sanford C. Bledsoe, Jr., Kresge Hearing Research Institute, University of Michigan, Ann Arbor, Michigan 48109, U.S.A.

*Richard P. Bobbin, Kresge Hearing Research Institute of the South, Department of Otorhinolaryngology and Biocommunication, Louisiana State University Medical Center, New Orleans, Louisiana 70119, U.S.A.

Stuart Braverman, Division of Otolaryngology, Department of Surgery, University of California-San Diego Medical School, La Jolla, California 92093, U.S.A.

Barbara Canlon, Kresge Hearing Research Institute, University of Michigan, Ann Arbor, Michigan 48109, U.S.A.

*Donald M. Caspary, Department of Pharmacology, Southern Illinois University School of Medicine, Springfield, Illinois 62708, U.S.A.

Andrew Catanzaro, Division of Otolaryngology, Department of Surgery, University of California-San Diego Medical School, La Jolla, California 92093, U.S.A.

*W. Ewart Davies, Department of Pharmacology, University of Birmingham Medical School, Birmingham B15 2TJ, England

David J. DeRosier, Graduate Program in Biophysics, and Department of Biology, Rosenstiel Basic Medical Sciences Research Center, Brandeis University, Waltham, Massachusetts 02254, U.S.A.

*Detlev Drenckhahn, Department of Anatomy and Cell Biology, Phillips-Universität, D-3550 Marburg, Federal Republic of Germany

*Dennis G. Drescher, Laboratory of Bio-otology, Department of Otolaryngology, and Department of Biochemistry, Wayne State University School of Medicine, Detroit, Michigan 48201, U.S.A.

Marian J. Drescher, Laboratory of Bio-otology, Department of Otolaryngology, Wayne State University School of Medicine, Detroit, Michigan 48201, U.S.A.

Jon D. Dunn, Department of Anatomy, Oral Roberts University, Tulsa, Oklahoma 74171, U.S.A.

Klaus Ehrenberger, I. Universität Klinik für HNO-Krankheiten, A-1090, Vienna, Austria

Carl L. Faingold, Department of Pharmacology, Southern Illinois University School of Medicine, Springfield, Illinois 62708, U.S.A.

Glenn R. Farley, Research Division, Boy's Town National Institute for Communication Disorders in Children, Omaha, Nebraska 68131, U.S.A.

*Dominik Felix, Division of Animal Physiology, University of Berne, CH-3012 Berne, Switzerland

*Jörgen Fex, Laboratory of Neuro-otolaryngology, National Institute of Neurological and Communicative Disorders and Stroke, National Institutes of Health, Bethesda, Maryland 20205, U.S.A.

*Åke Flock, Department of Physiology II, Karolinska Institutet, S-104 01 Stockholm, Sweden

*Taro Furukawa, Department of Physiology, Tokyo Medical and Dental University, 1-5-45 Yushima, Bunkyo-ku, Tokyo 113, Japan

*Donald A. Godfrey, Department of Physiology, Oral Roberts University, Tulsa, Oklahoma 74171, U.S.A.

*Paul S. Guth, Department of Pharmacology, Tulane University School of Medicine, New Orleans, Louisiana 70112, U.S.A.

*Cheng-Chun Huang, Department of Otolaryngology, College of Physicians and Surgeons, Columbia University, New York, New York 10032, U.S.A.

Eric Javel, Research Division, Boy's Town National Institute for Communication Disorders in Children, Omaha, Nebraska 68131, U.S.A.

Gary L. Jenison, Kresge Hearing Research Laboratory of the South, Department of Otorhinolaryngology and Biocommunication, Louisiana State University Medical Center, New Orleans, Louisiana 70119, U.S.A.

Doyle R. Jones, Department of Anatomy, University of Connecticut Health Center, Farmington, Connecticut 06032, U.S.A.

*Steven K. Juhn, Department of Otolaryngology, University of Minnesota Medical School, Minneapolis, Minnesota 55455, U.S.A.

Timothy T. K. Jung, Department of Otolaryngology, University of Minnesota Medical School, Minneapolis, Minnesota 55455, U.S.A.

Thomas P. Kerr, Laboratory of Bio-otology, Department of Otolaryngology, Wayne State University School of Medicine, Detroit, Michigan 48201, U.S.A.

Daniel C. Marcus, Department of Otolaryngology, Washington University School of Medicine, St. Louis, Missouri 63110, U.S.A.

*Michael R. Martin, Laboratory of Neuro-otolaryngology, National Institute of Neurological and Communicative Disorders and Stroke, National Institutes of Health, Bethesda, Maryland 20205, U.S.A.

Shiushi Matsuura, Department of Physiology, Osaka City University Medical School, Abeno-ku, Osaka 545, Japan

*Graciela Meza, Departamento de Neurosciencias, Centro de Investigaciones en Fisiologia Celular, Universidad Nacional Automoma de Mexico, 04510 Mexico, D. F., Mexico

D. Kent Morest, Department of Anatomy, University of Connecticut Health Center, Farmington, Connecticut 06032, U.S.A.

Hiroshi Moriyama, Department of Otolaryngology, College of Physicians and Surgeons, Columbia University, New York, New York 10032, U.S.A.

*Barbara J. Morley, Research Division, Boy's Town National Institute for Communication Disorders in Children, Omaha, Nebraska 68131, U.S.A.

Charles H. Norris, Department of Otolaryngology, Tulane University School of Medicine, New Orleans, Louisiana 70112, U.S.A.

Douglas R. Oliver, Department of Anatomy, University of Connecticut Health Center, Farmington, Connecticut 06032, U.S.A.

Christopher Owen, Department of Pharmacology, University of Birmingham Medical School, Birmingham B15 2TJ, England

Jami L. Park, Department of Physiology, Oral Roberts University, Tulsa, Oklahoma 74171, U.S.A.

Fulvio Perini, Department of Pharmacology, University of Michigan Medical School, Ann Arbor, Michigan 48109, U.S.A.

*Steven J. Potashner, Department of Anatomy, University of Connecticut Health Center, Farmington, Connecticut 06032, U.S.A.

Kenneth G. Pote, Department of Anatomy, University of Michigan Medical School, Ann Arbor, Michigan 48104, U.S.A.

Michael Prinz, Department of Anatomy and Cell Biology, Phillips-Universität, D-3550 Marburg, Federal Republic of Germany

C. David Ross, Department of Physiology, Oral Roberts University, Tulsa, Oklahoma 74171, U.S.A.

*Muriel D. Ross, Department of Anatomy, University of Michigan Medical School, Ann Arbor, Michigan 48104, U.S.A.

*Allen F. Ryan, Division of Otolaryngology, Department of Surgery, University of California-San Diego Medical School, La Jolla, California 92093, U.S.A.

Leonard P. Rybak, Division of Otolaryngology, Department of Surgery, and Department of Pharmacology, Southern Illinois University School of Medicine, Springfield, Illinois 62708, U.S.A.

*Jochen Schacht, Kresge Hearing Research Institute, University of Michigan, Ann Arbor, Michigan 48109, U.S.A.

Thomas Schäfer, Department of Anatomy and Cell Biology, Phillips-Universität, D-3550 Marburg, Federal Republic of Germany

*Ilsa R. Schwartz, Department of Surgery, School of Medicine, University of California, Los Angeles, California 90024, U.S.A.

William F. Sewell, Department of Otolaryngology, Harvard Medical School, Boston, Massachusetts 02115, U.S.A.

Frank R. Sharp, Division of Otolaryngology, Department of Surgery, University of California-San Diego Medical School, La Jolla, California 92093, U.S.A.

*Karen P. Steel, Medical Research Council Institute of Hearing Research, University of Nottingham, Nottingham NG7 2RD, England

*Olivier Sterkers, Department of Physiology, Faculté Xavier Bichat, Université Paris, and Institut de la Santé et de la Recherche Médicale, 75018 Paris, France

*Ruediger Thalmann, Department of Otolaryngology, Washington University School of Medicine, St. Louis, Missouri 63110, U.S.A.

*Lewis G. Tilney, Department of Biology, University of Pennsylvania, Philadelphia, Pennsylvania 19104, U.S.A.

*Robert J. Wenthold, Laboratory of Neuro-otolaryngology, National Institute of Neurological and Communicative Disorders and Stroke, National Institutes of Health, Bethesda, Maryland 20205, U.S.A.

Stephen L. Winbery, Kresge Hearing Research Laboratory of the South, Department of Otorhinolaryngology and Biocommunication, Louisiana State University Medical Center, New Orleans, Louisiana 70119, U.S.A.

Nigel K. Woolf, Division of Otolaryngology, Department of Surgery, University of California-San Diego Medical School, La Jolla, California 92093, U.S.A

*Asterisks designate senior authors.

PREFACE

During the last fifteen years, a new discipline has emerged. Represented initially by a small group of scientific innovators (see *Biochemical Mechanisms in Hearing and Deafness*, M. M. Paparella, editor, Charles C Thomas, Publisher, 1970), this discipline has since expanded to claim as its subject matter the molecular organization and function of the auditory, vestibular, and lateral-line systems (acousticolateralis systems), both peripheral and central. For present purposes, we will refer to this research area as "auditory biochemistry," because, of the acousticolateralis component systems, a majority of biochemical investigation has been performed on the auditory system. Auditory biochemistry draws heavily from the fields of neurochemistry, neuropharmacology, immunology, and physiological chemistry, fields which themselves have expanded during the last decade.

This book is intended to be "textlike" in character and to consolidate present research. The need for such a book arises from the fact that the leading investigators are scattered throughout the world, and their research is published in a variety of journals and other sources not easily accessible simultaneously. The original impetus for the book was a symposium held February 5–7, 1984, as part of the meeting of the Association for Research in Otolaryngology in St. Petersburg Beach, Florida. However, the pages which follow are not intended primarily to be a "proceedings" of the meeting. Authors were asked to submit chapters based both on current research and related work performed during the last ten years. In addition, some material in the text was not presented at the symposium. Chapters were written after the conclusion of the conference, allowing authors to reflect and relate their research to that of others. It is hoped that *Auditory Biochemistry* will serve as a reference work to neuroscientist-investigators, their graduate students, academic and practicing otolaryngologists, and medical students.

There are many people to whom I am greatly indebted for helping to plan this publication, in particular, Drs. Ernest Moore, Jörgen Fex, Ruediger Thalmann, David Lim, Robert Wenthold, Jochen Schacht, Donald Godfrey, Ilsa Schwartz, Åke Flock, Murray Goldstein, Ralph Naunton, Robert Mathog, and Maxwell Abramson. Drs. Michael Whipple, Marian Drescher, and Thomas Kerr helped to review the manuscripts. Dr. Mary Smith provided excellent typing and Ms. Maureen Doyle painstakingly proofed the chapters. Many scientists who attended the meeting provided critical comments during the symposium, which subsequently aided the editor in his task. The National Institute of Neurological and Communicative Disorders and Stroke and the Fogarty International Center of the National Institutes of Health, U.S.A., made possible both the symposium and book through a conference grant (R13 NS 20033). I wish to thank the publisher, Charles C Thomas, Publisher, for help and counsel during the preparation of this work.

Dennis G. Drescher

INTRODUCTION

This book is divided into four sections: Biochemistry of Receptoneural Transmission, Biochemistry of Central Auditory Neurotransmission, Biochemistry of Auditory Structural Elements, and Biochemistry of Labyrinthine Support Systems. The first section deals largely with postulated transmitters of hair cells and peripheral efferents. The second section concerns neurotransmitters and neurotransmitter receptors in the auditory brainstem and the central auditory pathways. The third section encompasses a relatively new area of study: structural proteins and other macromolecules present in the hair cell and surrounding entities. The last section deals with "support" systems; "support" is used here to identify processes and structures that aid and maintain hair cells in their transfer of mechanoreceptive information.

In each chapter, a review of the author's own work and related literature are presented, in addition to his current findings, in order to provide the reader with background material. Microtechniques that are particularly important to the study of auditory biochemistry are highlighted. Controversies are not always reconciled in the book, but individual viewpoints are adequately expressed, for example, the candidacy of GABA and glutamate for hair cell transmitters (first section) and the candidacy of glutamate, aspartate, or peptides for transmitters of the auditory nerve (second section).

In most cases, chapters conclude with suggestions for future research. As in other areas of medically-relevant investigation, an understanding of inner-ear processes at the molecular level promises to provide a rational approach for the eventual treatment of a multitude of disorders of hearing and balance. Perhaps some of the readers of this book will be the future investigators who expand auditory and vestibular research in directions suggested on the following pages.

CONTENTS

Preface   ix
Introduction   xi

PART I
BIOCHEMISTRY OF RECEPTONEURAL TRANSMISSION

Chapter

1. IMMUNOCYTOCHEMISTRY OF THE MAMMALIAN COCHLEA: RESULTS AND EXPECTATIONS   5
   Jörgen Fex and Richard A. Altschuler
2. MODE OF OPERATION OF AFFERENT SYNAPSES BETWEEN HAIR CELLS AND AUDITORY FIBERS   31
   Taro Furukawa and Shiushi Matsuura
3. PRIMARY AFFERENT TRANSMISSION IN ACOUSTICOLATERALIS ORGANS   42
   Paul S. Guth, Charles H. Norris, and William F. Sewell
4. HPLC ANALYSIS OF PRESUMPTIVE NEUROTRANSMITTERS IN PERILYMPH   50
   Dennis G. Drescher and Marian J. Drescher
5. THE ACTION OF PUTATIVE NEUROTRANSMITTER SUBSTANCES IN THE MAMMALIAN LABYRINTH   68
   Dominik Felix and Klaus Ehrenberger
6. CHARACTERIZATION OF GABA-ERGIC AND CHOLINERGIC NEUROTRANSMISSION IN THE CHICK INNER EAR   80
   Graciela Meza
7. ACTIONS OF PUTATIVE NEUROTRANSMITTERS AND OTHER RELEVANT COMPOUNDS ON *XENOPUS LAEVIS* LATERAL LINE   102
   Richard P. Bobbin, Sanford C. Bledsoe, Jr., Stephen L. Winbery, and Gary L. Jenison

PART II
BIOCHEMISTRY OF CENTRAL AUDITORY NEUROTRANSMISSION

8. GLUTAMATE AND ASPARTATE AS NEUROTRANSMITTERS OF THE AUDITORY NERVE 125
   Robert J. Wenthold
9. IDENTIFICATION OF GLUTAMATERGIC AND ASPARTATERGIC PATHWAYS IN THE AUDITORY SYSTEM 141
   Steven J. Potashner, D. Kent Morest, Douglas L. Oliver, and Doyle R. Jones
10. CHOLINERGIC NEUROTRANSMISSION IN THE COCHLEAR NUCLEUS 163
    Donald A. Godfrey, Jami L. Park, Jon D. Dunn, and C. David Ross
11. THE PHARMACOLOGY OF AMINO ACID RECEPTORS AND SYNAPTIC TRANSMISSION IN THE COCHLEAR NUCLEUS 184
    Michael R. Martin
12. THE EFFECTS OF INHIBITORY AND EXCITATORY AMINO-ACID NEUROTRANSMITTERS ON THE RESPONSE PROPERTIES OF BRAINSTEM AUDITORY NEURONS 198
    Donald M. Caspary, Leonard P. Rybak, and Carl L. Faingold
13. PUTATIVE NEUROTRANSMITTER RECEPTORS IN THE CENTRAL AUDITORY SYSTEM 227
    Barbara J. Morley, Glenn R. Farley, and Eric Javel
14. THE NATURE OF NEUROTRANSMITTERS IN THE MAMMALIAN LOWER AUDITORY SYSTEM 244
    W. Ewart Davies and Christopher Owen
15. AUTORADIOGRAPHIC STUDIES OF AMINO ACID LABELING OF NEURAL ELEMENTS IN THE AUDITORY BRAINSTEM 258
    Ilsa R. Schwartz

PART III
BIOCHEMISTRY OF AUDITORY STRUCTURAL ELEMENTS

16. THE ORGANIZATION OF ACTIN FILAMENTS IN THE STEREOCILIA OF THE HAIR CELLS OF THE COCHLEA 261
    Lewis G. Tilney and David J. DeRosier

17.	CONTRACTILE AND STRUCTURAL PROTEINS IN THE AUDITORY ORGAN Åke Flock	310
18.	ACTIN, MYOSIN, AND ASSOCIATED PROTEINS IN THE VERTEBRATE AUDITORY AND VESTIBULAR ORGANS: IMMUNOCYTOCHEMICAL AND BIOCHEMICAL STUDIES Detlev Drenckhahn, Thomas Schäfer, and Michael Prinz	317
19.	AUDITORY NERVE PROTEINS Robert J. Wenthold	336
20.	COMPOSITION AND PROPERTIES OF THE MAMMALIAN TECTORIAL MEMBRANE Karen P. Steel	351
21.	COLLAGENASE AND BONE RESORPTION IN CHRONIC OTITIS MEDIA Cheng-Chun Huang, Hiroshi Moriyama, and Maxwell Abramson	366

PART IV
BIOCHEMISTRY OF LABYRINTHINE SUPPORT SYSTEMS

22.	NOISE-INDUCED CHANGES OF COCHLEAR ENERGY METABOLISM Jochen Schacht and Barbara Canlon	389
23.	DEOXYGLUCOSE UPTAKE PATTERNS IN THE AUDITORY SYSTEM: METABOLIC RESPONSE TO SOUND STIMULATION IN THE ADULT AND NEONATE Allen F. Ryan, Nigel K. Woolf, Andrew Catanzaro, Stuart Braverman, and Frank R. Sharp	401
24.	PERSPECTIVES IN THE PHYSIOLOGICAL CHEMISTRY OF THE COCHLEAR DUCT Ruediger Thalmann and Daniel C. Marcus	422
25.	Na^+, K^+-ACTIVATED ADENOSINE TRIPHOSPHATASE AND CARBONIC ANHYDRASE: INNER-EAR ENZYMES OF ION TRANSPORT Dennis G. Drescher and Thomas P. Kerr	436
26.	ORIGIN AND ELECTROCHEMICAL COMPOSITION OF ENDOLYMPH IN THE COCHLEA Olivier Sterkers	473

27. TRANSPORT CHARACTERISTICS
OF THE BLOOD-LABYRINTH BARRIER   488
Steven K. Juhn, Leonard P. Rybak, and Timothy T. K. Jung

28. ANALYTICAL STUDIES OF THE ORGANIC MATERIAL
OF OTOCONIAL COMPLEXES,
INCLUDING ITS AMINO ACID
AND CARBOHYDRATE COMPOSITION   500
Muriel D. Ross, Kenneth G. Pote, and Fulvio Perini

Index   515

AUDITORY BIOCHEMISTRY

PART I
BIOCHEMISTRY OF RECEPTONEURAL TRANSMISSION

Chapter 1

IMMUNOCYTOCHEMISTRY OF THE MAMMALIAN COCHLEA: RESULTS AND EXPECTATIONS

JÖRGEN FEX AND RICHARD A. ALTSCHULER

I. Introduction
II. Methods
III. Antisera
IV. Results
V. Discussion
VI. Expectations and Anticipations
References

I. INTRODUCTION

The principles of immunocytochemistry were outlined in 1942 by Coons et al. (1942), and, in the 1970's, immunocytochemistry emerged as a powerful method for identifying structures and tracing pathways in the nervous system. Successful studies of monoamine cell systems were carried out (for review, see Hökfelt et al., 1975) using immunofluorescence and light microscopy for visualization of antigen-antibody complexes. Later, other techniques were developed that were compatible with electron microscopy. Excellent monographs and reviews of immunocytochemistry are available (e.g., Cuello, 1983; Chan-Palay and Palay, 1982; Sternberger, 1979). Immunocytochemistry now plays a fundamental role in the neuroanatomical and histochemical analysis of the central nervous system, as anticipated by Hökfelt et al. (1975).

Immunocytochemical studies of the mammalian cochlea were first reported in 1980 (Fex, 1980; Flock, 1980; Zenner, 1980). Since then, many studies on cochlear immunocytochemistry have been carried out, concerned with questions about neurotransmitter candidates (Altschuler et al., 1984b, c; Eybalin, 1982; Eybalin et al., 1983, 1984; Eybalin and Pujol, 1984; Fex and Altschuler, 1981, 1984; Fex et al., 1981, 1982a, b, 1984; Hoffman et al., 1984; Rubio et al., 1983) or about structual proteins (Zenner, 1981; Flock et al., 1981a, b, 1982; Drenckhahn et al., 1982). In Chapters 17 and 18 of this volume, Flock and Drenckhahn *et al.*, respectively, describe results related to structural proteins.

This chapter is about the distribution of immunoreactivity of enkephalin, choline acetyltransferase (ChAT), glutamate decarboxylase (GAD), aspartate aminotransferase (AATase), and glutaminase (GLNase) in the organ of Corti. ChAT is the enzyme that catalyzes the synthesis of acetylcholine (ACh). GAD is the terminal enzyme in the biosynthesis of the inhibitory neurotransmitter, γ-aminobutyric acid (GABA). AATase and GLNase are two enzymes involved in the metabolism of the excitatory neurotransmitter candidates aspartate and glutamate. Throughout the chapter we will use expressions such as "ChAT-like immunoreactivity" and "enkephalin-like immunoreactivity" instead of writing "ChAT immunoreactivity" and "enkephalin immunoreactivity". This is because different substances, be they closely related or seemingly quite dissimilar, have been found to have epitopes (i.e., antigenic determinants) in common, which bind the same species of antibody. Thus, finding immunoreactivity does not give conclusive evidence for the presence of the substance that was used to raise the polyclonal antiserum (or the monoclonal antibody).

We will describe below how immunoreactivity has been found in efferent fibers in the organ of Corti through the application of antisera to enkephalin, ChAT, GAD, AATase, and GLNase. It should be noted that the different antisera give different distributions of immunoreactivity. Also, we have unpublished pilot studies in which antisera to a series of substances were applied to the organ of Corti without showing immunoreactivity in the efferents. To the best of our knowledge, all efferent neurons of the organ of Corti come from the brainstem. Originally described as crossed and uncrossed olivocochlear neurons (Rasmussen, 1946, 1953, 1960), these efferents have recently been classified into a medial and a lateral system (Warr, 1975; Warr and Guinan, 1979; Guinan *et al.*, 1983; White and Warr, 1983) predominantly innervating, respectively, the outer hair-cell region and the inner hair-cell region (Fig. 1). Our immunocytochemical findings, particularly on the distribution of GAD-like immunoreactivity (Fex and Altschuler, 1984), indicate that there may be more than two different systems of efferents in the organ of Corti, as previously suggested by Schwartz and Ryan (1983).

II. METHODS

Concerning methods, techniques, and materials, a few pointers will be given that bear on problems that arise when working with the cochlea. For visualization of antigen-antibody complexes, we have used immunofluorescence and immunoperoxidase procedures (Fig. 2). We have much relied on surface preparations that are stained for the reaction product of horseradish peroxidase.

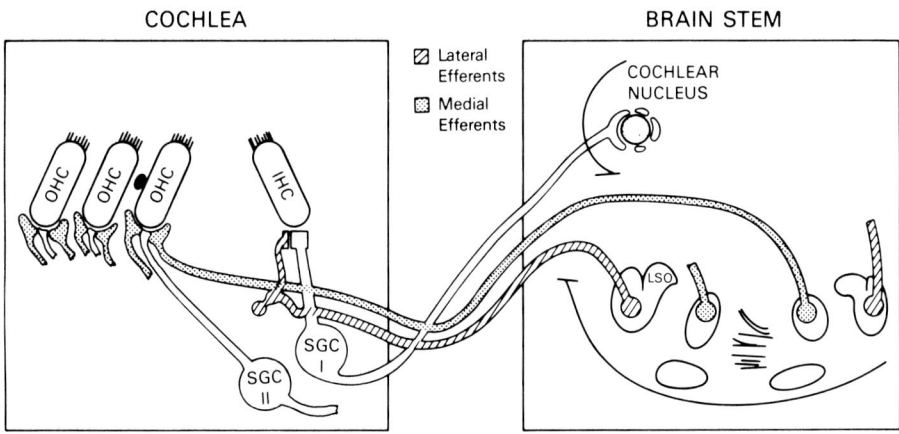

Figure 1-1. Schematic diagram of the efferent innervation of the guinea pig cochlea, based on the nomenclature of Warr, Guinan and co-workers (Warr, 1975; Warr and Guinan, 1975; Guinan et al., 1983; White and Warr, 1983). The *lateral* system of efferents (hatched) originates in the brainstem in the lateral superior olivary nuclei (LSO). Most of its nerve fibers run uncrossed. In the organ of Corti, the fibers run in the inner spiral bundle and the tunnel spiral bundle, terminating primarily at inner hair cells on afferent dendrites. The *medial* system of efferents (stippled) originates in medial olivary nuclei and many of its nerve fibers cross the midline. In the organ of Corti the nerve endings of this system synapse primarily at the bases of outer hair cells. The efferents ending on outer hair cells (shaded) at or above the level of the nucleus may be part of the medial and/or lateral systems.

Immunoperoxidase techniques have given us better signal-to-noise ratios for specific immunoreactivity in surface preparations of the organ of Corti than has immunofluorescence. Recently (Altschuler et al., 1984b), to achieve a maximal resolution of surface preparations in light-microscopy studies, we have used enhanced contrast video display (Inoue, 1981; Kachar, 1984). Cryostat sectioning of the cochlea has also been employed, with the advantage that a greater number of different antisera can be applied to comparable preparations from a single cochlea more easily than if surface preparations are used.

For light microscopy, we have generally used 4% paraformaldehyde in 0.1 M sodium cacodylate buffer for fixation. In most of our experiments, this fixative was applied systemically, the temporal bones were removed, and the cochleae were locally perfused with the same fixative. Recently, we have been given reasons to believe (Gulley, 1984) that results may be improved if

2

INDIRECT IMMUNOFLUORESCENCE TECHNIQUE

PAP TECHNIQUE

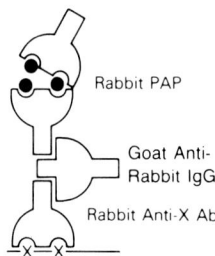

ABC TECHNIQUE

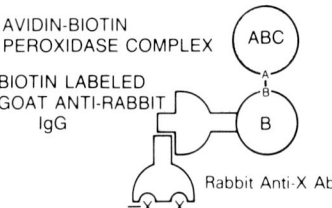

Figure 1-2. Immunocytochemical techniques used in studies described in this chapter. A primary antibody is applied either to cryostat sections of the cochlea, used only in combination with the indirect immunofluorescence technique (cf Coons, 1942), or to the cochlear spiral, remaining after that bony shell, vascular stria, and tectorial and Reissner's membrane have been removed from the cochlea. For indirect immunofluorescence, a second antibody with a fluorescent label, usually fluorescein, is applied. For the unlabeled-antibody-enzyme-PAP technique developed by Sternberger (1979), a second antibody is applied, followed by peroxidase anti-peroxidase complex. For the ABC technique developed by Hsu et al. (1981), a second antibody with a biotin label is applied, followed by an avidin-biotin-peroxidase complex. For both of the latter techniques, the peroxidase label is visualized with diaminobenzidine HCl as the chromogen.

cochleae are perfused during, and not after, the systemic application of the fixative. We have also used immersion of the previously unfixed cochlear spiral in this fixative with good results (Fex and Altschuler, 1984; Hoffman *et* al., 1984), and, although this method works well for the organ of Corti, fixation of such a spiral through immersion cannot be used for studying spiral ganglion cells. For most immuno-electron-microscopic studies, it is necessary to add glutaraldehyde to the fixative. In most cases, we use only small amounts of glutaraldehyde (0.05–0.2%), because this reagent decreases binding of most of our antibodies. Even with these small concentrations of glutaraldehyde, removing the tectorial membrane becomes a problem. It is more difficult to visualize the tectorial membrane after glutaraldehyde than after paraformaldehyde. Also, after glutaraldehyde, the tectorial membrane breaks apart instead of being relatively easily removed in one long, spiraling piece as after paraformaldehyde. Removal of the tectorial membrane is a critically important step. In surface preparations that otherwise correspond to our expectations, we have invariably found lack of immunoreactivity in areas under remaining pieces of the tectorial membrane. Much of Reissner's membrane should also be removed, so that no part of this tissue can cover the organ of Corti during the incubations. It is difficult to obtain surface preparations from the hook region and the beginning of the first turn of the organ of Corti of small mammals (guinea pig, rat, mouse); special dissection techniques must be developed. The rest of the spiraling organ of Corti is relatively easy to handle if extra care is taken so that the most apical part of the organ of Corti is not lost.

We have found when working with surface preparations that it is useful to add the nonionic detergent Triton X-100 to incubation solutions to make membranes penetrable to antibodies. Sometimes, as much as 0.3% of Triton X-100 is necessary to insure that the antibody has access to intracellular antigens (Altschuler et al., 1984c). Also, to suppress staining of intracellular endogenous peroxidase, particularly in the inner hair-cell region, we have been experimenting with treating the organ of Corti with H_2O_2 before incubation with antiserum. We have immersed the cochlear spiral in 3% H_2O_2 and 10% methoanol in phospate-buffered saline (PBS) for one min and then rinsed in PBS (cf Burns, 1975). This occasionally helps to increase the contrast between specific staining and nonspecific background. However, the regular structural organization of the organ of Corti, which serves as a useful, and sometimes even necessary, guide for determining the exact localization of immunoreactivity, is easily disrupted with H_2O_2. Background can also be decreased by pre-incubating with a 10% solution of normal serum from the animal species in which the second antibody is raised. Lately, an overriding priority in our techniques has been to keep cochleas steadily moving, albeit slowly, during the incubations. This came about when we realized that efferent subpopulations exist in the organ of Corti (Fex and Altschuler, 1984) that can be determined through the use of immunocytochemistry, if the techniques give good reproducibility. If cochleae remain motionless during incubations, staining of the organ of Corti will very likely be unevenly distributed, giving a distorted picture of the immunoreactivity; consequently, the staining will be far from optimally reproducible between cochleae. Further, we have not yet refined this technique to permit such incubation for cochleae that have been pre-treated with H_2O_2 and are more easily damaged by the movement.

In conclusion, the following steps are taken in a typical experiment. From a fixed cochlea, the bony shell, vascular stria, tectorial membrane, and Reissner's membrane are removed, leaving a spiral with the organ of Corti exposed. The spiral is rinsed in PBS for 16–20 h at 4°C. It is then preincubated in PBS with 10% normal goat serum for 1 h at 4°C, and placed in a BEEM embedding capsule for incubation in the primary antiserum diluted in PBS, with Triton X-100 at 0.3%. Dilutions range from 1:150 for some antisera raised to met-enkephalin, to 1:2,000 for antiserum to GLNase. This first incubation is for 16–20 h at 4°C with spiral moving in the BEEM capsule. The spiral is then rinsed in PBS, and, for the peroxidase-antiperoxidase (PAP) procedure of Sternberger (1979), placed in a BEEM capsule for the second incubation in goat anti-rabbit IgG (if the first antiserum was raised in rabbit), diluted 1:50 in PBS, with Triton X-100 at 0.3%. Alternatively, a solution of goat anti-rabbit antibodies with biotin label is used for the Vectastain ABC procedure of Hsu et al. (1981). Either incubation is performed for 60 min at room temperature with the spiral moving. The spiral is again rinsed in PBS and placed in a BEEM capsule for the third and final incubation, in the PAP complex diluted 1:60 for the PAP procedure, or in the avidin-biotin-peroxidase complex (ABC) for the ABC procedure. Either complex is diluted in PBS with Triton X-100 at 0.3% and either incubation is for 1 h at room temperature with the spiral moving. The spiral is then rinsed, reacted with diaminobenzidine HCl as the chromogen, which causes immunoreactivity to be visualized as the reaction product of peroxidase, and is rinsed in PBS again. Segments of the cochlear spiral are then dissected and usually mounted in well slides, as described before (Fex and Altschuler, 1981; Fex et al., 1982b), for viewing and photography. Some of the segments are mounted between two #0 coverslips and examined with an oil-immersion, 1.4 NA condenser and 1.3 NA Planapo 100X objective on a bright-field Zeiss Axiomat microscope equipped for video-enhanced (Inoue, 1981), Asymmetric Illumination (Kachar, 1984) contrast.

Besides normal cochleae, we used guinea-pig cochleae in which the olivocochlear fibers had degenerated after surgery. De-efferentation of the organ of Corti was carried out as described previously (Hoffman et al., 1984; Morrison et al., 1975; Rubio et al., 1983). The efferent neurons, which leave the brainstem with the vestibular nerve, close to where the efferents form an anastomosis between the vestibular nerve and the cochlear nerve in the inner auditory meatus. Usually four to six weeks later, immunocytochemistry was carried out on the cochleas, ipsilateral and contralateral to the lesion. It is difficult to de-efferent a cochlea without damaging its blood vessels, causing the organ of Corti to degenerate, and therefore the yield of successful de-efferentations is low.

III. ANTISERA

Enkephalin. Our antisera to methionine enkephalin (met-enkephalin) (TYR--GLY-GLY-PHE-MET) were generated in female New Zealand White rabbits by using met-enkephalin (Boehringer Mannheim) coupled with glutaraldehyde (Polysciences 70%, EM grade) to bovine thyroglobulin (Sigma type I), following procedures described elsewhere (Micevych and Elde, 1980; Skowsky and Fisher, 1972). Antiserum 163 has 1.5% crossreactivity and 164 has 1% cross-reactivity to leucine enkephalin (TYR-GLY-GLY-PHE-LEU). A major difference in cross-reactivity between the two antisera occurs with respect to their crossreactivity to des-tyrosyl met-enkephalin (GLY-GLY-PHE-MET); 163 has 5% cross-reactivity and 164 has 30% cross-reactivity. This indicates that antiserum 163 has a greater affinity for the N-terminal of the enkephalin moiety than does antiserum 164.

ChAT. We used: (1) a polyclonal antiserum to pig brain ChAT, provided by Dr. F. Eckenstein and described previously (Eckenstein and Sofroniev, 1983; Eckenstein and Thoenen, 1982), (2) a rat monoclonal antibody provided by Dr. F. Eckenstein (Eckenstein and Baugham, 1984), (3) a mouse polyclonal antibody provided by Dr. F. Eckenstein (Eckenstein and Sofroniev, 1983; Eckenstein and Thoenen, 1982), and (4) a commercially available rat monoclonal antibody to ChAT (Immuno Nuclear).

GAD. We used a sheep brain anti-GAD antiserum (1440) provided by Dr. Irwin J. Kopin and described previously (Oertel *et al.*, 1981a, b, c).

AATase. We used a polyclonal antiserum to cytoplasmic AATase (Boehringer Mannheim) from pig heart that was raised in rabbits and described previously (Altschuler *et al.*, 1981; Donoghue *et al.*, 1984; Wenthold and Altschuler, 1983).

GLNase. We used antisera that were raised in rabbits against phosphate-dependent glutaminase from rat kidney, provided by Drs. N. P. Curthoys and W. G. Haser and described previously (Altschuler *et al.*, 1984d; Clark and Curthoys, 1979; Curthoys and Weiss, 1974; Curthoys *et al.*, 1976; Donoghue *et al.*, 1984; Wenthold and Altschuler, 1983).

IV. RESULTS

In our first study of cochlear immunocytochemistry (Fex and Altschuler, 1981) we found enkephalin-like immunoreactivity in olivocochlear nerve fibers of the guinea pig and cat. We saw immunofluorescence at inner and outer hair cells and in the spiral bundle and tunnel bundle (Figs. 3 and 4). Later, we reported (Altschuler et al. 1983) that met-enkephalin antiserum 163 immunoreacted with efferents of both the medial and the lateral system, while met-enkephalin antiserum 164 immunoreacted with efferents of the lateral system only. Recently, using peroxidase reaction product for visualization, we examined, in a light and electron microscopic study, enkephalin-like immunoreactivity of the normal guinea pig cochlea and of the de-efferented cochlea (Altschuler et al., 1984c). We found immunoreactivity with antiserum 164 in small efferent endings high up on outer hair cells (Fig. 5a); this type of ending has been described previously (Bredberg, 1977a, b, c; Mangape and Harada, 1982; Nakai and Igarashi, 1974; Stopp and Comis, 1978; Wright and Preston, 1973). Also, we confirmed with electron microscopy that our antiserum 163, but not antiserum 164, shows immunoreactivity in large efferent terminals on outer hair cells (Fig. 5b).

The finding of enkephalin-like immunoreactivity in the efferents made it of interest to strengthen the evidence that cochlear efferents are cholinergic, particularly for the lateral system of efferents. Therefore, we studied the distribution of ChAT-like immunoreactivity in the organ of Corti (Fex et al., 1982a; Altschuler et al., 1984b), since ChAT is the marker of choice for cholinergic fibers (cf Rossier, 1984; Wainer et al., 1984). We found ChAT-like immunoreactivity in both the lateral and medial system of efferents in the organ of Corti (Fig. 6a and b), at all the sites where efferents are thought to be present. In the de-efferented organ of Corti, no ChAT-like immunoreactivity was found.

We complemented our studies by looking for signs of co-containment of ChAT and enkephalin in the brainstem. We found co-containment of ChAT-like immunoreactivity and enkephalin-like immunoreactivity in cells of origin of the lateral system of olivocochlear efferents (Altschuler et al., 1984a). This study was the first to show co-containment of enkephalin-like and ChAT-like immunoreactivities in neurons.

We have also been interested in the presence of GAD (Fex and Wenthold, 1976) and GABA (Gulley et al., 1979) in the organ of Corti. Recently, we applied antiserum to GAD (1440) to the organ of Corti. We found GAD-like immunoreactivity in the inner spiral bundle and tunnel spiral bundle, upper tunnel-crossing fibers, outer hair-cell synaptic regions and outer spiral bundles (Fig. 7). Most of the immunoreactivity was seen in the third and lower fourth turn of the guinea pig cochlea, but even there, many efferent fibers

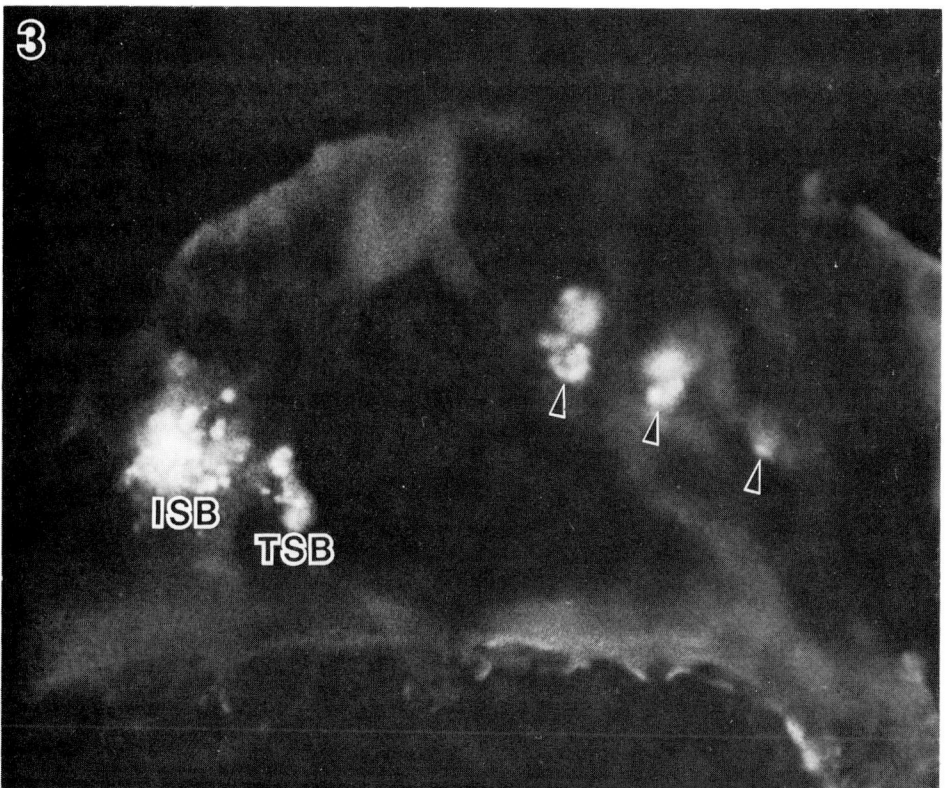

Figure 1-3. Enkephalin-like immunofluorescence is seen in the inner spiral bundle (ISB), tunnel spiral bundle (TSB), and at the bases of outer hair cells (arrowheads), in the second turn of a guinea pig cochlea on a cryostat section immunoreacted with antiserum 163 to methionine enkephalin.

and endings were unstained. Little immunoreactivity was seen in the basal turn.

The amino acids glutamate and aspartate are major candidates for excitatory neurotransmitters at many sites. Such sites include the auditory nerve synapses in the cochlear nucleus, with initial evidence, particularly for aspartate, presented in studies by Godfrey et al. (1977) and Wenthold and Gulley (1977). Much additional evidence on the auditory-nerve neurotransmitter has since accumulated, (for review, see Wenthold and Martin, 1984; also Chapter 8, this volume). We have found both AATase-like and GLNase-like immunoreactivity in axons and spiral ganglion cells of the auditory nerve (Altschuler et al., 1981; Fex et al., 1981). The hypothesis was proposed (Altschuler et al., 1981) that AATase and GLNase may serve as markers for glutamergic and aspartergic neurons. The question was raised as

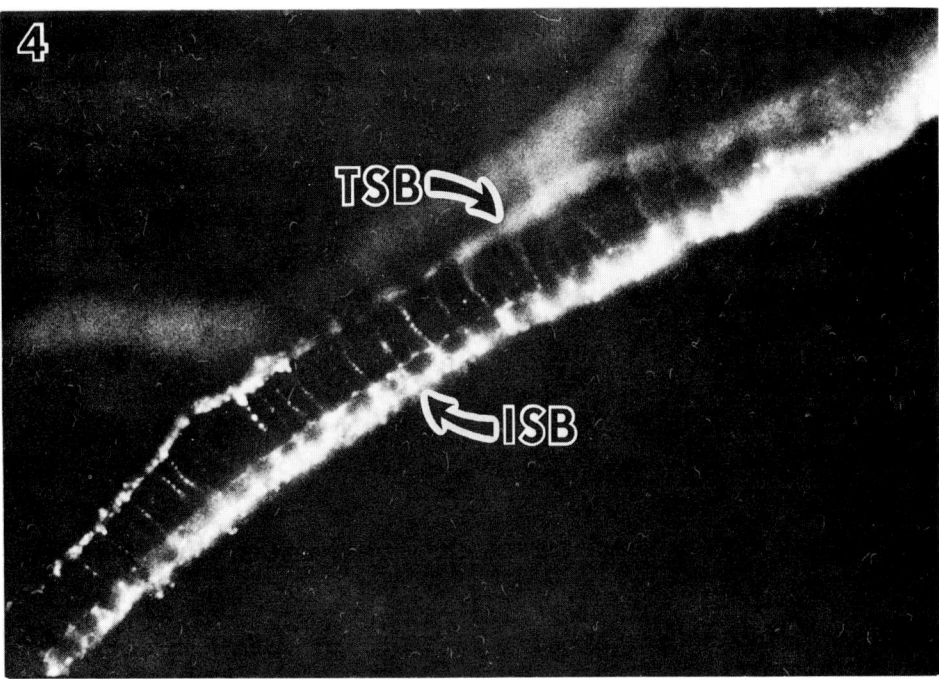

Figure 1-4. Enkephalin-like immunofluorescence is seen in the inner spiral bundle (ISB) and tunnel spiral bundle (TSB) in a surface preparation from the third turn of a cochlear spiral immunoreacted with antiserum 164 to methionine enkephalin.

to whether auditory dendrites at hair cells, and the hair cells themselves, would show immunoreactivity in response to antisera to the enzymes. We used antisera to AATase and studied the distribution of AATse-like immunoreactivity in the organ of Corti of the guinea pig (Fex *et al.*, 1982b). We found immunofluorescence in upper-tunnel crossing fibers and at the base of outer hair cells, distributed similarly to the efferent innervation of the outer hair cells. In some, but not all preparations, weak immunofluorescence was seen in the inner spiral bundle and tunnel bundle. No AATase-like immunoreactivity was seen in hair cells or in auditory dendrites. We later observed the same immunoreactivity distribution using the immunoperoxidase techniques (Fig. 8). We have also applied antisera to GLNase to the cochlea (Fex *et al.*, 1981, 1984) and found GLNase-like immunoreactivity at outer hair-cell bases (Figs. 9 and 10b), distributed similarly to the efferent innervation of outer hair cells. We saw immunoreactivity in upper-tunnel crossing fibers and in outer spiral bundles, particularly for upper-turn spiral bundles, where they contain the most efferents (Wright and Preston, 1973). We found weak and sparse immunoreactivity in the tunnel spiral bundle, similar to what we

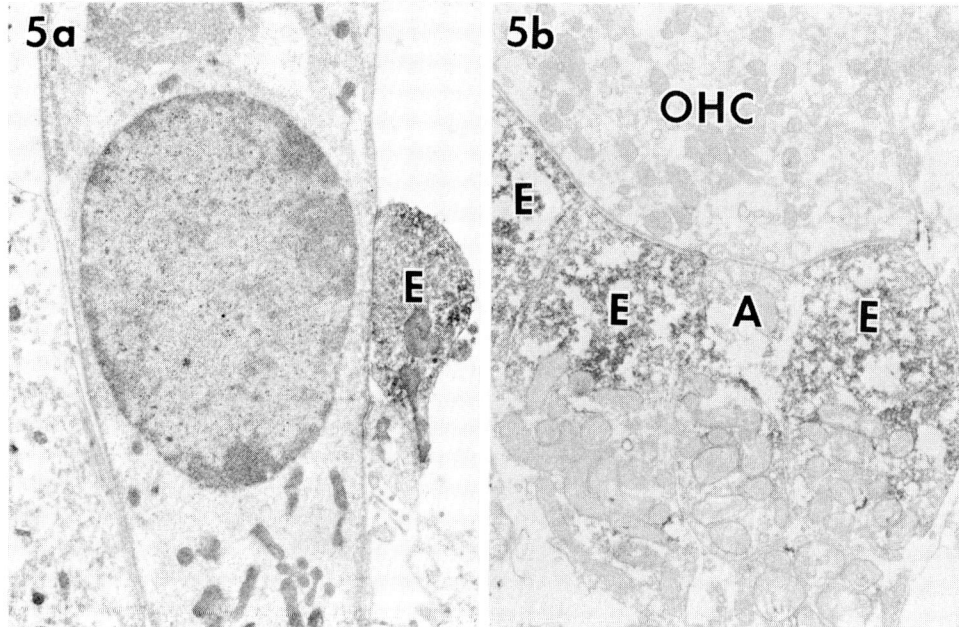

Figure 1-5. Immuno-electron-microscopic visualization of enkephalin-like immunoreactivity using the ABC (5a) and PAP (5b) immunoperoxidase techniques and antiserum 164 to methionine enkephalin (5a) and 163 (5b). 5a. Enkephalin-like immunoreactivity is seen in an efferent nerve ending (E) on an outer hair cell at the level of the nucleus. 5b. Enkephalin-like immunoreactivity is seen in efferent terminals (E) at the base of an outer hair cell (OHC). An auditory nerve dendrite (A) shows no immunoreactivity.

saw with AATase antiserum. There was a marked concentration of small puncta of immunoreactivity at inner hair cells (Figs. 9 and 10a). The concentration of such puncta in the de-efferented cochlea (Fig. 11) seemed only slightly less than in the normal cochlea. Such puncta, in all likelihood, represented afferent dendrites at the inner hair-cell base. Hair cells, particularly outer hair cells, inconsistently showed a fine, weakly staining network of immunoreactivity supranuclearly. This staining was much less intense than the staining of efferent endings on outer hair cells. Therefore, the hair cells were considered to contain GLNase-like material at much lower concentrations than the efferents.

V. DISCUSSION

It was initially surprising to find immunoreactivity in cochlear efferents in response to all the different antisera that we have mentioned here. However, finding this multitude of different immunoreactivities in a system of neurons

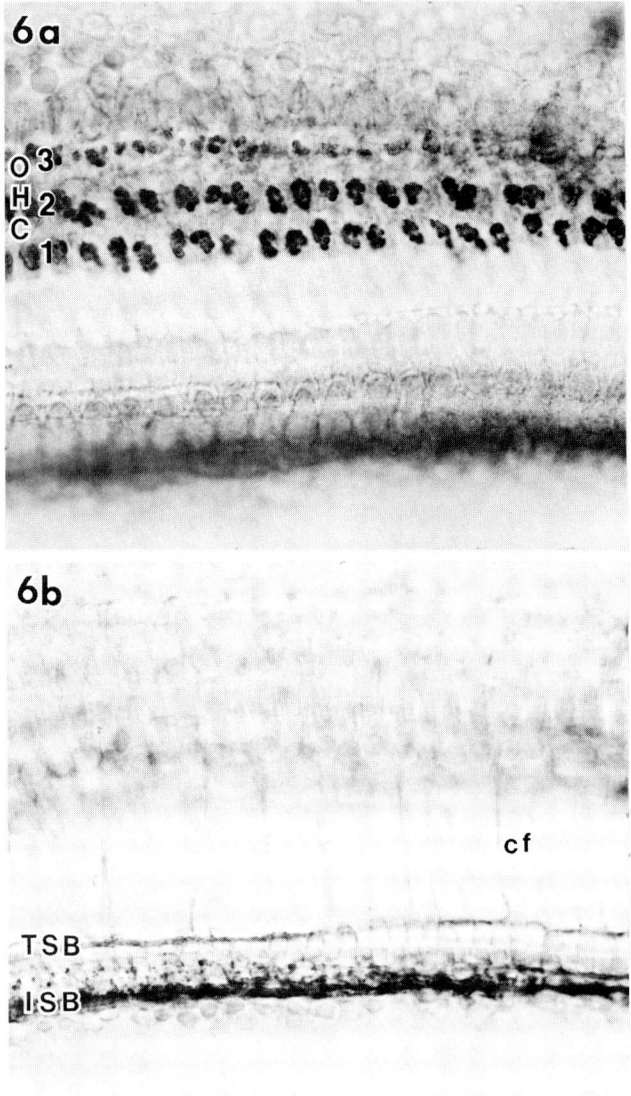

Figure 1-6. An area of a surface preparation from the second turn of a cochlear spiral immunoreacted with polyclonal antiserum to choline acetyltransferase, using the ABC immunoperoxidase technique visualized at two different focal planes. 6a. Choline acetyltransferase-like immunoreactivity is found at the bases of outer hair cells of rows 1, 2 and 3. 6b. Choline acetyltransferase-like immunoreactivity in inner spiral bundle (ISB), tunnel spiral bundle (TSB) and tunnel crossing fibers (cf).

is not unique. In a recent study, Chan-Palay *et al.* (1982) demonstrated the coexistence, in human and primate neuromuscular junctions, of enzymes catalyzing the synthesis of ACh (i.e., ChAT), catecholamines (i.e., tyrosine

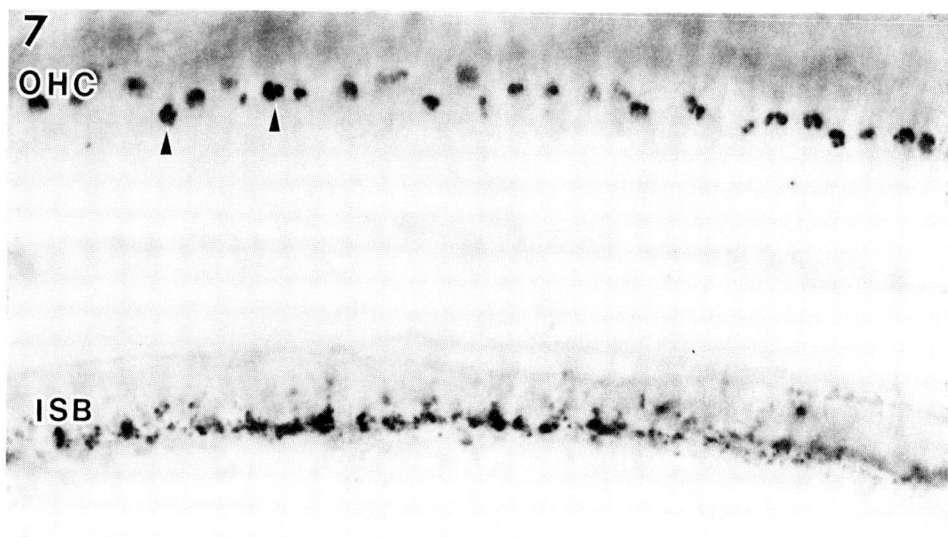

Figure 1-7. Glutamic acid decarboxylase (GAD)-like immunoreactivity in a surface preparation of the lower fourth turn of a cochlear spiral, immunoreacted with antiserum to GAD using the ABC technique. GAD-like immunoreactivity is seen by outer hair-cell bases (arrowheads) and in the inner spiral bundle (ISB).

hydroxylase), taurine (i.e., cysteine sulfinic acid decarboxylase, CSD) and GABA (i.e., GAD). Early evidence that cochlear efferents are cholinergic was provided by studies showing that they contained acetylcholinesterase (Churchill and Schuknecht, 1959; Schuknecht et al., 1959) and that d-tubocurarine in the cochlea blocked effects of efferent stimulation (Fex, 1968). Much evidence was later added (for references, see Altschuler et al., 1984a, b), but this evidence mostly concerned the medial system of efferents. Also, the lateral system of efferents in the organ of Corti very likely contains enkephalin (Fex and Altschuler, 1981), raising the possibility that the lateral efferent system is only partly cholinergic, if at all (cf Altschuler et al., 1983). Our results with a polyclonal antiserum to ChAT, as well as with two different monoclonal antibody species to this enzyme, show that most or all of the efferents of both the lateral and medial systems contain ChAT-like immunoreactivity. These results give strong evidence that both efferent systems are cholinergic. Godfrey et al. (1982, 1984, Chapter 10, this volume) have carried out quantitative histochemical studies of ChAT activity of the olivocochlear fibers at the point where these fibers form a bundle. Their findings are compatible with the assumption that all efferents in the organ of Corti are cholinergic. However, it is possible that not every efferent neuron in the organ of Corti contains ChAT.

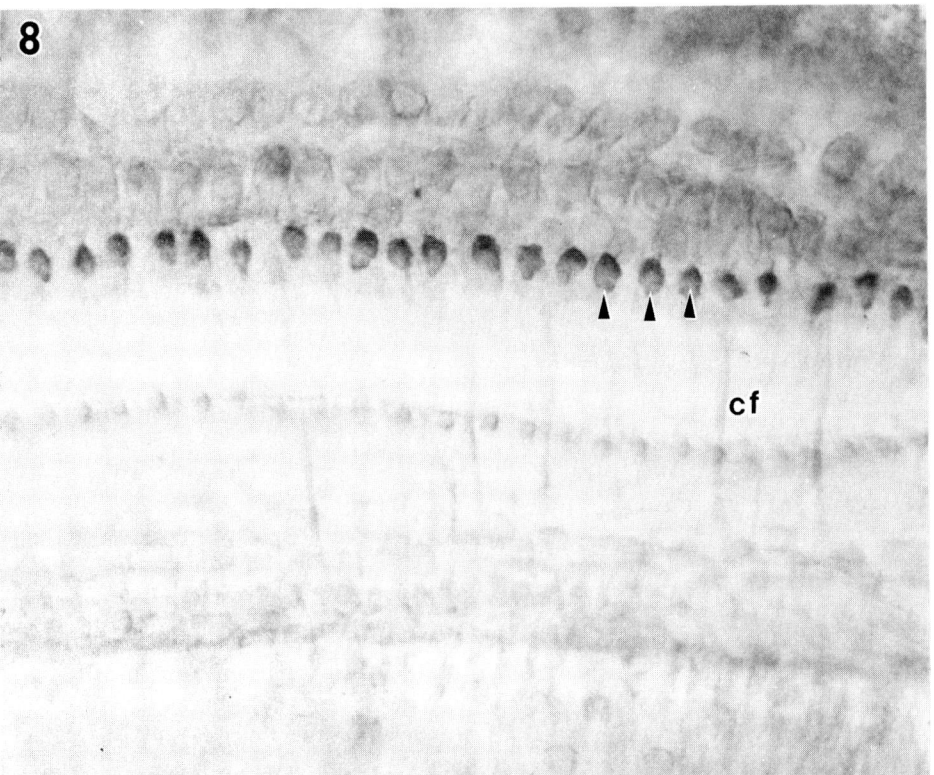

Figure 1-8. Aspartate aminotransferase-like immunoreactivity is seen at the bases of outer hair cells (arrowheads) and in tunnel-crossing fibers (cf) in a surface preparation from the second turn of a cochlear spiral, immunoreacted with antiserum to aspartate aminotransferase using the PAP technique.

Our findings of enkephalin-like immunoreactivity in efferents in the organ of Corti (Fex and Altschuler, 1981; Altschuler *et al.*, 1983, 1984c) have been confirmed by Eybalin, Cupo and Pujol (Eybalin *et al.*, 1983, 1984; Eybalin and Pujol, 1984). The latter workers also showed, by means of electron microscopy studies, that enkephalin-like immunoreactivity is present below inner hair cells in terminals that synapse with afferent dendrites (Eybalin *et al.*, 1983, 1984). Furthermore, with high-performance liquid chromatography (HPLC), combined with radioimmunoassay (RIA), evidence has been presented that the organ of Corti contains multiple enkephalin-related peptides (Eybalin *et al.*, 1984; Hoffman *et al.*, 1983, 1984), including metenkephalin and leucine enkephalin. It also has been shown that an HPLC component with the retention time of met-enkephalin increases in perilymph after sound stimulation (Drescher *et al.*, 1983, Chapter 4, this volume). These different findings, taken together, give strong evidence that

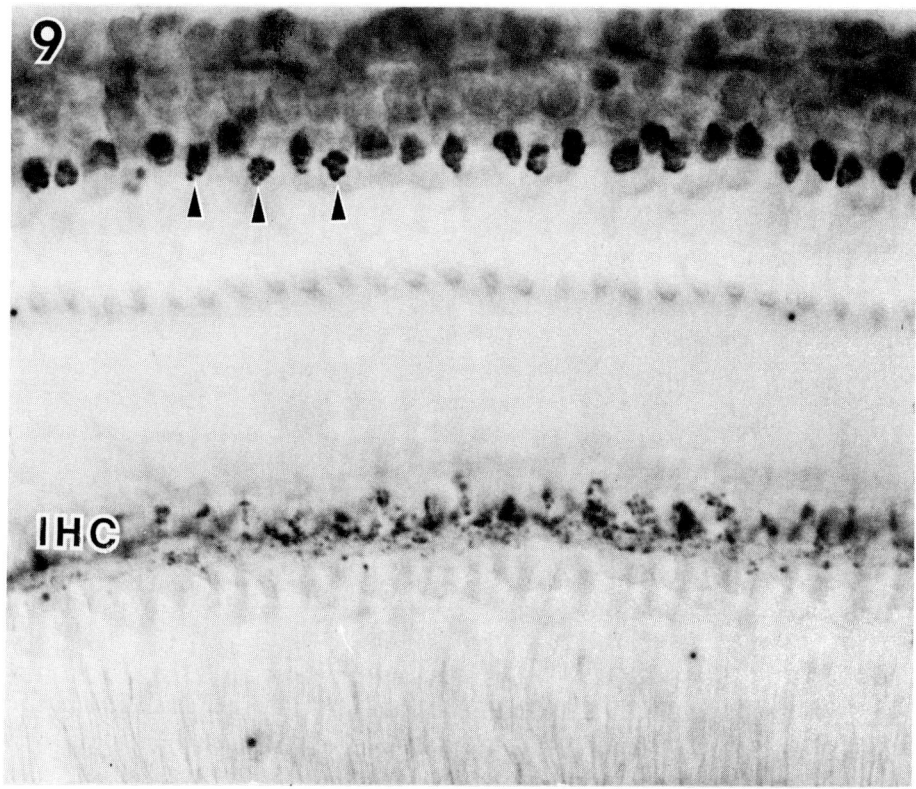

Figure 1-9. Glutaminase-like immunoreactivity is seen as puncta at the bases of outer hair cells (arrowheads) and in the inner hair cell region (IHC) in a surface preparation from the second turn of a cochlear spiral, immunoreacted with antiserum to glutaminase using the ABC technique.

met-enkephalin is present in the lateral system of efferents. Our findings of co-containment of ChAT-like and enkephalin-like immunoreactivity in cells of origin in the lateral system of efferents (Altschuler *et al.*, 1984a) makes it appear probable that there is a corresponding co-containment in the organ of Corti. Thus, there may be cholinergic efferents in the organ of Corti whose actions are modified by the met-enkephalin which they may contain. The medial system of efferents probably contains no met-enkephalin, but rather, some other opioid(s), since the medial system shows no immunoreaction with antiserum 164; however, the small endings high up on outer hair cells may constitute an exception (cf Fig. 5a).

The pattern of immunoreactivity evoked with anti-GAD antiserum is totally different from that seen with other antisera (Fex and Altschuler, 1984). Even where the immunoreactivity was at its most dense, with

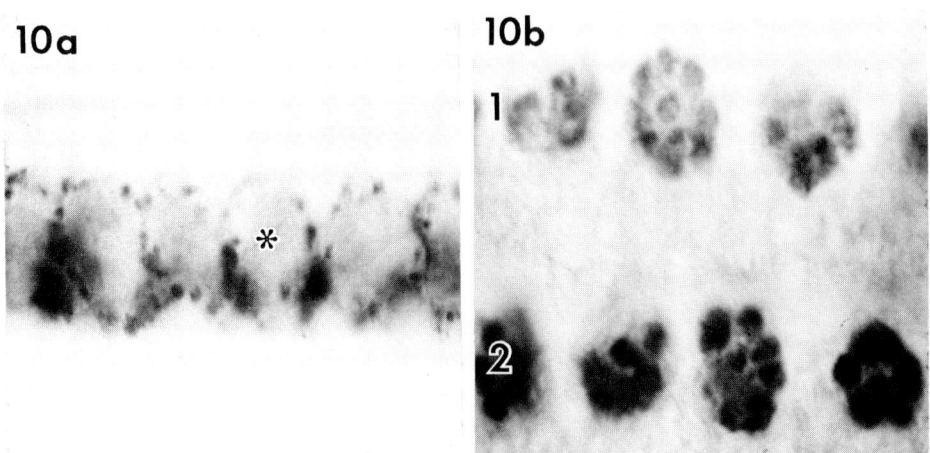

Figure 1-10. Video-enhanced contrast microscopy with asymmetric illumination shows: (10a) glutaminase-like immunoreactivity around the bases of inner hair cells (asterisk), and (10b) by the bases of outer hair cells in rows 1 and 2. From a surface preparation of the second turn of a cochlear spiral, immunoreacted with antiserum to glutaminase using the PAP technique.

immunoreactive components of both the medial and the lateral efferent system, there were many efferent endings and fibers with no GAD-like immunoreactivity. This is strong evidence the two efferent systems each have a subpopulation of neurons that are chemically different from the other efferents. Eventually, it may prove useful to discuss efferents of the organ of Corti in terms of three systems, the third system being different from the others through its GAD-like immunoreactivity and through other characteristics that are still undefined. There is good evidence that GAD-like immunoreactivity is a marker of choice for visualizing GABAergic neurons (Oertel et al., 1982; Wu, 1983), as first shown by Saito et al. (1974). Our results indicate that if there are GABAergic efferent neurons in the organ of Corti, they may be present in small numbers. This may explain why previous evidence is tenuous, but not lacking, for GABA as a neurotransmitter of cochlear efferents. The concentration of GABA in the cochlea is very low (Godfrey et al., 1976; Kuriyama and Kimura, 1976; Tachibana and Kuriyama, 1974; Thalmann et al., 1975), as is enzymatic GAD activity (Fex and Wenthold, 1976). On the other hand, efferents are marked through incubation of the cochlea with ^{3}H–GABA (Richrath et al., 1974; Gulley et al., 1979; Eybalin, 1982; Schwartz and Ryan, 1983). Substances blocking GABA-sensitive receptors partially block effects of efferent stimulation in the guinea-pig cochlea (Klinke and Oertel, 1977). As determined through HPLC, a GABA-like component becomes elevated in perilymph during sound stimulation (Drescher et al., 1983, Chapter 4, this volume). As noted above, GAD-

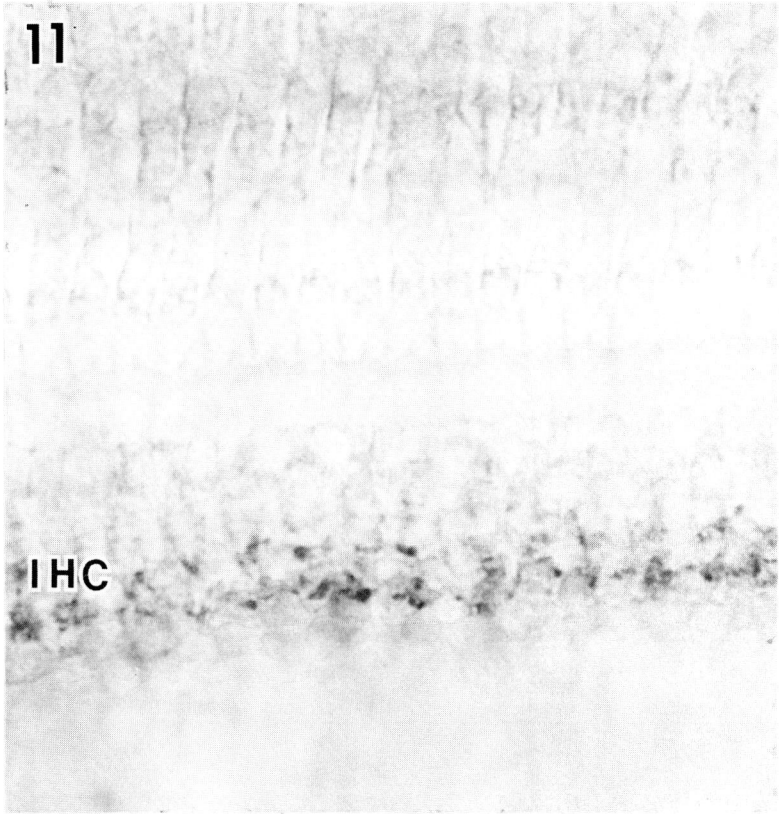

Figure 1-11. Glutaminase-like immunoreactivity is seen in the inner hair-cell region (IHC) of a surface preparation immunoreacted with antiserum to glutaminase using the ABC technique. From the second turn of a cochlear spiral de-efferented four weeks prior to immunocytochemistry.

like immunoreactivity can also be present in cholinergic neurons (Chan-Palay et al., 1982). Whether efferents in the organ of Corti with GAD-like immunoreactivity also contain ChAT has not yet been determined. If such efferents are few and are lacking ChAT, this lack may not, or could not, have been noted in our experiments on ChAT immunoreactivity (Altschuler et al., 1984b) and in the experiments of Godfrey et al., (1982, 1984a, b) on ChAT enzymatic activity. On the other hand, the GAD-like immunoreactivity that we have observed could have been due to the taurine-synthesizing enzyme, CSD, since the antiserum (1440) also reacts with CSD (Oertel et al., 1981b, c). In addition, the GAD-like immunoreactivity could have been due to a macromolecule different from both GAD and CSD but sharing an epitope with both, or due to some other cross reactivity between antiserum 1440 and the reactive efferents. Therefore, we are continuing to study the issue of GABA-ergic efferents in the organ of Corti. In a preliminary series of experiments, we have recently applied to the organ of Corti anti-GABA antiserum (Immuno Nuclear) that has been developed according to the protocol of Storm-Mathisen et al. (1983). It is possible that anti-GABA antisera will be a valuable, and even necessary, complement to anti-GAD antisera for marking GABA-ergic neurons (Storm-Mathisen et al., 1983).

The studies using antisera to GLNase and AATase are of special interest because the findings can be related to glutamate and aspartate as neurotransmitter candidates of immunoreactive cells. Auditory dendrites at the base of inner hair cells contain GLNase-like immunoreactivity. No such immunoreactivity was seen at outer hair cells in those apical regions where efferent endings probably are absent. Consequently, auditory nerve dendrites at outer hair cells probably contain little or no GLNase. This agrees with the following: There is evidence that Type I spiral ganglion cells of the auditory nerve exclusively innervate inner hair cells and that Type II spiral ganglion cells exclusively innervate outer hair cells (Kiang et al., 1982). Type I spiral ganglion cells show intense GLNase-like immunoreactivity, and Type II cells show such immunoreactivity weakly or not at all (Altschuler et al., 1984d). AATase-like immunoreactivity also is intense in Type I spiral ganglion cells (Altschuler et al., 1981; Fex et al., 1981, 1982b). The few, small spiral ganglion cells that regularly show no AATase immunoreactivity may be type II cells (Fex et al., 1982b); this remains to be demonstrated. On the other hand, our findings in the organ of Corti indicate there there is little or no AATase-like immunoreactivity in auditory dendrites. This suggest that in Type I spiral ganglion cells a substance with GLNase-like immunoreactivity is transported from the cell body out into the dendrite, while the substance causing AATase-like immunoreactivity is not. Furthermore, the findings of intense GLNase-like and AATase-like immunoreactivity in the Type I cells, and the probably low activities of both enzymes in the Type II cells, may indicate that Type I cells use glutamate or aspartate as neurotransmitter, and that Type II spiral ganglion cells do not. Antisera to both GLNase (Fex et al., 1981, 1984) and AATase (Fex et al., 1982b) demonstrate immunoreactivity in the medial system of efferents. There is no evidence that these efferents use glutamate or aspartate as a neurotransmitter. Uptake in the organ of Corti of these and other amino acids in tritiated form has been studied, in attempts to obtain evidence concerning neurotransmission at synapses in the organ of Corti, using light microscopy to determine labeling with silver grains in autoradiographs. Different laboratories have obtained somewhat different results, perhaps due to different methods being used. Gulley et al. (1979) found label over outer hair-cell efferent ending after incubation with tritiated glutamate, but found stronger labeling over the inner spiral bundle and tunnel spiral bundle with both glutamate and aspartate. Schwartz and Ryan (1983) found that efferent endings under inner hair cells and small fibers in the inner spiral bundle were preferentially labeled with glutamate and aspartate; glutamate labeling was the least dense. Neither of these amino acids gave labeling at outer hair cells. Eybalin and Pujol (1983), using tritiated glutamate and glutamine, found labeling of the inner spiral bundle and tunnel spiral bundle and no labeling at outer hair cells. Thus, these uptake studies do not indicate why GLNase-like and AATase-like immunoreactivity are found in the medial system of efferents and give no evidence for aspartate or glutamate as a neurotransmitter for the medial system of efferents. Therefore, findings regarding the cochlear efferents show that more research is needed to clarify the status of immunocytochemically-demonstrated GLNase-like and AATase-like immunoreactivity as markers of glutamatergic or aspartatergic neurons. A similar view is expressed in two recent studies (Donoghue et al., 1984; Wenthold and Altschuler, 1983). Wenthold and Altschuler (1983) found, much on the basis of a study of the cerebellum, "that AAT may be present in other neurons, perhaps some GABAergic neurons, while GLNase appears more likely to be enriched only in those neurons which are believed to release glutamate or aspartate". In a study of GLNase-like immunoreactivity and AATase-like immunoreactivity in neurons of cerebral neocortex (Donoghue et al., 1984), results were taken to suggest that GLNase marks neurons that may use an excitatory amino-acid neurotransmitter. On the other hand, the form and distribution of cells marked by AATase antiserum were taken to suggest that such cells were GABAergic neurons (Donoghue et al., 1984). Future studies with the continued use of antisera to amino acids (cf Storm-Mathisen et al., 1982) may help to determine which markers best define glutamatergic and aspartatergic neurons.

VI. EXPECTATIONS AND ANTICIPATIONS

We have great expectations for the future development of the immunocytochemistry of the mammalian cochlea, much because of exciting recent developments concerning monoclonal antibodies and light microscopy. It has been demonstrated that monoclonal antibodies can be raised *in vitro* (for references, see Reading, 1982). Apart from inherent beauty, the technique has a major advantage. Much less antigen is needed for raising antibodies *in vitro* than in an animal. It has been reported that tens of nanograms of antigenic substances can lead to the formation of antibodies *in vitro*, while micrograms are needed *in vivo*. Consequently, the number of cochleae needed for raising antibodies to a species of molecule in a mammalian hair cell or spiral ganglion cell may no longer be prohibitive. Therefore, it seems certain that soon papers will appear describing new kinds of results with monoclonal antibodies that have been raised to molecules from hair cells and spiral ganglion cells. Such antibodies will be used for immunocytochemistry, purification of proteins, developmental studies, pharmacological studies and physiological studies. Monoclonal antibodies, whether they are raised *in vitro* or *in vivo*, can be internally labeled by using radioactive amino acids in the cell culture medium containing the hybridomas that produce the antibody (Cuello et al., 1982). Using such radioactive antibodies, immunoreactive sites can be localized with light or electron microscopy of silver grain distribution in autoradiographs (radioimmunocytochemistry). No secondary antibody is needed, but can be used if a complementing means of visualization is wanted, such as fluorescence, peroxidase reaction product, ferritin, or gold.

A recent, exciting development in light microscopy is the technique of video-enhanced light microscopy (Inoue, 1981), which we (Altschuler et al., 1984b) have used with Asymmetric Illumination contrast (Kachar, 1984). Such microscopy can give a stunningly powerful resolution and beautiful visualization of living structures under tissue culture conditions, as well as of fixed surface preparations of the organ of Corti. We expect that with the help of this powerful technique it will be possible to test the working hypothesis that events at efferent ending-outer hair cell synapses cause structural changes in the cell, which in their turn are important enough to change the micromechanics of the organ of Corti. In a recent study (Brownell, 1983), this hypothesis is reviewed and new evidence is provided with findings of electrically and pharmacologically-induced shape changes of outer hair cells of the guinea pig cochlea. We expect that a series of new studies of a particular kind will soon appear. Thus, experiments with isolated inner-ear sensory cells will be carried out with video-enhanced light microscopy in which simultaneously transduction, synaptic events, and structural changes will be studied and controlled, using labeled monoclonal antibodies that serve a dual purpose as immunocytochemical markers and as excitatory or blocking ligands. Recordings from single hair cells using patch clamp techniques (for references, see Sakman and Neher, 1983) will be combined with injection into the cell of monoclonal antibodies that can be traced through non-toxic labels and that may be designed to monitor and control specific functions of the cell.

We also anticipate the use of monoclonal antibodies for immunocytochemistry in developmental studies of the inner ear in otocyst cultures, such as those mastered by Van de Water and Ruben (Van de Water and Ruben, 1971; for references see Van de Water, 1983) and in surface preparation-like cultures, such as those pioneered by Sobkowicz (Sobkowicz et al., 1975; for references, see Sobkowicz and Rose, 1983). The Sobkowicz type of culture has already proven to be excellent for the study of synapse development under visual control. Undoubtedly immunocytochemistry, combined with video enhanced light microscopy, will be used for continued study of auditory nerve-hair cell synapses in tissue culture, marking selected molecules on inner and outer hair cells with labeled monoclonal antibodies. Furthermore, we expect attempts will be made to study how auditory nerve synapses are formed in cultures of cochlear tissue and brainstem tissue; immunocytochemistry will superbly serve to identify and localize key cellular components. If successful, such attempts may indicate where the Type II spiral ganglion cells have their synaptic targets in the adult brainstem. It is also expected that judiciously selected and marked antibodies will be used in more direct studies of where the Type II spiral ganglion cell synapses are. Suitable antibodies for this are expected to become available through the continued study of the spiral ganglion cells, using combined immunocytochemical and biochemical techniques.

Our expectations thus concern questions of great interest and importance, for both auditory research in particular, and for neuroscience in general. We hope that these expectations will prove to be good prophesies.

ACKNOWLEDGMENTS

We thank Drs. N. P. Curthoys and W. G. Haser for providing us with antisera against glutaminase and Dr. F. Eckenstein for providing antiserum and antibodies against choline acetyltransferase. We are grateful to the Laboratory of Clinical Science, NIMH, for providing us with the antiserum against glutamic acid decarboxylase. This antiserum (1440) was developed in the Laboratory of Clinical Science, NIMH, under the supervision of Dr. I. J. Kopin with Drs. W. Oertel, D. E. Schmechel, and M. Tappaz, with its effective use in immunocytochemistry greatly aided through the laboratory of Dr. E. Mugnaini (University of Connecticut, Storrs). We thank Drs. M. R. Martin and R. J. Wenthold for suggestions to improve the manuscript and Ms. M. L. Adams for expert typing.

REFERENCES

Altschuler, R. A.; Fex, J., Parakkal, M. H., and Eckenstein, F.: Colocalization of enkephalin-like and choline acetyltransferase-like immunoreactivities in olivocochlear neurons of the guinea pig. *J. Histochem. Cytochem.* In press, 1984a.

Altschuler, R. A.; Kachar, B., Rubio, J. A., Parakkal, M. H., and Fex, J.: Immunocytochemical localization of choline acetyltransferase-like immunoreactivity in the guinea pig cochlea. Submitted, 1984b.

Altschuler, R. A.; Neises, G. R., Harmison, G. G., Wenthold, R. J., and Fex, J.: Immunocytochemical localization of aspartate aminotransferase immunoreactivity in the cochlear nucleus of the guinea pig. *Proc. Natl. Acad. Sci. U.S.A.* 78: 6553-6557, 1981.

Altschuler, R. A.; Parakkal, M. H., and Fex, J.: Localization of enkephalin-like immunoreactivity in acetylcholinesterase positive cells in the guinea pig lateral superior olivary complex that project to the cochlea. *Neuroscience 9:* 621-630, 1983.

Altschuler, R. A.; Parakkal, M. H., Rubio, J. A., Hoffman, D. W., and Fex, J.: Enkephalin-like immunoreactivity in the guinea pig organ of Corti: Ultrastructural and lesion studies. Submitted, 1984c.

Altschuler, R. A.; Wenthold, R. J., Schwartz, A. M., Haser, W. G., Curthoys, N. P., Parakkal, M. H., and Fex, J.: Immunocytochemical localization of glutaminase-like immunoreactivity in the auditory nerve. *Brain Res. 291:* 173-178, 1984d.

Bredberg, G.: Innervation of the organ of Corti. *Acta Otolaryngol. 83:* 71-78, 1977a.

Bredberg, G.: Innervation of the organ of Corti as revealed in the scanning electron microscope. In Evans, E. F., and Wilson, J. P. (Eds.): *Psychophysics and Physiology of Hearing.* New York, Academic Press, 1977b, pp. 3-11.

Bredberg, G.: Ultrastructural features of the small nerve endings high up on the outer hair cells. In Evans, E. F., and Wilson, J. P. (Eds.): *Psychophysics and Physiology of Hearing.* New York, Academic Press, 1977c, pp. 12-13.

Brownell, W. E.: Observations on a motile response in isolated outer hair cells. In Webster, W. R., and Atkin, L. M. (Eds.): Mechanisms of Hearing. Clayton, Victoria, Australia, Monash University Press, 1983, pp. 5-10.

Burns, J.: Background staining and sensitivity of the unlabelled antibody-enzyme (PAP) Method. Comparison with the peroxidase labelled antibody sandwich method using formalin fixed paraffin embedded material. *Histochemistry 43:* 291-294, 1975.

Chan-Palay, V.; Engel, A. G., Wu, J.-Y., and Palay, S. L.: Coexistence in human and primate neuromuscular junctions of enzymes synthesizing acetylcholine, catecholamine, taurine and γ-aminobutyric acid. *Proc. Natl. Acad. Sci. U.S.A. 79:* 7027-7030, 1982.

Chan-Palay, V., and Palay, S. L. (Eds.): *Cytochemical Methods in Neuroanatomy.* New York, Alan R. Liss, Inc., 1982.

Churchill, J. A., and Schuknecht, H.: The relationship of acetylcholinesterase in the cochlea to the olivocochlear bundle. *Henry Ford Hosp. Med. Bull. 7:* 202-205, 1959.

Clark, V. M., and Curthoys, N. P.: Cause of subunit heterogeneity in purified rat renal phosphate dependent glutaminase. *J. Biol. Chem. 254:* 4939-4941, 1979.

Coons, A. H.; Creech, H. J., Jones, R. N., and Berliner, E.: The demonstration of pneumococcal antigen in tissues by the use of fluorescent antibody. *J. Immunol. 45:* 159-170, 1942.

Cuello, A. C. (Ed.): Immunohistochemistry. New York, John Wiley & Sons, 1983.

Cuello, A. C.; Priestly, J. V., and Milstein, C.: Immunocytochemistry with internally labeled monoclonal antibodies. *Proc. Natl. Acad. Sci. U.S.A. 79:* 665-669, 1982.

Curthoys, N. P.; Kuhlenschmidt, T., Godfrey, S. S., and Weiss, R. F.: Phosphate dependent glutaminase from rat kidney. *Arch. Biochem. Biophys. 172:* 162-167, 1976.

Curthoys, N. P., and Weiss, R. F.: Regulation of renal ammoniagenesis; subcellular localization of rat kidney glutaminase isoenzymes. *J. Biol. Chem. 249:* 3261-3266, 1974.

Donoghue, J. P.; Wenthold, R. J., and Altschuler, R. A.: Localization of glutaminase-like and aspartate aminotransferase-like immunoreactivity in neurons of cerebral neocortex. Submitted, 1984.

Drenckhahn, D.; Kellner, J., Mannherz, H. G., Gröschel-Stewart, U., Kendrick-Jones, J., and Scholey, J.: Absence of myosin-like immunoreactivity in stereocilia of cochlear hair cells. *Nature (Lond.) 300:* 531-532, 1982.

Drescher, M. J.; Drescher, D. G., and Medina, J. E.: Effect of sound stimulation at several levels of concentrations of primary amines, including neurotransmitter candidates, in perilymph of the guinea pig inner ear. *J. Neurochem. 41:* 309-320, 1983.

Eckenstein, F., and Baughman, R. W.: Two types of cholinergic innervation in cortex, one co-localized with vasoactive intestinal polypeptide. *Nature 309:* 153-155, 1984.

Eckenstein, F., and Sofroniew, M. V.: Identification of central cholinergic neurons containing both choline acetyltransferase and acetylcholinesterase and of central neurons containing only acetylcholinesterase. *J. Neurosci. 3:* 2286-2291, 1983.

Eckenstein, F., and Thoenen, H.: Production of specific antisera and monoclonal antibodies to choline acetyltransferase: characterization and use for identification of cholinergic neurons. *The EMBO J. 1:* 363-368, 1982.

Eybalin, M.: Approche morphologique des neurotransmetteurs de la cochlée. Thesis, Université de Droit, d'Economie et des Sciences d'Aix, Marseille, 1982.

Eybalin, M.; Cupo, A., and Pujol, R.: Localisation ultrastuctale des immunoréactions a un anticorps met-enképhaline dans l'organe de Corti. *C. R. Acad. Sci., Paris Sér III, 296:* 1125-1128, 1983.

Eybalin, M.; Cupo, A., and Pujol, R.: Met-enkephalin characterization in the cochlea: High performance liquid chromatography and immunoelectron microscopy. *Brain Res.* In press, 1984.

Eybalin, M., and Pujol, R.: A radioautographic study of (^{3}H)L-glutamate and (^{3}H)L-gultamine uptake in the guinea pig cochlea. *Neuroscience 9:* 863-871, 1983.

Eybalin, M., and Pujol. Immunofluorescence with met-enkephalin and leu-enkephalin antibodies in the guinea cochlea. *Hearing Res. 12:* 135-140, 1984.

Fex, J.: Efferent inhibition in the cochlea by the olivocochlear bundle. In Reuck, V. S., and Knight, J. (Eds.): *Ciba Found. Symp. Hearing Mechanisms in Veterbrates.* London, Churchill, 1968, pp. 169-181.

Fex, J.: Immunoreactivity in the mammalian organ of Corti. *Abstr. 17th Workshop on Inner Ear Biology,* p. 43, 1980.

Fex, J., and Altschuler, R. A.: Enkephalin-like immunoreactivity of olivocochlear nerve fibers in cochlea of guinea pig and cat. *Proc. Natl. Acad. Sci. U.S.A. 78:* 1255-1259, 1981.

Fex, J., and Altschuler, R. A.: Glutamic acid decarboxylase immunoreactivity of olivocochlear neurons in the organ of Corti of guinea pig and rat. *Hearing Res.* In press, 1984.

Fex, J.; Altschuler, R. A., Harmison, G. G., Neises, G. R., Parakkal, M. H., and Wenthold, R. J.: Aspartate aminotransferase (AAT) and glutaminase-like immunoreactivity in the cochlea and auditory nerve of the guinea pig. *Abstr. 18th Workshop on Inner Ear Biology,* 1981.

Fex, J.; Altschuler, R. A., Parakkal, M. H., and Eckenstein, F.: Immunocytochemical localization of choline acetyltransferase-like immunoreactivity in olivocochlear fibers in the guinea pig cochlea. *Soc. Neurosci. Abstr. 8:* 41, 1982a.

Fex, J.; Altschuler, R. A., Wenthold, R. J., and Parakkal, M. H.: Aspartate aminotransferase immunoreactivity in cochlea of guinea pig. *Hearing Res. 7:* 149-160, 1982b.

Fex, J.; Kachar, B., Rubio, J. A., Parakkal, M. H., and Altschuler, R. A.: Immunocytochemical localization of glutaminase-like immunoreactivity in the guinea pig cochlea. In preparation, 1984.

Fex, J., and Wenthold, R. J.: Choline acetyltransferase, glutamate decarboxylase and tyrosine hydroxylase in the cochlea and cochlear nucleus of the guinea pig. *Brain Res. 109:* 575-585, 1976.

Flock, A.: Contractile proteins in hair cells. *Hearing Res. 2:* 411-412, 1980.

Flock, A.; Bretscher, A., and Weber, K.: Immunohistochemical localization of several cytoskeletal proteins in inner ear sensory and supporting cells. *Hearing Res. 6:* 75-89, 1982.

Flock, A.; Cheung, C., Flock, B., and Utter, G.: Three sets of actin filaments in sensory cells of the inner ear. Identification and functional orientation determined by gel electrophoresis, immunofluorescence and electron microscopy. *J. Neurocytol. 10:* 133-147, 1981a.

Flock, A.; Hoppe, Y., and Wei, X.: Immunofluorescence localization of proteins in semithin 0.2-1 μm frozen sections of the ear. A report of improved techniques including gelatin encapsulation of cryoultramicrotomy. *Arch. Otorhinolaryngol. 233:* 55-66, 1981b.

Godfrey, D. A.; Carter, J. A., Berger, S. J., Lowry, O. H., and Matschinsky, F. M.: Quantitative histochemical mapping of candidate transmitter amino acids in cat cochlear nucleus. *J. Histochem. Cytochem. 25:* 417-431, 1977.

Godfrey, D. A.; Carter, J. A., Berger, S. J., and Matschinsky, F. M.: Levels of putative transmitter amino acids in the guinea pig cochlea. *J. Histochem. Cytochem. 24:* 468-470, 1976.

Godfrey, D. A.; Park, J. L., Dunn, J. D., and Ross, L. D.: Enzymes of acetylcholine metabolism in centrifugal fibers to the inner ear of the rat. *Soc. Neurosci. Abstr. 8:* 346, 1982.

Godfrey, D. A.; Park, J. L., and Ross, L. D.: Choline acetyltransferase and acetylcholinesterase in centrifugal labyrinthine bundle of rats. *Hearing Res.* In press, 1984.

Guinan, J. J.; Warr, W. B., and Norris, B. E.: Differential olivocochlear projections from lateral versus medial zones of the superior olivary complex. *J. Comp. Neurol. 221:* 358-370, 1983.

Gulley, R. L.: Personal communication, 1984.

Gulley, R. L.; Fex, J., and Wenthold, R. J.: Uptake of putative neurotransmitters in the organ of Corti. *Acta Otolaryngol. 88:* 177-182, 1979.

Hoffman, D. W.; Altschuler, R. A., and Fex, J.: High performance liquid chromatographic identification of enkephalin-like peptides in the cochlea. *Hearing Res. 9:* 71-78, 1983.

Hoffman, D. W.; Rubio, J. A., Altschuler, R. A., and Fex, J.: Several distinct receptor binding enkephalins in olivocochlear fibers and terminals in the organ of Corti. *Brain Res.* In press, 1984.

Hökfelt, T.; Fuxe, K., and Goldstein, M.: Applications of immunohistochemistry to studies on monoamine cell systems with special reference to nervous tissues. *Ann. New York Acad. Sci. 254:* 407-432, 1975.

Hsu, S-M.; Raine, L., and Fanger, H.: Use of avidin-biotin-peroxidase complex (ABC) in immunoperoxidase techniques: a comparison between ABC and unlabeled antibody (PAP) procedures. *J. Histochem. Cytochem. 29:* 577-580, 1981.

Inoue, S.: Video image processing greatly enhances contrast, quality, and speed in polarization-based microscopy. *J. Cell Biol. 89:* 346-356, 1981.

Kachar, B.: Asymmetric illumination contrast: A method of image formation for video light microscopy. Submitted, 1984.

Kiang, N. Y. S.; Rho, J. M., Northrop, C. C., Liberman, M. C., and Ryugo, D. K.: Hair-cell innervation by spiral ganglion cells in adult cats. *Science 217:* 175-177, 1982.

Klinke, R., and Oertel, W.: Evidence that GABA is not the afferent transmitter in the cochlea. *Exp. Brain Res. 28:* 311-314, 1977.

Kuriyama, K., and Kimura, H.: Distribution and possible functional roles of GABA in the retina, lower auditory pathway, and hypothalamus. In Roberts, E.; Chase, T. N., and Tower, D. B. (Eds.): *GABA in Nervous System Function*. New York, Raven Press, 1976, pp. 203-216.

Mangape, D., and Harada, Y.: Scanning electron microscopy of the nerve endings of the outer hair cells in the organ of Corti. *J. Laryngol. Otol. 96:* 591-597, 1982.

Micevych, P., and Elde, R.: Relationships between enkephalinergic neurons and the vasopressin-oxytocin neuroendocrine system of the cat: An immunohistochemical study. *J. Comp. Neurol. 190:* 135-146, 1980.

Morrison, D.; Schindler, R. A., and Wersäll, J.: A quantitative analysis of the afferent innervation of the organ of Corti in the guinea pig. *Acta Otolaryngol. 79:* 11-23, 1975.

Nakai, Y., and Igarashi, M.: Distribution of the crossed olivo-cochlear bundle terminals in the squirrel monkey cochlea. *Acta Otolaryngol. 77:* 393-404, 1974.

Oertel, W. H.; Mugnaini, E., Schmechel, D. E., Tappaz, M. L., and Kopin, I. J.: The immunocytochemical demonstration of gammaaminobutyric acidergic neurons—methods and application. In Chan-Palay, V., and Palay, S. L. (Eds.): *Cytochemical Methods in Neuroanatomy.* New York, Alan R. Liss, Inc., 1982, pp. 297-329.

Oertel, W. H.; Schmechel, D. E., Mugnaini, E., Tappaz, M. L., and Kopin, I. J.: Immunocytochemical localization of glutamate decarboxylase in rat cerebellum with a new antiserum. *Neuroscience 6:* 2715-2735, 1981a.

Oertel, W. H.; Schmechel, D. E., Tappaz, M. L., and Kopin, I. J.: Production of a specific antiserum to rat brain glutamic acid decarboxylase by injection of an antigen-antibody complex. *Neuroscience 6:* 2689-2700, 1981b.

Oertel, W. H.; Schmechel, D. E., Weise, V. K., Ransom, D. H., Tappaz, M. L., Krutsch, H. C., and Kopin, I. J.: Comparison of cysteine sulphinic acid decarboxylase isoenzymes and glutamic acid decarboxylase in rat liver and brain. *Neuroscience 6:* 2701-2714, 1981c.

Rasmussen, G. L.: The olivary peduncle and other fiber projections of the superior olivary complex. *J. Comp. Neurol. 84:* 141-219, 1946.

Rasmussen, G. L.: Further observations on the efferent cochlear bundle. *J. Comp. Neurol. 99:* 61-74, 1953.

Rasmussen, G. L.: Efferent fibers of the cochlear nerve and cochlear nucleus. In Rasmussen, G. L., and Windle, W. F. (Eds.): *Neural Mechanisms of the Auditory and Vestibular Systems.* Springfield, Charles C. Thomas, 1960, pp. 105-115.

Reading, C. L.: Review Article. Theory and methods for immunization in culture and monoclonal antibody production. *J. Immunol. Methods 52:* 261-291, 1982.

Richrath, W.; Kraus, H., and Fromme, H. G.: Lokalisation von ^{3}H-γ-Aminobuttersaure in der Cochlea. Licht- und Elektronenoptische Autoradiographie. *Arch. Otorhinolaryngol. 208:* 283-293, 1974.

Rossier, J.: On the mapping of the cholinergic neurons by immunocytochemistry. *Neurochem. Int. 6:* 183-184, 1984.

Rubio, J. A.; Altschuler, R. A., Kachar, B., Eckenstein, F., and Fex, J.: Immunocytochemical localization in the guinea pig organ of Corti after peripheral efferent lesions. *Soc. Neurosci. Abstr. 9:* 43, 1983.

Saito, K.; Barber, R., Wu, J-Y., Matsuda, T., Roberts, E., and Vaughn, J. E.: Immunohistochemical localization of glutamic acid decarboxylase in rat cerebellum. *Proc. Natl. Acad. Sci. U.S.A. 71:* 269-273, 1974.

Sakmann, B., and Neher, E. (Eds.): *Single-Channel Recording.* New York, Plenum Press, 1983.

Schuknecht, H. G.; Churchill, J. A., and Doran, R.: The localization of acetylcholinesterase in the cochlea. *Arch. Otolaryngol. 69:* 549-559, 1959.

Schwartz, I. R., and Ryan, A. F.: Differential labeling of sensory cell and neural populations in the organ of Corti following amino acid incubations. *Hearing Res. 9:* 185-200, 1983.

Skowsky, W. R., and Fisher, D. A.: The use of thyroglobulin to induce antigenicity to small molecules. *J. Lab. Clin. Med. 80:* 134-144, 1972.

Sobkowicz, H. M.; Bereman, B., and Rose, J. E.: Organotypic development of the organ of Corti in culture. *J. Neurocytol. 4:* 543-572, 1975.

Sobkowicz, H. M., and Rose, J. E.: Innervation of the organ of Corti of the fetal mouse in culture. In Romand, R. (Ed.): *Development of Auditory and Vestibular Systems.* New York, Academic Press, 1983, pp. 27-45.

Sternberger, L. A.: *Immunocytochemistry.* New York, John Wiley & Sons, 1979.
Stopp, P. E., and Comis, S. D.: Afferent and efferent innervation of the guinea pig cochlea: A light microscopic and histochemical study. *Neuroscience 3:* 1197-1206, 1978.
Storm-Mathisen, J.; Leknes, A. K., Bore, A. T., Vaaland, J. L., Edminson, P., Haug, F.-M.S., and Ottersen, O. P.: First visualization of glutamate and GABA in neurons by immunocytochemistry. *Nature (Lond.) 301:* 517-520, 1983.
Tachibana, M, and Kuriyama, K.: Gamma-aminobutyric acid in the lower auditory pathway of the guinea pig. *Brain Res. 69:* 370-374, 1974.
Thalmann, R.; Miyoshi, T., Kusakari, J., and Ise, I.: Normal and abnormal energy metabolism of the inner ear. *Otolaryngol. Clin. North Am. 8:* 313-333, 1975.
Van de Water, T. R.: Embryogenesis of the inner ear: "*in vitro* studies". In Romand, R. (Ed.): *Development of Auditory and Vestibular Systems.* New York, Academic Press, 1983, pp. 337-374.
Van de Water, T. R., and Ruben, R. J.: Organ culture of the mammalian inner ear. *Acta Otolaryngol. 71:* 303-312, 1971.
Wainer, B. H.; Levey, A. I., Mufson, E. J., and Mesulam, M. M.: Cholinergic systems in mammalian brain identified with antibodies against choline acetyltransferase. *Neurochem. Int. 6:* 163-182, 1984.
Warr, W. B.: Olivocochlear and vestibular efferent neurons of the feline brain stem: their location, morphology and number determined by retrograde axonal transport and acetylcholinesterase histochemistry. *J. Comp. Neurol. 161:* 159-182, 1975.
Warr, W. B., and Guinan, J. J., Jr.: Efferent innervation of the organ of Corti: two separate systems. *Brain Res. 173:* 152-155, 1979.
Wenthold, R. J., and Altschuler, R. A.: Immunocytochemistry of aspartate aminotransferase and glutaminase. In Hertz, L.; Kbamme, E., McGeer, E. G., and Schoushoe, A. (Eds.): *Glutamine, Glutamate and GABA in the Central Nervous System.* New York, Alan R. Liss, Inc., 1983, pp. 33-50.
Wenthold, R. J., and Gulley, R. L.: Aspartic acid and glutamic acid levels in the cochlear nucleus after auditory nerve lesion. *Brain Res. 138:* 111-113, 1977.
Wenthold, R. J., and Martin, M. R.: Neurotransmitters of the auditory nerve and central auditory system. In Berlin, C. (Ed.): *Recent Advances Series in Speech, Hearing and Language: Hearing Sciences.* San Diego, College Hill Press. In press, 1984.
White, J. S., and Warr, W. B.: The dual origins of the olivocochlear bundle in the albino rat. *J. Comp. Neurol. 219:* 203-214, 1983.
Wright, C. G., and Preston, R. E.: Degeneration and distribution of efferent nerve fibers in the guinea pig organ of Corti. A light and scanning electron microscopic study. *Brain Res. 58:* 37-59, 1973.
Wu, J-Y.: Preparation of glutamic acid decarboxylase as immunogen for immunocytochemistry. In Cuello, A. C. (Ed.): *Immunohistochemistry.* New York, John Wiley & Sons, 1983, pp. 159-191.
Zenner, H. P.: Antibodies against tubulin and actin: specific visualization of cytoskeletal elements in the inner ear. *Abstr. 17th Workshop on Inner Ear Biology,* p. 42, 1980.
Zenner, H. P.: Cytoskeletal and muscle-like elements in cochlear hair cells. *Arch. Otorhinolaryngol. 230:* 81-92, 1981.

Chapter 2

MODE OF OPERATION OF AFFERENT SYNAPSES BETWEEN HAIR CELLS AND AUDITORY FIBERS

Taro Furukawa and Shiushi Matsuura

I. Introduction
II. Some Basic Findings
   A. Graded Synaptic Action
   B. Successive Decline of EPSP's
   C. Incremental and Decremental Responses
III. Explanations of Adaptive Phenomena
   A. Multiple Release Sites Model
   B. Number of Release Sites Available to Each S1 Fiber
   C. Release Parameters n and p
   D. The Case of the Cochlear Nerve
IV. Summary and Conclusions
References

I. INTRODUCTION

The chemical nature of afferent transmission between hair cells and auditory fibers can be hypothesized from ultrastructual evidence. Physiological support is provided by the presence of a synaptic delay and its dependence on temperature, and by the virtual absence of electrically-mediated postsynaptic potentials. Most of our evidence for hair-cell chemical transmission has been obtained from studies on the goldfish saccule, i.e., its inner ear (Furukawa & Ishii, 1967; Furukawa et al., 1972a, b, 1978, 1982; Furukawa, 1978, 1981; Furukawa and Matsuura, 1978). For the saccule system, low temperature or anoxia produces a marked reduction in the amplitude of the excitatory postsynaptic potentials (EPSP's) leaving the microphonic potentials unaffected (Furukawa, 1983; Kuno et al., 1983).

The ear of the goldfish has a relatively simple structure, and intracellular recording from auditory fibers is facilitated by the presence of large fibers (S1 fibers). An unique feature of the hair cell-afferent fiber synapse is the graded release of transmitter. This corresponds to the fact that the activity of hair cells is graded to the sound intensity. The afferent synapse is the site where various adaptive phenomena are produced. As distinct from the retina, where adaptive reduction in light sensitivity follows a reduction in the magnitude of the receptor potentials, microphonic potentials of hair cells do not show adaptation. Rather, adaptation in the rate of afferent firings, as observed in the cochlea and other hair-cell organs, is based on events at the afferent synapse, such as depletion and replenishment of transmitter quanta.

In this chapter, adaptive phenomena observed in the goldfish ear are briefly described and our interpretations of the related mechanisms are outlined.

II. SOME BASIC FINDINGS

The ear of the fish responds predominantly to low-frequency sounds, and each sound wave serves as an independent stimulus. Figure 1 shows a typical sample record of the EPSP's produced by a moderately strong sound stimulus. The EPSP's are largest at the onset of the sound, but thereafter successively decline in amplitude. Normally, afferent spikes are elicited when the amplitude of the EPSP's exceeds a certain level. (Spike potentials were suppressed in these experiments with locally applied tetrodotoxin.) One can see that there is a delay between the microphonic deflections and the initiation of the EPSP's. Figure 1 may give the impression that the mode of operation of the hair cell-afferent fiber synapse is similar to that of the neuromuscular junction. However, such is not the case.

A. GRADED SYNAPTIC ACTION

Figure 2A (curve 1) shows how the amplitudes of the first EPSP's are related to sound intensity. The curve rises rather steeply with the sound intensity to the point of saturation. This saturation seems to be produced mostly postsynaptically, by a nonlinear summation of the EPSP's. The latter interpretation is supported by the finding that the amplitude of the EPSP's increases in a straightforward manner to a much higher level when the afferent fiber is hyperpolarized by passing a DC current through the recording microelectrode (Fig. 2B).

Theoretically, saturation could occur at more peripheral levels (i.e., presynaptically), due to limitations in the amount of transmitter released, limitations of transduction processes, and so forth. In our preparation, however, the magnitude of the EPSP's was so large, and the rate of growth with the sound intensity so steep, that saturation at the postsynaptic level was much more marked than events occurring at other sites.

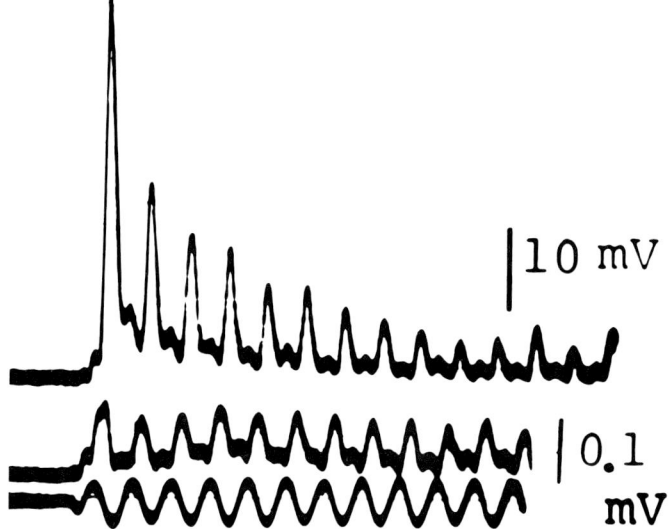

Figure 2-1. Sample records of EPSP's and microphonics from the goldfish saccule. Top: EPSP's recorded intracellularly from large S1 fibers; spike potentials were blocked with a local application of tetrodotoxin. Middle: microphonic potentials (extracellular). Bottom: sound monitor (500 Hz, 95 dB SPL).

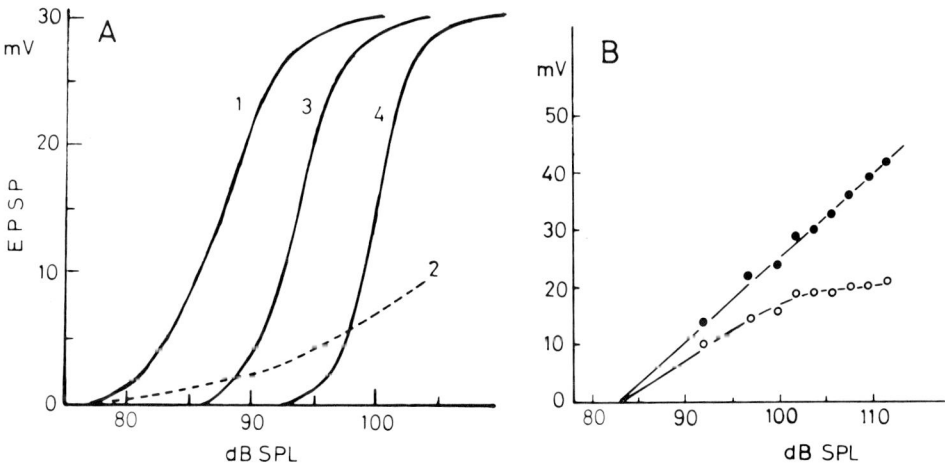

Figure 2-2. A. Relation between the amplitude of the EPSP's and the sound intensity for the onset of sound (curve 1), for the quasi-steady state (curve 2), and for dynamic responses (curves 3 and 4). B. Absence of postsynaptic saturation during hyperpolarization applied to the afferent fiber. Open circles, control; filled circles, measurements during hyperpolarization.

B. SUCCESSIVE DECLINE OF EPSP'S

The decline ("rundown") in the EPSP size, such as observed in Fig. 1, can most probably be attributed to a depletion of the transmitter quanta at the presynaptic release site. Similar phenomena have been noted for other synapses (Elmqvist and Quastel, 1965; Kusano and Landau, 1975), but for the present case, the rundown (i.e., synaptic depression) takes place more markedly than at other synapses. Also, the recovery from the depression occurs much more rapidly (the time constant of recovery: 15-20 ms, compared to 4-5 s at the frog neuromuscular junction and at other synapses). These differences may relate to differences in the fine morphological structure of the presynaptic sites.

Reduction in the amplitude of the EPSP's is initially exponential, but it soon slows down, due to the intervention of a replenishing process. Eventually, a quasi-steady state is reached where the rate of transmitter release is balanced by the rate of replenishment. The amplitudes of the EPSP's at this quasi steady state, shown in curve 2 of Fig. 2A, represent the rate of replenishment for different intensities of sound.

C. INCREMENTAL AND DECREMENTAL RESPONSES

It was mentioned previously that the adaptive reduction in amplitude of the EPSP's is attributable to a depletion of available transmitter quanta. Here, we use the term "depletion" to designate a functional concept, because even in a depressed state, an increment in the sound intensity readily gives rise to new release of transmitter (Fig. 3A, top). The EPSP's, thus increased in amplitude, run down again, in the same manner as at the start of the sound.

On the other hand, a decrease (decrement) in intensity is found to produce a temporary reduction in the amplitude of the EPSP's. The EPSP's, thus reduced in amplitude, gradually revert toward the level appropriate for the reduced sound intensity (Fig. 3A, second to fourth traces). Under certain conditions, this decremental response is similar to the "off-suppression" of spontaneous firings, as observed in the cochlear nerve. In either case, responses, once suppressed by the reduction of sound intensity or by cessation of sound, gradually recover toward the level of activity maintained by a continuous drive from the hair cells (Sewell, 1984).

Another approach to testing "off-suppression" is to observe the change in the response magnitude for test sounds delivered at various intervals after cessation of the (conditioning) sound. This procedure is analogous to the method for testing for forward masking (Smith, 1977; Harris & Dallos, 1979). A typical result obtained with this mode of testing is shown in Fig. 3B.

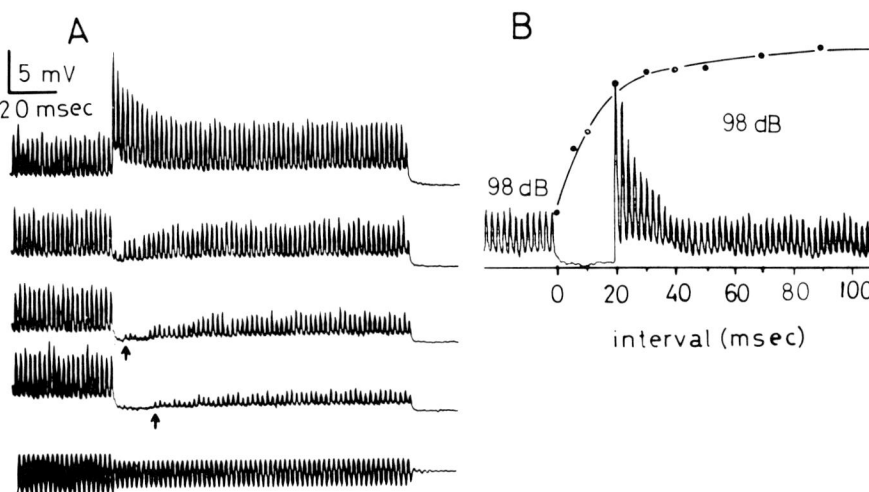

Figure 2-3. Incremental and decremental responses (A) and recovery from depression (B), observed as changes in the amplitude of the EPSP's. A. Sound intensity was increased by 3 dB (from 98 dB SPL) in the top trace, and decreased by 1, 3 and 5 dB in the second, third, and fourth traces, respectively. EPSP's were completely suppressed until the time shown with arrows. B. Test tone (same intensity as the adapting tone) was delivered at various intervals (dots) after cessation of the adapting tone.

The EPSP's produced by test sounds increased to a much higher level than for the decremental response, because the recovery occurs after *cessation* of the conditioning sound.

Due to the occurrence of incremental and decremental responses, one can draw dynamic response curves for different intensities of adapting sound. Curves 3 and 4 of Fig. 2A represent dynamic response curves for an adapting sound of 88 and 98 dB, respectively. Note that the rate of maximum rise of these dynamic response curves is very steep.

Adaptive phenomena, including incremental and decremental responses, as observed in the rate of unitary firing of the cochlear nerve, have been studied in detail by several investigators (Smith & Zwislocki, 1975; Smith, 1979; Smith & Brachman, 1982). These workers point out that the adaptive phenomena are additive, not multiplicative, in nature. By "additive," it is meant that the process underlying the adaptive decline in the magnitude of the response does not produce any change in the gain of the system. Rather, the response magnitude is reduced during adaptation. The possibility of gain changes was excluded in the above studies by the finding that a small increment in the stimulus intensity gave rise to approximately the same amount of increase in the response magnitude, either in the adapted or non-adapted state. The results of the previous workers pertain to the rate of unitary afferent firings (spikes) of the cochlear nerve, and ours pertain to the amplitude of the EPSP's recorded from single auditory fibers of goldfish. Despite the differences, all of the previous results are in fairly good agreement. In particular, the time course of the adaptive decline in the response and that of recovery from off-suppression (or decremental response) are in parallel. There is one significant difference, relating to the origin of saturation in the response magnitude: Whereas saturation mostly occurs at the postsynaptic level in goldfish, it occurs at more peripheral levels (presynaptically) in the cochlea.

III. EXPLANATIONS OF ADAPTIVE PHENOMENA

A fundamental assumption of the quantal release hypothesis can be stated as $m = n \cdot p$, where m is the mean number of the quanta released upon stimulation, n is the number of available quanta at the presynaptic site (see below), and p is the release fraction. The value of these release parameters can be determined experimentally, for example, by the use of binomial statistics (Furukawa *et al.*, 1978, 1982).

A. MULTIPLE RELEASE SITES MODEL

Our "multiple release sites model" was designed to explain adaptive phenomena in the hair cell-afferent fiber synapse (Furukawa & Matsuura, 1978). It is assumed in the model that (1) there are numerous release sites on the presynaptic membrane, and (2) each release site has its own threshold, so that release can occur from the site only when threshold is reached. It is further assumed that (3) at the most, a single synaptic vesicle is allotted to each release site, and (4) when release occurs from an occupied release site, the site becomes empty (unoccupied) and remains so until a replenishment is made from the store.

It follows from these assumptions that depletion would take place only at release sites whose threshold had been reached. The incremental response can thus readily be explained, because an increment in the sound intensity would trigger a new release from the release sites whose threshold is above the level of the adapting sound, sites that would have remained fully occupied. It also follows that the response curve, expressed in terms of the amount of transmitter released, would advance in a straight-line fashion without saturation, up to a fairly strong sound intensity.

B. NUMBER OF RELEASE SITES AVAILABLE TO EACH S1 FIBER

The model, as depicted above, should conform to the actual anatomical situation in the goldfish saccule. How many release sites could be available to each S1 fiber? The problem is important in view of very large size of the EPSP's recorded and the small value of the quantum size (about 0.4 mV in S1 fibers). Further, in order to explain dynamic response curves such as those shown in Fig. 2A, one must assume that the amplitude of the EPSP's (in the absence of the effect of non-linear summation) should increase with the sound intensity to a value at least as great as 100 mV. If p, the release fraction, is set at 0.5, then there must be at least 500 release sites per each S1 fiber.

Calculations based on morphological findings seem to satisfy the requirement. We started from the results of Hama and Saito (1977), which indicated that synaptic vesicles at each presynatic site were arranged as parallel rows beneath the spherical dense body. Since there are about five such rows, each containing about five synaptic vesicles, the number of release sites existing at each presynaptic site is about 25. The next problem is to determine the number of hair cells which are connected to single S1 fibers. We looked into this problem with an intraaxonal injection of Lucifer Yellow (Fig. 4, from a previously-unpublished observation of S. Sento). Each S1 axon entering the saccular macula gives off fine dendritic processes to an average of ten hair cells located within an area that is 25 to 50 μm across (based on observations on 31 injected S1 fibers). There is another possibility that a single dendritic process may cover two presynaptic sites (Nakajima & Wang, 1974). If the latter factor is estimated as 2, then a figure of 500 can be obtained for the number of release sites per single S1 fiber ($25 \times 10 \times 2 = 500$).

C. RELEASE PARAMETERS n AND p

Supporting evidence for multiple release sites was obtained by calculating release parameters n and p from statistical analyses of the amplitude of the EPSP's (Furukawa et al., 1978, 1982). Results of these studies indicated that it was the value of parameter n that changed under different conditions, while the value of parameter p remained largely unchanged.

It must be noted here that n does not represent the number of physically existing synaptic vesicles. It is appropriate to our multiple release sites model to define n as the number of occupied release sites which are activated by the sound intensity used. Further, the statement made above to the effect that parameter p remains unchanged for different stimulus conditions is somewhat of a simplification. Results of statistical analyses showed that the value of p increased with the sound intensity (Fig. 10 of Furukawa et al.,

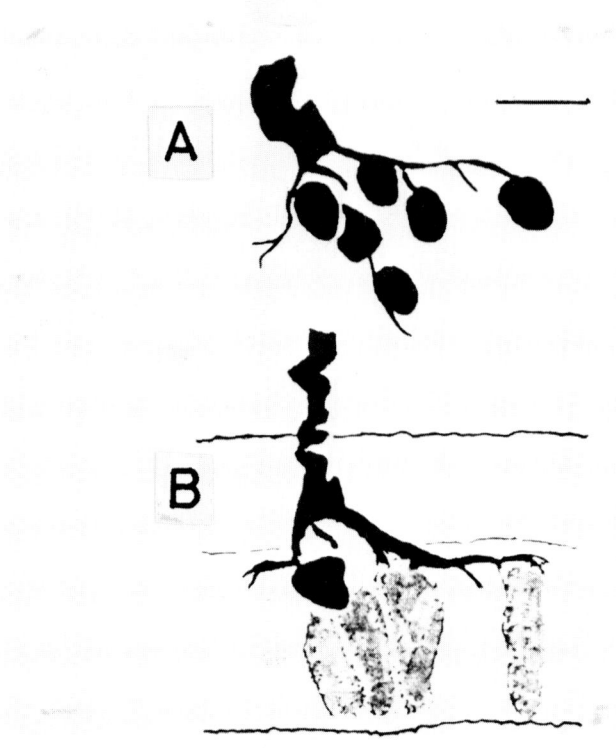

Figure 2-4. Dendritic branchings of an S1 fiber, revealed with an intraaxonal injection of Lucifer Yellow. A. View from the luminal surface of the saccular macula. Scale, 10 μm. B. Side view of the same place as in A. Note that the fiber sends several dendritic branches and that the dye spreads transsynaptically to about five hair cells. Such transsynaptic "dye-coupling" was observed fairly frequently.

1978). We also were under the impression that the rate of rundown of the EPSP's for a near-threshold sound was much smaller than for a moderately strong sound. Generally speaking, it is difficult to estimate the value of p for the onset of the sound because (1) sound waves often show an irregularity for a few cycles at the onset, although we tried to select a frequency which would minimize this sort of distortion, and (2) more fundamentally, the amplitude of large EPSP's can best be estimated in an approximate manner, for non-linear summation would cause a large error, even with the use of compensatory measures. On the other hand, changes in the value of n and p for incremental or decremental responses can be estimated more easily when the extent of the increment or decrement is moderate.

D. THE CASE OF THE COCHLEAR NERVE

In the case of the cochlear afferent fibers terminating on inner hair cells, the number of release sites available to each afferent fiber must be much smaller than for fish S1 fibers, because there is no branching before the cochlear afferents synapse on single inner hair cells (Liberman, 1982). The number of release sites for the cochlear case would be ten or so. It follows from this that (1) the quantal size, i.e., the amplitude of the EPSP a single quantum can produce, would be much larger, (2) the amplitude of the EPSP's must show stepwise changes, and (3) a release of one (or two) quanta would be enough to set up an afferent spike. Thus, the situation at afferent synapses of cochlear inner hair cells is different from that observed at synapses of S1 fibers. Despite these differences, it may be postulated that basically similar mechanisms underlie the graded release and adaptive phenomena for both of these systems.

An interesting picture emerges if one assumes that release of a single quantum is sufficient to set up an afferent impulse. For example, the rate of spontaneous firings would then be set by the rate of spontaneous release of transmitter quanta, and the rate of sustained firings during sound stimulation would correspond to the rate of replenishment of quanta to releasing sites. Namely, afferent firings at a rate of 200/s could be accounted for by the rate of replenishment of 200 quanta/s, which does not seem excessively high, because a comparable value can be assumed without much difficulty for S1 afferent synapses. (Similar assumptions were made by Schroeder-Hall and by Ohno; see Geisler *et al.*, 1979.)

Observations on the fish S1 fiber lead to interesting suggestions when considering the origin of the random nature of firings in cochlear afferent fibers. Kuno (1983) showed that (1) the rate of afferent firings cannot exceed that of sound frequency in a strongly phase-locked state, and (2) the rate of afferent firings is linearly related to the mean amplitude of the EPSP's for a certain range of sound intensity. Randomness in the release process, namely the random fluctuations in the EPSP amplitude, are thus found to convert the mean EPSP amplitude to the firing rate.

IV. SUMMARY AND CONCLUSIONS

Action of hair cell-afferent fiber synapses is characterized by a graded release, adaptive rundown in the amount of release, a quick recovery, and incremental and decremental responses. These features reflect the structural differentiation at the presynaptic sites, i.e., a very limited number of available transmitter quanta at the presynaptic site, and the presence of presynaptic organelles such as synaptic ribbons or spherical dense bodies. The limited number of available quanta seems to relate to the marked development of adaptive decline in the EPSP amplitude, while the presynaptic organelles serve as instruments for a rapid replenishment of transmitter quanta by providing routes of transport of synaptic vesicles. However, differences in the threshold among different release sites also have to be taken into account when considering various adaptive phenomena.

REFERENCES

Elmqvist, D., and Quastel, D. M. J.: A quantitative study of endplate potentials in isolated human muscle. *J. Physiol. 178:* 505–529, 1965.

Furukawa, T.: Sites of termination on the saccular macula of auditory nerve fibers in the goldfish as determined by intracellular injection of procion yellow. *J. Comp. Neurol. 180:* 807–814, 1978.

Furukawa, T.: Effects of efferent stimulation on the saccule of goldfish. *J. Physiol. 315:* 203–215, 1981.

Furukawa, T.: Mechanism of transmission at afferent auditory synapse. In Webster, W. R., and Aitkin, L. M. (Eds.): *Mechanisms of Hearing.* Melbourne, Monash University Press, 1983, pp. 40–45.

Furukawa, T.; Hayashida, Y., and Matsuura, S.: Quantal analysis of the size of excitatory post-synaptic potentials at synapses between hair cells and afferent nerve fibres in goldfish. *J. Physiol. 276:* 211–226, 1978.

Furukawa, T., and Ishii, Y.: Neurophysiological studies on hearing in goldfish. *J. Neurophysiol. 30:* 1377–1403, 1967.

Furukawa, T.; Ishii, Y., and Matsuura, S.: An analysis of microphonic potentials of the sacculus of goldfish. *Jap. J. Physiol. 22:* 603–616, 1972a.

Furukawa, T.; Ishii, Y., and Matsuura, S.: Synaptic delay and time course of postsynaptic potentials at the junction between hair cells and eighth nerve fibers in the goldfish. *Jap. J. Physiol. 22:* 617–635, 1972b.

Furukawa, T.; Kuno, M., and Matsuura, S.: Quantal analysis of a decremental response at hair cell-afferent fibre synapses in the goldfish sacculus. *J. Physiol. 322:* 181–195, 1982.

Furukawa, T., and Matsuura, S.: Adaptive rundown of excitatory postsynaptic potentials at synapses between hair cells and eighth nerve fibres in the goldfish. *J. Physiol. 276:* 193–209, 1978.

Geisler, C. D.; Le, S., and Schwid, H.: Further studies on the Schroeder-Hall hair-cell model. *J. Acoust. Soc. Am. 65:* 985–990, 1979.

Hama, K., and Saito, K.: Fine structure of the afferent synapse of the hair cells in the saccular macula of the goldfish, with special reference to the anastomosing tubules. *J. Neurocytol. 6:* 361–373, 1977.

Harris, D., and Dallos, P.: Forward masking of auditory nerve fiber responses. *J. Neurophysiol. 42:* 1083–1107, 1979.

Kuno, M.: Adaptive changes in firing rates in goldfish auditory fibers as related to changes in mean amplitude of excitatory postsynaptic potentials. *J. Neurophysiol. 50:* 573–581, 1983.

Kuno, M.; Matsuura, S., and Kyogoku, I.: Effects of temperature changes on the peripheral auditory organ of the goldfish. In Webster, W. R., and Aitkin, L. M. (Eds.): *Mechanisms of Hearing.* Clayton, Victoria, Australia, Monash University Press, 1983, p. 61.

Kusano, K., and Landau, E. M.: Depression and recovery of transmission at the squid giant synapse. *J. Physiol. 245:* 13–32, 1975.

Liberman, M. C.: The cochlear frequency map for the cat: Labelling auditory-nerve fibers of known characteristic frequency. *J. Acoust. Soc. Am. 72:* 1441–1449, 1982.

Nakajima, Y., and Wang, D. W.: Morphology of afferent and efferent synapses in hearing organ of the goldfish. *J. Comp. Neurol. 156:* 403–416, 1974.

Sewell, W. F.: The relation between the endocochlear potential and spontaneous activity in auditory nerve fibres of the cat. *J. Physiol. 347:* 685–696, 1984.

Smith, R. L.: Short-term adaptation in single auditory nerve fibers: Some poststimulatory effects. *J. Neurophysiol. 40:* 1098–1112, 1977.

Smith, R. L.: Adaptation, saturation, and physiological masking in single auditory-nerve fibers. *J. Acoust. Soc. Am. 65:* 166–178, 1979.

Smith, R. L., and Brachman, M. L.: Adaptation in auditory-nerve fibers: A revised model. *Biol. Cybernetics 44:* 107–120, 1982.

Smith, R. L., and Zwislocki, J. J.: Short-term adaptation and incremental responses of single auditory-nerve fibers. *Biol. Cybernetics 17:* 169–182, 1975.

Chapter 3

PRIMARY AFFERENT TRANSMISSION IN ACOUSTICOLATERALIS ORGANS

Paul S. Guth, Charles H. Norris, and William F. Sewell

I. Introduction
II. Experimental Strategies for Identifying Neurotransmitters
III. The Release-Detection Paradigm
IV. Identification of Transmitters in Various Acousticolateralis Systems
References

I. INTRODUCTION

The identity of the primary afferent transmitter(s) of audition is not known. In fact, none of the afferent transmitters in the labyrinthine organs (e.g., saccule, utricle, semicircular canals, lagena/cochlea) is known. However, research efforts in this regard are not only continuing, but gaining momentum, as more laboratories become involved in what now appears to be a problem that can be solved.

Although we do not know what the primary, afferent auditory transmitter is, we do know what it is not. It seems not to be any of the standard, small-molecule transmitters, such as acetylcholine (ACh), the monoamines (5-hydroxytryptamine, norepinephrine, and dopamine) and γ-aminobutyric acid (GABA) (Guth and Melamed, 1982). The frontrunner candidates are excitatory amino acids, such as glutamate, aspartate, or some closely-related substance. As will be clear later in this chapter, circumspection in this regard is appropriate until the evidence is more convincing.

We should also be cautious in making the assumption that the transmitters of all hair cell-neuronal synapses are the same. Although this hypothesis appears to be reasonable, accumulating pharmacological evidence casts doubt upon it.

Sensation, sensory neurotransmission, and the effects of drugs on sensory processes have not been of classical concern to pharmacologists. Only in very recent times has there been a body of literature which can be identified as auditory or labyrinthine pharmacology. Clearly, the pharmacology of the sensitive mechanosensory systems is a rapidly growing area of interest.

Gisselson (1950) was the first to awaken a general interest in transmitters of the auditory system. Based on the observation that the anticholinesterase agent, physostigmine increased the latency of the N_1 component of the compound action potential, Gisselson suggested that ACh might be involved in cochlear neurotransmission. Similar suggestions were later made by Davis (1957) and by Vinnikov and Titova (1964) concerning the auditory role of ACh.

A decade earlier, Martini (1941) had detected ACh-like biological activity in cochlear fluids during sound stimulation. Norris and Guth (1974) also found ACh in cochlear perilymph. However, in company with Gisselson (1952) and Ranke (1953), Norris and Guth (1974) found no increase in ACh content in perilymph following sound stimulation. Using the technique of cochlear perilymphatic collection pioneered by Konishi and Kelsey (1968), Norris and Guth (1974) found that perilymphatic ACh did increase upon stimulation of the so-called crossed olivo-cochlear bundle (COCB). The latter experiments helped to substantiate the idea that perilymph is in diffusional contact with the hair-cell synaptic region. This idea is basic to pharmacologic studies, in which one hopes to obtain a putative transmitter by collecting perilymph, or where one wants to apply substances to an auditory synapse by way of the perilymph. The release of ACh upon COCB stimulation helped to establish that at least some of the cochlear efferents are cholinergic. There had already been clues pointing to that possibility, such as the association of acetylcholinesterase with the efferents (Schuknecht et al., 1959).

II. EXPERIMENTAL STRATEGIES FOR IDENTIFYING NEUROTRANSMITTERS

Until recently, there were no clues pointing toward the nature of the primary afferent transmitter of audition or of other acousticolateralis senses. In this regard, there are three possible strategies for discovering the identity of such a transmitter: (1) the *biochemical strategy,* (2) the *pharmacological strategy,* and (3) the *release-detection strategy* (also known as the Loewi paradigm).

The *biochemical strategy* employs labeled transmitter precursors to determine the ability of a given cell or organ to take up a putative transmitter or, possibly, to localize histochemically the transmitter candidate within a cell. Labeled antibodies can also be used to determine the presence of an enzyme capable of synthesizing the suspected transmitter (see Fex et al., Chapter 1, this volume).

The *pharmacological strategy* depends upon the mimicking of natural neurotransmission by the application of transmitter candidates, or the influencing of natural neurotransmission by drugs (such as antagonists or potentiators) in a manner predictable from the known interactions with the putative transmitter.

The biochemical and the pharmacological strategies must presuppose, or guess at, the identity of the transmitter before deciding on the appropriate precursor, antibody, mimetic, or antagonist to use. If the true transmitter is on the list of possible, identified transmitters, then these approaches are useful. If, as may be the case, the true transmitter is not on such a list, then these approaches will fail. If the true transmitter is not thought to be on the list of transmitter candidates, then one may resort to the strategy of release-detection.

The *release-detection strategy* makes no presuppositions as to the identity of the transmitter(s), but only that it is possible to cause release of a sufficiently large amount of transmitter(s) that terminating mechanisms are saturated and some natural transmitter overflows into the extracellular fluid. The transmitter, having overflowed, can now be detected *biologically,* i.e., by its ability to cause the postsynaptic element to respond appropriately. This simple but powerful stratagem was first used by Otto Loewi (1921) to detect what was later identified as ACh from the vagus nerve.

III. THE RELEASE-DETECTION PARADIGM

We applied the strategy of release-detection to the study of the primary afferent transmitter of audition in the frog (Sewell *et al.,* 1978) and were able to demonstrate the presence of an "auditory nerve-activating substance" (ANAS). ANAS appeared in the perilymph of frogs and guinea pigs upon stimulation with intense sound. Its presence was determined by the ability of perilymph samples containing it to cause an increase in the firing rate of afferent single units derived from the frog auditory papillae. ANAS was found only in perilymph collected during sound stimulation, and not in perilymph collected during quiet. Guinea-pig perilymph samples, either containing ANAS or devoid of ANAS, were analyzed for the presence of several obvious constituents, such as electrolytes (Table I) and amino acids (Table II) (Melamed *et al.,* 1982).

There were no differences between ANAS-containing and ANAS-devoid samples in any of the constituents measured. Thus, the conclusion had to be that ANAS was not an electrolyte, glutamate (GLU), aspartate (ASP), or γ-aminobutyric acid (GABA).

This same release-detection paradigm has been used successfully by Obara's group (Umekita *et al.,* 1980) in searching for the primary afferent transmitter released from electroreceptor cells. Obara has likewise concluded that the afferent transmitter cannot be GLU, because the concentration of exoge-

TABLE 3-I
CONSTITUENTS OF GUINEA-PIG PERILYMPH COLLECTED EITHER
DURING ACOUSTIC STIMULATION OR DURING SILENCE

	Acoustic Stimulation Mean ± SE	N	Silence Mean ± SE	N
Na^+ (meq/L)	138 ± 4	20	141 ± 4	16
K^+ (meq/L)	2.2 ± 0.15	20	2.2 ± 0.05	16
Ca^{++} (meq/L)	2.24 ± 0.15	20	2.04 ± 0.1	28
Osmolarity (mosmol/kg)	327 ± 12.4	5	328 ± 12.1	6
Glucose (mg/100 ml)	77 ± 25	12	81 ± 19	18

Acoustic stimulation used in these studies released auditory nerve-activating substance (see Sewell et al., 1978).

TABLE 3-II
AMINO ACID CONTENT OF PERILYMPHATIC PERFUSATE

Amino Acid	(μM) Sound	(μM) Quiet
Glutamine	30.6 ± 5.8	25.3 ± 4.2
Glycine	21.8 ± 3.9	12.6 ± 2.9
Aspartate	0.08 ± 0.001	0.09 ± 0.05
Glutamate	0.18 ± 0.06	0.21 ± 0.07
Serine	0.99 ± 1.1	1.1 ± 0.22
Alanine	23.1 ± 3.3	14.5 ± 2.8
GABA	0.37 ± 0.07	0.38 ± 0.07
Proline	0.71 ± 0.08	0.89 ± 0.17
Methionine	0.56 ± 0.09	0.86 ± 0.2
Valine	9.6 ± 0.09	8.4 ± 1.9
Tryptophan	0.11 ± 0.02	0.13 ± 0.02

From Melamed et al. (1982). See text.

nous GLU needed to activate the afferent dendrites is far greater than the concentration normally present (endogenously) during the release and detection experiment.

Sewell and Mroz (1984) are employing the release-detection stratagem with some success in searching for the transmitter released from fish saccular hair cells. They are using K^+-induced release, as is Bobbin's group (Jenison et al., 1984) in the mammalian cochlea. Bobbin (Bledsoe et al., 1980) has also used mechanical release and chemical detection in searching for the primary afferent transmitter of the lateral line (see Chapter 7, this volume).

Having successfully demonstrated the release of a substance with biological activity, the next phase of research must involve the isolation, purification and identification of the biologically active substance. Drescher *et al.* (1983) have produced exciting results in this regard, using chemical separation and detection techniques, after release of substances from the cochlea (see Chapter 4, this volume).

IV. IDENTIFICATION OF TRANSMITTERS IN VARIOUS ACOUSTICOLATERALIS SYSTEMS

Several review articles (Guth *et al.*, 1976; Guth *et al.*, 1981; Klinke, 1981; Guth and Melamed, 1982), in covering the subject of the nature of the primary afferent transmitter in acousticolateralis organs, have circumspectly, and correctly, stated that the identities are as yet unknown. These articles have considered the small-molecule transmitters in detail; Table III simply summarizes those results.

TABLE 3-III
STUDIES WITH MAJOR PRIMARY, AFFERENT AUDITORY-TRANSMITTER CONDIDATES

Candidates	Biochemical Findings	Pharmacological Findings
Acetylcholine	AChE, ChAT associated with OCB; not synthesized by amphibian basilar papilla	Antagonists do not affect afferent transmission
Catecholamines	Not synthesized; not present by histochemistry	Mimetics and antagonists do not consistently interfere with afferent transmission; reserpine and guanethidine affect electron density of synaptic body
5-Hydroxytryptamine	Not synthesized	Mimetics and antagonists have some effect on CAP, but show no consistent pattern
GABA	Synthesized from precursors (may be nonspecific); little in cochlea	No effect on cochlea or semicircular canal; picrotoxin inactive in lateral line; picrotoxin and bicuculline inactive in cochlea; GABA excites saccule, bicuculline and picrotoxin block
Glutamate	AAT not found in inner hair cells; glutamate found in organ of Corti; glutamine taken up by hair cells; glutamate taken up by glia	Excitatory on all synapses studied; NMDA inactive in cochlea, active in lateral line; DAA inactive in cochlea, active in lateral line and semicircular canal

Experimental results summarized above are described in cited review articles (see text). Picrotoxin and bicuculline are GABA antagonists. Abbreviations: AAT = aspartate aminotransferase; AChE = acetylcholinesterase; CAP = compound action potential; ChAT = choline acetyltransferase (ACh-synthesizing enzyme); DAA = D-α-aminoadipate (a glutamate antagonist); NMDA = N-methyl-D-aspartate (a glutamate mimetic); OCB = olivocochlear bundle.

TABLE 3-IV
ACTIVITIES OF SMALL-MOLECULE TRANSMITTERS IN ACOUSTICOLATERALIS ORGANS

Transmitters	Lateral Line[1]	Cochlea[2]	Papilla[3]/Saccule[4]	Semicircular Canal[5]
L-Glutamate	+	+	nt	+
L-Aspartate	+	+	nt	+
Acetylcholine	+/−	−	nt	+/−
GABA	−	0	+	0
Serotonin	0	0	nt	nt
DOPA/Dopamine	0	0	nt	nt
Epinephrine/Norepinephrine	0	0	nt	nt
Histamine	0	0	nt	−

[1]Bobbin et al. (1984; [2]As reviewed in Guth and Melamed (1982); [3]Personal observations (1983); [4]Felix and Ehrenberger (1981); [5]Personal observations (1984) and Valli et al. (1984). 0 = inactive; − = inhibitory; + = excitatory; +/− = biphasic effect; nt = not tested. Serotonin is the same as 5-hydroxytryptamine, and DOPA refers to 3,4-dihydroxyphenylalanine.

As the pharmacology of the acousticolateralis organs is studied in greater depth, the realization emerges that these organs do not respond similarly to the application of drugs (Table IV). Consequently, the operative assumption that the transmitter(s) might be the same for all acousticolateralis synapses is probably wrong. For example, the basilar-papillar and saccular afferents respond to the application of GABA by increasing their firing rates, whereas cochlear and ampullar afferents fail to respond, and lateral-line afferents are inhibited (Table IV).

GABA has, in fact, been suggested as the primary afferent transmitter in the saccule of mammals (Felix and Ehrenberger, 1977, 1981, Chapter 5, this volume), based on the above-cited excitatory effect and the antagonism of afferent transmission by the GABA antagonists, picrotoxin and bicuculline (Felix and Ehrenberger, 1981).

At other afferent synapses in acousticolateralis organs, GABA, ACh, catecholamines, and 5-hydroxytryptamine may be removed from consideration as transmitter candidates. Of the small-molecule transmitters, only the excitatory amino acids are currently under consideration. The problems with their candidacy are obvious: they are found in all cells and are almost universally excitatory. Even in a junction generally thought to be glutamatergic, namely, the crayfish neuromuscular junction, Ishida and Shinozaki (1980) have produced some pharmacological evidence which is not supportive of a transmitter role for GLU. In fact, it may be said that advocates for GLU are engaged in producing evidence supportive of GLU's transmitter role, when, what should be sought, if the possibility is to be stringently tested, is an incisive disproof. If GLU's candidacy stands up to that test, the candidacy is much more likely to be a valid one.

The present authors have run into the problem of the candidacy of GLU as the afferent transmitter in the frog semicircular canal. According to our data, GLU causes an increase in the firing rate of ampullar afferents by releasing the natural transmitter (Valli et al., 1984). This release of the natural transmitter is blocked presynaptically by the GLU antagonist, α-aminoadipate. Furthermore, other manipulations which prevent release of the natural ampullar transmitter also prevent GLU from increasing ampullar afferent firing rates. Thus, GLU's candidacy as the primary afferent transmitter, at least in the semicircular canal, cannot be comfortably accommodated.

In summary, the primary afferent transmitters of acousticolateralis organs are still not known. The transmitters may be different at different synapses. The front-running candidates are the excitatory amino acids and GABA. Although evidence for the identification of acousticolateralis transmitters is still incomplete, this is a most exciting time for sensory pharmacologists and biochemists. Once the identities of the sensory transmitter(s) are known, rational therapy for sensory disorders is likely to become a reality.

REFERENCES

Bledsoe, S.; Bobbin, R. P.; Thalmann, R., and Thalmann, I.: Stimulus-induced release of endogenous amino acids from skins containing the lateral line organ in *Xenopus laevis*. *Exp. Brain Res. 40:* 97-101, 1980.

Bobbin, R. P.; Bledsoe, S., Jr., Winbery, S., and Jenison, G.: Actions of putative neurotransmitters and other relevant compounds on Xenopus laevis lateral line. *Assoc. Res. Otolaryngol. Abstr. 7:* 131-132, 1984.

Davis, H.: Biophysics and physiology of the inner ear. *Physiol. Rev. 37:* 1-49, 1957.

Drescher, M. J.; Drescher, D. G., and Medina, J. E.: Effect of sound stimulation of several levels on concentrations of primary amines, including neurotransmitter candidates, in perilymph of the guinea pig inner ear. *J. Neurochem. 41:* 301-320, 1983.

Felix, D., and Ehrenberger, K.: The action of GABA and acetylcholine in the labyrinth of the cat. In Portmann, M., and Aran, J.-M (Eds.): *Inner Ear Biology, Vol. 68*, Paris, INSERM, 1977, pp. 147-154.

Felix, D., and Ehrenberger, K.: The action of putative neurotransmitter substances in the cat labyrinth. *Acta. Otolaryng. 93:* 101-109, 1981.

Gisselson, L.: Experimental investigation into the problem of humoral transmission in the cochlea. *Acta. Otolaryng. Suppl. 82:* 3-78, 1950.

Gisselson, L.: The effect of acetylcholine-esterase inhibiting substances on the muscles of the middle ear and on the latency of the cochlear potentials. *Acta. Otolaryng. 42:* 208-218, 1952.

Guth, P. S., and Melamed, B.: Neurotransmission in the auditory system: A primer for pharmacologists. *Annu. Rev. Pharmacol. Toxicol. 22:* 383-412, 1982.

Guth, P. S.; Norris, C. H., and Bobbin, R. P.: The pharmacology of transmission in the perilymph auditory system. *Pharmacol. Rev. 28:* 95-127, 1976.

Guth, P. S.; Sewell, W. F., and Tachibana, M.: The pharmacology of the cochlear afferents and cochlear nucleus. In Brown, R. D. (Ed.): *Pharmacology of Hearing: Experimental and Clinical Bases.* New York, John Wiley and Sons, 1981, pp. 99-136.

Ishida, M., and Shinozaki, H.: Differential effects of diltiazem on glutamate potentials and excitatory junctional potentials at the crayfish neuromuscular junction. *J. Physiol. 298:* 301-319, 1980.

Jenison, G. L.; Bobbin, R. P., and Thalmann, R.: K^+-induced release of glutamate and taurine into perilymph is calcium-dependent. *J. Acoust. Soc. Am. 75:* S82, 1984.

Klinke, R.: Neurotransmitters in the cochlea and cochlear nucleus. *Acta. Otolaryng. 91:* 541-554, 1981.

Konishi, T., and Kelsey, E.: Effect of cyanide on cochlear potentials. *Acta. Otolaryng. 65:* 381-390, 1968.

Loewi, O.: Uber Humorale ubertragbarkeit der Herznervenwerkung. *Arch. F. Ges. Physiologie 189:* 239-242, 1921.

Martini, V.: Liberazione di sostanza acetilcolinosimile nel-l'orecchio interno durante la stimolazione sonora. *Arch. Sci. Biol. 27:* 94-100, 1941.

Melamed, B.; Norris, C., Bryant, G., and Guth, P.: Amino acid content of guinea pig perilymph collected under conditions of quiet or sound stimulation. *Hearing Res. 7:* 13-18, 1982.

Norris, C. H., and Guth, P. S.: The release of acetylcholine (ACh) by the crossed olivo-cochlear bundle (COCB). *Acta. Otolaryng. 77:* 318-326, 1974.

Ranke, O. F.: Physiologie des Gehörs. In Trendelenburg, W., and Schütz, E. (Eds.): *Lehrbuch der Physiologie. Gehör, Stimme, Sprache* (Vol. by Ranke, O. F., and Lullies, H.). Berlin, Springer-Verlag, 1953, pp. 3-162.

Schuknecht, H. F.; Churchill, J. A., and Doran, R.: The localization of acetylcholinesterase in the cochlea. *Arch. Otolaryng. 69:* 549-559, 1959.

Sewell, W. F., and Mroz, E.: Personal communication, 1984.

Sewell, W. F.; Norris, C. H., Tachibana, M., and Guth, P. S.: Detection of an auditory nerve activating substance. *Science 202:* 910-912, 1978.

Umekita, S.-I.; Matsumoto, Y.; Abe, T., and Obara, S.: The afferent neurotransmitter in the ampullary electroreceptors: stimulus-dependent release experiments refute the transmitter role of L-glutamate. *Neurosci. Lett. Suppl. 4:* 5-7, 1980.

Valli, P.; Zucca, G., Prigioni, I., Botta, L., Casella, C., and Guth, P.: The effect of glutamate on the frog semicircular canal. *Brain Res.*, 1984. In press.

Vinnikov, Ya. A., and Titova, L. K.: The organ of Corti. New York, Consultants Bureau, 1964.

Chapter 4

HPLC ANALYSIS OF PRESUMPTIVE NEUROTRANSMITTERS IN PERILYMPH

Dennis G. Drescher and Marian J. Drescher

I. Introduction
II. HPLC Methods
III. Primary Amine Composition of Perilymph
IV. Effect of Sound Stimulation on Concentrations of Primary Amines in Perilymph
V. Glutamic Acid as a Hair-Cell Transmitter
VI. GABA as an Acousticolateralis Transmitter
VII. Summary
References

I. INTRODUCTION

Perilymph is an extracellular fluid that bathes the sensory cells and nerve terminals of the cochlea (Schuknecht, 1970). Evidence that presumptive auditory neurotransmitters are secreted into perilymph was first presented by Fex (1973) and Norris and Guth (1974), who demonstrated an increase in acetylcholine bioactivity in perilymph after electrical stimulation of crossed olivocochlear fibers. Acetylcholine is thought to be an efferent, inhibitory neurotransmitter in the cochlea (Fex, 1968; Jasser and Guth, 1973; Fex and Wenthold, 1976; Godfrey *et al.*, 1976b). The subsequent electrophysiological detection of a perilymphatic "auditory nerve-activating substance" (Sewell *et al.*, 1978), a candidate for the primary, afferent auditory transmitter, further suggested that perilymph reflects cochlear activity. Therefore, we applied high-performance-liquid chromatographic (HPLC) techniques to the analysis of mammalian perilymph, in order to identify chemically presumptive neurotransmitters in this fluid.

II. HPLC METHODS

Because many presumed neurotransmitters, such as the amino acids and peptides, have primary amine groups, automated amino acid analysis is a technique which is well suited to scan for neurotransmitters in perilymph. However, for classical amino acid analysis (Spackman et al. 1958; Moore and Stein, 1963), large volumes of sample are required and sensitivity is typically in the micromole-to-nanomole range. Cation-exchange HPLC with fluorescence detection, a relatively recent development (Benson, 1975; Hare, 1975; Drescher and Lee, 1978; Lee and Drescher, 1978, 1979), allows analysis of primary amines at picomole-to-femtomole concentrations, and, in addition, permits analyses of high resolution.

In the present studies, we used an Aminco amino-acid analyzer. The stainless-steel chromatography column, 50 cm × 0.24 cm (internal diameter), contained a sulfonated polystyrene cation-exchange resin, 8.75% crosslinked with divinylbenzene, with bead diameter of 5 ± 0.5 μm. Lithium-based buffers were used for elution of compounds (Drescher et al., 1983). The primary amine groups were detected fluorometrically as isoindole adducts, which were formed by the reaction of primary amines with ortho-phthalaldehyde (OPA) and 2-mercaptoethanol (2-ME) in basic solution (Fig. 1). Detection was by means of a ratio fluorometer, with filtered excitation peak at 360 nm, emission peak at 500 nm, and sufficient overlap in spectra to allow detection of isoindole adducts, which have a peak of excitation of 340 nm and a peak of emission at 455 nm (Roth, 1971). The ratio fluorometer reduced baseline instability by comparing the fluorescence intensity of the effluent with that of the lamp. The output from the fluorometer was fed simultaneously to three strip-chart recorders, which produced plots 1/6, 1/2, and 1 times full-scale (25 cm), respectively.

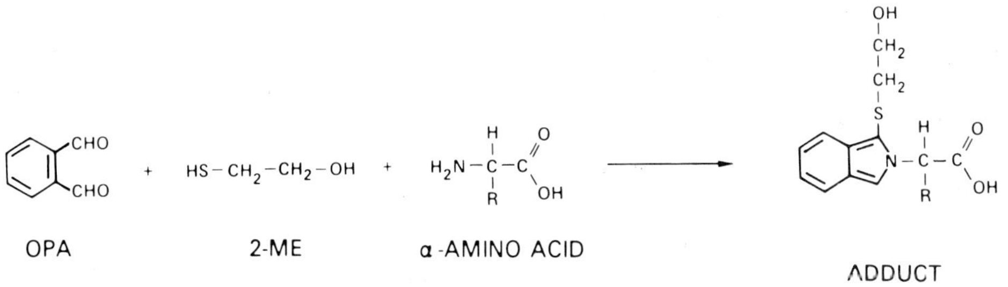

Figure 4-1. Scheme for combination of o-phthalaldehyde (OPA), 2-mercaptoethanol (2-ME) and α-amino acids to yield highly fluorescent adducts (Roth, 1971; Simons and Johnson, 1976, 1977). (From Lee and Drescher, 1978.)

Using six buffer changes, two changes in column-pump pressure, and a temperature change, we were able to achieve high resolution and sensitivity, suitable for the analysis of primary amines in mammalian perilymph, as shown in Fig. 2. In all, 81 different primary-amine compounds, indicated by the small dots in the illustration, were resolved in 10 µl of guinea-pig perilymph. The limit of sensitivity for the analysis of Fig. 2 was about 100 femtomoles. Ten neurotransmitter candidates could be analyzed simultaneously: taurine, aspartic acid, glutamic acid, glycine, γ-aminobutyric acid (GABA), methionine-enkephalin, leucine-enkephalin, carnosine, norepinephrine, and dopamine. Peaks were quantitated by measuring peak heights in comparison to standards, or by integrating peak areas and making similar comparisons. Norleucine, included as an internal standard, indicated standard deviations between runs of $\pm$ 3% (N = 10). The present analytical method is, to our knowledge, the most sensitive aqueous, cation-exchange procedure available today. Further increases in sensitivity of the present analysis appear to be limited only by such variables as purity of available buffer reagents, steadiness of pump pressure, and magnitude of signal/noise ratio of electronic components.

III. THE PRIMARY AMINE COMPOSITION OF PERILYMPH

Cochlear perilymph probably originates from several different sources. One source is the cerebrospinal fluid (CSF) that reaches the cochlea by way of the cochlear aqueduct. The cochlear aqueduct is a bony channel that connects the scala tympani of the basal cochlear turn with the subarachnoid space of the posterior cranial cavity; thus CSF can contact and mix with perilymph (Schuknecht and Seifi, 1963). Another, probably minor, source of perilymph is the CSF that enters the cochlea through the modiolus, the channel through which the auditory nerve reaches the hair cells. A third source is likely to be the system of blood vessels that supplies the inner ear itself (Axelsson, 1968). The hair cells, supporting cells, and nerve fibers of the cochlea are in contact with perilymph (Schuknecht and Seifi, 1963; Masuda et al., 1971) and probably also exchange metabolites and oxygen with this fluid.

Crifò and Crifò (1971), using classical, automated liquid-chromatographic methods, first determined amino acids in perilymph obtained postmortem from horses. However, the study of the amino acid content of living animals received little attention until 1981, when Medina and Drescher first quantitated major primary amine fractions in cochlear perilymph and CFS of living, anesthetized guinea pigs by HPLC with fluorescence detection. The results of these initial analyses are shown in Table 1. The chemical composition of perilymph in Table I indicates that CSF may contribute to the formation of perilymph, but that perilymph is nevertheless a fluid with its own specific properties. Glycine and alanine concentrations are markedly elevated in perilymph compared to CSF, a finding subsequently confirmed by Thalmann et al. (1982). This observation implies that glycine and alanine, and possibly

Presumptive Neurotransmitters in Perilymph

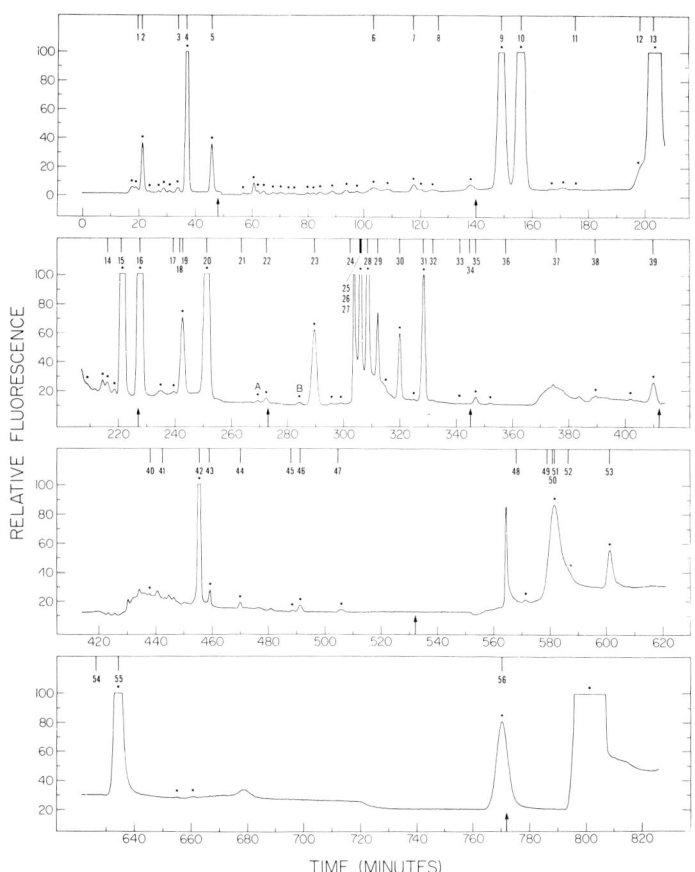

Figure 4-2. Expanded HPLC analysis of 10 μl of perilymph collected from control guinea pigs. Relative fluorescence of OPA/2-ME adducts of primary amines is plotted on the ordinate. Vertical arrows on the abscissa show timed changes in lithium citrate buffer, temperature, and pressure (Drescher et al., 1983). Small dots mark perilymphatic peaks that are present in the sample chromatographs and absent in the blank. Elution positions of standard compounds are numbered at the top of the chromatogram: 1, L-cysteic acid; 2, O-phospho-DL-serine; 3, L-α-glycerophosphorylethanolamine; 4, taurine; 5, O-phosphoethanolamine; 6, γ-L-glutamyl-L-glutamic acid; 7, L-aspartic acid; 8, reduced glutathione; 9, L-threonine; 10, L-serine; 11, L-asparagine; 12, L-glutamic acid; 13, L-glutamine; 14, L-α-aminoadipic acid; 15, L-glycine; 16, L-alanine; 17, L α amino-n-butyric acid; 18, D-(+)-glucosamine; 19, L-citrulline; 20, L-valine; 21, DL α aminopimelic acid, 22, L-cystine; 23, L-methionine; 24, L-alanylglycylglycine; 25, L-(+)-cystathionine; 26, L-isoleucine; 27, tetraglycine; 28, L-leucine; 29, L-norleucine; 30, L-tyrosine; 31, L-phenylalanine; 32, DL-γ-amino-β-hydroxybutyric acid; 33, DL-β-amino-n-butyric acid; 34, L-homocystine; 35, β-alanine; 36, DL-β-aminoisobutyric acid; 37, 5-hydroxy-L-tryptophan; 38, γ-amino-n-butyric acid (GABA); 39, L-tryptophan; 40, methionine-enkephalin; 41, leucine-enkephalin; 42, L-histidine; 43, L-3-methylhistidine; 44, L-1-methylhistidine; 45, L-carnosine; 46, L-homocarnosine; 47, L-anserine; 48, L-norepinephrine; 49, α amino-β-guanidinopropionic acid; 50, DL-hydroxylysine; 51, 2-aminoethanol; 52, ammonia; 53, L-ornithine; 54, dopamine; 55, L-lysine; 56, L-arginine (From Drescher et al., 1983.)

other amino acids co-chromatographing with them, arise from sources other than CSF, sources that are probably located within the cochlea.

TABLE 4-I
AMINO ACID ANALYSIS OF PERILYMPH AND CEREBROSPINAL FLUID

	Perilymph	Cerebrospinal Fluid
Taurine	5.1 ± 0.7	6.0 ± 1.1
Aspartic Acid	0.8 ± 0.4	0.8 ± 0.2
Glutamic Acid	11.7 ± 12.4	9.3 ± 4.8
Glycine	45.1 ± 14.0	11.6 ± 2.6
Alanine	78.0 ± 33.2	21.6 ± 5.9
Valine	15.0 ± 1.7	12.3 ± 2.6
Methionine	5.5 ± 0.8	6.7 ± 1.2
Isoleucine	10.3 ± 2.0	7.8 ± 0.5
Leucine	18.5 ± 3.9	16.6 ± 1.9
Tyrosine	10.6 ± 3.9	9.5 ± 2.7
Phenylalanine	9.4 ± 1.7	10.8 ± 1.3
Tryptophan	7.1 ± 1.9	5.6 ± 0.8
Lysine	54.4 ± 5.9	69.3 ± 3.3
Ammonia	81.4 ± 74.2	43.0 ± 19.1
Arginine	34.5 ± 7.5	36.3 ± 3.1

Table I lists the concentrations of free amino acids and ammonia (μM; mean ± SD, N = 3) in perilymph and cerebrospinal fluid obtained from the guinea pig. Perilymph was collected from the apical turn. (From Medina and Drescher, 1981.)

When guinea pig perilymph is analyzed by HPLC at higher resolution, the number of primary amine compounds which can be detected is larger than that revealed by simpler analyses. Table II shows the normative values of primary amine compounds in perilymph from animals kept in silence (Drescher et al., 1983), as obtained by an expanded analysis illustrated in Fig. 2. Table II lists the concentration of each component in perilymph, analyzed in this way, and identified with respect to the standard that eluted in the same position. Standards were chromatographed singly, and mixed with sample perilymph, to define with certainty positions of coelution. In all, concentrations of 35 identified primary-amine components are listed in Table II. The concentrations of about 46 unidentified primary amine components are not listed in the table; these ranged in concentration from 20 nM to 2 μM, relative to the fluorescence of the leucine-OPA/2-ME adduct.

The perilymph analyzed in our studies was collected from a small hole drilled in the scala tympani of either the basal or the apical cochlear turn. The relative contribution of CSF to perilymph formation for closed vs. open

TABLE 4-II
AMINO ACIDS AND RELATED COMPOUNDS IN PERILYMPH
FROM GUINEA PIGS MAINTAINED IN SILENCE,
DETERMINED BY EXPANDED HPLC ANALYSIS

Components	Concentration (μM)	
O-Phospho-DL-serine	0.63 ± 0.18	(3)
L-α-Glycerophorylethanolamine	0.29 ± 0.02	(4)
Taurine	13.26 ± 1.30	(4)
O-Phosphoethanolamine	4.42 ± 0.25	(4)
γ-L-Glutamyl-L-glutamic acid	0.56 ± 0.06	(4)
L-Aspartic acid	0.80 ± 0.19	(4)
L-Threonine	28.66 ± 1.16	(4)
L-Serine	46.34 ± 4.00	(4)
L-Asparagine	0.06 ± 0.01	(4)
L-Glutamic acid	3.59 ± 0.60	(4)
L-Glycine	39.15 ± 13.34	(4)
L-Alanine	68.48 ± 6.26	(4)
L-α-Amino-n-butyric acid	0.38 ± 0.09	(4)
L-Citrulline	6.35 ± 0.73	(4)
L-Valine	30.20 ± 1.06	(4)
L-Cystine	1.79 ± 0.27	(4)
L-Methionine	10.84 ± 1.11	(4)
L-Isoleucine	12.92 ± 0.42	(4)
L-Leucine	24.29 ± 1.52	(4)
L-Tyrosine	9.47 ± 0.48	(4)
L-Phenylalanine	10.02 ± 0.86	(4)
β-Alanine	0.92 ± 0.15	(4)
γ-Amino-n-butyric acid	0.14 ± 0.01	(4)
L-Tryptophan	2.30 ± 0.17	(4)
Methionine-enkephalin	4.30 ± 0.58	(4)
L-Histidine	21.72 ± 2.58	(4)
L-3-Methylhistidine	1.70 ± 0.31	(4)
L-1-Methylhistidine	0.26 ± 0.05	(4)
L-Carnosine	0.05 ± 0.04	(4)
L-Homocarnosine	0.92 ± 0.15	(4)
L-Anserine	0.28 ± 0.07	(4)
2-Aminoethanol	26.96 ± 6.61	(4)
L-Ornithine	7.42 ± 0.54	(4)
L-Lysine	50.56 ± 4.40	(4)
L-Arginine	21.78 ± 1.11	(4)

Concentration values expressed as mean ± SEM; number of animals given in parentheses. Glutamine and ammonia were not quantified in this analysis; estimates of their concentrations in normative perilymph (Drescher et al., 1981) are 408.5 and 13.4 μM, respectively. (From Drescher et al., 1983).

cochlea (in the latter case where CSF pressure is relieved) is at present

IV. EFFECT OF SOUND STIMULATION ON THE CONCENTRATION OF PRIMARY AMINES IN PERILYMPH

In experiments aimed at identifying auditory neurotransmitters in perilymph, Drescher *et al.* (1983) exposed guinea pigs to sound and analyzed their perilymph. The acoustic stimulus was presented at three different levels in order to identify compounds that increase in perilymph in proportion to stimulus intensity, and thus identify those that are most likely to have neurotransmitter roles. This procedure was based on the premise that induced release by, and in proportion to, physiological stimuli of the neurosecretory cells is one of the primary criteria for identification of neurotransmitters (Orrego, 1979). Error due to variation in the composition of perilymph between animals was minimized by exposing single animals to both noise and silence in different periods and comparing the perilymph composition across the periods. Fluid was sampled from the scala tympani of the first cochlear turn in a manner similar to that described in studies in which auditory bioactivity was detected in perilymph (Sewell *et al.*, 1978).

Figure 3 outlines the experimental procedures employed. After surgery,

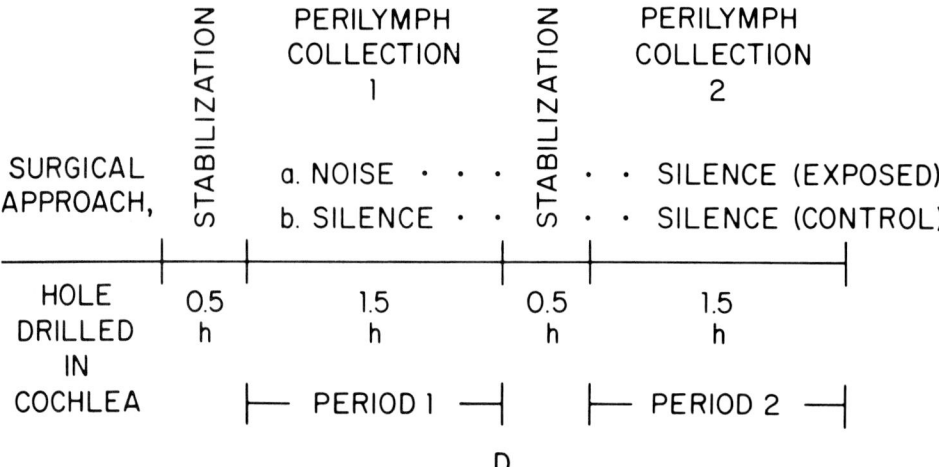

Figure 4-3. Schedule for collection of perilymph as described in the text. Axis indicates time elapsed; the surgical procedure required 2–3 h. D marks time of disarticulation of the stapes. (From Drescher *et al.*, 1983.)

each animal was allowed to stabilize for 0.5 h in the silence of a soundproof booth (50 dB SPL overall) and perilymph was subsequently collected during a 1.5-h exposure to wide-band noise at 80, 90, or 115 dB SPL. Following exposure, the stapes was disarticulated and the animal was stabilized for 0.5 h, after which perilymph was again collected, in silence, for 1.5 h. The first 1.5 h collection is termed "period 1" and the second period, "period 2." For the control animals, the schedule for collection of perilymph was similar, except that silence replaced noise in period 1.

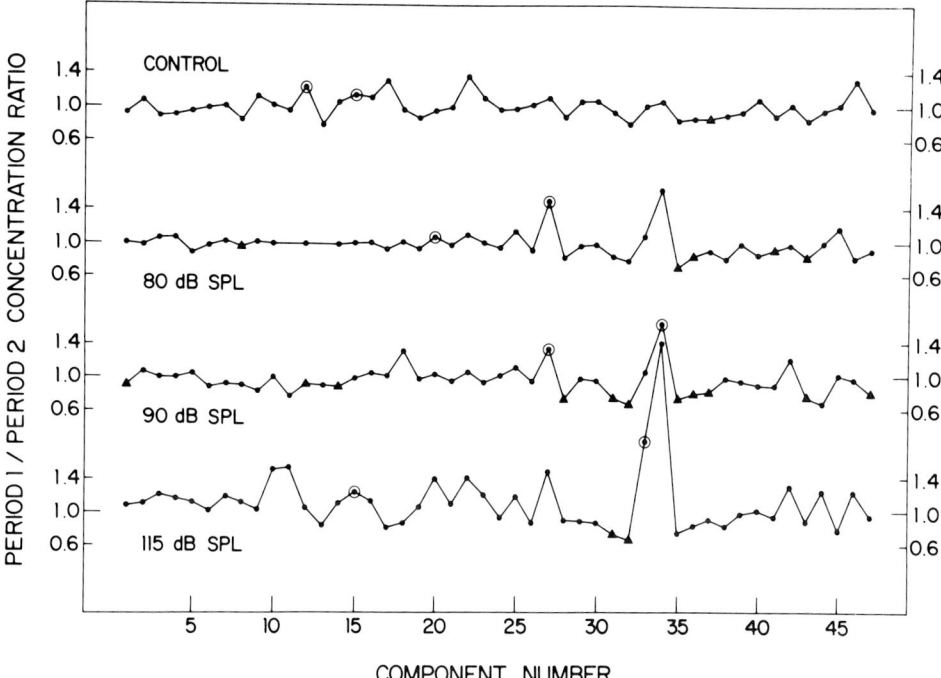

Figure 4-4. Plots of mean period 1:period 2 concentration ratios for perilymph components, numbered according to elution position in shortened HPLC analysis (Drescher et al., 1983). The mean ratios were obtained from three control animals, four animals at 80 dB SPL, eight animals at 90 dB SPL, and three animals at 115 db SPL. Ratio points are connected by lines in the illustration. Circled dots indicate components that are present in greater concentration during period 1 than during period 2, significant by paired-variate analysis at p = 0.05 (two-tailed test); triangles indicate components that are present in greater concentration in period 2 than in period 1 at the same significance level. (From Drescher et al., 1983.)

Figure 4 shows plots of the ratios of the concentration of each component in perilymph collected during period 1 to the concentration of the same component in perilymph collected during period 2, across 47 components determined by HPLC analysis (Drescher et al., 1981, 1983). In general, the concentrations of the primary amines for control animals (upper plot) did not change from the first to the second collection period, as reflected in the ratios, indicating that the animal preparations were stable. At 80 dB SPL, there was overall little deviation from 1.0 of the period 1:period 2 (noise:silence) concentration ratios, but components 20 and 27, unidentified primary amines, were significantly elevated in concentration in period 1 compared with period 2, by paired variate analysis ($p<0.05$). At 90 dB SPL, a general increase in the deviation of the ratios from 1.0 occurred, compared with the ratios at 80 dB SPL. However, only two components, peaks 27 and 34, were significantly elevated during the 90-dB-SPL sound exposure compared with the subsequent silence, as indicated by the ratios. At 115 dB SPL, the largest concentration ratios corresponded to peak 33 (β-alanine position) and peak 34 (the GABA position). In all, relatively few components showed elevated noise:silence ratios, significant by paired-variate analysis (Drescher et al., 1983). Another feature of Fig. 4 is noteworthy: the plots show marked similarities of ratio pattern in the region of components 19–37 for the 80-, 90-, and 115-dB-SPL plots, and in the region of components 41–43 for all plots, indicating high reproducibility of the preparation.

Table III shows the significance of regression coefficients relating exposure

TABLE 4-III
SIGNIFICANCE FOR LINEAR REGRESSION ANALYSIS OF THE RELATION
BETWEEN EXPOSURE LEVEL AND CONCENTRATION
OF PERILYMPHATIC COMPONENTS

Component Assignment	(1) Significance for period 1 – period 2 concentration difference	(2) Significance for period 1 concentration	(3) Significance for period 2 concentration
GABA	$p < 0.005$	$p < 0.05$	ns
β-ALA	$p < 0.025$	$0.05 < p < 0.10$	ns
Peak 20	$p < 0.025$	ns	ns
Peak 27	ns	ns	ns
ASP	ns	ns	$0.05 < p < 0.10$
GLU	ns	ns	ns
TAU	ns	ns	ns
GLY	ns	ns	ns
LEU	ns	ns	ns

Table III reflects the extent of linearity of the relation between exposure levels (50, 80, 90, and 115 dB SPL) and concentration of perilymphatic components. Levels of significance for regression coefficients (which are not listed) were determined from linear least-squares regression analysis of the relation between (1) exposure level and the difference in concentration between periods 1 and 2, (2) exposure level and period 1 concentration, and (3) exposure level and period 2 concentration, shown in the corresponding columns above. The letters ns indicate not significant at $p = 0.10$. The t-test ($n - 2$) for zero regression was used. (From Drescher et al., 1983.)

level to concentrations of components and to differences in concentrations of components between the periods. Putative amino acid neurotransmitters, and the non-transmitter leucine, are included in the table for comparison. The regression coefficient relating the exposure level to differences in concentration between noise and silence was highly significant for the GABA-like component (peak 34). Absolute concentrations of the GABA-like component also increased significantly during period 1, as the SPL was raised. For β-alanine, the regression coefficient relating concentration differences and SPL was significant, with $p<0.025$, but the regression coefficient for period 1 concentrations vs. exposure level was only significant at $0.05 < p < 0.10$. Results for peak 20 were similar to those for β-alanine, except that the regression coefficient for period 1 concentrations was not significant ($p>0.10$). For peak 27, significant concentration differences between periods 1 and 2 occurred only at 80 and 90 dB SPL (not shown), and neither the concentration differences between period 1 and 2, nor the absolute concentrations during period 1 or period 2, were significantly related to the SPL of the exposure. For aspartic acid, concentrations in period 2 following sound stimulation (but not in period 1 during sound stimulation) increased with SPL ($0.05 < p < 0.10$). In the case of the neurotransmitter candidates glutamic acid, taurine, and glycine, there was no statistical evidence for dependence of perilymphatic concentration on sound exposure.

In a second series of experiments with the expanded HPLC, we determined concentrations of primary amines in perilymph at 115 dB SPL, the upper limit of sound stimulation, which presumably would release the maximum amount of neurotransmitters. In the period during acoustic stimulation there was a significant increase, compared with controls, in the concentration of the GABA component, two unidentified components A and B eluting in position close to peak 27 of the shortened HPLC analysis, glutamic acid, and asparagine (Table IV). As in the first experimental series, aspartic acid was elevated in concentration following sound exposure in period 2, relative to controls. A methionine-enkephalin-like component, distinct from leucine-enkephalin and resolved with the expanded analysis (Fig. 5), was also present in higher concentration in period 2 following sound simulation, compared with controls.

In the present studies, the strongest evidence, from the simultaneous analysis of 47 primary amine components in perilymph relating concentration to sound pressure level, was obtained for the GABA-like component (Table III). Differences in concentration of this component between period 1 (stimulation period) and period 2 (silent period) were statistically related to sound-pressure level, with a regression coefficient significant at $p<0.005$.

The higher concentration of aspartic acid in period 2 following sound stimulation compared with controls may reflect a sound-induced release of

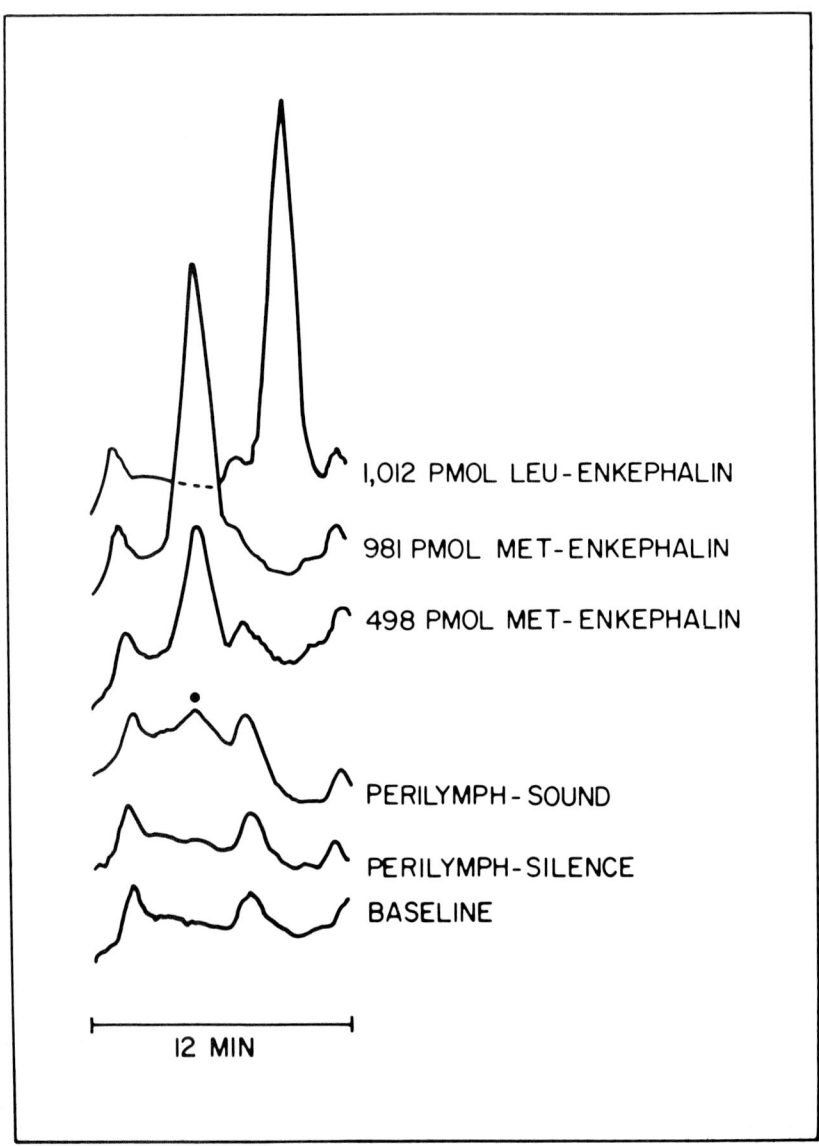

Figure 4-5. Separation of standard methionine-enkephalin from standard leucine-enkephalin by HPLC as described in the text (upper three traces) and detection of a methionine-enkephalin-like component in guinea-pig perilymph. Dot marks position of the perilymphatic component enhanced by sound at 115 dB SPL.

aspartic acid at the level of the cochlear nucleus (Wenthold and Gulley, 1977; Martin, 1980). The latter structure is situated in close anatomical proximity to the internal auditory canal, which in turn communicates with

TABLE 4-IV
RATIOS OF PRIMARY AMINES IN PERILYMPH, DETERMINED BY EXPANDED HPLC ANALYSIS,
FOR EXPOSURE TO NOISE AT 115 dB SPL

Component	Control[1]	115 dB SPL[1]	Period 1, 115 dB to Period 1, control[2]	Period 2, 115 dB, to Period 2, control[3]
GABA	1.10 (ns)[4]	1.37 ($p < 0.02$)[4]	2.54 ($p < 0.005$)[5]	2.05 ($p < 0.01$)[5]
β-ALA	1.07 (ns)	1.30 ($p < 0.06$)	1.21 (ns)	0.99 (ns)
Peak A	1.14 (ns)	1.24 (ns)	1.87 ($p < 0.05$)	1.78 ($p < 0.10$)
Peak B	1.11 (ns)	1.90 ($p < 0.01$)	1.60 ($p < 0.05$)	0.81 (ns)
ASP	1.51 ($p < 0.10$)	0.90 ($p < 0.10$)	1.43 (ns)	2.49 ($p < 0.005$)
GLU	1.08 (ns)	1.12 (ns)	2.10 ($p < 0.05$)	2.00 ($p < 0.005$)
TAU	1.03 (ns)	1.19 (ns)	0.93 (ns)	0.79 ($p < 0.05$)
GLY	1.02 (ns)	1.17 ($p < 0.025$)	1.29 (ns)	1.17 (ns)
ASN	1.00 (ns)	0.78 (ns)	4.69 ($p < 0.05$)	4.94 ($p < 0.005$)
ALA	1.01 (ns)	1.12 ($p < 0.05$)	1.17 (ns)	1.04 (ns)
MET-enkephalin	1.11 (ns)	1.10 (ns)	1.27 (ns)[6]	2.88 ($0.05 < p < 0.10$)[6]
LEU	1.15 (ns)	1.10 (ns)	0.87 (ns)	0.90 (ns)

The letters ns designate not significant at $p = 0.10$. [1]Ratio listed is the average of individual ratios of the period 1 concentration to the period 2 concentration. [2]Ratio of the mean concentration during period 1 at 115 dB SPL to the mean concentration during period 1 for control (silence). [3]Same as footnote 2, except for period 2. [4]Values in parentheses in these columns indicate significance, by paired-variate analysis, of differences in concentration between periods 1 and 2, using the two-tailed t-test. [5]Values in parentheses in these columns indicate significance of the difference between the mean from the numerator and the mean from the denominator, using the two-tailed t-test, and assuming independent groups with equal variances. [6]Same as footnote 5, except assuming independent groups with unequal variances.

the scala tympani via the modiolar spaces and the perilymphatic canaliculae. In addition, the cochlear aqueduct connects the cranial cavity with scala tympani. Thus, aspartic acid, released by the auditory nerve at the cochlear nucleus in response to sound stimulation, may appear in the perilymph and at later times than neurotransmitter substances released within the cochlea.

The increase in concentration of the methionine-enkephalin-like component detected in period 2 following 115 db SPL exposure may have resulted from release from olivocochlear fibers. Fex and Altschuler (1981) found enkephalin-like immunoreactivity in the efferent, olivocochlear nerve fibers of the cochlea (see Chapter 1, this volume). Fex (1962) had previously demonstrated firing of efferent fibers in response to sound stimulation to the contralateral ear.

In contrast to the relatively few components whose concentrations increased in response to sound stimulation, it should be emphasized that the majority of perilymphatic compounds did not change in concentration. Over 40 quantified components in one series of experiments and over 50 quantified components in another series of experiments were not elevated in concentration during sound stimulation relative to their concentrations in the following silence, nor were they elevated in period 1 or 2 relative to corresponding control concentrations (Drescher et al., 1983). Thus, the present experimental approach incorporates many internal controls, which validate the relatively few changes in perilymphatic compounds that were observed.

V. GLUTAMIC ACID AS A HAIR-CELL TRANSMITTER

Glutamic acid has been suggested as the transmitter between the hair cell and its primary afferent fibers in acousticolateralis systems. We found by HPLC methods no change in the perilymphatic concentration of glutamic acid in proportion to sound stimulation for the guinea pig (Drescher et al., 1983). However, at 115 dB SPL, a maximum exposure level, which is borderline for the disruption of hair cells (Bohne, 1976), we found that glutamic acid was significantly elevated (the noise:silence ratio for period 1 was 2.10, $p<0.05$; Table IV). Melamed et al. (1982) did not see an elevation of glutamate in perilymph with exposure to noise at 95 dB SPL. Sewell et al. (1978) had also previously measured glutamic acid, by non-HPLC methods, in both perilymph for guinea pigs exposed to sound and in perilymph from animals kept in silence, under conditions for which biological activity was detected in perilymph from exposed animals, and found no change in the concentration of glutamate. Thus, in all, there is little evidence of an increase in the concentration of glutamic acid in perilymph as a result of low-to-moderate levels of sound exposure.

Steinbach and Bennett (1971) and Teeter and Bennett (1981) suggested that glutamic acid or a related compound may be the transmitter at the primary afferent synapse of electroreceptor organs of electric fish. Electroreceptor cells are closely related to hair cells (Bennett, 1970). Electroreceptor sensory fibers are excited by low concentrations of applied glutamic acid, a finding which appears to support a neurotransmitter role for glutamate at this synapse. However, recent evidence suggests that the primary, afferent transmitter for electroreceptors may not be glutamic acid. Obara et al. (1981) reported that electrical stimulation of electroreceptors released an excitatory substance that was chromatographically different from glutamic acid, and was active at a much lower concentration than glutamate. Nagai et al (1984) were able to block electroreceptor afferent discharges, induced by iontophoretically-applied glutamic acid, with a spider toxin known to block glutamate receptors specifically (Abe et al., 1983), but were unable to block the natural synaptic response by means of the toxin. To the extent that electroreceptor cells are similar to hair cells, this evidence may contradict the hypothesis that glutamic acid is an acousticolateralis sensory transmitter.

Bobbin et al. (Chapter 7, this volume) strongly support glutamic acid as the hair-cell transmitter for both auditory and lateral-line systems. Their conclusions are based partly on experiments in which glutamic acid, applied exogenously to the cochlea, produced an excitatory response of the auditory nerve (Bobbin, 1979), and on experiments where stimulated efflux of glutamate from frog skins containing lateral-line organs was significantly greater than the stimulated efflux from non-lateral-line skins (Bledsoe et al., 1980). Because glutamic acid is a general neural excitant (Curtis et al., 1972) and because relatively large amounts of glutamate were reported to be specifically released from lateral-line tissue that contained few hair cells (Bledsoe et al., 1980), further work will be necessary to decide whether glutamic acid is the natural acousticolateralis sensory transmitter or an agonist of this compound.

VI. GABA AS AN ACOUSTICOLATERALIS TRANSMITTER

Since Flock and Lam (1974) first suggested that GABA might be an excitatory transmitter used by hair cells, the idea of GABA as a peripheral acousticolateralis transmitter has been both supported and opposed, across several systems of study. GABA has gained some experimental support as a transmitter, possibly excitatory, in the vestibular periphery of warmblooded animals. Felix and Ehrenberger (1977, 1982, Chapter 5, this volume) reported that GABA increased the firing rate of afferent nerve fibers of the cat saccule when applied iontophoretically, and the GABA antagonists bicuculline and picrotoxin blocked both the spontaneous and GABA-induced activity in the latter studies. Meza et al. (1982; Chapter 6, this volume) have demonstrated both GABA synthesis and the presence of GABA receptors in vestibular sensory tissue of the chick inner ear. In the lateral line, GABA has been reported to be inhibitory (Bledsoe et al., 1983).

The role of GABA in the cochlea is less clear. The content of GABA (Tachibana and Kuriyama, 1974; Godfrey et al., 1976a) and its synthesizing enzyme, glutamate decarboxylase (Fex and Wenthold, 1976), has been reported to be low in the organ of Corti of the cochlea. Richrath et al. (1974) concluded that GABA was present in the cochlea, and was localized to the efferent nerve endings, after seeing silver grains over cochlear efferent endings in autoradiographs following perfusion of the cochlea with solutions containing [^{3}H]-GABA. Gulley et al. (1979) and Schwartz et al. (1983) with light and electron microscopic autoradiography, respectively, observed accumulation of silver grains due to ^{3}H-GABA over efferent olivocochlear nerve fibers to the outer hair cells. Klinke and Oertel (1977) applied picrotoxin and bicuculline to the cochlea and found that these drugs did not affect the afferent, compound action potential of the cochlear nerve, but apparently did block the efferents at high doses. Recently, Fex et al. (Chapter 1, this volume) demonstrated the presence of a subpopulation of cochlear efferent endings on outer hair cells which shows staining for antiserum to glutamate decarboxylase, raising the possibility that some cochlear efferents may use GABA as a neurotransmitter or neuromodulator.

In the present studies, the increase in concentration of the GABA-like component in perilymph in proportion to intensity of the sound stimulus appears to support a neurotransmitter role for this component in the cochlea. The question of the source of the GABA-like component in perilymph, whether from cochlear efferent endings or from hair cells, will require further investigation. Also, additional analytical techniques may be necessary to establish with certainty the identity of the GABA-like component.

VII. SUMMARY

By means of HPLC analysis of mammalian perilymph and related fluids, we have found that:

1. Glycine and alanine are present in perilymph in greater concentrations than in CSF. This finding indicates that while CSF may represent a "carrier" fluid, perilymph is nevertheless chemically unique.

2. A GABA-like HPLC component increases in perilymph in proportion to sound-stimulus intensity. Assuming a synaptic origin of this component, it is possible that either (a) the component is GABA itself, originating from a subpopulation of cochlear efferents, or (b) it is an unidentified chromatographic component comigrating with GABA, originating from the hair cells. Other permutations exist, requiring a further increase in experimental specificity to determine the identity of the component and its source.

3. There is no consistent increase in concentration of glutamic acid in perilymph in proportion to the intensity of sound stimulation. This lack of change does not support glutamate as an acousticolateralis transmitter. However, at high levels of exposure (115 dB SPL), the concentration of glutamate is increased in perilymph compared with controls.

4. Aspartic acid, although not immediately elevated in perilymph after sound exposure, is increased in concentration relative to controls 2–3.5 hours after onset of the stimulus. This observation is compatible with the release of aspartic acid from central auditory synapses.

5. A methionine-enkephalin-like component, distinct from leucine-enkephalin, is detected in perilymph from control animals and is elevated in response to noise exposure at 115 dB SPL. The elevation of the methionine-enkephalin-like component may result from stimulation of cochlear efferents, which have been reported to possess enkephalin-like immunoreactivity.

Future directions for research on neurotransmitters in perilymph include both the development of more sensitive HPLC procedures and the bioassay of HPLC fractions. The present sensitivity of HPLC can be improved by removing impurities from buffer reagents, lowering pump pressure variation by damping, and increasing the signal:noise ratio of the electronic constituents.

Unknown HPLC components showing biological activity can be further identified by mass spectrometry. Knowing the chemical identity of a suspected transmitter, one can then obtain the pure substance in large amounts and characterize it by detailed bioassay.

ACKNOWLEDGMENTS

This work was supported by NIH grant NS 16166. Permission to reproduce previously-published illustrations was given by Raven Press and Pergamon Press.

REFERENCES

Abe, T.; Kawai, N., and Miwa, A.: Effects of a spider toxin on the glutaminergic synapse of lobster muscle. *J. Physiol. 339:* 243–252, 1983.

Axelsson, A.: The vascular anatomy of the cochlea in the guinea pig and in man. *Acta Otolaryng. Suppl. 243:* 1–134, 1968.

Bennett, M. V. L.: Comparative physiology: electric organs. *Ann. Rev. Physiol. 32:* 471–528, 1970.

Benson, J. R.: Some recent advances in amino acid analysis. In Perham, R. N. (Ed.): *Instrumentation in Amino Acid Sequence Analysis.* New York, Academic Press, 1975, pp. 1–39.

Bledsoe, S. C., Jr.; Bobbin, R. P., Thalmann, R., and Thalmann, I: Stimulus-induced release of endogenous amino acids from skins containing the lateral-line organ in *Xenopus laevis. Exp. Brain Res. 40:* 97–101, 1980.

Bledsoe, S. C., Jr.; Chihal, D. M., Bobbin, R. P., and Morgan, D. N.: Comparative actions of glutamate and related substances on the lateral line of *Xenopus laevis. Comp. Biochem. Physiol. 75C:* 199–206, 1983.

Bobbin, R. P.: Glutamate and aspartate mimic the afferent transmitter in the cochlea. *Exp. Brain Res. 34:* 389–393, 1979.

Bohne, B. A.: Mechanisms of noise damage in the inner ear. In Henderson, D.; Hamernik, R. P., Dosanjh, D. S., and Mills, J. H. (Eds.): *Effects of Noise on Hearing.* New York, Raven Press, 1976, pp. 41–68.

Crifò, S., and Crifò, C.: Free amino acid content of perilymph. *Arch. Otolaryngol. 93:* 65–67, 1971.

Curtis, D. R.; Duggan, A. W., Felix, D., Johnston, G. A. R., Tebecis, A. K., and Watkins, J. C.: Excitation of mammalian central neurones by acidic amino acids. *Brain Res. 41:* 283–301, 1972

Drescher, D. G., and Lee, K. S.: Extraction of fixed, stained protein bands from gels for micro amino acid analysis using o-phthaldialdehyde. *Anal. Biochem. 84:* 559–569, 1978.

Drescher, M. J.; Drescher, D. G., and Medina, J. E.: Effect of sound stimulation at several levels on concentrations of primary amines, including neurotransmitter candidates, in perilymph of the guinea pig inner ear. *J. Neurochem. 41:* 309–320, 1983.

Drescher, M. J.; Medina, J. E., and Drescher, D. G.: High-resolution analysis of physiological amino acids and related compounds in ten-microliter samples of guinea pig perilymph by the use of high-performance liquid chromatography. *Anal. Biochem. 116:* 280–286, 1981.

Felix, D., and Ehrenberger, K.: The action of GABA and acetylcholine in the labyrinth of the cat. In Portmann, M., and Aran, J. M. (Eds.): *Inner Ear Biology.* Paris, INSERM, 1977, pp. 147-154.

Felix, D., and Ehrenberger, K.: The action of putative neurotransmitter substances in the cat labyrinth. *Acta Otolaryng. 93:* 101-105, 1982.

Fex, J., and Altschuler, R. A.: Enkephalin-like immunoreactivity of olivocochlear nerve fibers in cochlea of guinea pig and cat. *Proc. Natl. Acad. Sci. U.S.A. 78:* 1255-1259, 1981.

Fex, J.: Auditory activity in centrifugal and centripetal cochlear fibres in cat. *Acta Physiol. Scand. (Suppl. 189) 55:* 1-68, 1962.

Fex, J.: Efferent inhibition in the cochlea by the olivocochlear bundle. In de Reuck, A. V. S., and Knight, J. (Eds.): *Hearing Mechanisms in Vertebrates.* London, Churchill, 1968, pp. 169-181.

Fex, J.: Neuropharmacology and potentials of the inner ear. In Moller, A. R. (Ed.): *Basic Mechanisms in Hearing.* New York, Academic Press, 1973, pp. 377-421.

Fex, J., and Wenthold, R. J.: Choline acetyltransferase, glutamate decarboxylase and tyrosine hydroxylase in the cochlea and cochlear nucleus of the guinea pig. *Brain Res. 109:* 575-585, 1976.

Flock, A., and Lam, D. M. K.: Neurotransmitter synthesis in inner ear and lateral line sense organs. *Nature 249:* 142-144, 1974.

Godfrey, D. A.; Carter, J. A., Berger, S. J., and Matschinsky, F. M.: Levels of putative transmitter amino acids in the guinea pig cochlea. *J. Histochem. Cytochem. 24:* 468-470, 1976a.

Godfrey, D. A.; Krzanowski, J. J., and Matschinsky, F. M.: Activities of enzymes of the cholinergic system in the guinea pig cochlea. *J. Histochem. Cytochem. 24:* 470-472, 1976b.

Gulley, R. L.; Fex, J., and Wenthold, R. J.: Uptake of putative neurotransmitters in the organ of Corti. *Acta Otolaryng. 88:* 177-182, 1979.

Hare, P. E.: Amino acid composition by column chromatography. In Needleman, S. B. (Ed.): *Protein Sequence Determination.* New York, Springer, 1975, pp. 204-231.

Jasser, A., and Guth, P. S.: The synthesis of acetylcholine by the olivocochlear bundle. *J. Neurochem. 20:* 45-53, 1973.

Klinke, R., and Oertel, W.: Evidence that GABA is not the afferent transmitter in the cochlea. *Exp. Brain Res. 28:* 311-314, 1977.

Lee, K. S., and Drescher, D. G.: Derivatization of cysteine and cystine for fluorescence amino acid analysis with the o-phthaldialdehyde/2-mercaptoethanol reagent. *J. Biol. Chem. 254:* 6248-6251, 1979.

Lee, K. S., and Drescher, D. G.: Fluorometric amino-acid analysis with o-phthaldialdehyde (OPA). *Int. J. Biochem. 9:* 457-467, 1978.

Martin, M. R.: The effects of iontophoretically applied antagonists on auditory nerve and amino acid evoked excitation of anteroventral cochlear nucleus neurons. *Neuropharmacol. 19:* 519-528, 1980.

Masuda, Y.; Sando, I., and Hemenway, W. G.: Perilymphatic communication routes in guinea-pig cochlea. *Arch. Otolaryngol. 94:* 240-245, 1971.

Medina, J. E., and Drescher, D. G.: The amino acid content of perilymph and cerebrospinal fluid from guinea pigs and the effect of noise on the amino-acid composition of perilymph. *Neuroscience 6:* 505-509, 1981.

Melamed, B.; Norris, C., Bryant, G., and Guth, P. S.: Amino acid content of guinea pig perilymph collected under conditions of quiet or sound stimulation. *Hearing Res. 7:* 13-18, 1982.

Meza, G.; Carabez, A., and Ruiz, M.: GABA synthesis in isolated vestibular tissue of chick inner ear. *Brain Res. 241:* 157-161, 1982.

Moore, S., and Stein, W. H.: Chromatographic determination of amino acids by the use of automatic recording equipment. In Colowick, S. P., and Kaplan, N. O. (Eds.): *Methods in Enzymology,* Vol. 6. New York, Academic Press, 1963, pp. 819-831.

Nagai. T.; Obara, S., and Kawai, N.: Differential blocking effects of a spider toxin on synaptic and glutamate responses in the afferent synapse of the acoustico-lateralis receptors of *Plotosus. Brain Res. 300:* 183-187, 1984.

Norris, C. H., and Guth, P. S.: The release of acetylcholine (ACh) by the crossed olivocochlear bundle (COCB). *Acta Otolaryng. 77:* 318-326, 1974.

Obara, S.; Umekita, S., and Matsumoto, Y.: Stimulus-induced release experiments disprove L-glutamate as the afferent neurotransmitter in ampullary electroreceptors of the marine catfish, *Plotosus. Eighth Int. Congr. Pharmacol. (Tokyo) Abstracts 8:* 391, 1981.

Orrego, F.: Criteria for the identification of central neurotransmitters, and their application to studies with some nerve tissue preparations *in vitro. Neuroscience 4:* 1037-1057, 1979.

Richrath, W.; Kraus, H., and Fromme, H. G.: Lokalisation von ^{3}H-γ-aminobuttersäuer in der cochlea. *Arch. Otorhinolaryngol. 208:* 283-293, 1974.

Roth, M.: Fluorescence reaction for amino acids. *Analyt. Chem. 43:* 880-882, 1971.

Schuknecht, H. F.: Pathophysiology of the fluid systems of the inner ear. In Neff, W. D. (Ed.): *Contributions to Sensory Physiology.* New York, Academic Press, 1970, pp. 75-93.

Schuknecht, H. F., and Seifi, A. E.: Experimental observations on the fluid physiology of the inner ear. *Ann. Otol. Rhinol. Laryng. 72:* 687-712, 1963.

Schwartz, I. R., and Ryan, A. F.: Differential labeling of sensory cell and neural populations in the organ of Corti following amino acid incubations. *Hearing Res. 9:* 185-200, 1983.

Sewell, W. F.; Norris, C. H., Tachibana, M., and Guth, P. S.: Detection of an auditory nerve-activating substance. *Science 202:* 910-912, 1978.

Simons, S. S., Jr., and Johnson, D. F.: Preparation of a stable, fluorescent l-alkylthio-2-alkylisoindole. *J. Chem. Soc. Chem. Commun. 11:* 374-375, 1977.

Simons, S. S., Jr., and Johnson, D. F.: The structure of the fluorescent adduct formed in the reaction of o-phthalaldehyde and thiols with amines. *J. Am. Chem. Soc. 98:* 7098-7099, 1976.

Spackman, D. H.; Stein, W. H., and Moore, S.: Automatic recording apparatus for use in the chromatography of amino acids. *Anal. Chem. 30:* 1190-1206, 1958.

Steinbach, A. B., and Bennett, M. V. L.: Effects of divalent ions and drugs on synaptic transmission in phasic electroreceptors in a Mormyrid fish. *J. Gen. Physiol. 58:* 580-598, 1971.

Tachibana, M., and Kuriyama, K.: Gamma-aminobutyric acid in the lower auditory pathway of the guinea pig. *Brain Res. 69:* 370-374, 1974.

Teeter, J. H., and Bennett, M. V. L.: Synaptic transmission in the ampullary electroreceptor of the transparent catfish, *Kryptopterus. J. Comp. Physiol. 142:* 371-377, 1981.

Thalmann, R.; Comegys, T. H., and Thalmann, I.: Amino acid profiles in inner ear fluids and cerebrospinal fluid. *Laryngoscope 92:* 321-328, 1982.

Wenthold, R. J., and Gulley, R. L.: Aspartic acid and glutamic acid levels in the cochlear nucleus after auditory nerve lesion. *Brain Res. 138:* 111-123, 1977.

Chapter 5

THE ACTION OF PUTATIVE NEUROTRANSMITTER SUBSTANCES IN THE MAMMALIAN LABYRINTH

Dominik Felix and Klaus Ehrenberger

I. Introduction
II. Integrating Functions of Primary Vestibular Afferents
III. Modulation of the Irregular Firing Pattern
IV. Modulation of the Regular Firing Pattern
V. Summary and Conclusions
  References

I. INTRODUCTION

Our knowledge of synaptic transmission centers on the nature of the molecules which transmit neuronal information and how these molecules can be identified. Fundamentals of the chemical transmission process are the synthesis and release of the transmitter in response to presynaptic excitation and the postsynaptic effect produced by the substance. The identification of a so-called "neurotransmitter" depends largely upon interdisciplinary methods of research (Klinke, 1981). For the electrophysiologist, the most important criterion is that of "identity of action" (Werman, 1966). This implies that exogenous application of a suspected transmitter to the region of the synapse should mimic the response elicited by release of the natural transmitter into the synaptic cleft. The method of administering such a substance is of the utmost importance. Microiontophoresis, a technique by which drugs can be applied locally and the membrane response recorded, serves as a useful tool for assessing the chemical sensitivity of neurons, which is preliminary to identifying transmitter substances (Curtis, 1964; Kelly, 1975; Martin, Chapter 11, this volume). Furthermore, the use of multibarreled microelectrodes makes it possible to test substances, sequentially and at a common locus, as antagonists to synaptic transmission.

The question as to the identity of the neurotransmitter acting between hair cells and afferent, as well as efferent, terminals has attracted a good deal of interest in recent years. Hair cells in the acousticolateralis system, including those of the vestibular sensory organ, receive efferent fibers which make contact with the hair cells at their bases (Engström and Wersäll, 1958; Flock, 1971). Furthermore, hair cells are innervated by afferent nerve fibers, to which they transmit information about mechanical stimuli. There is evidence that synaptic transmission at afferent, as well as at efferent synapses, is chemical in nature. Excitatory postsynaptic potentials have been recorded in afferent nerve fibers (Furukawa and Ishii, 1967; Flock and Russell, 1973) and inhibitory postsynaptic potentials have been recorded from hair cells in response to efferent fiber stimulation (Flock and Russell, 1973). The results of several investigations suggest that acetylcholine acts as the efferent transmitter. Its effect is blocked by curare (Fex, 1973), and acetylcholinesterase and choline acetyltransferase have been shown to be present in efferent fibers (Hilding and Wersäll, 1962; Russell, 1971; Jasser and Guth, 1973). The nature of the afferent transmitter, however, is less clear. Based on results obtained using radioactive precursors, Flock and Lam (1974) proposed that GABA mediates afferent transmission in the acousticolateralis system of cold-blooded animals. The bullfrog basilar papilla, which has only afferent innervation, appeared to synthesize GABA but not acetylcholine, dopamine, or noradrenaline. On the other hand, the skate ampulla, which has both afferent and efferent innervation, synthesized GABA as well as acetylcholine (Flock and Lam, 1974). The data for glutamate and the catecholamines are controversial. Higuchi *et al.* (1980) provided evidence that L-glutamate mimics the natural transmitter in ampullary electroreceptors, whereas data from stimulus-dependent release experiments by Umekita *et al.* (1980) refute the role of this excitatory amino acid as the afferent neurotransmitter in ampullary electroreceptors. In elasmobranches, electroreceptor cells generate action potentials. Inward currents that generate such action potentials are transmitted by calcium ions. Frequency is controlled by outward repolarizing K^+ current, which is activated by the increase in cytoplasmic Ca^{++} (Bennett, 1984). Thus, calcium ions may play a role both in transmitter release and electrogenesis. Using an isolated lateral line preparation, Bobbin *et al.* (1984; Chapter 7, this volume) found that *Xenopus laevis* afferent fibers respond to each of the three amino acid analogs currently being used to classify vertebrate excitatory amino acid receptors: N-methyl-D-aspartate (NMDA), quisqualate, and kainate. The data imply that the receptors activated by NMDA are the amino acid receptors activated by the endogenous afferent transmitter. The blockade of the responses to glutamate and aspartate by the NMDA antagonist, D-α-amino-adipate, is consistent with the hypothesis that these two amino acids act at the NMDA-preferring receptors.

In the skate, the fact that spontaneous and evoked afferent responses were inhibited by picrotoxin after stimulation of the semicircular canal (Flock and

Goldstein, 1973) suggests the existence of a postsynaptic GABA receptor in this system. Furthermore, GABA appeared to be synthesized from glutamate in vestibular tissue of fish, implying the presence of glutamate decarboxylase (Flock and Lam, 1974). All of these studies were performed in lower vertebrates. In 1982, Meza *et al.* presented evidence for enzyme-mediated GABA synthesis in the vestibular tissue of the chick inner ear. Using two different methods, [^{14}C]-GABA isolation by paper chromatography and measurement of $^{14}CO_2$ from [^{14}C]-glutamate, they showed that glutamate decarboxylase is active in isolated chick vestibular cristae. This activity, demonstrated by the two different assay methods, was affected to the same extent by amino-oxyacetic acid (AOAA), an inhibitor of brain glutamate decarboxylase. These data indicate that GABA is synthesized in the chick labyrinth, with glutamate decarboxylase acting as the synthesizing enzyme. However, whether this activity is located in the hair cells, thus implying that GABA is the mediator between these cells and the afferent terminals in the vestibular system of higher vertebrates, has yet to be demonstrated.

Since there is some evidence that GABA and acetylcholine are involved in synaptic processes at afferent, as well as at efferent, synapses in the mammalian labyrinth, we have carried out experiments using microiontophoretic techniques, with the aim of testing the effects of these transmitter candidates in the saccular macula of the cat. Before we deal with these kinds of experiments, some electrophysiological properties of the labyrinth will be described.

II. INTEGRATING FUNCTIONS OF PRIMARY VESTIBULAR AFFERENTS

The discharge pattern of fibers recorded in the subsynaptic region of the saccular macula is either regular or irregular, depending on the recording site (Goldberg and Fernandez, 1971; Walsh *et al.*, 1972). Type I hair cells are joined to the first neuron singly or in groups of up to five (Spoendlin, 1975). In contrast, the population of afferents of vestibular type II receptor cells is formed by integrating neurons. The mammalian vestibular nerve contains central neurites of the two populations. Figure 1 illustrates the two distinct recordings in the saccular macula. The multibarreled electrodes were inserted into the subsynaptic region of the saccular macula and, when both the endolymphatic sac and the hair cell layer had been penetrated, it was possible to record spontaneous, irregular nerve activity. The frequency of fiber discharge remained at a relatively low level of 0.5–10/s. It was often possible to observe a rhythmic discharge with a periodicity of 6–10/min. The characteristic activity of a fiber is shown in Fig. 1A. The unimodal, asymmetric distribution evident in the on-line, recorded interspike-interval histogram shows the irregularity of fiber discharge in this region. Fiber activity was also observed 40–90 μm below the first recording sites. The pattern of this spontaneous discharge was regular, in contrast to the irregular pattern of activity recorded immediately and subsynaptically. This regularity can be seen from the interspike histogram shown in Fig. 1B. The fibers corresponded to those action potentials which had a constant interval length. The regular firing sequence varied between 40 and 70 spikes/s and was clearly distinguishable from 50 Hz alternating current.

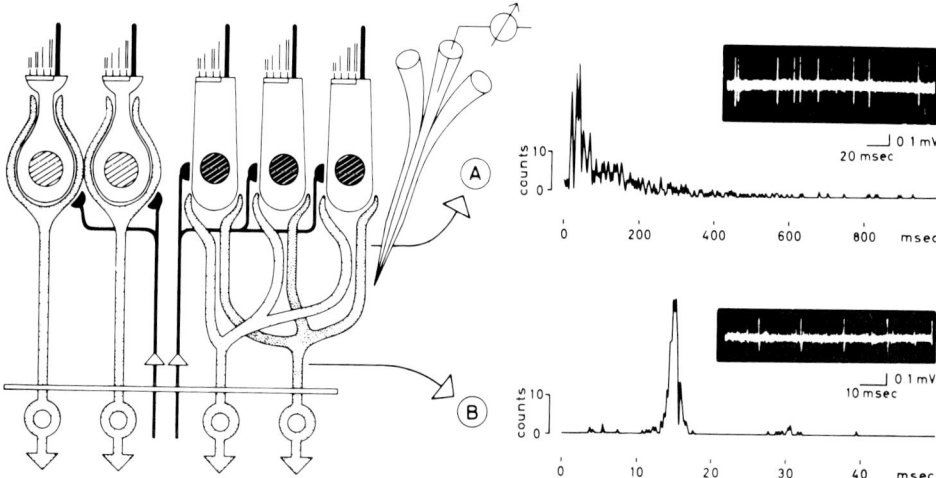

Figure 5-1. Schematic diagram showing the two distinct recordings in the saccular macula. Irregular (A) and regular (B) firing pattern, as revealed by interspike-interval histograms and oscilloscope tracing. (From Felix and Ehrenberger, 1973, with permission.)

Ehrenberger (1974) found that the spontaneous phasic activity in the lateral labyrinthine ampulla of the cat, recorded immediately subsynaptically, is always irregular and indicates a stochastic process. In one population of neurons, irregular spontaneous activity must therefore be transformed into regular spontaneous activity. Segundo et al. (1968) were able to explain such a transformation on an integrating neuron using a general model simulated on a computer. They found the following: 1) as the number of independent inputs increases, the impulse sequence of outputs becomes more and more regular, irrespective of irregular or regular impulse sequences in the individual input channels; 2) when the input channels become dependent or interdependent, however, regular input trains lead to regular output trains and irregular input trains to irregular output trains.

According to the model, the vestibular type II afferents should be capable of changing from irregular to regular spontaneous activity if their inputs are numerous and independent. This is, indeed, the case, as shown by morphological investigations (Spoendlin, 1975). Primary, vestibular type II afferents have contact with efferent nerve endings. So far, an inhibitory influence has been attributed to these efferents (Precht, 1974; Sala, 1965). Under the influence of efferent activity following the stimulation of the contralateral labyrinth, integrating cell inputs lose their independence, and, as a consequence, output activity follows input activity and the regular spontaneous activity becomes irregular (Ehrenberger *et al.*, 1979). Figure 2 illustrates such modification of the spontaneous activity of afferents by efferent stimulation. After on-line identification, the contralateral ear is irrigated with cold water in order to activate the efferents. Under the conditions prevailing in this test, the irrigation was observed to cause the spike rate to decrease and the regular spontaneous activity to become irregular. This effect was checked off-line, using several statistical methods (Fig. 2). It can be seen from the interspike interval histogram that a unimodal asymmetric distribution replaces the "quasi-Gaussian" distribution, a development which evolves through intermediate stages of multi-frequency distribution. This is due to repeated failure of spikes in still-regular spike trains. A state is finally reached where the discharge pattern is practically stochastic, as has been confirmed by autocorrelograms. After a certain period of time, the efferent influence completely disappears and spontaneous activity returns, while the spike rate rises again and the sequence of spike trains becomes highly regular once more. Analysis confirms a high functional interdependence of both labyrinths.

III. MODULATION OF THE IRREGULAR FIRING PATTERN

As described above, the main aim of this study was to investigate the action of the neurotransmitter candidates, GABA and acetylcholine, on fiber discharge in the labyrinth. We shall now describe the influence of these substances on the irregular spontaneous firing frequency, as observed in experiments using microiontophoretic techniques. Iontophoretic application of GABA was seen to enhance the firing rate in the majority of fibers. The latency of the effect of GABA varied between 5 s and 1 min, and it was not possible to establish a correlation between this and the amount of current applied. The latency rate fell, however, following repeated ejection of the amino acid at short intervals. Figure 3A shows an original recording of a unit during the control period, 30 s after the onset of GABA ejection and 1 min after GABA application ceased. It was observed that the application of this substance led to a 2.5-fold increase in activity without, however, affecting the spike amplitude. A continuous recording of the integrated firing frequencies

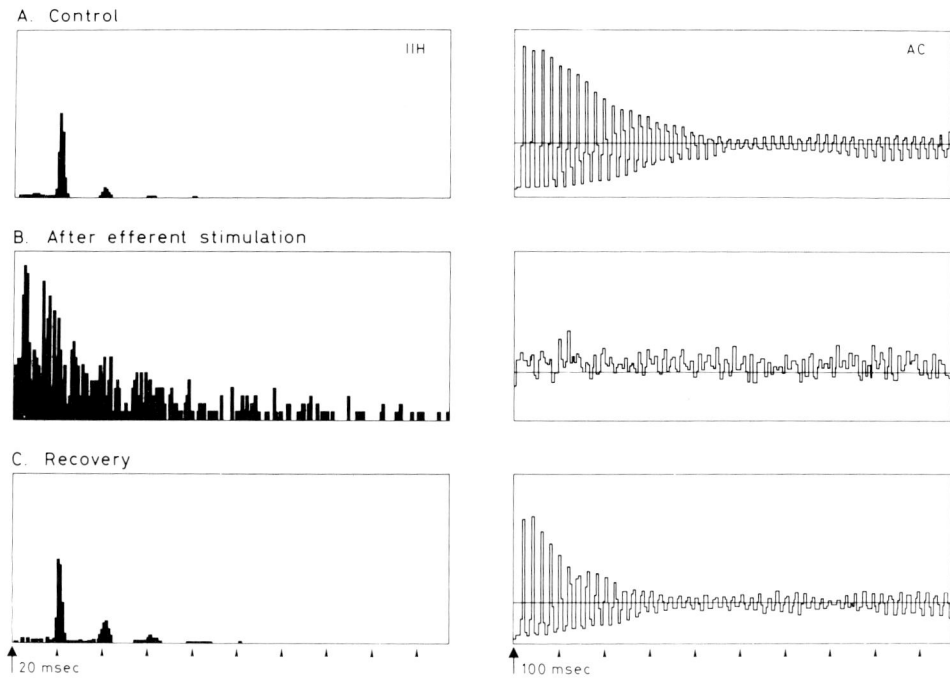

Figure 5-2. Modification of (A) spontaneous activity of afferents by (B) efferent stimulation, followed by (C) recovery. IIH, interspikeinterval histogram; AC, autocorrelogram. (From Ehrenberger *et al.*, 1979, with permission.)

of another fiber can be seen in Fig. 3B; here a definite excitatory effect due to GABA can be seen. Sodium chloride was used as a current control. Acetylcholine ejected with 100 nA clearly depressed spontaneous activity. In comparison to the effect produced by GABA, it was necessary to use higher amounts of current to produce a distinct acetylcholine effect. Unit discharge was depressed in two-thirds of the fibers (Felix and Ehrenberger, 1977, 1982).

The question subsequently arose as to whether it was possible to block GABA-induced activity using the well-known GABA antagonists, bicuculline (Curtis *et al.*, 1971) and picrotoxin. An experiment was carried out where these agents were applied in doses of 50 nA over a period of 1–2 min. Bicuculline (Fig. 4A) was seen to reduce or even completely block the spontaneous firing rate for several minutes. Both agents had varied effects on GABA-induced activity, but there was a noticeable reduction in the action of GABA in all fibers tested. Where antagonism was 100%, recovery from the action of GABA occurred at the same time as when spontaneous activity returned to the control value. Another unit which responded to GABA in a characteristic way is shown in Fig. 4B. A dose of 50 nA of picrotoxin

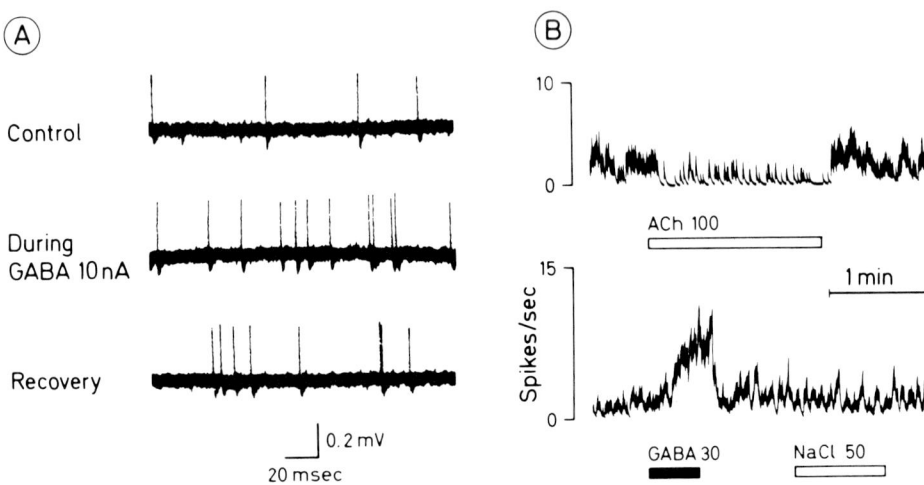

Figure 5-3. Effect of GABA and acetylcholine (ACh) on irregular fiber activity. A. Original oscilloscope tracing of discharge before (control), during, and after application of GABA. B. Continuously-recorded, integrated firing frequency; ejected substances are indicated below the traces. (From Felix and Ehrenberger, 1982, with permission.)

antagonized the action of GABA completely and recovery occurred 3 min after picrotoxin ejection ceased (Felix and Ehrenberger, 1982).

IV. MODULATION OF THE REGULAR FIRING PATTERN

Because it had been observed that GABA had a clearly excitatory effect on spontaneous activity in irregularly-discharging units, the consequent question was whether a similar effect would be observed in units which had a regular firing pattern. Such units, with average frequencies of 45–72/s, were tested. Application of GABA caused the firing frequency to rise, although there was no change in the regularity of the discharge (see Fig. 5A). As can be seen from the interspike-interval histogram, the average length of the interval decreased and the firing rate increased. The variability of the discharge subsequently fell, as can be seen from the distribution in the histogram. The effect of the GABA antagonist bicuculline was to shift the histogram data to the right, i.e., to reduce the firing frequency (Fig. 5B). As for the case of GABA, there was no change in the regularity of the discharge pattern (Felix and Ehrenberger, 1977, 1982).

Putative Neurotransmitter Action in the Labyrinth

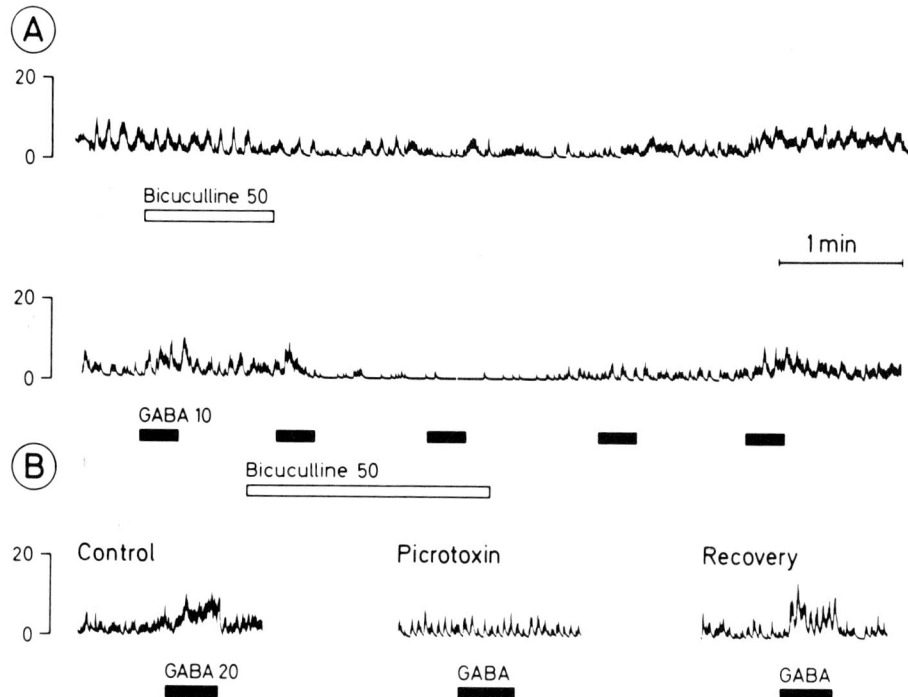

Figure 5-4. Effects of (A) bicuculline and (B) picrotoxin on spontaneous irregular activity and GABA-induced activity. (From Felix and Ehrenberger, 1982, with permission.)

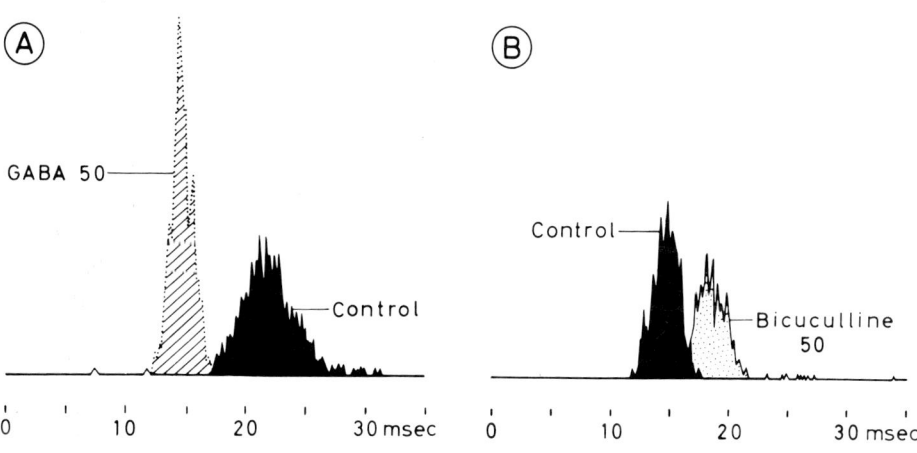

Figure 5-5. Interspike-interval histogram showing the effect of 50 nA GABA and 50 nA bicuculline on regular discharge patterns. Two different fibers were used in A and B. (From Felix and Ehrenberger, 1982, with permission.)

V. SUMMARY AND CONCLUSIONS

Our findings confirm that the regularity of the discharge pattern of fibers recorded in the subsynaptic region of the saccular macula depends on the site chosen for the recording (Goldberg and Fernandez, 1971; Walsh et al., 1972; Ehrenberger, 1974). The irregular spontaneous input activity is transformed into regular output by vestibular type II afferents, which are integrating neurons, and this output is controlled by efferents (Ehrenberger et al., 1979).

Our data on possible neurotransmitters show that microiontophoretically applied GABA and acetylcholine affect fiber discharge (Felix and Ehrenberger, 1977, 1982), and that this effect is more marked in regions of irregular activity. Our results confirm the theory that acetylcholine is the inhibitory transmitter used by efferent fibers of the acousticolateralis system (Flock and Russell, 1973; Fex, 1973; Fex and Adams, 1978) and that GABA is the excitatory transmitter between the hair cells and the afferent fibers (Flock and Lam, 1974). This is an interesting possibility, because GABA acts in most neuronal systems as an inhibitor, i.e., it reduces excitability by increasing subsynaptic membrane conductance, mainly to chloride ions (for a review, see Curtis and Johnston, 1974). There is evidence, however, that GABA mediates the synaptic depolarization of primary afferent terminals (Schmidt, 1963; Levy, 1975; Davidson and Southwick, 1971) and that sodium is the principle ion involved (Barker and Nicoll, 1972). GABA is, therefore, a true excitatory agent and can mimic the natural transmitter in that it depolarizes the membrane (Davidoff, 1972), an effect which can be blocked by bicuculline, a specific GABA-antagonist, in dorsal root terminals (Davidoff, 1972; Curtis et al., 1971). This antagonistic action of bicuculline was observed during our experiments, where this alkaloid not only blocked the action of GABA, but also caused a long-lasting drop in the firing frequency.

One unsatisfactory aspect of the results we obtained is the time course of action of the substances used. Onset usually occurred relatively slowly, after a lapse of up to a minute. It is therefore possible that GABA, applied to the region of the sensory cells, has an indirect, modulatory effect on the response. This hypothesis should now be investigated by means of experiments involving simultaneous intracellular recording of the receptor potentials and microiontophoretic ejection of the substances in question.

In a recent clinical study carried out at the University Hospital in Vienna, the GABA antagonist picrotoxin was shown to suppress labyrinthine spontaneous nystagmus and vertigo in man (Ehrenberger et al., 1982). A total of 45 patients, divided into two groups, received picrotoxin in concentrations of between 1 and 10 mg% (total dose, 1–50 mg, usually 1–5 mg). One group comprised patients with true peripheral labyrinthine spontaneous nystagmus and corresponding labyrinthine vertigo of Menière's disease, acute loss of vestibular function and other symptoms. The other group of patients were suffering from central spontaneous nystagmus and vertigo, with a known neurological diagnosis. At specific dosages, picrotoxin can block the excitability of the vestibular labyrinth, the peripheral spontaneous nystagmus, and thus labyrinthine vertigo. In contrast, manifest central spontaneous nystagmus remained unaffected by picrotoxin.

The clinical results correspond to the evidence obtained from experiments on animals, indicating that GABA acts as an excitatory transmitter in the mammalian labyrinth. The mechanism underlying the action of picrotoxin, however, remains unclear. Picrotoxin might raise the threshold of excitability of both right and left labyrinths and thereby decrease the absolute value of a tonic difference. This would then result in a reduction of the peripheral spontaneous nystagmus. Hypothetical accumulation of picrotoxin in the inner ear would explain the suprathreshold peripheral vestibular effect, as well as the prolonged, dose-dependent peripheral action, without any manifest central analeptic effect.

ACKNOWLEDGMENTS

This work was supported by grants 3.017-0.81 and 3.627-0.84 from the Swiss National Science Foundation and by grant 3411 of the Austrian Science Research Foundation. We thank Mrs. R. Bandi for expert technical assistance.

REFERENCES

Barker, J. L., and Nicoll, R. A.: Gamma-aminobutyric acid: role in primary afferent depolarization. *Science 176:* 1043–1045, 1972.

Bennett, M. V. L.: Electroreceptors as model acoustico-lateralis receptors. *Assoc. Res. Otolaryngol. Abstr. 7:* 130–131, 1984.

Bobbin, R. P.; Bledsoe, S. Jr., Winbery, S., and Jenison, G.: Action of putative neurotransmitters and other relevant compounds on Xenopus laevis lateral line. *Assoc. Res. Otolaryngol. Abstr. 7:* 131–132, 1984.

Curtis, D. R.: Microelectrophoresis. In Nastuk, W. L. (Ed.): *Physical Technique in Biological Research.* New York, Academic Press, 1964, pp. 145–190.

Curtis, D. R.; Duggan, A. W., Felix, D., and Johnston, G. A. R.: Bicuculline, an antagonist of GABA and synaptic inhibition in the spinal cord of the cat. *Brain Res. 32:* 69–96, 1971.

Curtis, D. F., and Johnston, G. A. R.: Amino acid transmitters in the mammalian central nervous system. *Ergebn. Physiol. 69:* 94–188, 1974.

Davidoff, R. A.: Gamma-aminobutyric acid antagonism and presynaptic inhibition in the frog spinal cord. *Science 175:* 331–333, 1972.

Davidson, N., and Southwick, C. A. P.: Amino acids and presynaptic inhibition in the rat cuneate nucleus. *J. Physiol. (Lond.) 219:* 689–708, 1971.

Ehrenberger, K.: Klassifizierung und morphologische Zuordnung spontanaktiver Einheiten der lateralen Bogengangsampulle der Katze. *Montsschr. Ohrenheilkd. Laryngorhinol. 108:* 485-495, 1974.

Ehrenberger, K.; Benkoe, E., and Felix, D.: Suppressive action of picrotoxin, a GABA antagonist, on labyrinthine spontaneous nystagmus and vertigo in man. *Acta Otolaryngol. 93:* 269-273, 1982.

Ehrenberger, K.; Felix, D., and Wyss, U.: Efferent controlled integrating functions of primary vestibular afferents. *Acta Otolaryngol. 87:* 472-476, 1979.

Engström, H., and Wersäll, J.: The ultrastructural organization of the organ of Corti and the vestibular sensory epithelia. *Exp. Cell Res. Suppl. 5:* 460-492, 1958.

Felix, D., and Ehrenberger, K.: The action of GABA and acetylcholine in the labyrinth of the cat. In Portmann, M., and Aran, J.-M. (Eds.): *Inner Ear Biology.* Paris, INSERM, 1977, pp. 147-154.

Felix, D., and Ehrenberger, K.: The action of putative neurotransmitter substances in the cat labyrinth. *Acta Otolaryngol. 93:* 101-105, 1982.

Fex, J.: Neuropharmacology and potentials of the inner ear. In Moller, A. R. (Ed.): *Basic Mechanisms in Hearing.* New York, Academic Press, 1973, pp. 377-421.

Fex, J., and Adams, J. C.: α-Bungarotoxin blocks reversibly cholinergic inhibition in the cochlea. *Brain Res. 159:* 440-444, 1978.

Flock, A.: Sensory transduction in hair cells. In Loewenstein, W. R. (Ed.): *Handbook of Sensory Physiology, Vol. 1, Principles of Receptor Physiology.* Berlin, Springer-Verlag, 1971, pp. 396-441.

Flock, A., and Goldstein, H.: Preliminary studies of the skate's semicircular canal. *Biol. Bull. 145:* 433, 1973.

Flock, A., and Lam, D. M. K.: Neurotransmitter synthesis in inner ear and lateral line sense organs. *Nature 249:* 142-144, 1974.

Flock, A., and Russell, I.: Postsynaptic action of efferent fibres on hair cells. *Nature. 243:* 89-91, 1973.

Furukawa, T., and Ishii, Y.: Neurophysiological studies on hearing in goldfish. *J. Neurophysiol. 30:* 1377-1403, 1967.

Goldberg, J. M., and Fernandez, C.: Physiology of peripheral neurons innervating semicircular canals of the squirrel monkey. I. Resting discharge and response to contant angular accelerations. *J. Neurophysiol. 34:* 635-660, 1971.

Higuchi, T.; Nagai, T., Umekita, S. H., and Obara, S.: The afferent neurotransmitter in the ampullary electroreceptors: L-glutamate mimics the natural transmitter. *Neurosci. Lett. Suppl. 4:* S7, 1980.

Hilding, D. A., and Wersäll, J.: Cholinesterase and its relation to the nerve endings in the inner ear. *Acta Otolaryngol. 55:* 205-217, 1962.

Jasser, A., and Guth, P. S.: The synthesis of acetylcholine by the olivo-cochlea bundle. *J. Neurochem. 20:* 45-53, 1973.

Kelly, J. S.: Microiontophoretic application of drugs onto single neurons. In Iversen, L. L.; Iversen, S. D., and Snyder, S. H. (Eds.): *Handbook of Psychopharmacology, Vol. 2.* New York, Plenum Press, 1975, pp. 29-67.

Klinke, R.: Neurotransmitters in the cochlea and the cochlear nucleus. *Acta Otolaryngol. 91:* 541-554, 1981.

Levy, R. A.: The effect of intravenously administered α-aminobutyric acid on afferent fiber polarization. *Brain Res. 92:* 21-34, 1975.

Meza, G.; Carabez, A., and Ruiz, M.: GABA synthesis in isolated vestibular tissue of chick inner ear. *Brain Res. 241:* 157-161, 1982.

Precht, W.: Physiological aspects of the efferent vestibular system. In Kornhuber, H. H. (Ed.): *Handbook of Sensory Physiology. Vol. VI/1: Vestibular System, Part 1, Basic Mechanisms.* Berlin, Springer-Verlag, 1974, pp. 221–236.

Russell, I. J.: The pharmacology of efferent synapses in the lateral line system of *Xenopus laevis. J. Exp. Biol. 54:* 643–658, 1971.

Sala, I.: The efferent vestibular system. *Acta Otolaryngol. Suppl. 197:* 1–34, 1965.

Schmidt, R. F.: Pharmacological studies on the primary afferent depolarization of the toad spinal cord. *Pflügers Arch. ges. Physiol. 277:* 325–346, 1963.

Segundo, J. P.; Perkel, D. H., Wyman, H., Hegstad, H., and Moore, G. P.: Input-output relations in computer-simulated nerve cells. *Kybernetik 4:* 157–171, 1968.

Spoendlin, H.: Relation entre structure et fonction du récepteur vestibulaire. *Acta Otorhinolaryng. Belg. 29:* 75–91, 1975.

Umekita, S. H.; Matsumota, Y., Abe, T., and Obara, S.: The afferent neurotransmitter in ampullary electroreceptors: Stimulus-dependent release experiments refute the transmitter role of L-glutamate. *Neurosci. Lett. Suppl. 4:* S7, 1980.

Walsh, B. T.; Miller, J. B., Gacek, R. R., and Kiang, N. Y.-S.: Spontaneous activity in the eighth cranial nerve of the cat. *Int. J. Neurosci. 3:* 221–235, 1972.

Werman, R.: Criteria for identification of a central nervous system transmitter. A review. *Comp. Biochem. Physiol. 18:* 745–766, 1966.

Chapter 6

CHARACTERIZATION OF GABA-ERGIC AND CHOLINERGIC NEUROTRANSMISSION IN THE CHICK INNER EAR

GRACIELA MEZA

   I. Introduction
  II. The Chemical Nature of Neurotransmission in the Vestibular System of Vertebrates
 III. Brief Description of the Chick Inner Ear and Dissection Procedure
 IV. Neurotransmission Mediated by GABA
  V. Acetylcholine Participation in Neurotransmission
 VI. Cellular Localization of GABA and Acetylcholine-Synthesizing Enzymes: The Ontogenetic Approach
VII. Summary and Conclusions
     References

I. INTRODUCTION

In vertebrates, the vestibular part of the inner ear provides the major source of sensory information utilized by the CNS in order to maintain balance in the organism.

The vestibular membranous labyrinth consists of five main organs: the utricle, the saccule, and, in most vertebrates, three semicicular canals. These structures are interconnected with each other as well as with the auditory portion of the inner ear, together forming the statoacoustic system. The membranous labyrinth is encased in a bony structure of much the same shape as the membranous labyrinth, called the bony labyrinth, which is embedded in the petrous part of the temporal bone. A fluid called endolymph fills the membranous labyrinth, whereas the perilymph (of different composition than the endolymph) is found in the space between the membranous and the bony labyrinth.

The sensory regions are located in specialized areas of sensory epithelium, comprised of mechanoreceptor cells (hair cells), supporting cells, afferent and efferent fiber connections, and a system for the coupling of movement.

The vestibular sensory epithelia are highly specialized and organized. Two types of hair cells, called types I and II, are found in all of the sensory areas of the vestibular labyrinth of warm-blooded animals. The type I hair cells are flasked-shaped and each is surrounded by a nerve chalice from one of the terminal branches of the vestibular nerve. The type II hair cells are shaped like cylinders. Both types of hair cells have at their free surface cytoplasmic elongations called stereocilia and one true cilium, or kinocilium. In its basal part, the type II hair cell is contacted by multiple synapses along the plasma membrane at different levels, both below and above the nucleus. Both kinds of cells receive efferent innervation, but the main difference between type I and type II cells is that the former are in direct connection with the afferent termination or calyx, whereas, in the type II, efferent connections make direct contact with the basal part of the cell. Cold-blooded animals have only type II hair cells in their epithelia, while the other structures are similar to those for warm-blooded animals.

The sensory epithelia are located in discrete areas of the various organs of the membranous labyrinth, and are named differently, according to their morphology. For the utricle and saccule, the sensory epithelium is called the macula (from latin maculare, spot); these sensory epithelia are covered with a gelatinous substance in which calcium carbonate crystals (otoconia; see Chapter 28) are embedded, forming the otolithic membrane.

Each of the semicircular canals widens at one end to form an ampulla, from whose floor projects a ridge composed of connective tissue which is lined by the sensory epithelium; this structure is called the crista. A gelatinous substance, the cupula, covers it. The otolithic membrane and the cupula constitute the movement-coupling system, as will be explained below.

In the utricle and the saccule, shearing takes place between the otolithic membrane and the sensory epithelium with a change in linear motion of the head, which disturbs the equilibrium between the otolithic membrane and the hair cells. This shearing motion, which is the result of linear acceleration, is thought to be the stimulus for the receptor cells.

In the case of the semicircular canals, angular movements of the head produce a shift of the endolymph within the semicircular canals lying in the plane of the acceleration. The pressure from the movement of the endolymph induces the cupula to move, which, in turn, results in the movement of cytoplasmic extensions (stereocilia) of the receptor hair cells, thus stimulating them.

Regardless of how the stimulus reaches the hair cell, this stimulus is thought to change the permeability in the apical part of the cell membrane, producing a receptor potential, probably due to the entrance of K^+ ions. This, in turn, causes the basal part of the cell to admit calcium ions, which presumably release neurotransmitter into the synaptic cleft. Information is thus transmitted by the hair cell to the central nervous system (CNS), through afferent fibers. In turn, higher neural centers render information to the hair cells via efferent connections.

The molecular level at which all these events take place is presently unknown. In our laboratory, we are interested in unravelling these mysteries by biochemical means. Specifically, we wish to uncover the mechanisms by which hair cells and the CNS communicate, i.e., the identity of the neurotransmitter molecules responsible for carrying mechanoreceptive information.

II. THE CHEMICAL NATURE OF NEUROTRANSMISSION IN THE VESTIBULAR SYSTEM OF VERTEBRATES

Ample experimental evidence, both morphological and physiological, indicates that synaptic transmission between the hair cell and the afferent fiber, as well as the communication between the efferent fiber and the hair cell, is chemical in nature.

Ultrastructural studies have revealed that the synaptic cleft between the receptor cell and the afferent fiber is approximately 200 Å wide, and the presynaptic cell (in this case, the hair cell) contains a typical presynaptic body or ribbon, whereas the postsynaptic terminal (afferent fiber) shows membrane thickening. Presynaptic bodies were first described by Sjöstrand (1958) in the retina, and such bodies were found thereafter in vestibular receptors (Wersäll et al., 1956; Wersäll and Flock, 1965; Nakajima and Wong, 1974; Wersäll et al., 1975; Flock, 1967).

Physiological investigations have demonstrated excitatory postsynaptic potentials in the primary afferent fibers of the goldfish saccule (Furukawa and Ishii, 1967; Chapter 2, this volume), and in the frog semicircular canal (Rossi et al., 1977). Furukawa et al. (1972) also demonstrated the existence of a delay in the synaptic potential at the receptoneural junction of the eighth nerve of the goldfish after stimulation of the receptor cell, and a quantal build-up of postsynaptic potentials in the same preparation (Ishii et al., 1971).

Extensive morphological reports at the electron-microscopic level have characterized the efferent synapse. Generally, the efferent endings (boutons) contain many vesicles, some of them dense-core, whereas a typical subsynaptic cistern is usually visualized in the hair cell. Also, the distance between the efferent ending and the sensory cell is about 200 Å (Gacek, 1974; Wersäll and Bagger-Sjöbäck, 1974). All of these features conform to the structure of a typical chemical synapse.

Inhibitory postsynaptic potentials, induced by stimulation of efferent fibers, have been described in the frog labyrinth (Gleissner and Henriksson, 1963;

Rossi *et al.*, 1980). Furthermore, overwhelming physiological evidence exists that hair-cell-efferent communication is chemical in nature (see Precht, 1974, for review).

However, all of the foregoing information says nothing about the identity of the chemical mediators involved, either in afferent or efferent neurotransmission. Obviously, the identification of chemical excitatory or inhibitory substances at specific synapses is a very challenging task.

Sophisticated methodology has been developed in order to solve this problem for specific neuronal pathways, either in the CNS or in sensory receptors such as the inner ear. However, for practical purposes, homogeneous rules should exist so that a comparison can be made between the results of different investigations. Based on the characteristics shown by a neurotransmitter at well-defined synapses, several authors have proposed criteria that must be fulfilled for a compound to be called a neurotransmitter. Among them, Werman's (1966) criteria have been especially useful because they are succinct, comprehensive, and can be satisfied with currently available techniques. The criteria can be divided in "presynaptic" and "postsynaptic" criteria. The presynaptic criteria include: (1) the presence of the substance in nerve endings, (2) the availability of the transmitter precursor, (3) the presence of a synthesizing enzyme, (4) the existence of a specific release mechanism and collectability of the substance upon stimulation of the terminal, and (5) the existence of inactivating mechanisms. Among the postsynaptic criteria are: (1) the application of the substance must have the same effects as presynaptic stimulation, (2) drugs which block effects of the test compound must block the effects of stimulation, and (3) the substance must produce similar changes in the electrical properties of the membrane as does stimulation. (The latter criterion calls for the need of intracellular recording.)

In general, presynaptic criteria can be studied by biochemical means, whereas postsynaptic criteria require electrophysiological methods, although lately, the identification of the postsynaptic receptors can also be approached biochemically.

III. BRIEF DESCRIPTION OF THE CHICK INNER EAR AND DISSECTION PROCEDURE

The anatomy and embryology of the chick inner ear are well known and, therefore, this system is particularly appropriate for biochemical studies of vestibular function.

The inner ear is embedded in three porous bones—the prootic, the opisthotic, and the epiotic (Jollie, 1957)—which together are equivalent to the mammalian petrous part of the temporal bone.

The membranous labyrinth (Fig. 1) is formed by three semicircular canals, the utricle, and the saccular cavities, which connect through the ductus reuniens to the corresponding structures of the mammalian cochlea, namely, the cochlear duct and the lagena. As in other vertebrates, the sensory structures are called cristae in the ampullae of the semicircular canals and maculae in the otolithic organs (the utricle, saccule, and lagena). The sensory organ of the cochlear duct is termed the basilar papilla. In birds, the osseous labyrinth conforms closely to the shape of the membranous labyrinth.

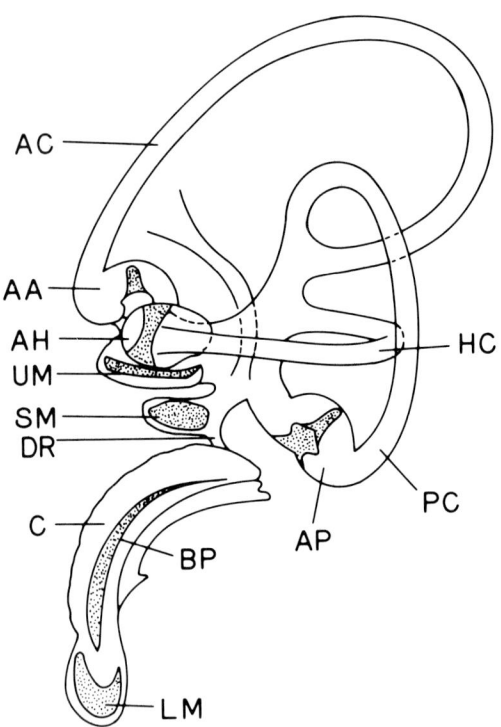

Figure 6-1. Schematic representation of the main parts of the chick inner ear. AC, anterior canal; HC, horizontal canal; PC, posterior canal; AA, anterior ampulla; AH, horizontal ampulla; AP, posterior ampulla; UM, ultricular macula; SM, saccular macula; DR, ductus reuniens; C, cochlear duct; BP, basilar papilla; LM, lagenar macula. Stippled areas in the ampullae are the sensory regions (cristae) of the semicircular canals.

The vestibular sensory epithelium, as previously described, consists of supporting cells and type I and II hair cells innervated by the vestibular branch of the eighth nerve. Because of easy access, the vestibular cristae were chosen for our biochemical studies.

One-day-old male, Rhode Island Red chicks were used throughout our experiments. The animals were sacrificed by decapitation and their craniums were cut sagitally along the dorsal midline. The brain mass was removed to expose the cranial cavity. The inner-ear-containing bones, localized by their proximity to the cerebellum, were trimmed from the inside to reach the periotic capsule. The membranous semicircular canals, which are not visible through the bone, were pulled out by inserting a fine forceps into their bony ducts, which are localized as ridges lateral to the cerebellum. All of these operations were performed under the dissecting microscope.

Immediately after removal, the semicircular canals containing the ampullary cristae (Fig. 2) were placed in chilled phosphate buffer. The dissection time for each canal was about 15 s. Organs from both ears of 10 to 30 chicks (36–100 mg wet tissue) were needed for biochemical studies.

IV. NEUROTRANSMISSION MEDIATED BY GABA

The chemical identity of the neurotransmitter in the receptoneural junction of the vestibular sensory epithelia in vertebrates is nowadays a matter of controversy: glutamate or a glutamate-like substance has been implicated (Higuchi et al., 1980; Umekita et al., 1980; Annoni et al., 1983). Also, catecholamines have been suggested (Thornhill, 1972). Notwithstanding, the present author believes that the most coherent evidence involves GABA as the afferent neurotransmitter candidate, since a number of the neurotransmitter criteria mentioned earlier have been fulfilled for this amino acid. In lower vertebrates, GABA synthesis from ^{3}H-glutamate was reported for the fish vestibule, implying the presence of the GABA-synthesizing enzyme, glutamate decarboxylase (GAD) (Flock and Lam, 1974). In addition, spontaneous and evoked afferent responses in the skate's labyrinth (Flock and Lam, 1974) were found to be inhibited by picrotoxin, a blocker of GABA receptors (Oja et al., 1977).

Our laboratory reported the synthesis of GABA in chick vestibular cristae when either uniformly-labeled ^{14}C-glutamate or 1-^{14}C-glutamate were used as precursors (Meza et al., 1981b). A high-affinity, Na$^+$- and energy-dependent uptake system for ^{3}H-GABA, possibly representing a GABA-inactivating mechanism, was described (Fig. 3) (Meza et al., 1981a). The blocking by aminooxyacetic acid (AOAA), a well-known, in vitro GAD inhibitor, of the enzyme present in chick vestibular tissue was later reported; thus an enzyme-mediated GABA synthesis was established (Meza et al., 1982a) (Fig. 4). In the same year, additional, physiological evidence for GABA as a vestibular neurotransmitter became available: (a) spontaneous and evoked labyrinthine responses were blocked by bicuculline and picrotoxin applied to the isolated cat labyrinth (Felix and Ehrenberger, 1982) and (2) systemically-administered picrotoxin blocked vertigo crises in patients suffering from peripheral vestibu-

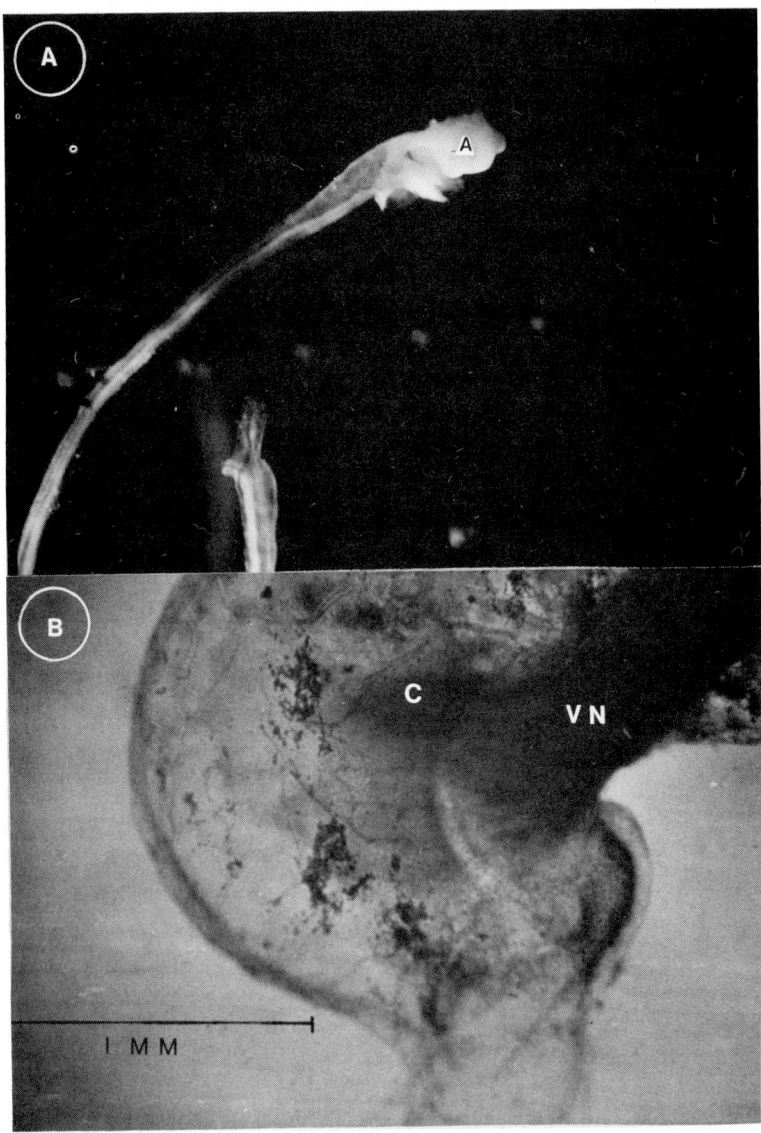

Figure 6-2: A. Photograph, with dark-field illumination, of an isolated posterior semicircular canal of the chick inner ear, showing the open ampulla (A). B. Picture of the ampulla of the posterior canal, showing the branch of the vestibular nerve (VN) and the cristae (C), stained with 2% osmium tetroxide.

lar disorders (Ehrenberger et al., 1982). Thus, because bicuculline or picrotoxin, both blockers of GABA receptors (Takeuchi and Takeuchi, 1969; Curtis et al., 1971), apparently blocked vestibular afferent responses, the presence of a post-synaptic GABA receptor was supported.

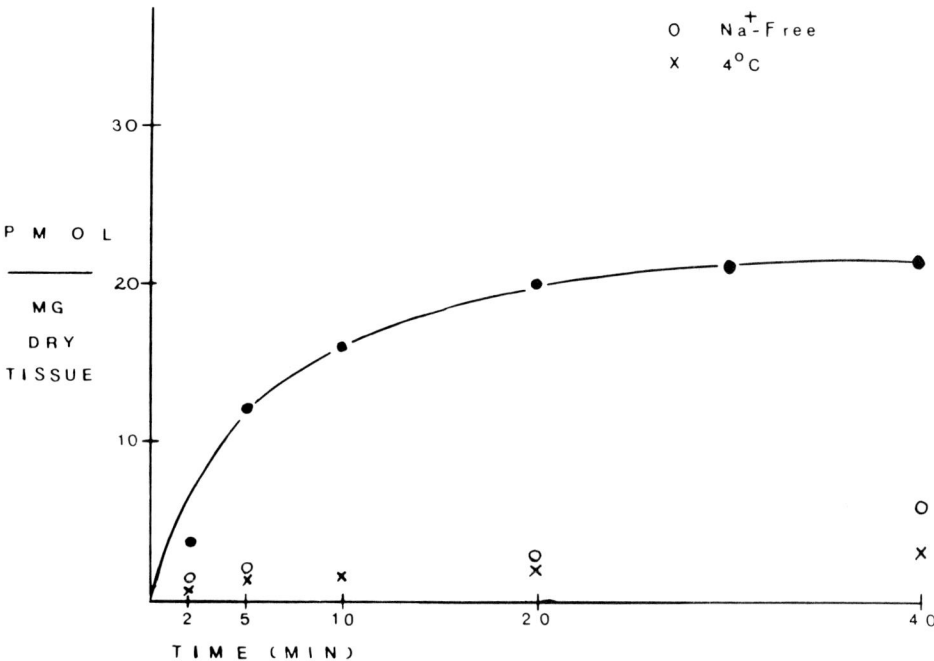

Figure 6-3. ^{3}H–GABA uptake by isolated vestibular cristae of the chick inner ear. After dissection, isolated semicircular canals were incubated in a Krebs-Ringer-bicarbonate buffer at pH 7.4 and 37°C, in the presence of 0.5 µM ^{3}H–GABA for the time indicated. The reaction was stopped by filtration and the tissue was thoroughly washed with buffer. Samples were allowed to dry to constant weight and were digested in 0.1% NCS solubilizer with gentle heating. Tritosol-based scintillation fluid was added and the samples were counted in a Packard Tri-Carb liquid scintillation spectrometer. For Na$^+$-free experiments, the same procedure was followed except that NaCl present in the medium was replaced by choline chloride. An ice-water bath was used to incubate the samples for the 4°C experiments. This illustration indicates that the GABA transport is carrier-mediated, showing saturation with time and depression both in the absence of Na$^+$ and at low temperature.

Following the same biochemical scheme, and using the chick ampullary cristae as a model, the characterization of GAD was carried out. Interesting information emerged regarding the resemblance of vestibular GAD to its homologous enzyme in nervous tissue, for which a role in the regulation of neural excitability has been demonstrated. Our results revealed properties of the vestibular GAD which were very similar to those of brain-tissue GAD, at least with regard to the effect of activators and inhibitors such as pyridoxal phosphate and AOAA (Table 1). On the grounds of the latter findings, a possible regulatory character for GAD in the inner ear was postulated (Meza, 1984a).

The presence of postsynaptic GABA receptors (Zukin et al., 1974) in the

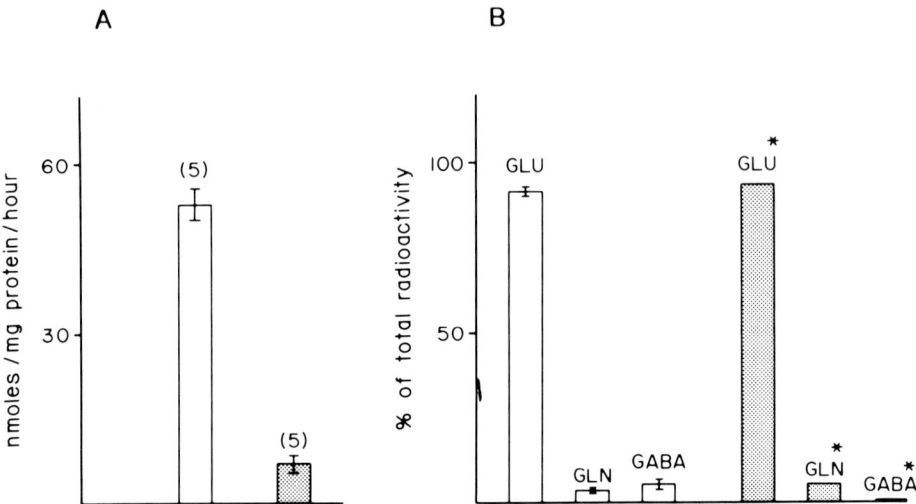

Figure 6-4. A. GAD activity, measured as $^{14}CO_2$ evolution, after incubating chick vestibular cristae homogenates with 1-^{14}C-glutamic acid. The empty bar represents control GAD activity, and the shaded bar shows GAD activity when 1 mM aminooxyacetic acid (AOAA) was included in the incubation medium (an 86.2% decrease). GAD activity is expressed as nmol of glutamate decarboxylated/mg protein/h incubation, ± SEM. The number of experiments are shown in parentheses. B. GABA synthesis, when measured by incubation of intact chick vestibular cristae with uniformly-labeled ^{14}C-glutamic acid. After making an alcoholic extract of the incubated tissue, paper chromatography was performed. Three peaks of radioactivity were obtained, corresponding to glutamate (GLU), glutamine (GLN), and GABA. When the same experiment was performed in the presence of 1 mM aminooxyacetic acid, the GABA peak was decreased by 84.3%. Results were calculated as percent of the radioactivity in the whole chromatogram present in each peak of the amino acids indicated. Controls were repeated 3 times in duplicate. Bars indicate SEM. Experiments with AOAA were performed twice, also with duplicate samples.

vertebrate labyrinth was also investigated by biochemical means. When measuring specific ^{3}H–GABA binding to labyrinthine membranes, the binding should be displaced by an excess of the unlabeled GABA and by specific receptor blockers, both interactions being Na^+-independent. The binding of ^{3}H–GABA to a crude membrane preparation of chick ampullary cristae was measured. It was shown that ^{3}H–GABA bound to labyrinthine membrane fractions could be displaced by 0.033 μM unlabeled GABA and by the specific GABA antagonist bicuculline (105 μM) (Table 2). In addition, this binding was independent of Na^+. The magnitude of the binding can be compared with what has been found for either crude or purified synaptic membranes of the mouse brain (results not shown). ^{3}H–GABA binding to vestibular fractions, measured at different GABA concentrations, showed

TABLE 6-I

COMPARISON OF SOME PROPERTIES OF CHICK INNER-EAR
GLUTAMIC ACID DECARBOXYLASE WITH THOSE OF
THE HOMOLOGOUS ENZYME IN NERVOUS TISSUE

GAD Source	Percent Activation by Pyridoxal Phosphate (PLP)			Percent Inhibition by Aminooxyacetic Acid (AOAA)	
	0.1 mM	0.5 mM	1.0 mM	0.1 mM	1.0mM
	(PLP concentration)			(AOAA concentration)	
Chick ampullary cristae[1]	42.5	59.6	76.6	48.2	88.0
Rat brain[2]	41.0, 40.5, 40.6	—	—	—	—
Mouse brain [3,4]	—	—	93.1	100.0	100.0

[1]Meza et al. (1984a); [2]Itoh and Ichimura (1981) for caudate nucleus, amigdala basal nucleus, and ventral tegmental areas, respectively; [3]Tapia and Awapara (1969); [4]Roberts and Simonsen (1963).

typical saturation kinetics (Fig. 5). Binding constants, calculated from the saturation curves, compared favorably with those reported for neural-tissue membranes (Table III). All of these results can be interpreted as reflecting the existence of a GABA receptor, of the postsynaptic type, in crude membrane preparations of the chick inner ear.

TABLE 6-II

INHIBITION OF SPECIFIC ^{3}H-GABA BINDING TO
A MEMBRANE FRACTION OF CHICK VESTIBULAR CRISTAE

Compound	$IC_{50}(\mu M)$
GABA	0.033
Bicuculline methiodide	105.000

The displacement of specific binding of 8 nM ^{3}H-GABA was determined for each of the above unlabeled compounds at 4–6 different concentrations (0.200–1.0 μM). The concentrations of the compounds inhibiting 50% of the specific binding (IC_{50} values) were calculated by log-probit analysis. Samples were run in triplicate.

Further information regarding GABA participation in acousticolateralis neurotransmission has been recently reported. A GABA-like substance was detected in guinea-pig cochlear perilymph after sound stimulation (Drescher et al., 1983; Chapter 1, this volume). These results may or may not be related to previous autoradiographic evidence locating ^{3}H-GABA in efferent terminals in the cochlea (Richrath et al., 1974; Klinke and Oertel, 1977). For no other putative neurotransmitter has so much evidence been gathered as for GABA, particularly in the vertebrate vestibular system.

Although the precise cellular localization of the synthesis and binding of GABA has not been elucidated, the above-cited lines of evidence can be taken together to support GABA participation in neurotransmission in the

TABLE 6-III
^{3}H-GABA BINDING DATA FOR CHICK INNER-EAR MEMBRANES
COMPARED WITH VALUES OBTAINED FOR NERVOUS TISSUES

Preparation	K_D (nM)	Bmax (pmol/mg protein)	Reference
Chick inner ear	19.4	0.59	Our work
Human anterior pituitary[1]	40	0.65	Grandison et al. (1982)
Rat retina[2]	18	0.60	Enna and Snyder (1976)
Rat brain[2]	16	0.60	Enna and Snyder (1977)
Rat brain[2]	32	1.17	Greenlee et al. (1978)
Rat cerebellum[2]	32	1.4	Browner et al. (1981)
Rat brain[2,3]	20	0.48	Jordan et al. (1982)

[1] Only one high-affinity site was found.
[2] Although two sites were present, data refer only to the high-affinity site.
[3] Highly purified, rat-brain synaptic plasma membranes were used.

vertebrate vestibular system. Later in this chapter, evidence for a possible afferent nature of GABA will be presented.

V. ACETYLCHOLINE PARTICIPATION IN NEUROTRANSMISSION

Physiological and histochemical evidence implicates acetylcholine (ACh) in vestibular efferent neurotransmission of some vertebrates (Dohlman et al., 1958; Brunetti et al., 1963; Dohlman, 1965; Iurato et al., 1971; Warr, 1975). Retrograde transport of ^{3}H-choline by efferent synapses in the cat labyrinth (Dememes et al., 1983) further supports this implication.

Ample evidence indicates that the vestibular sensory periphery of the bird receives efferent innervation from higher centers (Dohlman et al., 1958; Dohlman, 1965; Jahnke, et al., 1969; Joergensen, 1970; Boord and Karten, 1974; Hirokawa, 1978b; Schwarz et al., 1978, 1981; Tanaka and Smith, 1978; Eden and Correia, 1979; Strutz and Schmidt, 1982) and its nerve terminals possess AchE activity, as shown by histochemical techniques (Dohlman et al., 1958; Dohlman, 1965).

The demonstration of an enzyme of synthesis is one of the criteria which a putative neurotransmitter should fulfill. Regarding the synthesis of ACh, choline acetyltransferase (ChAT; E.C. 2.3.1.6.) has been reported to be present in the isolated labyrinth of amphibians and fish (Flock and Lam, 1974). Therefore, investigations of ChAT in the chick labyrinth are particularly important in trying to implicate ACh in neurotransmission in the chick inner ear. Specific activity of choline acetyltransferase in chick vestibular cristae homogenates, as quantified by a rapid radiochemical method (Fonnum, 1975), was shown to be 4.1 nmol/mg protein/min. This value is of the same

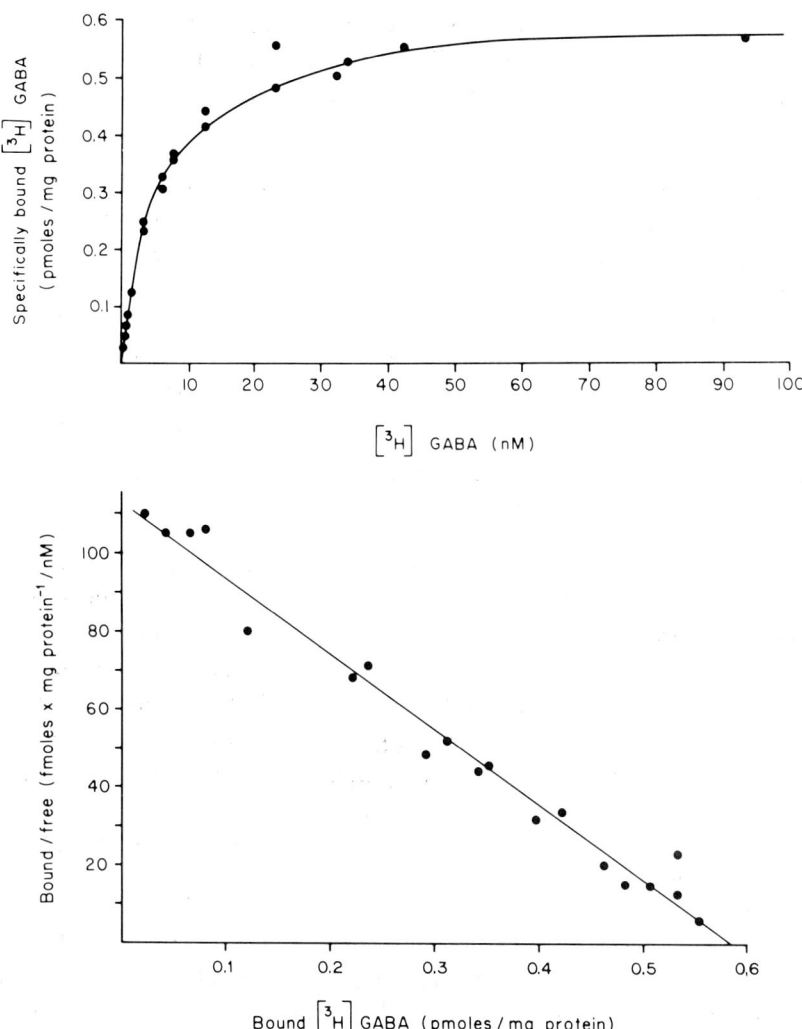

Figure 6-5. ^{3}H–GABA binding at different concentrations of unlabeled amino acid when incubation was performed in a Na$^+$-free medium. Specifically-bound ^{3}H–GABA was calculated by subtracting the ^{3}H–GABA bound to a crude membrane preparation of the chick inner ear in the presence of 1 mM unlabeled GABA from the total ^{3}H–GABA bound in the absence of the unlabeled amino acid. Results are expressed as pmol/mg protein for the crude membrane preparation. Each point in the curve represents the mean of 2–3 independent experiments run in duplicate. B. The same data as in A, graphed by the method of Scatchard; K_D for GABA was found to be 19.4 mM and B_{max} was 0.58 pmol/mg protein. Both graphs were calculated with the aid of a PDP-11/34 computer and drawn with a Hewlett-Packard 7225 plotter.

order of magnitude as that obtained for the homologous enzyme in chick neural tissue (Ross and McDougal, 1976; Marchi et al., 1980) (Table IV). When properties of vestibular ChAT activity, with regard to its sensitivity to phosphate and its dependence on chloride were examined, they were found to be identical to those described for ChAT of vertebrate neural tissue (Fonnum, 1975; Rossier et al., 1977; Hersh, 1980) (Fig. 6).

TABLE 6-IV
INNER-EAR CHOLINE ACETYLTRANSFERASE ACTIVITY AS COMPARED WITH ACTIVITY OF THE SAME ENZYME IN OTHER AREAS OF THE CHICK NERVOUS SYSTEM

ChAT Origin	nmol/mg Protein/min	Reference
Chick cristae ampullares	4.2	Meza et al. (1984b)
Retina	2.5[1]	Ross and McDougal (1976)
Iris	5.0	Marchi et al. (1980)
Lumbar sympathetic ganglion	8.3	Marchi et al. (1980)
Ciliary ganglion	50.0	Marchi et al. (1980)

[1] Value calculated assuming a 10% protein, 90% water content in this tissue.

The presence of an inactivating mechanism is another prerequisite for a compound to be called a neurotransmitter. For ACh, the degradative enzyme acetylcholinesterase (AChE; E. C. 3.1.1.7) provides such a mechanism. Therefore, the activity of this enzyme was investigated by the method of Ellman (1961) in a crude membrane fraction of chick ampullary cristae.

The value for AChE activity was found to fall within the range found for a similar enzyme in neural tissue (Table V). Physostigmine sulfate, a specific, reversible, *in-vivo* (Bullock et al., 1946) and *in-vitro* (Augustinsson and Nachmanson, 1949) inhibitor of AChE, completely blocked the enzyme activity at 10 μM (Fig. 7).

The existence of enzymes of synthesis and degradation for ACh in chick vestibular tissue, together with histochemical localization of AChE, may be taken to support ACh as a neurotransmitter in the chick labyrinth. Evidence to be presented later lends support to ACh being an efferent neurotransmitter.

VI. CELLULAR LOCALIZATION OF GABA- AND ACETYLCHOLINE-SYNTHESIZING ENZYMES: THE ONTOGENETIC APPROACH

Various studies performed on developing stato-acoustic organs of certain vertebrates indicates that there exists a differential maturation of the various cell types composing the sensory epithelium. Apparently, hair cells and their afferent synapses mature early in development, as judged by characteristic morphological features observed under the electron microscope (Kikuchi and Hilding, 1965; Vazquez-Nin and Sotelo, 1968; Van de Water et al., 1977; Hirokawa, 1978a; Ginzberg and Gilula, 1980; Jones and Eslami, 1983).

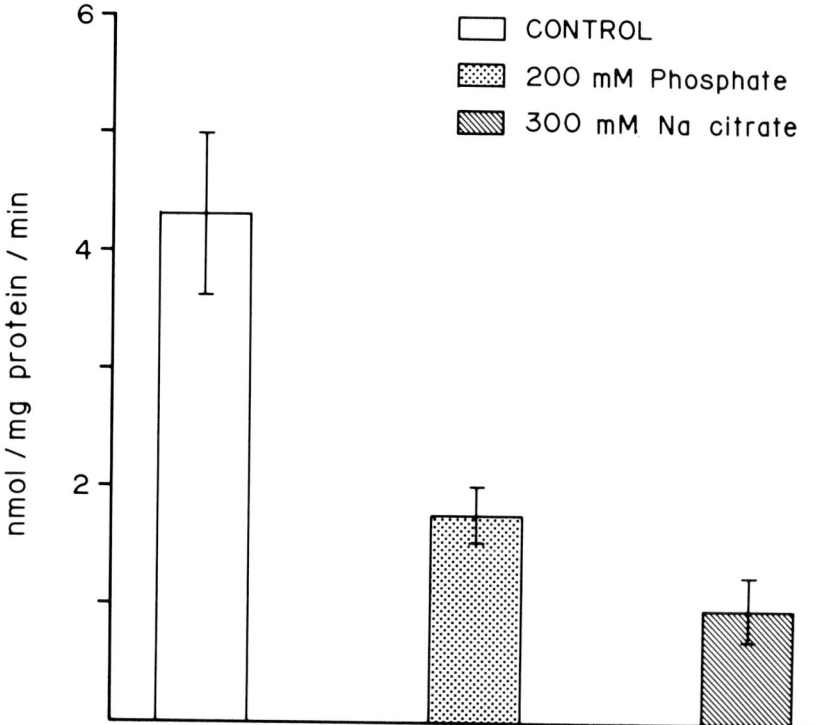

Figure 6-6. Activity of ChAT obtained in homogenates of ampullary cristae from the chick inner ear, by the method of Fonnum (1975), measured either in the presence of 10 mM phosphate, 200 mM phosphate, or in standard incubation medium (10 mM phosphate) in which NaCl was replaced by sodium citrate. As shown, the enzyme is inhibited by elevated concentrations of phosphate and its activity is highly dependent on the presence of chloride. These properties are similar to those for ChAT in other nervous tissues (Fonnum, 1975). Results are expressed as nmol of ^{3}H-acetyl CoA deacetylated/mg protein/min. Three to five independent experiments were performed in duplicate. Bars indicate SEM. (Data from Meza et al., 1984b.)

In contrast, the appearance of mature efferent synapses seems to be a late event in development. It has been observed that the number of efferent terminals or the area occupied by the terminals suddenly increases in late stages of ontogenesis which probably indicates the establishment of functional efferent innervation (Nakai, 1970; Hirokawa, 1977; Cohen and Fermin, 1978; Lenoir et al., 1980; Pujol et al., 1980; Shnerson et al., 1982; Jones and Eslami, 1983; Rebillard and Pujol, 1983; Fermin and Cohen, 1984).

The experimental evidence from our laboratory and the laboratories of others, described earlier in this chapter, seems to point towards GABA as being the afferent neurotransmitter, at least in the warm-blooded vertebrate

TABLE 6-V
ACTIVITY OF ACETYLCHOLINESTERASE IN A CRUDE MEMBRANE PREPARATION OF
CHICK INNER EAR, AS COMPARED WITH VALUES OBTAINED FOR
THE SAME ENZYME FROM DIFFERENT SOURCES[1]

Tissue	$\mu mol/(L \times min \times g\ tissue)$	Reference
Chick vestibular membranes	31.2	Our work
Whole pigeon brain	22.5	Wächtler (1981)
Whole rat brain	10.3	Ellman et al. (1961)
Guinea pig crude synaptosomal fraction, water suspension	1.3	Whittaker et al. (1964)

[1]Although the preparations are not the same and, therefore, not quite comparable, at least values of the same order of magnitude were obtained.

vestibular systems, and thus GABA is probably synthesized in the hair cell. The data also implicate ACh as the most likely efferent mediator; therefore, its synthesizing enzyme (ChAT) would have to be located in the efferent terminal, as described for other cholinergic areas of the CNS (Hebb and Whittaker, 1958; Hebb, 1963). In addition, the presence of ChAT has been claimed to represent a good correlate of neural activity (Ekstrom, 1978).

Assuming that GAD and ChAT are located in distinct cell types whose maturation occurs at different times in ontogenesis, we decided to measure these enzymes in developing chick ampullary cristae and to compare the values obtained with those described for the one-day-old chick, whose inner ear is considered to be mature.

Indeed, when GAD and ChAT activities were measured in ampullary cristae of embryonic chicks at different stages (days 13, 15, 17, 18, 19, and 20; Fig. 8), it was found that the values of GAD activity were practically the same at the earliest stage (day 13 of embryogenesis) as at the stage of inner-ear maturity (one-day-old chicks). In contrast, ChAT activity was practically undetectable up to day 17 of embryonic development, but rose suddenly at the eighteenth day, reaching the one-day-old level at the nineteenth day of embryonic age.

These data obtained for both GAD and ChAT in the developing inner ear of the chick indeed seem to indicate that each enzyme is located in a different cell type (or compartment), whose maturation occurs at a different time during development, in accord with the hypothesis that GAD may be located in the hair cell, whereas ChAT is found in the efferent terminal.

VII. SUMMARY AND CONCLUSIONS

That synaptic transmission in the vertebrate vestibular system is chemical in nature is presently accepted; however, the corresponding neurotransmitters

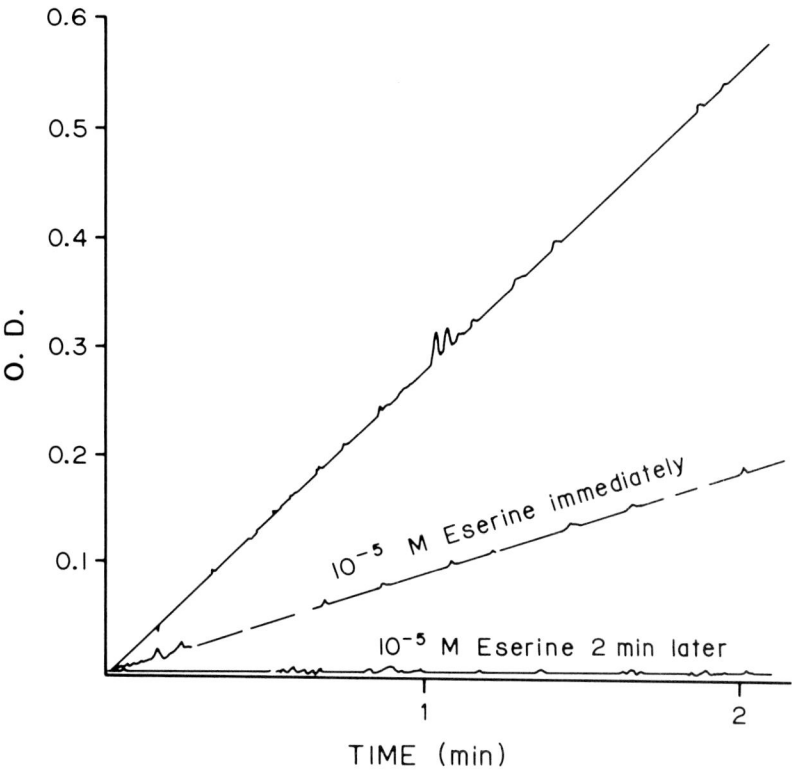

Figure 6-7. Drawing of the strip-chart record of a typical experiment for AChE determination by the Ellman (1961) method. The first trace from the left represents a record of a control experiment when the reaction was started with acetylthiocholine as substrate. The lower traces show the recordings taken either immediately, or 2 min after, the addition of 10^{-5}M eserine sulfate to the control curvette. The reaction rates were followed and recorded from an Aminco double-beam spectrophotometer equipped with a stopped-flow device.

have not been identified. Some evidence implicates GABA as the afferent neurotransmitter and acetylcholine as the efferent neuromediator. Data in support of these assumptions have been obtained using several vestibular preparations.

A number of criteria must be fulfilled before a neurotransmitter candidate can be called a neurotransmitter. For GABA, the synthesizing enzyme appears to be present in the vestibular organs of amphibians, fish, and birds. Sodium and energy-dependent GABA transport, possibly representing a GABA-inactivating mechanism, has been demonstrated in intact chick vestibular cristae. Mimicking of the natural response by GABA in the cat labyrinth and blocking by picrotoxin and bicuculline of the effects of GABA on afferent fibers in fish and cat labyrinths speak of the possible existence of a postsynap-

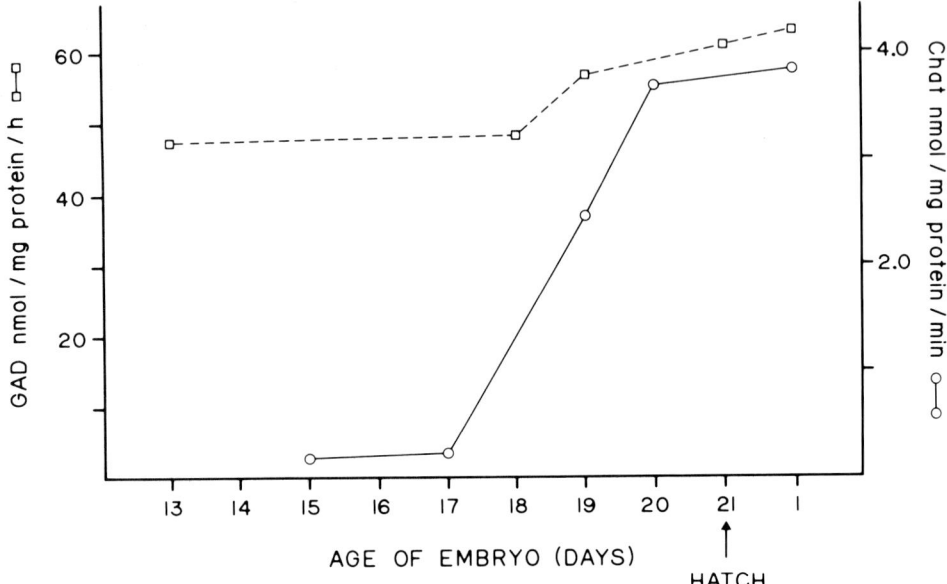

Figure 6-8. GAD and ChAT activities measured in isolated ampullary cristae of the embryonic chick. Activities were determined in homogenates of 8-10 animals (approximately 18 mg wet tissue) of the ages indicated. For GAD, the $^{14}CO_2$ evolution method was used (Meza et al., 1984a); data are expressed as nmol of glutamate decarboxylated/mg protein/h. For ChAT (Meza et al., 1984b), the data are expressed as nmol ^{3}H-acetyl CoA deacetylated/mg protein/min. Each point on the graph, for either enzyme, represents the mean of 2-3 experiments run in duplicate. Results show a differential increase in activity with time for the two enzymes, possibly indicating their localization in different cell types or compartments which mature differently during ontogenesis.

tic GABA receptor. Biochemical experiments measuring ^{3}H-GABA binding in a crude membrane preparation of chick ampullary cristae tend to confirm the idea of a GABA receptor. Successful treatment with picrotoxin of patients suffering from vertigo crises of peripheral origin may also be correlated with the presence of a postsynaptic GABA receptor.

For acetylcholine, the enzyme of synthesis, choline acetyltransferase, has been found in isolated vestibular tissue from the frog and chick. Acetylcholinesterase, the degradative (inactivating) enzyme for acetylcholine, can be demonstrated in crude membrane preparations of chick vestibular tissue and in most other vertebrate vestibular sensory tissue at the efferent level, as shown by histochemical techniques. Both enzymes have properties which are similar to those described for neural tissue. Acetylcholine mimics the natural efferent stimulus in the isolated frog labyrinth, and pharmacological studies suggest the presence of a nicotinic receptor in this preparation.

Experiments performed in the developing chick labyrinth show measurable GAD and ChAT activities at times which can be correlated with maturation of afferent and efferent connections.

Considered as a whole, these data are consistent with the possible role of GABA as the afferent neurotransmitter and ACh as the probable efferent neuromediator in the sensory periphery of the vestibular system in warm-blooded vertebrates.

It is hoped that the investigations described above will contribute to a better understanding of the normal function of the human inner ear and thus eventually lead to the development of drugs which can alleviate a variety of vestibular disorders in man.

ACKNOWLEDGMENTS

A number of persons assisted in performing the work described in this chapter. Dr. Mariano Ruiz, Dept. Histología, Facultad de Medicina, UNAM designed and performed the dissection procedures; Miss Patricia Cuadros, Miss Teresa González, and Mr. Ivan López, respectively, carried out the GAD chromatographic experiments, the GABA-receptor determinations, and the ChAT and AChE measurements. Mrs. Virginia Godínez provided excellent secretarial help and Mr. Arturo Franco prepared the illustrations and photographs. Grant PCCBBNA 020897, from Consejo Nacional de Ciencia y TecnologíA (CONACyT), México, partially financed this investigation.

REFERENCES

Annoni, J. M.; Cochran, S. L., and Precht, W.: Synaptic and amino acid-induced excitation and antagonism at the vestibular hair cell primary afferent synapse of the frog. *Neurosci. Lett. Suppl. 14:* S9, 1983.

Augustinsson, K. B., and Nachmansohn, D.: Studies on cholinesterase. VI. Kinetics of the inhibition of acetylcholine esterase. *J. Biol. Chem. 179:* 543–549, 1949.

Boord, R. L., and Karten, H. J.: The distribution of primary lagenar fibers within the vestibular nuclear complex of the pigeon. *Brain Behav. Evol. 10:* 228–235, 1974.

Browner, M.; Ferkany, J. W., and Enna, S. J.: Biochemical identification of pharmacologically and functionally distinct GABA receptors in rat brain. *J. Neurosci. 1:* 514–518, 1981.

Brunetti, F.; Rossi, G., Vocnna, G., Buonogiovanni, S., and Cortesina, G.: L'action locale de l'acetycholine de substances anticholine-esterasiques sur la fonction vestibulaire. *Acta Otolaryng. 57:* 294–298, 1963.

Bullock, T. H.; Nachmansohn, D., and Rothenberg, M. A.: Effect of inhibitors of choline esterase on the nerve action potential. *J. Neurophysiol. 9:* 9–22, 1946.

Cohen, G. M., and Fermin, C. D.: The development of hair cells in the embryonic chick's basillar papila. *Acta Otolaryng. 86:* 342–358, 1978.

Curtis, D. R.; Duggan, A. W., Felix, D., and Johnston, G. A. R.: Bicuculline, an antagonist of GABA and synaptic inhibition in the spinal cord of the cat. *Brain Res. 32:* 69–96, 1971.

Dememes, D.; Raymond, J., and Sans, A.: Selective retrograde labelling of vestibular efferent neurons with ^{3}H-choline. *Neuroscience* 8: 285-290, 1983.

Dohlman, G. F.: Histochemical studies of vestibular mechanism. In Rasmussen, G. L., and Windl, W. F. (Eds.): *Neural mechanisms of the Auditory and Vestibular Systems.* Springfield, Charles C. Thomas, 1965, pp. 258-275.

Dohlman, G. F.; Farkashidy, J., and Salonna, F.: Centrifugal nerve fibers to the sensory epithelium of the vestibular labyrinth. *J. Laryng.* 72: 984-991, 1958.

Drescher, M. J.; Drescher, D. G., and Medina, J. E.: Effect of sound stimulation at several levels on concentrations of primary amines, including neurotransmitter candidates, in perilymph of the guinea pig inner ear. *J. Neurochem.* 41: 309-320, 1983.

Eden, A. R., and Correia, M. J.: Horseradish peroxidase identification of four separate groups of vestibular efferent neurons in the adult pigeon. *Soc. Neurosci. Abstr.* 5: 690, 1979.

Ehrenberger, K.; Benkoe, E., and Felix, D.: Suppressive action of picrotoxin, a GABA antagonist, on labyrinthine spontaneous nystagmus and vertigo in man. *Acta Otolaryng.* 93: 269-273, 1982.

Ekstrom, J.: Acetylcholine synthesis and its dependence on nervous activity. *Experientia* 34: 1247-1253, 1978.

Ellman, G. L.; Courtney, K. D., Andres, V., Jr., and Featherstone, R. M.: A new and rapid colorimetric determination of acetylcholinesterase activity. *Biochem. Pharmacol.* 7: 88-95, 1961.

Enna, S. J., and Snyder, S. H.: Gamma-aminobutyric acid (GABA) receptor binding in mammalian retina. *Brain Res.* 115: 174-179, 1976.

Enna, S. J., and Snyder, S. H.: Influences of ions, enzymes and detergents on γ-aminobutyric acid-receptor binding in synaptic membranes of rat brain. *Mol. Pharmacol.* 13; 442-453, 1977.

Felix, D., and Ehrenberger, K.: The action of putative neurotransmitter substances in the cat labyrinth. *Acta Otolaryng.* 93: 101-105, 1982.

Fermin, C. D., and Cohen, G. M.: Developmental gradients in the embryonic chick's basillar papilla. *Acta Otolaryng.* 97: 39-51, 1984.

Flock, A.: Ultrastructure and function in the lateral line organs. In Cahn, P. (Ed.): *Lateral line detectors.* Indiana, University Press, 1967, pp. 163-197.

Flock, A., and Lam, D. M. K.: Neurotransmitter synthesis in inner ear and lateral line sense organs, *Nature* 249: 142-144, 1974.

Fonnum, F.: A rapid radiochemical method for the determination of choline acetyltransferase. *J. Neurochem.* 24: 407-409, 1975.

Furukawa, T., and Ishii, Y.: Neurophysiological studies on hearing in goldfish: *J. Neurophysiol.* 30: 1377-1403, 1967.

Furukawa, T.; Ishii, Y., and Matsumura, S.: Synaptic delay and time course of the postsynaptic potential at the junction between hair cells and eighth nerve fibers in the goldfish. *Jap. J. Physiol.* 22: 617-635, 1972.

Gacek, R. R.: Morphological aspects of the afferent vestibular system. In Kornhuber, H. H. (Ed.): *Handbook of Sensory Physiology,* Vol. VI/I, New York, Springer-Verlag, 1974, pp. 213-220.

Ginzberg, R. D., and Gilula, N. B.: Synaptogenesis in the vestibular sensory epithelium of the chick embryo. *J. Neurocytol.* 9: 405-424, 1980.

Gleissner, L., and Henrikson, N. G.: Efferent and afferent activity pattern in the vestibular nerve of the frog. *Acta Otolaryng. Suppl.* 192: 90-103, 1963.

Grandison, L.; Cavagnini, F., Schmid, R., Invitti, C., and Guidotti, A.: γ-Aminobutyric acid and benzodiazepine binding sites in human anterior pituitary tissue. *J. Clin. Endocrinol. Metab.* 54: 597-601, 1982.

Greenlee, D. V.; Vanness, P. C., and Olsen, R. W.: Endogenous inhibitor of GABA binding in mammalian brain. *Life Sci. 22:* 1653–1662, 1978.

Hebb, C. O.: Formation, storage and liberation of acetylcholine. In Koelle, G. B. (Ed.): *Handbuch der experimentellen Pharmacologie. Vol. 15,* New York Springer-Verlag, 1963, pp. 55–88.

Hebb, C. O., and Whittaker, V.: Intracellular distributions of acetylcholine and choline acetylase. *J. Physiol. 142:* 187–196, 1958.

Hersh, L. B.: Studies on the kinetic mechanism and salt activation of bovine brain choline acetyltransferase, *J. Neurochem. 34:* 1077–1081, 1980.

Higuchi, T.; Nagai, T., Umekita, S. H., and Obara, S.: The afferent neurotransmitter in the ampullary electroreceptors: L-glutamate mimics the natural transmitter. *Neurosci. Lett. Suppl. 4:* S7, 1980.

Hirokawa, N.: Disappearance of afferent and efferent nerve terminals in the inner ear of the chick embryo after chronic treatment with α-bungarotoxin. *J. Cell. Biol. 73:* 27–46, 1977.

Hirokawa, N.: Synaptogenesis in the basilar papilla of the chicken. *J. Neurocytol. 7:* 283–300, 1978a.

Hirokawa, N.: The ultrastructure of the basilar papilla of the chick. *J. Comp. Neurol. 181:* 361–374, 1978b.

Ishii, Y.; Matsuura, S., and Furukawa, T.: Quantal nature of transmission at the synapse between hair cells and the eighth nerve fibers, *Jap. J. Physiol. 19:* 79–89, 1971.

Itoh, M., and Ichimura, H.: Regional differences in cofactor saturation of glutamate decarboxylase (GAD) in discrete brain nuclei of the rat. *Neurochem. Res. 6:* 1283–1289, 1981.

Iurato, S.; Luciano, L., Pannese, E., and Reale, E.: Histochemical localization of acetylcholinesterase (AChE) activity in the inner ear, *Acta Otolaryng. Suppl. 279:* 1–50, 1971.

Jahnke, V.; Lundquist, P. G., and Wersall, J.: Some morphological aspects of sound perception in birds. *Acta Otolaryng. 67:* 583–601, 1969.

Jollie, M. T.: The head skeleton of the chicken and remarks on the anatomy of this region in other birds. *J. Morphol. 100:* 389–436, 1957.

Jones, D. G., and Eslami, H.: An ultrastructural study of the development of afferent and efferent synapses on outer hair cells of the guinea pig organ of Corti. *Cell Tissue Res. 231:* 533–539, 1983.

Jordan, C.; Matus, A. I., Piotrowski, W., and Wilkinson, D.: Binding of ^{3}[H]-aminobutyric acid and ^{3}[H]-muscimol in purified rat brain synaptic plasma membranes and the effects of bicuculline. *J. Neurochem. 39:* 52–58, 1982.

Joergensen, J. M.: On the structure of the macula lagenae in birds with some notes on the avian maculae utriculi and sacculi. *Vidensk. Medd. Dan. Naturh. Foren. 133:* 121–147, 1970.

Kikuchi, K., and Hilding, D.: The development of the organ of Corti in the mouse. *Acta Otolaryng. 69:* 207–222, 1965.

Klinke, R., and Oertel, W.: Evidence that GABA is not the afferent transmitter in the cochlea. *Exp. Brain Res. 30.* 141–143, 1977.

Lenoir, M.; Shnerson, A., and Pujol, R.: Cochlear receptor development in the rat with emphasis on synaptogenesis. *Anat. Embryol. 160:* 253–262, 1980.

Marchi, M.; Hoffman, D. W., Giacobini, E., and Fredrickson, T.: Acetyltransferase activities in autonomic ganglia and iris of the chick. *Develop. Neurosci. 3:* 235–247, 1980.

Meza, G.: Some characteristics of glutamic acid decarboxylase of chick ampullary cristae. *J. Neurochem. 43:* 634–639, 1984a.

Meza, G.; Cuadros, P., Lopez, I., and Ruiz, M.: Neurotransmitter synthesizing enzymes in the developing chick inner ear: a model for cellular localization of putative chemical mediators in the vestibular sensory periphery. *Soc. Neurosci. Abstr. 8:* 42, 1982a.

Meza, G.; Hernandez, C., and Ruiz, M.: GABA synthesis in isolated vestibullary tissue of chick inner ear. *Brain Res. 241:* 157-161, 1982b.

Meza, G.; Hernandez, C., and Ruiz, M.: ^{3}H-GABA uptake in isolated vestibullary cristae of chick inner ear. *Soc. Neurosci. Abstr. 7:* 147, 1981a.

Meza, G.; Lopez, I., and Ruiz, M.: Possible cholinergic neurotransmission in cristae ampullares of chick inner ear. *Neurosci. Lett.* In press. 1984b.

Meza, G.; Ruiz, M., and Cuadros, P.: GABA synthesis in isolated ampullar cristae of chick inner ear. *Trans. Am. Soc. Neurochem. 12:* 256, 1981b.

Nakai, Y.: The development of the sensory epithelium of the cristae ampullares in the rabbit. *Pract. Otorhinolaryngol. 32:* 268-278, 1970.

Nakajima, Y., and Wong, D. W.: Morphology of afferent and efferent synapses in the hearing organ of the goldfish. *J. Comp. Neurol. 156:* 403-416, 1974.

Oja, S. S.; Kontro, P., and Löhdesmäki, P.: *Neurotransmitters Progress in Pharmacology*, Stuttegard, Fisher-Verlag, 1977.

Precht, W.: Physiological aspects of the efferent vestibular system. In Kornhuber, H. H. (Ed.): *Handbook of Sensory Physiology, Vol. VI/I*, Berlin, New York, Springer-Verlag, 1974, pp. 221-236.

Pujol, R.; Carlier, E., and Lenoir, M.: Ontogenetic approach to inner and outer hair cell function. *Hearing Res. 2:* 423-430, 1980.

Rebillard, M., and Pujol, R.: Innervation of the chicken basillar papilla durings its development. *Acta Otolaryng. 96:* 379-388, 1983.

Richrath, W.; Kraus, H., and Fromme, H. G.: Lokalisation von ^{3}H-γ-Aminobuttersaure in der Cochlea. *Arch. Otolaryngol. 208:* 283-286, 1974.

Roberts, E., and Simonsen, D. G.: Some properties of L-glutamic decarboxylase in mouse brain. *Biochem. Pharmacol. 12:* 113-134, 1963.

Ross, C. D., and McDougal, D. B.: The distribution of choline acetyltransferase in vertebrate retina. *J. Neurochem. 26:* 521-526, 1976.

Rossi, M.; Prigioni, I., Valli, P., and Casella, C.: Activation of the efferent system in the isolated frog labyrinth, effects on the afferent EPSPs and spike discharge recorded from single fibers of the posterior nerve. *Brain Res. 185:* 125-137, 1980.

Rossi, M. L.; Valli, P., and Casella, C.: Post-synaptic potentials recorded from afferent nerve fibers of the posterior semicircular canal in the frog. *Brain Res. 135:* 67-75, 1977.

Rossier, J.; Spanditakis, Y., and Benda, P.: The effect of Cl$^-$ on choline acetyltransferase kinetic parameters and a proposed role for Cl$^-$ in the regulation of acetylcholine synthesis. *J. Neurochem. 29:* 1007-1012, 1977.

Schwarz, D. W. F.; Schwarz, I. E., and Tomlinson, R. D.: Avian efferent vestibular neurons identified by axonal transport of ^{3}H-adenosine and horseradish-peroxidase. *Brain Res.* 155: 103-107, 1978.

Schwarz, I.; Schwarz, D., Fredrickson, J., and Landolt, J.: Efferent vestibular neurons. A study employing retrograde tracer-methods in the pigeon (Columbia livia). *J. Comp. Neurol. 196:* 1-12, 1981.

Shnerson, A.; Devigne, C., and Pujol, R.: Age related changes in the C57BL/65 mouse cochlea. II. Ultrastructural findings. *Dev. Brain Res. 2:* 77-88, 1982.

Sjöstrand, F. S.: Ultrastructure of the retinal rod synapses of the guinea pig eye as revealed by three-dimensional reconstructions from serial sections. *J. Ultrastruct. Res. 2:* 122-170, 1958.

Strutz, J., and Schmidt, L.: Acoustic and vestibular efferent neurons in the chicken (Gallus domesticus). *Acta Otolaryng. 94:* 45-51, 1982.

Takeuchi, A., and Takeuchi, N.: A study of the action of picrotoxin on the inhibitory neuromuscular junction of the crayfish. *J. Physiol. (Lond.)* 205: 377–391, 1969.

Tanaka, T., and Smith, C. A.: Structure of the chicken's inner ear: SEM and TEM study. *Am. J. Anat.* 153: 251–272, 1978.

Tapia, R., and Awapara, I.: Effects of various substituted hydrazones and hydrazines of pyridoxal-5'-phosphate on brain glutamate decarboxylase. *Biochem. Pharmacol.* 18: 145–152, 1969.

Thornhill, R. A.: The effect of catecholamine precursors and related drugs on the morphology of the synaptic bars in the vestibular epithelia of the frog, *Rana temporaria*. *Comp. Gen. Pharmacol.* 3: 89–97, 1972.

Umekita, S. H.; Matsumoto, Y., Abe, T., and Obara, S.: The afferent neurotransmitter in ampullary electroreceptors: stimulus-dependent release experiments refute the transmitter role of L-glutamate. *Neurosci. Lett. Suppl.* 4: S7, 1980.

Van de Water, T. R.; Anniko, M., Nordemar, H., and Wersäll, J.: Embryonic development of the sensory cells in macula utriculae of mouse. In Portmann, M., and Aran, J. M. (Eds.): *Inner Ear Biology*, Vol. 68, 1977, pp. 25–36.

Vaquez-Nin, G. H., and Sotelo, J. R.: Electron microscope study of the developing nerve terminals in the acoustic organs of the chick embryo. *Zeitsch. Zellforsch.* 92: 325–338, 1968.

Wächtler, K.: The regional distribution of acetylcholine, cholineacetyltransferase and acetylcholinesterase in vertebrate brains of different phylogenetic levels. In Pepeu, G., and Ledinsky, H. (Eds.): *Cholinergic Mechanisms*. New York, Plenum, 1981, pp. 59–72.

Warr, W. B.: Olivocochlear and vestibular efferent neurons of the feline brain stem: their location, morphology and number determined by retrograde axonal transport and acetylcholinesterase histochemistry. *J. Comp. Neurol.* 161: 159–183, 1975.

Werman, R.: Criteria for identification of a central nervous system transmitter. A review. *Comp. Biochem. Physiol.* 18: 745–766, 1966.

Wersäll, J.: Studies on the structure and innervation of the sensory epithelium of the cristae ampullares in the guinea pig. A light and electronmicroscopic investigation. *Acta Otolaryng.* 126: 1–85, 1956.

Wersäll, J., and Bagger-Sjöbäck, D.: Morphology of the vestibular sense organs. In Kornhuber, H. H. (Ed.): *Handbook of Sensory Physiology*. Vol VI/I, Berlin, New York, Springer-Verlag, 1974, pp. 123–170.

Wersäll, J., and Flock, A.: Functional Anatomy of the Vestibular and Lateral Line Organs. In Neff, W. D. (Ed.): *Contributions to Sensory Physiology*, Vol. 1. New York, Academic Press, 1965, pp. 39–61.

Wersäll, J.; Flock, A., and Lundquist, P. G.: Structural basis for directional sensitivity in cochlear and vestibular sensory receptors. *Cold Spr. Harb. Symp. Quant. Biol.* 155: 115–132, 1975.

Whittaker, V. P.; Michaelson, I. A., and Kirkland, R. J. A.: The separation of synaptic vesicles from nerve-ending particles (synaptosomes). *Biochem. J.* 90: 293–303, 1964.

Zukin, S. R.; Young, A. B., and Snyder, S. H.: Gamma-aminobutyric acid binding to receptor sites in the rat central nervous system. *Proc. Natl. Acad. Sci. U.S.A.* 71: 4802–4807, 1974.

Chapter 7

ACTIONS OF PUTATIVE NEUROTRANSMITTERS AND OTHER RELEVANT COMPOUNDS ON *XENOPUS LAEVIS* LATERAL LINE

Richard P. Bobbin, Sanford C. Bledsoe, Jr.,
Stephen L. Winbery, and Gary L. Jenison

I. Introduction
II. Action
III. Pharmacological Antagonism
IV. Release
V. Summary and Conclusions
References

I. INTRODUCTION

This chapter concerns neurotransmitters in the lateral line of *Xenopus laevis*. We will focus on three criteria that must be satisfied in order to establish a chemical as a transmitter (Siegel *et al.*, 1981): (1) action, i.e., what chemicals have an effect at a given synapse, (2) pharmacological antagonism, i.e., what chemicals antagonize this action and the action of the endogenous transmitter, and (3) release, i.e., what chemicals are released by natural stimulation. In addition, because we are interested in the transmitters of mammals, we will compare results in the lateral line to those obtained in the cochlea.

The *Xenopus* lateral line has long been used as a model for understanding sensory transduction in hair-cell systems. For example, it has been studied by others interested in afferents (Bauknight *et al.*, 1976), efferents (Russell, 1968, 1971, 1976), ototoxic antibiotics (Kroese and Bercken, 1982), and the function of the cupula (Russell and Sellick, 1976). The reasons for this are many, and include the ease of experimental methods and the viability of *in vitro* preparations for long periods of time. Figure 1 shows the distribution of the sensory structures, the stitches, comprising the *Xenopus* lateral line. Stitches are located in the skin and contain five to twelve clumps of hair cells; each clump of hair cells is called a neuromast. As illustrated in Fig. 2, each stitch is innervated by two myelinated afferent nerve fibers, which, in turn, innervate most of the neuromasts (Harris and Flock, 1967).

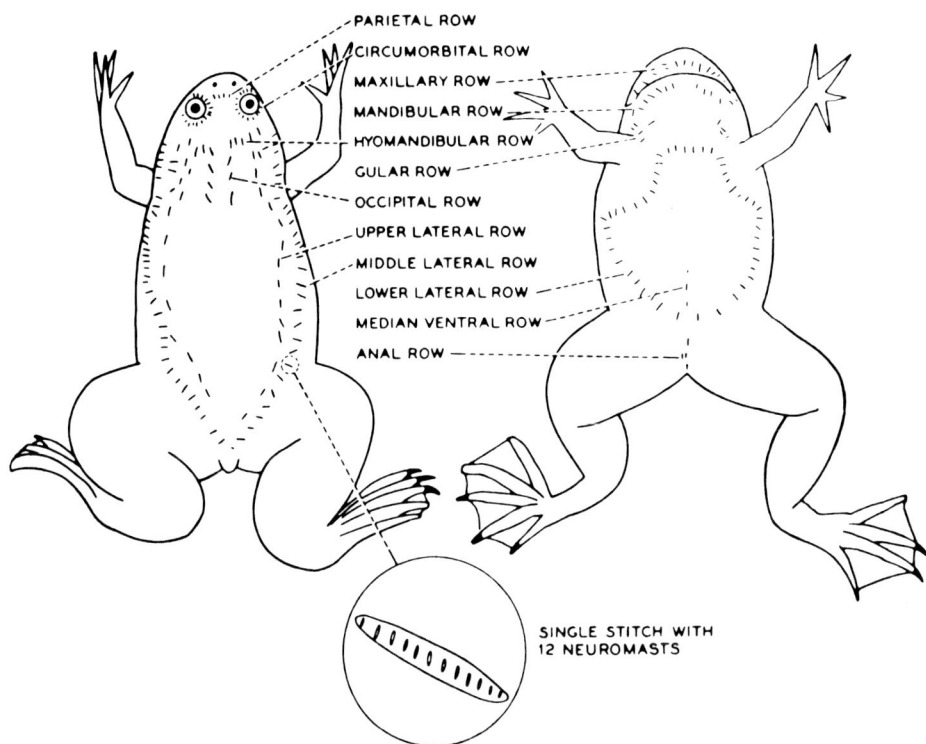

Figure 7-1. An illustration of the distribution of the lateral-line organ in the Xenopus laevis. From Harris and Milne (1966), with permission.

The methods we used to prepare an isolated stitch for recording are similar to those of others (e.g., Bauknight *et al.*, 1976; Harris and Milne, 1966; Harris and Flock, 1967). The animal is anesthetized by cooling to about 0°C in water from its own tank, decapitated and pithed. A portion of skin containing the medial lateral row of stitches is removed from the animal's side and pinned to a wax board on a piece of filter paper. A schematic of the experimental set-up is shown in Fig. 3. A single stitch is isolated by sectioning the lateral-line nerve trunk distal to the selected stitch and sectioning all nerves leading to stitches proximal to the one selected for study. The response of the two afferent nerve fibers is recorded with a suction electrode, which holds the nerve bundle. The preparation is maintained in an oxygenated Ringer solution that washes continuously over the serosal surface of the skin. The wax board is positioned on a slant to allow fluid to

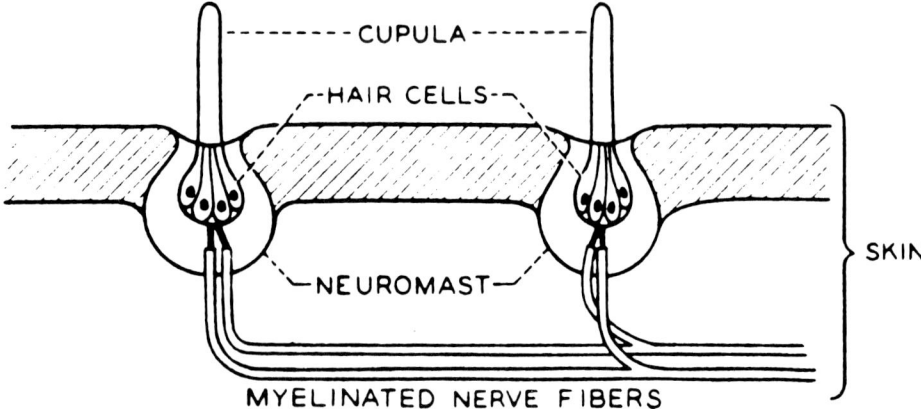

Figure 7-2. Drawing of the afferent fibers innervating two neuromasts. From Harris and Milne (1966), with permission.

drain off. Drugs are applied to the stitch in several ways. In one method, a drug is added to the wash solution, wherein it remains in contact with the skin for a long time. In another, the drug is injected into the wash-delivery tube so that a small volume of chemical contacts the stitch briefly and is then washed away. A third method involves turning the wash off and allowing the skin to drain for a time before squirting microliter amounts of the drug solution onto the stitch. The wash is turned back on after waiting a few minutes for the chemical to have an effect. In addition, we have adapted the lateral line to a bath preparation to study the effects of chemical agents on water-motion induced activity and also to examine the chemicals released into the bath (Bledsoe et al., 1980).

II. ACTION

A limited number of chemicals, which are known or are suspected to be transmitters elsewhere, have been shown to possess activity in the lateral line. Many investigators have contributed to our knowledge in this area, with the work of Russell (1976) probably being most recognized. He showed that acetylcholine was a powerful inhibitor of afferent spontaneous activity in the *Xenopus* lateral line and a candidate for the inhibitory transmitter released by efferent fibers. Recently, we have confirmed these results but, in addition, have observed that acetylcholine also increases spontaneous activity (Winbery and Bobbin, 1983; Bobbin et al., 1984b). In Fig. 4, results reported by Russell (1971) have been redrawn (top portion), along with some of our results (bottom portion). As shown (Fig. 4, bottom), acetylcholine, when applied in eserine (physostigmine) to prevent its breakdown by cholinesterase,

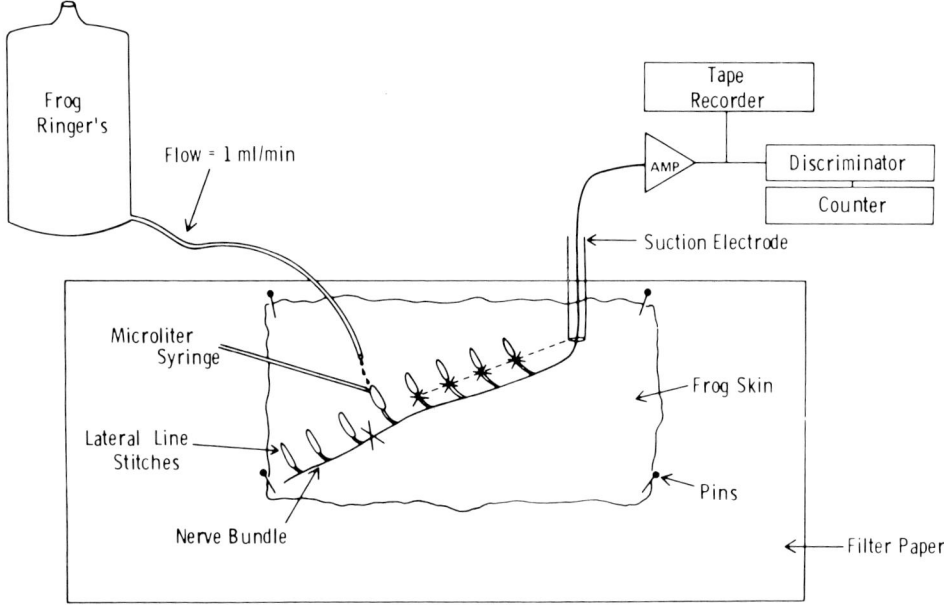

Figure 7-3. Drawing of the frog skin with associated lateral-line neuromasts, pinned to a wax board, illustrating the points at which the nerve fibers are sectioned (X's), the recording apparatus, and methods of drug application.

produces an excitation of afferent fibers before the suppression described by Russell.

In Fig. 4, we applied the drugs in the wash solution for a long time to approximate the method used by Russell. What happens when acetylcholine is applied in low concentrations for a short period of time? In Fig. 5, the drug solution (50 μl) was squirted onto the stitch, with the Ringer wash turned off 1 min before application, and the solution was allowed to remain in contact for 4 min before the wash was turned back on. Again, acetylcholine applied in eserine increased spontaneous activity and some of the increases in activity were found not to be followed by a large suppression. Russell (1971) has shown that the suppression was duplicated by the acetylcholine analogue, carbachol, which is chemically and pharmacologically similar to acetylcholine but resists breakdown by cholinesterase. Therefore, this analogue can be applied without eserine in the test solution or in the Ringer wash. We have tested carbachol and found it to duplicate both the excitation and suppression produced by acetylcholine (Fig. 5).

How does this increase in spontaneous rate come about? Because the efferents appear to synapse solely on hair cells in *Xenopus* lateral line (Bledsoe and Zajic, 1984), the most likely hypothesis is that acetylcholine

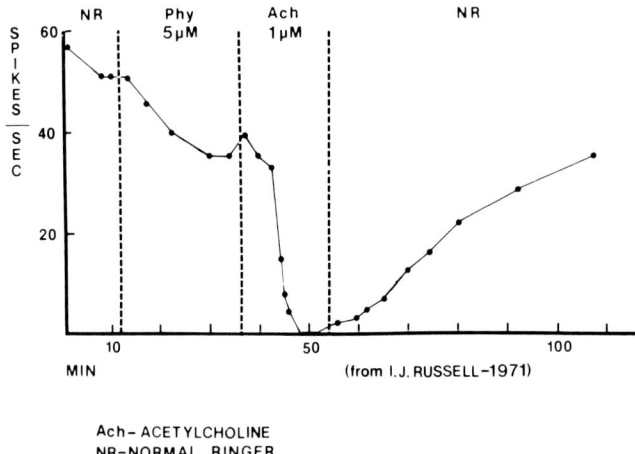
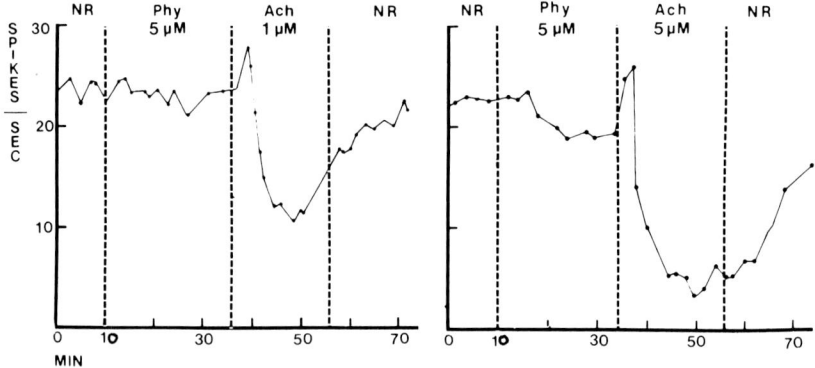

Figure 7-4. Action of normal Ringer solution (NR), eserine (Phy), and acetylcholine with eserine (ACh) on the spontaneous unit activity of lateral-line afferents when applied in a large volume and over a long period of time. The top portion is a figure redrawn from Russell (Fig. 1, 1971, by permission). The bottom portion is our data from experiments where the drugs were dissolved in the Ringer solution and washed continuously over the stitch for the period between the broken vertical lines.

acts presynaptically (on the hair cells) to release afferent transmitter which, in turn, results in the increase in afferent activity we observe. Others have shown that transmitter release is dependent on the presence of calcium ions and that high concentrations of magnesium ions antagonize the action of calcium (e.g., Eccles, 1964). Therefore, to test whether acetylcholine or carbachol acts on the hair cells, we applied them in a Ringer solution containing high magnesium (10 mM) and low calcium (0.1 mM). Spontaneous activity was abolished, as was the response to acetylcholine and carbachol, but the

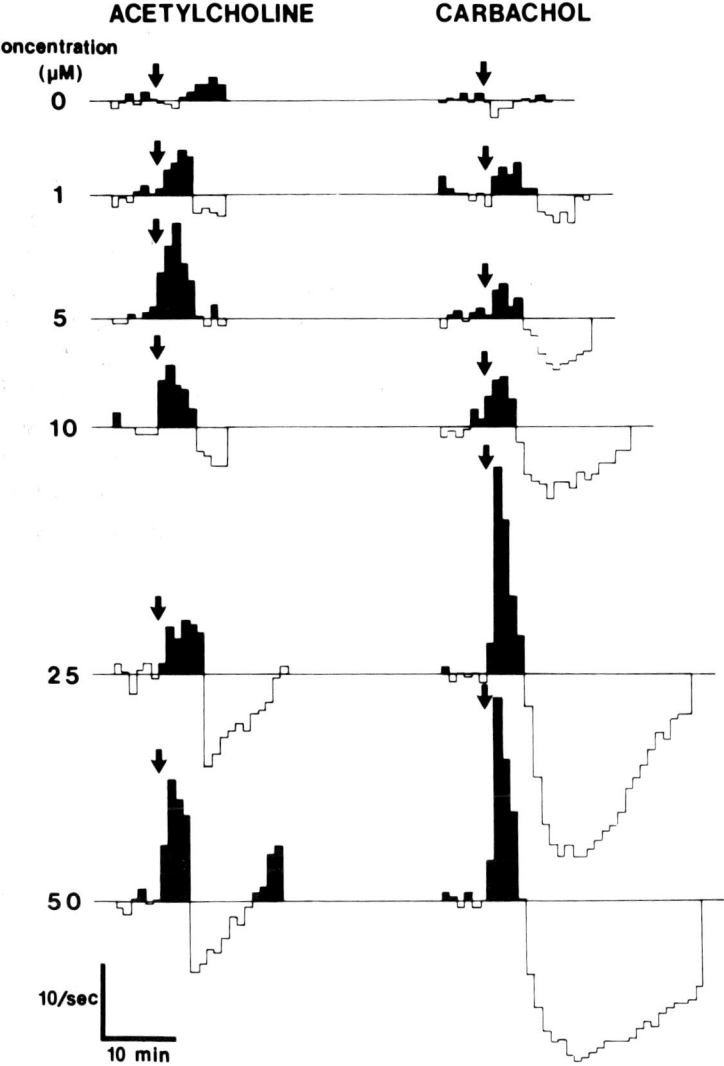

Figure 7-5. Action of normal Ringer solution (0 μM) and increasing concentrations of acetylcholine (with 5 μM eserine) and carbachol on the spontaneous activity (1-min time bins) of lateral-line afferents. The drugs were applied (arrows) by squirting the drug solution (50 μl) onto the stitch after turning off the wash and allowing the preparation to drain for 1 min. The wash was turned back on 4 min after drug application. The horizontal line indicates mean rate (5 min) before drug application. Increases in rate above baseline have been shaded.

response to glutamate was, on the average, only reduced slightly (Bobbin et al., 1984b). This supports the notion that acetylcholine and carbachol increase spontaneous activity indirectly by releasing afferent transmitter and that to a large extent glutamate acts directly on the afferents. Since activation of the efferents appears to result solely in suppression of afferent activity (Russell, 1971), the physiological role of the increases that we observe remains to be determined.

With regard to the identity of the afferent transmitter, we have studied a number of other agonists in the lateral line. The group of chemicals known as the excitatory amino acids may be major excitatory transmitters in the animal kingdom (Watkins, 1978). Though there are many naturally occurring amino acids, the most studied compounds include glutamic acid and aspartic acid. Thus, glutamate was one of the first compounds studied by us (Bobbin and Morgan, 1980) and others (Russell, 1976). Figure 6 illustrates the response to glutamate. Again, like acetylcholine (Fig. 6), glutamate produces an increase in spontaneous activity. At larger concentrations (not shown) the increase is greater, but like the acetylcholine response, it is not sustained, being rapidly followed by a suppression which presumably results from a depolarization blockade (Bobbin et al., 1981b; Bledsoe et al., 1983). We will return to expand on the subject of the glutamate response.

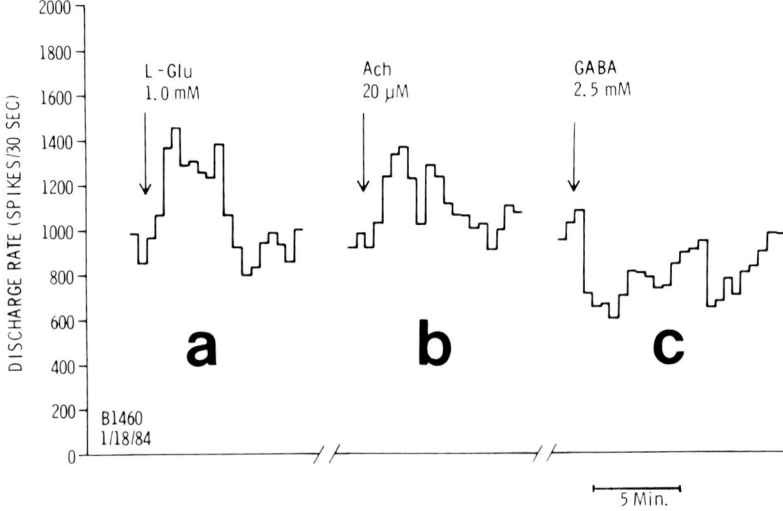

Figure 7-6. Lateral-line afferent nerve fiber responses produced by glutamate (L–Glu), acetylcholine (ACh) in 5 μM eserine, and GABA in one experiment. Drugs were applied (arrows) in 1 ml of Ringer solution at the indicated concentrations and time by injection into the Ringer solution-wash-delivery tube so that they washed over the stitch and were in contact with it for only 1 min. Spontaneous activity (30-s time bins) is shown.

Gamma-aminobutyric acid (GABA) is a compound which has received major attention as an inhibitory transmitter in the brain and elsewhere (e.g., DeFeudis and Mandel, 1981). The response to GABA, applied in the same manner as glutamate and acetylcholine, is shown in Fig. 6. Unlike glutamate and acetylcholine, GABA produces a response that is inhibitory and bimodal, i.e., at high concentrations the response enters a second inhibitory phase following some recovery from the initial suppression (Bledsoe et al., 1983).

TABLE 7-I
POTENCIES OF VARIOUS PUTATIVE TRANSMITTERS ON AFFERENT NERVE ACTIVITY IN THE LATERAL LINE AND COCHLEA

Chemical	Lateral Line	Cochlea
L-Glutamate	+[1]	+[6,7]
L-Aspartate	+[1]	+[6,7]
Acetylcholine	+++/−−−[2,3]	−−−[7,8]
GABA	−−−[1]	0[6]
Glycine	0[1]	0[6]
Serotonin	0[4]	0[6]
Dopamine	0[1]	0[6]
Epinephrine	0[5]	0[6]
Norepinephrine	0[5]	0[6]
Histamine	0[4]	0[6]

Results are expressed as excitation (+), inhibition (−), and no activity (0) with potency indicated by the number of symbols. [1]Bledsoe et al. (1983); [2]Winbery and Bobbin (1983); [3]Russell (1971, 1976); [4]Unpublished observations; [5]Chihal et al. (1980); [6]Bobbin and Thompson (1978); [7]Comis and Leng (1979); [8]Robertson and Johnstone (1978).

Table 1 summarizes the information on substances tested in the lateral line and compares these results to those for the cochlea. Aspartate, an analogue of glutamate, is equal to glutamate in potency, and the two are indistinguishable in the lateral line (Bobbin et al., 1981b; Bledsoe et al., 1983) and cochlea (Klinke and Oertel, 1977a; Bobbin and Thompson, 1978; Bobbin, 1979; Comis and Ling, 1979). In contrast, the response to acetylcholine in the cochlea has been reported as strictly inhibitory (Robertson and Johnstone, 1978; Comis and Ling, 1979), not excitatory/inhibitory as we have found in the lateral line. In addition, we and others (Bobbin and Guth, 1970; Klinke and Oertel, 1977b; Bobbin and Thompson, 1978) have failed to detect a response to GABA in the cochlea, though, it is quite active in the lateral line. To date, we have not observed significant responses to other putative transmitter substances, such as glycine, serotonin, dopamine, epinephrine, norepinephrine, or histamine, in either the lateral line (Chihal et al., 1980; Bledsoe et al., 1983) or cochlea (Klinke and Oertel, 1977c; Bobbin and Thompson, 1978; Klinke, 1981; Bobbin et al., 1984a).

Others have proposed that excitatory amino acids, such as glutamate and aspartate, can act on at least three types of receptors, which can be distinguished pharmacologically (Watkins, 1981; Martin, Chapter 11, this volume). Figure 7 shows glutamate and aspartate acting on three receptors named by the drugs that preferentially activate them: quisqualate, kainate, and N-methyl-D-aspartate (NMDA). These drugs do not occur naturally in nervous tissue. Table II summarizes our results with these compounds in the lateral line (Bledsoe et al., 1983) and cochlea (Bledsoe et al., 1981b; Jenison and Bobbin, 1983a). All three are active in the lateral line but only two, quisqualate and kainate, are active in the cochlea. Interestingly, their potencies are approximately parallel in the two preparations, with quisqualate being the most potent. Kainate is ten times less potent that quisqualate but ten times more potent than glutamate. NMDA is approximately as potent as glutamate in the lateral line, but not active in the cochlea. Thus, it appears that all three subtypes of excitatory amino acid receptors are present in the lateral line, but only two exist in the cochlea.

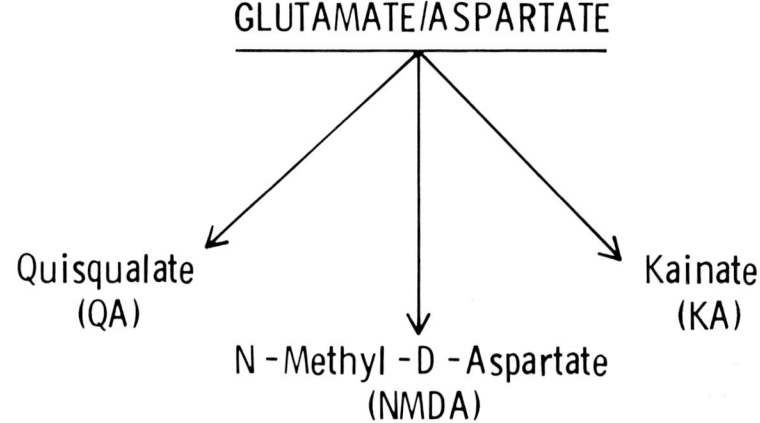

Figure 7-7. Types of receptors which may be activated by glutamate and/or aspartate.

TABLE 7-II
POTENCIES OF GLUTAMATE AND GLUTAMATE ANALOGUES ON AFFERENT NERVE ACTIVITY IN THE LATERAL LINE AND COCHLEA

Chemical	Lateral Line	Cochlea
Kainate (KA)	++[1]	++[2]
Quisqualate (QA)	+++[1]	+++[3]
N-Methyl-D-aspartate (NMDA)	+[1]	0[3]
Glutamate	+[1]	+[3]

Results are expressed as excitation (+), inhibition (−), and no activity (0), with potency indicated by the number of symbols. [1]Bledsoe et al. (1983); [2]Bledsoe et al. (1981b); [3]Jenison and Bobbin (1983a).

III. PHARMACOLOGICAL ANTAGONISM

The above results have revealed several substances to be active in the lateral line and therefore potential transmitters there. To gain further insight into synaptic mechanisms, we have evaluated the effects of antagonists to these various substances. For example, D-α-aminoadipic acid (DAA) has been described as a selective NMDA antagonist (Watkins and Evans, 1981), glutamic acid diethyl ester (GDEE) as a quisqualate antagonist (Watkins and Evans, 1981), bicuculline as a reversible antagonist of GABA (Curtis et al., 1971), and atropine and curare as selective acetylcholine antagonists (Siegel et al., 1981).

Figure 8 illustrates our results in the lateral line with DAA (Bledsoe and Bobbin, 1982b). Responses to glutamate, NMDA, and kainate are shown before, during, and after the application of a Ringer solution control and Ringer solution containing 0.25 mM and 0.5 mM DAA. DAA selectively blocks the action of NMDA, suppresses glutamate, and has no effect on responses to kainate. Figure 9 extends these results to natural stimulation, demonstrating that DAA suppresses natural stimulation at the same concentrations that it antagonizes the action of NMDA. This strongly suggests that NMDA, DAA, and the unknown endogenous transmitter released by hair cells all act on the same receptor.

Russell (1971) reported that curare and atropine, two antagonists of acetylcholine, readily block the suppressive action of acetylcholine in the lateral line. We have examined the effect of atropine (1 μM) on the cholinergic excitatory response (Bobbin et al., 1983) and found that it selectively blocks the response to carbachol without affecting the excitation to glutamate (Fig. 10a-d). Others have suggested that acetylcholine may be involved in afferent transmission (e.g., Guth et al., 1976) and, since we have found acetylcholine to be excitatory, it seemed appropriate to test atropine against natural stimulation. At the same concentration that blocks the carbachol response, atropine has no effect on responses to natural stimulation (Bobbin et al., 1983, 1984b). Thus, we conclude that acetylcholine is not the afferent transmitter.

Bicuculline is a selective GABA antagonist in other systems (Curtis et al., 1971). We have examined GABA further by showing that its response is blocked selectively by bicuculline, as shown in Fig. 10e-h (Bobbin et al., 1983, 1984b). In addition, bicuculline has no effect on activity evoked by water motion, suggesting that GABA is also not the afferent transmitter (Bobbin et al., 1983, 1984b).

Table III summarizes the current data with regard to various antagonists tested and reported, both by us and others, in the lateral line and cochlea. Experiments by Fex and Martin (1980) and by ourselves (Bobbin et al.,

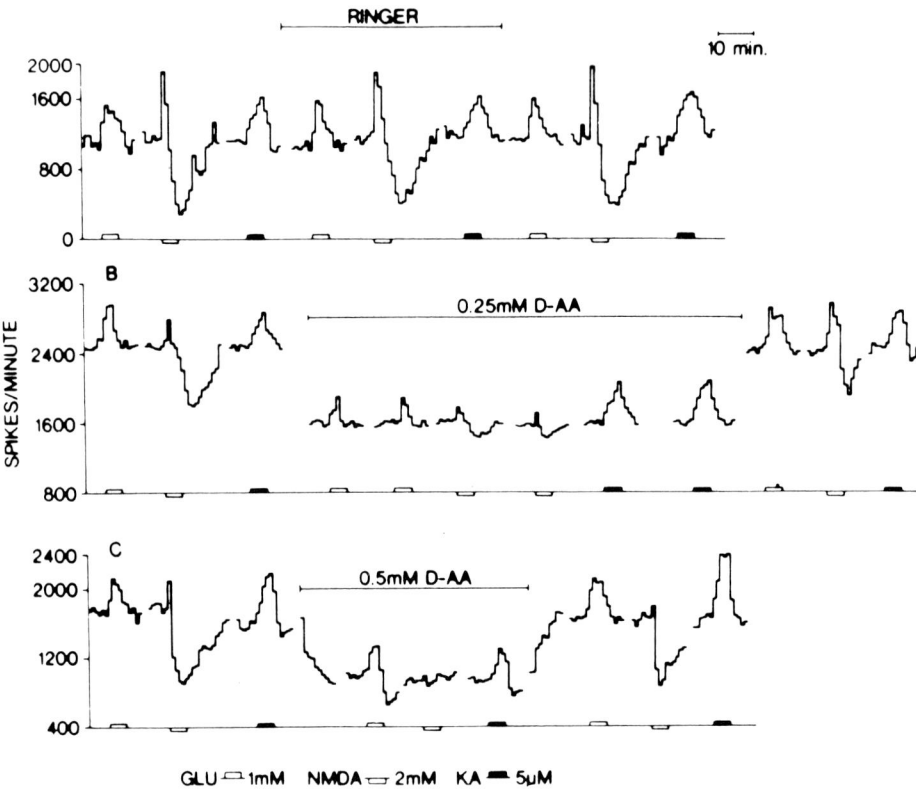

Figure 7-8. Responses of lateral-line afferent fibers to exogenously applied L-glutamate (GLU), N-methyl-D-aspartate (NMDA), and kainate (KA) in the presence of drug-free Ringer solution, 0.25 mM DAA (B), and 0.5 mM DAA (C). Results are from three different experiments for which spontaneous activity was recorded. Open and closed bars indicate agonist concentrations and the time of contact with the preparation. From Bledsoe and Bobbin (1982b), with permission.

1981a) have shown that DAA has no effect on afferents in the cochlea. To the best of our knowledge, DAA has not been examined for an effect on the efferents in either the lateral line or the cochlea. GDEE, which is thought of as quisqualate blocker, is less potent than DAA in the lateral line (Bledsoe and Bobbin, 1982a). In contrast, GDEE in the cochlea is a powerful blocker of the compound action potential of the auditory nerve (Bobbin et al., 1981a), although the drug also has marked effects on the cochlear microphonics and some effects on the efferents. Atropine does not affect afferent activity in either the lateral line (Bobbin et al., 1983, 1984b) or the cochlea (Bobbin and Konishi, 1974). On the other hand, this blocker does antagonize the effects of efferent stimulation in both the lateral line and cochlea (e.g., Guth et al., 1976). This is consistent with the thought that acetylcholine is

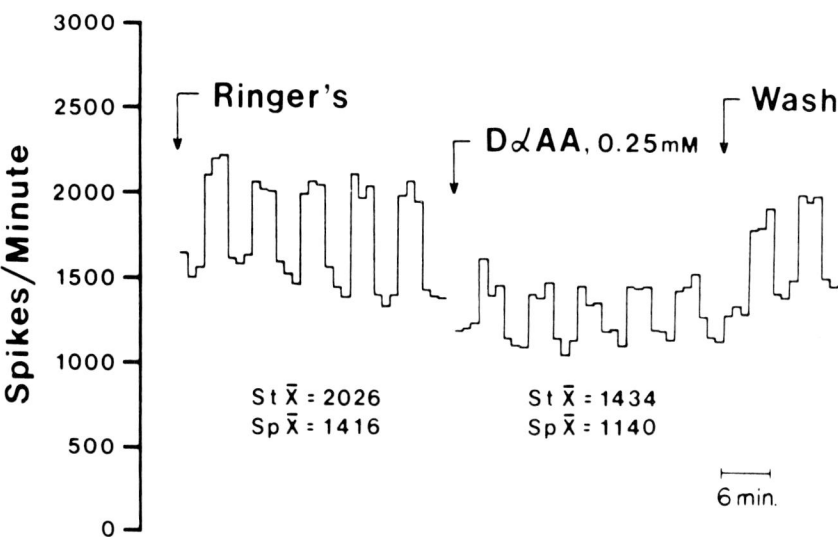

Figure 7-9. Suppression of spontaneous activity and water-motion-induced excitation of afferent nerve fibers in the *Xenopus* lateral line by 0.25 mM DAA. At the arrow designated "Ringer's", the drug-free Ringer wash was turned off and the remaining fluid on the serosal surface of the skin rapidly exchanged for a 400-μl volume of fresh, drug-free Ringer solution. An initial 3 min of spontaneous activity, followed by five 3-min periods of stimulated and spontaneous activity, were recorded. At the arrow designated DAA, the 400 μl of Ringer solution in the chamber was rapidly exchanged for 400 μl of Ringer solution containing 0.25 mM DAA. The stimulated-then-spontaneous sampling sequence was repeated following a 2-min break in the recording, and, at the arrow designated "Wash", the drug-free Ringer was was turned back on and the preparation permitted to recover. Mean values for the stimulated (St) and spontaneous (Sp) conditions obtained in each of the solutions are given. From Bledsoe and Bobbin (1982b), with permission.

the transmitter within efferent endings. Bicuculline, a GABA antagonist, has essentially no effect on afferent transmission in the lateral line (Bobbin et al., 1983, 1984b) and the cochlea (Bobbin and Guth, 1970; Klinke and Oertel, 1977b; Bobbin and Thompson, 1978). Bicuculline has not been examined on the lateral-line efferents and has only a slight effect on the efferents in the cochlea (Klinke and Oertel, 1977b).

IV. RELEASE

There is little doubt that acetylcholine is the endogenous substance acting on the acetylcholine-sensitive receptors (e.g., Guth et al., 1976; Daigneault, 1981). However, the identities of both the endogenous substance acting on the presumptive NMDA receptor in the lateral line and the analogous substance acting on the presumptive quisqualate receptor in the cochlea are

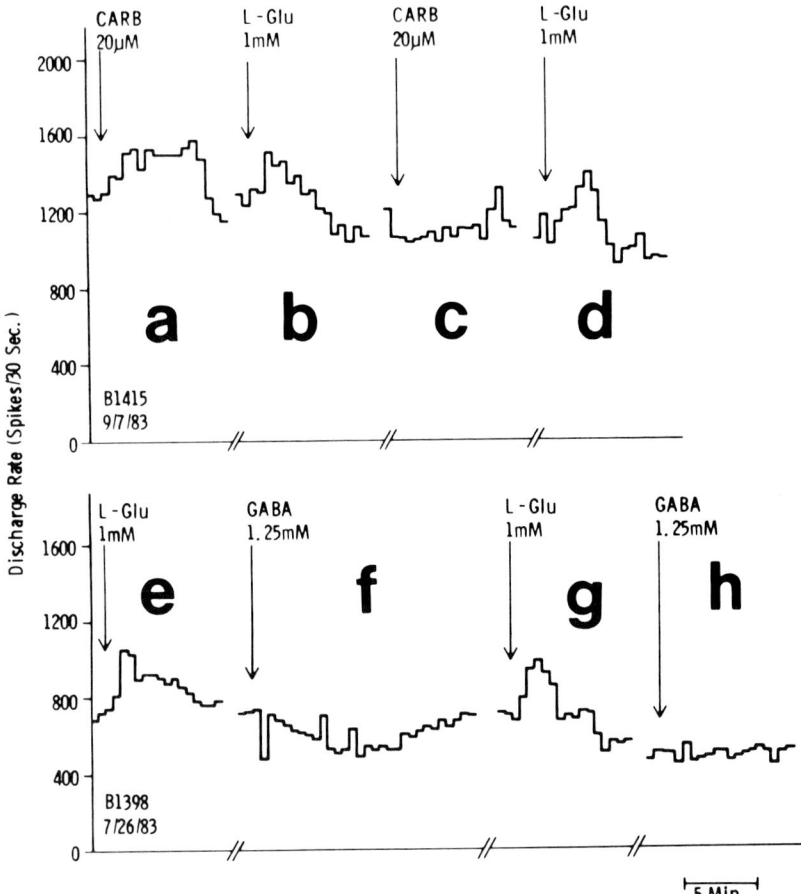

Figure 7-10(a–d): Responses of lateral-line afferent fibers to exogenously applied carbachol (CARB) and glutamate (L–Glu), in the presence of drug free Ringer solution (a, b) and in the presence of 1 µM atropine sulfate (c, d), showing that carbachol but not glutamate is blocked by the atropine. Figure 10(e–h): Responses of lateral-line afferent fibers to exogenously applied glutamate (L–Glu) and GABA in the presence of drug-free Ringer solution (e, f) and in the presence of 100 µM bicuculline methiodide (g, h), showing that GABA but not glutamate is blocked by the bicuculline. Spontaneous activity is shown; agonists were applied as in Fig. 6.

unknown. Therefore, to obtain further information, we have attempted to detect the release of the chemical into the fluids bathing the hair cells after natural stimulation. Figure 11 shows the chamber used to study release in the lateral line (Bledsoe et al., 1980). The skin was held on a chamber with tap water on the outside and Ringer solution bathing the inside. The stimulus was water motion. We compared the release of glutamate, aspartate, and

TABLE 7-III
POTENCIES OF ANTAGONISTS OF PUTATIVE TRANSMITTERS ON
AFFERENT AND EFFERENT NERVE ACTIVITY IN
THE LATERAL LINE AND COCHLEA

		Natural Systems			
		Lateral Line		Cochlea	
Antagonist	Receptor Type	Afferent	Efferent	Afferent	Efferent
DAA	NMDA	++++[1]	?	0[5,6]	?
GDEE	QA	++[2]	?	+++[6]	+[6]
Atropine	ACh	0[3]	+++++[4]	0[7]	+++++[7]
Bicuculline	GABA	0[3]	?	0[8]	+[8]

Results are expressed as active (+), not active (0), and not tested (?), with potency indicated by the number of symbols. [1]Bledsoe and Bobbin (1982b); [2]Bledsoe and Bobbin (1982a); [3]Bobbin et al. (1983); [4]Russell (1968, 1971, 1976); [5]Fex and Martin (1980); [6]Bobbin et al. (1981a); [7]Bobbin and Konishi (1974); [8]Kline and Oertel (1977b).

glycine (not shown) from skins containing lateral line stitches with release occurring from skins without stitches under stimulated (S) and non-stimulated (N) conditions. The results (Fig. 12) indicated a significantly greater release of glutamate from skins with stitches (Bledsoe et al., 1980). This suggests that the amino acid may have been released from hair cells, and is consistent with the hypothesis that glutamate is the substance that acts on the NMDA receptor in the lateral line. As mentioned earlier, calcium is necessary for transmitter release in most systems and magnesium ions act as presynaptic blockers preventing the action of calcium (Eccles, 1964). We have not determined the effects of presynaptic blockade on release to determine what proportion of the amino acid release is synaptic in origin.

Given the success of detecting an increase in glutamate in the fluids bathing the lateral line (Bledsoe et al., 1980) and an unknown excitatory substance in the cochlear fluids (Sewell et al., 1978) we (Bledsoe et al., 1981a), and others (Medina and Drescher, 1981; Melamed et al., 1982; Drescher et al., 1983) have looked for a sound-induced change in the level of amino acids in cochlear fluids. Drescher et al. (1983) and Drescher and Drescher (Chapter 4, this volume) detected an increase in a GABA-like substance in cochlear fluids that was proportional to sound stimulus intensity, but, to date no one has detected an increase in glutamate at moderate exposure levels, except for a preliminary study reported by us (Bledsoe et al., 1981a). One interpretation for the lack of ability to detect a change in glutamate levels may be that not enough transmitter is released from the hair cells for detection, due to efficient degradation or uptake. Sound stimulation is in fact very difficult to shape such that both apical and basal regions of the cochlea are stimulated maximally. (This is not the situation for mechanical

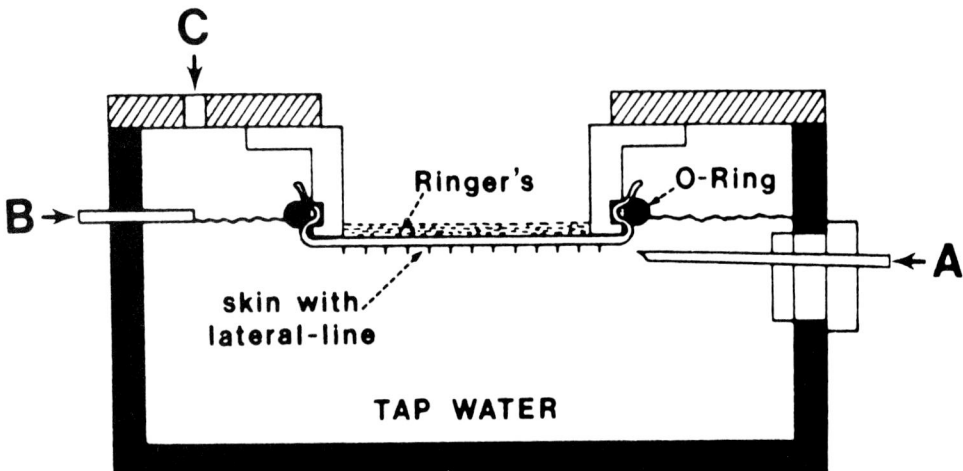

Figure 7-11. Schematic cross-section of the experimental apparatus to evoke release of the transmitter. A, inlet needle for pulsatile water application to the stitches; B, suction application to remove water; C, pressure relief hole. From Bledsoe et al. (1980), with permission.

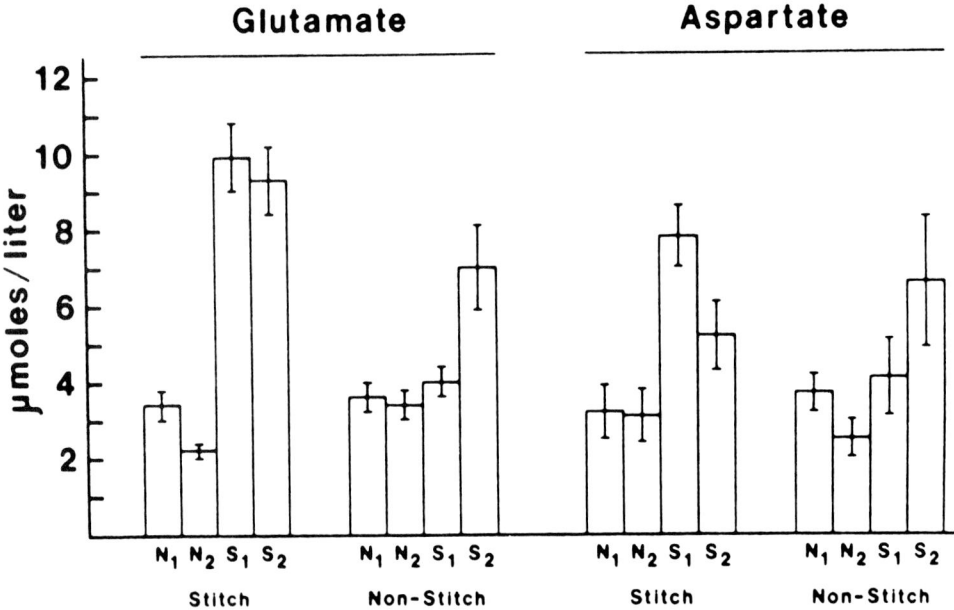

Figure 7-12. Release of endogenous glutamate and aspartate from stitch and non-stitch skins. Values are given as means ± SEM for 13 skins of each type. Data are presented for two 30-min periods of the stimulated (S_1, S_2) and non-stimulated (N_1, N_2) conditions. From Bledsoe et al. (1980), with permission.

stimulation in the lateral line.) We have recently turned to high potassium to induce release in the cochlea (Jenison and Bobbin, 1983b; Jenison et al., 1984). High potassium has been used by investigators in other systems to maximally depolarize a large number of presynaptic nerve endings in concert, and thus induce a relatively greater release of transmitter substance (Wenthold, 1979). In addition, we physically blocked the cochlear aqueduct to limit the entry of cerebrospinal fluid, and animals were run for which Ringer solutions were altered to contain high magnesium and low calcium.

Of the seventeen amino acid fractions assayed, only those identified as glutamate and taurine were significantly released in a calcium-dependent manner during the perfusion of the high potassium solution (Jenison et al., 1984). The attenuation in this release in the presence of the high magnesium-low calcium solution suggests that the release originated from synaptic pools. If the release is synaptic, then it could have originated from at least four sources: inner hair cells, outer hair cells, medial efferents and lateral efferents. One other possibility is that the amino acids may be degradation products of the transmitter, as suggested by others (Koller and Coyle, 1984; Ryan and Schwartz, 1984).

V. SUMMARY AND CONCLUSIONS

Figure 13 is a summary of our current view of the two hair-cell systems, based on the data discussed. Atropine does not affect afferent activity in either the cochlea (Bobbin and Konishi, 1974) or lateral line (Bobbin et al., 1983, 1984a, b). Therefore, we conclude that the afferent transmitter does not act on acetylcholine receptors. Since efferent fibers synapse on the hair cells in the lateral line and acetylcholine is thought to be the transmitter at these efferent endings, we have depicted acetylcholine receptors on the hair cells. Likewise in the cochlea, the acetylcholine receptors are associated with efferent synapses: on hair cells for the outer hair cells and on nerve fibers for the inner hair cells.

Figure 13 also shows the GABA receptor as being absent in the cochlea, but present on the afferent nerve fiber in the lateral line, for want of a more clearly defined site. The GABA receptor is probably not involved in afferent transmission, since bicuculline has no effect on afferent transmission at concentrations which block the effects of exogenously applied GABA (Bobbin et al., 1983, 1984a, b). It therefore remains an unsolved problem for future research.

In the lateral line, since DAA blocks both the NMDA response and the effects of natural stimulation (Bledsoe and Bobbin, 1982b), we speculate that the afferent transmitter acts on the same receptor as NMDA. Note that in the cochlear system the NMDA receptors are depicted as absent. In the

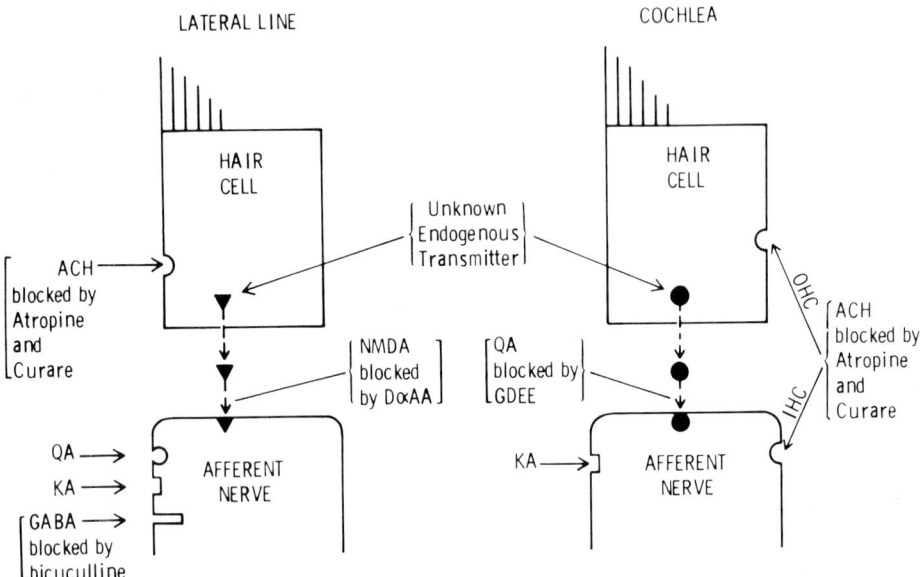

Figure 7-13. Proposed sites of action of drugs at the hair cell-nerve fiber terminals in the lateral line and cochlea. In the lateral line, N-methyl-D-aspartate (NMDA) is shown acting at the afferent transmitter site on the afferent nerve, where both substances appear to be blocked by D-α-aminoadipate (DαAA). In the cochlea, the NMDA–DAA sites are absent, and instead, quisqualate (QA) and glutamate diethyl ester (GDEE) are suggested as acting at the afferent transmitter site. Acetylcholine, a proposed efferent transmitter, is shown acting at only a hair-cell site in the lateral line, but at both hair-cell (OHC, outer hair cell) and nerve-fiber sites (IHC, inner hair cell) in the cochlea. Gamma-aminobutyric acid (GABA) is depicted as acting at sites in the lateral line which are absence in the cochlea. Kainate (KA) is shown as acting at nerve-fiber sites in both systems.

lateral line, the roles of the kainate and quisqualate receptors in transmission are at present not known. Kainate is not affected by DAA in any system including the lateral line (Bledsoe and Bobbin, 1982b) and, to date, quisqualate has not been studied with DAA in the lateral line. Agents which are thought of as quisqualate blockers are less potent than DAA in the lateral line (Bledsoe and Bobbin, 1982a). Thus, for the lateral line, the kainate and quisqualate receptors in Fig. 13 are positioned aside. In contrast, we have found GDEE, a quisqualate blocker, to be a powerful antagonist of the compound action potential in the cochlea (Bobbin et al., 1981a). Therefore, we have tentatively depicted the quisqualate receptor as the one used by the cochlear afferent transmitter.

What is the identity of the endogenous chemical that acts on the afferent transmitter receptors? In the lateral line, the suppression of glutamate by DAA is consistent with the hypothesis that this substance is the endogenous transmitter (Bledsoe and Bobbin, 1982b). In addition, our results in both the cochlea and the lateral line, which detected the release of glutamate into the surrounding fluid, constitute further evidence for the assertion that the unknown substance in both systems is glutamate.

Obviously, we have learned much about both systems and have generated further questions, to be answered by future research. Probably the two most pressing issues are the origin of the released glutamate in the lateral line and the origin of the released glutamate and taurine in the cochlea. In addition, the roles of the GABA and kainate receptors and the released taurine need defining. In addition, the lateral line may very well prove to be one of the best bioassays for examining the chemistry of the NMDA receptor and substances which act on it.

ACKNOWLEDGMENTS

Thanks to Gail Ceasar, M. S., for technical assistance and to Cindy Frazier and Vibha Flax for their aid in the preparation of this manuscript. Supported at Kresge Hearing Research Laboratory of the South, New Orleans, Louisiana, by NIH grants NS 16080 and NS 07058, the Deafness Research Foundation, the Kresge Foundation, and the Louisiana Lions Eye Foundation, and at Kresge Hearing Research Institute, Ann Arbor, Michigan, by grants from NIH (NS 05785) and the University of Michigan.

REFERENCES

Bauknight, R. S.; Strelioff, D., and Honrubia, V.: Effective stimulus for the *Xenopus laevis* lateral-line hair-cell system. *Larynogoscope 86:* 1836–1844, 1976.

Bledsoe, S. C., Jr., and Bobbin, R. P.: Effects of antagonists of excitatory amino acids on the activity of afferent fibers in the *Xenopus laevis* lateral line. *Assoc. Res. Otolaryngol. Abstr. 5:* 88, 1982a.

Bledsoe, S. C., Jr., and Bobbin, R. P.: Effects of D-α-aminoadipate on excitation of afferent fibers in the lateral line of *Xenopus laevis. Neurosci. Lett. 32:* 315–320, 1982b.

Bledsoe, S. C., Jr.; Bobbin, R. P., and Chihal, D. M.: Technique for studying sound-induced release of endogenous amino acids from the guinea pig cochlea. *Assoc. Res. Otolaryngol. Abstr. 4:* 24, 1981a.

Bledsoe, S. C., Jr.; Bobbin, R. P., and Chihal, D. M.: Kainic acid: An evaluation of its action on cochlear potentials. *Hearing Res. 4:* 109–120, 1981b.

Bledsoe, S. C., Jr.; Bobbin, R. P., Thalmann, R., and Thalmann, I.: Stimulus-induced release of endogenous amino acids from the *Xenopus laevis* lateral-line organ. *Exp. Brain Res. 40:* 97–101, 1980.

Bledsoe, S. C., Jr.; Chihal, D. M., Bobbin, R. P., and Morgan, D. N.: Comparative actions of glutamate and related substances on the lateral line of *Xenopus laevis. Comp. Biochem. Physiol. 75C:* 119–206, 1983.

Bledsoe, S. C., Jr., and Zajic, G.: Personal communication, 1984.

Bobbin, R. P.: Glutamate and aspartate mimic the afferent transmitter in the cochlea. *Exp. Brain Res. 34:* 389–393, 1979.

Bobbin, R. P.; Bledsoe, S. C., Jr., and Chihal, D. M.: Effects of various excitatory amino acid antagonists on guinea pig cochlear potentials. *Assoc. Res. Otolaryngol. Abstr. 4:* 27, 1981a.

Bobbin, R. P.; Bledsoe, S. C., Jr., Chihal, D. M., and Morgan, D. N.: Comparative actions of glutamate and related substances on the *Xenopus laevis* lateral line. *Comp. Biochem. Physiol. 69C:* 145–147, 1981b.

Bobbin, R. P.; Bledsoe, S. C., Jr., and Jenison, G. L.: Neurotransmitters of the cochlea and lateral line organ. In Berlin, C. I. (Ed.): *Recent Advances in Hearing Science.* San Diego, College-Hill Press, 1984a. In press.

Bobbin, R. P.; Bledsoe, S. C., Jr., Jenison, G. L. Winbery S., and Caesar, G.: Actions of atropine and bicuculline on the activity of afferent fibers in the *Xenopus laevis* lateral line. *Neuroscience 9:* 739, 1983.

Bobbin, R. P.; Bledsoe, S. C., Jr., Winbery, S., Ceasar, G., and Jenison, G. L.: Comparative actions of GABA and acetylcholine on the *Xenopus laevis* lateral line, 1984b. In preparation.

Bobbin, R. P., and Guth, P. S.: Evidence that gamma-aminobutyric acid is not the inhibitory transmitter at the crossed olivocochlear nerve-hair cell junction. *Neuropharmacol. 9:* 567–574, 1970.

Bobbin, R. P., and Konishi, T.: Action of cholinergic and anticholinergic drugs at the crossed olivocochlear bundle-hair cell junction. *Acta Otolaryng. (Stockh.) 77:* 56–65, 1974.

Bobbin, R. P., and Morgan, D. N.: Glutamate mimics the afferent transmitter in the Xenopus laevis lateral line. In Gorlin, R. J. (Ed.): *Morphogenesis and Malformation of the Ear, Vol. 16.* New York, Alan R. Liss, 1980, pp. 107–109.

Bobbin, R. P., and Thompson, M. H.: Effects of putative transmitters on afferent cochlear transmission. *Ann. Otol. Rhinol. Laryngol. 87:* 185–190, 1978.

Chihal, D. M.; Bledsoe, S. C., Bobbin, R. P., and Morgan, D. N.: The glutamate receptor site in the lateral line of *Xenopus laevis:* Structure activity relationships (SAR). *Assoc. Res. Otolaryngol. Abstr. 3:* 7, 1980.

Comis, S. D., and Leng, G.: Action of putative neurotransmitters in the guinea pig cochlea. *Exp. Brain Res. 36:* 119–128, 1979.

Curtis, D. R.; Duggan, A. W., Felix, D., Johnston, G. A. R., and McLennan, H.: Antagonism between bicuculline and GABA in the cat brain. *Brain Res. 33:* 57–73, 1971.

Daigneault, E. A.: Pharmacology of the cochlear efferents. In Brown, R. D., and Daigneault, E. A. (Eds.): *Pharmacology of Hearing.* New York, John Wiley & Sons, 1981, pp. 137–151.

DeFeudis, F. V., and Mandel, P.: Amino acid neurotransmitters. In Costa, E., and Greengard, P. (Eds.): *Advances in Biochemical Psychopharmacology.* New York, Raven Press, 1981.

Drescher, M. J.; Drescher, D. G., and Medina, J. E.: Effect of sound stimulation at several levels on concentrations of primary amines, including neurotransmitter candidates, in perilymph of the guinea pig inner ear. *J. Neurochem. 41:* 309–320, 1983.

Eccles, J. C.: *The Physiology of Synapses.* Berlin, Springer-Verlag, 1964.

Fex, J.; and Martin, M. R.: Lack of effect of DL-α-aminoadipate, an excitatory amino acid antagonist, on cat auditory nerve responses to sound. *Neuropharmacol. 19:* 809–811, 1980.

Guth, P. S.; Norris, C. H., and Bobbin, R. P.: The pharmacology of transmission in the peripheral auditory system. *Pharmacol. Rev. 28:* 95–125, 1976.

Harris, G. G., and Flock, A.: Spontaneous and evoked activity from the *Xenopus laevis* lateral line. In Cahn, P. (Ed.): *Lateral Line Detectors.* Bloomington, Indiana University Press, 1967, pp. 135–161.

Harris, G. G., and Milne, D. C.: Input-output characteristics of the lateral-line sense-organs of *Xenopus laevis. J. Acoust. Soc. Am.* 40: 32-42, 1966.

Jenison, G. L., and Bobbin, R. P.: Effects of quisqualate on guinea pig cochlear potentials. *Assoc. Res. Otolaryngol. Abstr.* 6: 47, 1983a.

Jenison, G. L., and Bobbin, R. P.: Potassium-induced changes in the levels of endogenous amino acids of guinea pig perilymph. *Neurosci. Abstr.* 9: 41, 1983b.

Jenison, G. L.; Bobbin, R. P., and Thalmann, R.: Potassium-induced release of glutamate and taurine into perilymph is calcium-dependent. *J. Acoust. Soc. Am.* 75: S82, 1984.

Klinke, R.: Neurotransmitters in the cochlea and the cochlear nucleus. *Acta Otolaryng.* 91: 541-554, 1981.

Klinke, R., and Oertel, W.: Amino acids putative afferent transmitter in the cochlea? *Exp. Brain Res.* 30: 145-148, 1977a.

Klinke, R., and Oertel, W.: Evidence that GABA is not the afferent transmitter in the cochlea. *Exp. Brain Res.* 28: 311-314, 1977b.

Klinke, R., and Oertel, W.: Evidence that 5-HT is not the afferent transmitter in the cochlea. *Exp. Brain Res.* 30: 141-143, 1977c.

Koller, K. J., and Coyle, J. T.: Characterization of the interactions of N-acetyl-aspartyl-glutamate with [^{3}H]L-glutamate receptors. *Eur. J. Pharmacol.* 98: 193-199, 1984.

Kroese, A. B. A., and van den Bercken, J.: Effects of ototoxic antibiotics on sensory hair cell functioning. *Hearing Res.* 6: 183-194, 1982.

Medina, J. E., and Drescher, D. G.: The amino-acid content of perilymph and cerebrospinal fluid from guinea pigs and the effect of noise on the amino acid composition of perilymph. *Neuroscience,* 6: 505-509, 1981.

Melamed, B.; Norris, C., Bryant, G., and Guth, P.: Amino acid content of guinea pig perilymph collected under conditions of quiet or sound stimulation. *Hearing Res.* 7: 13-18, 1982.

Robertson, D., and Johnstone, B. M.: Efferent transmitter substance in the mammalian cochlea: single neuron support for acetylcholine. *Hearing Res.* 1: 31-34, 1978.

Russell, I. J.: Amphibian lateral line receptors. In Llinas, R., and Precht, W. (Eds.): *Frog Neurobiology, A Handbook.* New York, Springer-Verlag, 1976.

Russell, I. J.: Influence of efferent fibers on a receptor. *Nature.* 2: 177-178, 1968.

Russell, I. J.: The pharmacology of efferent synapses in the lateral line system of *Xenopus laevis. J. Exp. Biol.* 54: 643-658, 1971.

Russell, I. J., and Sellick, P. M.: Measurement of potassium and chloride ion concentrations in the cupulae of the lateral lines of *Xenopus laevis. J. Physiol. (Lond.)* 257: 245-255, 1976.

Ryan, A. F., and Schwartz, I. R.: Preferential glutamine uptake by cochlear hair cells: implications for the afferent cochlear transmitter. *Brain Res.* 290: 376-379, 1984.

Sewell, W.; Norris, C. H., Tachibana, M., and Guth, P.: Detection of an auditory nerve-activating substance. *Science.* 202: 910 912, 1978.

Siegel, G. J.; Albers, R. W., Agranoff, B. W., and Katzman, R. K.: *Basic Neurochemistry* Boston, Little, Brown and Co., 1981.

Watkins, J. C.: Excitatory amino acids. In McGear, E. G., Olney; J. W., and McGeer, P. L. (Eds.): *Kainic Acid as a Tool in Neurobiology.* New York, Raven Press, 1978.

Watkins, J. C.: Pharmacology of excitatory amino acid receptors. In Roberts, P. J.; Storm-Mathisen, J., and Johnston, G. A. R. (Eds.): *Glutamate: Transmitter in the Central Nervous System.* New York, John Wiley & Sons Ltd., 1981.

Watkins, J. C., and Evans, R. H.: Excitatory amino acid transmitters. *Annual Review of Pharmacology and Toxicology.* Palo Alto, Annual Reviews, Inc., 1981.

Wenthold, R. J.: Release of endogenous glutamic acid, aspartic acid and GABA from cochlear nucleus slices. *Brain Res. 162:* 338–343, 1979.

Winbery, S. L., and Bobbin, R. P.: Actions of acetylcholine and carbachol on the spontaneous activity of afferent fibers in the Xenopus laevis lateral line. *Assoc. Res. Otolaryngol. Abstr. 6:* 47, 1983.

PART II
BIOCHEMISTRY OF
CENTRAL AUDITORY NEUROTRANSMISSION

Chapter 8

GLUTATMATE AND ASPARTATE AS NEUROTRANSMITTERS OF THE AUDITORY NERVE

Robert J. Wenthold

I. Introduction
II. Presence of Glutamate and Aspartate in Presynaptic Terminals of the Auditory Nerve
III. Release of Glutamate and Aspartate from the Cochlear Nucleus
IV. Biosynthesis of Glutamate and Aspartate in the Cochlear Nucleus
V. Uptake and Retrograde Transport of D-Aspartate in the Auditory Nerve
VI. Pharmacological Studies on the Auditory-Nerve Neurotransmitter
VII. Conclusions
References

I. INTRODUCTION

Since the discovery of chemically-mediated synaptic transmission, the identification and characterization of neurotransmitters has occupied a major position in neurobiological research. These research efforts have provided us with a list of several substances believed to act as neurotransmitters, in addition to information on their synthesis, release, and receptor interactions. We have gained insights into roles that certain neurotransmitters may play in neurological disorders, as well as the treatment of these disorders through manipulation of neurotransmitter systems. The study of neurotransmitters, however, is becoming increasingly complicated. Recent findings have shown that two or more neuroactive substances often co-exist in the same presynaptic terminal and that some neurotransmitters may have several different postsynaptic receptors. Furthermore, many of the neuro-active substances may be acting as neuromodulators rather than neurotransmitters. Such findings, coupled with the growing list of putative neurotransmitters, make the identification of the neurotransmitter of a particular synapse a formidable undertaking.

Neurotransmitters of the auditory system have received only modest attention, with the auditory nerve being one of the most studied pathways.

Several lines of evidence point to glutamate (GLU) or aspartate (ASP) as a neurotransmitter of the auditory nerve. This evidence includes the presence of GLU and ASP in auditory nerve terminals, the presence of enzymes capable of synthesizing GLU and ASP in auditory nerve terminals, the release of GLU and ASP from auditory nerve terminals, the presence of receptors for GLU and ASP on postsynaptic neurons in the cochlear nucleus, and the uptake and retrograde transport of D-aspartate in the auditory nerve. As with other systems which are thought to use GLU or ASP, studies on the auditory nerve are complicated by the ubiquitous distribution of these amino acids in neuronal tissue and the widespread nature of apparently non-synaptic receptors for GLU and ASP. Due to the difficulty of adequately demonstrating the criteria of specific release and postsynaptic identity of action, the support for GLU and ASP as neurotransmitters at the auditory nerve, or at any mammalian synapse, is circumstantial. However, the data supporting GLU or ASP as an auditory-nerve neurotransmitter are multifaceted and come from several laboratories. Essentially no results are inconsistent with GLU or ASP being the neurotransmitter of the auditory nerve and, of the relatively large number of putative neurotransmitters surveyed, only GLU and ASP appear as promising candidates. In this review, the strengths and weaknesses of the evidence supporting GLU and ASP as neurotransmitters of the auditory nerve are presented and discussed.

II. PRESENCE OF GLUTAMATE AND ASPARTATE IN PRESYNAPTIC TERMINALS OF THE AUDITORY NERVE

One of the most basic, and in many cases the most easily satisfied, criteria for transmitter identification is the demonstration of its presence in sufficient quantity in the presynaptic terminal. For neurotransmitters such as acetylcholine, GABA, and the catecholamines, a lesion study or immunocytochemical localization is usually sufficient to satisfy this criterion. In the case of amino acids, however, the obvious difficulty is distinguishing between the neurotransmitter pool and non-neurotransmitter pool of the suspected neurotransmitter. Simply demonstrating that an amino acid is present in the presynaptic terminal is not sufficient to support a neurotransmitter role. However, research on several pathways has shown that certain populations of neurons are enriched in amino acids which are commonly thought to act as neurotransmitters. One example is glycine (GLY), which is present at much higher concentrations in the spinal cord and brainstem than in the cortex, and this distribution seems to correlate with glycine's proposed neurotransmitter roles (Curtis and Johnston, 1974). The distributions of GLU and ASP are not as regionally well-defined as that of GLY, but several neuronal populations have been identified, primarily with lesion studies, which contain high concentrations of GLU and ASP in their presynaptic terminals (Cotman et al., 1981).

The first suggestion that GLU or ASP may be the neurotransmitter of the auditory nerve came from the studies of Godfrey et al. (1977), who measured the levels of GLU, ASP, GABA, and GLY in subdivisions of the cochlear nucleus and auditory nerve of the cat. It was found that the distribution of aspartate somewhat followed that of auditory nerve terminals and fibers, while the distribution of glutamate was more uniform. Similar distributions of these amino acids were subsequently reported in the cochlear nuclei of the rat and guinea pig (Godfrey et al., 1978; Wenthold, 1978).

Evidence which more directly links GLU and ASP with auditory-nerve terminals is that both amino acids decrease in the cochlear nucleus with degeneration of the auditory nerve. This has been demonstrated for degeneration after cochlear ablation (Wenthold and Gulley, 1977) and for the waltzing guinea pig, a genetically-deaf mutant which exhibits an age-dependent decrease in spiral ganglion cells and auditory-nerve terminals (Wenthold and Gulley, 1978). In both cases, the decrease in GLU and ASP in the cochlear nucleus coincides with the morphological degeneration of auditory nerve terminals. No other amino acid or neurotransmitter was found to decrease significantly, and loss of GLU and ASP in cochlear-nucleus subdivisions was greatest in those having the most dense innervation by the auditory nerve (Wenthold, 1978).

The results outlined above could be explained in several ways. They could indicate that GLU and ASP are neurotransmitters of the auditory nerve and are, therefore, concentrated in the presynaptic terminal where they are stored for release. Alternatively, the findings are consistent with GLU and ASP being metabolic precursors or products of the auditory-nerve neurotransmitter, which may also be concentrated in the presynaptic terminal. A third interpretation is that GLU and ASP are concentrated in the auditory-nerve terminals, but that they serve a function unrelated to the neurotransmitter role. Finally, the results may indicate that GLU and ASP are enriched in neurons or glia within the cochlear nucleus in a pool that is sensitive to the loss of auditory-nerve input. While none of these possibilities can be ruled out, the fact that auditory-nerve lesion does not affect other amino acids and that the decrease in GLU and ASP closely parallels the morphological degeneration of auditory-nerve terminals, tends to rule out a major metabolic disruption of the cochlear nucleus as the explanation for the decrease in GLU and ASP. Perhaps a definitive answer regarding the localization of GLU and ASP can be obtained with the promising new technique of immunocytochemistry with a GLU-specific antibody (Storm-Mathisen et al., 1983). This, however, would not indicate if the GLU or ASP is acting as a neurotransmitter, which would need to be addressed by testing other criteria.

A remaining question is why both GLU and ASP decrease in the cochlear nucleus with loss of the auditory nerve. If we assume that one of these amino acids is acting as a neurotransmitter, a likely explanation is that the other is a precursor or product. Lesion studies done on other putative GLU/ASP systems show that in many cases, both GLU and ASP decrease together, while in some cases only one amino acid decreases. If one postulates a metabolic relation as an explanation for the loss of both GLU and ASP in the cochlear nucleus, it would also have to be suggested that GLU and ASP are synthesized differently in different glutamatergic and aspartatergic neurons. This, in fact, may be the case, because the distribution of enzymes involved in the metabolism of these amino acids seems to be variable among neurons (Wenthold and Altschuler, 1983). Alternatively, both GLU and ASP could be neurotransmitters for the auditory nerve, either released together or released independently from different populations of presynaptic terminals. Iontophoretic studies show GLU and ASP have similar effects on receptors and some suggest GLU and ASP may be co-released and together interact with the same postsynaptic receptor (Freeman *et al.*, 1981).

III. RELEASE OF GLUTAMATE AND ASPARTATE FROM THE COCHLEAR NUCLEUS

One of the most specific criteria for demonstrating that a substance is a neurotransmitter at a particular synapse is to demonstrate its release from the presynaptic terminal with stimulation. This is also one of the most difficult aspects to show, however, especially in the central nervous system. After its interaction with the postsynaptic receptor, a neurotransmitter is rapidly inactivated. In attempting to measure a release, it must be assumed that this inactivation system is not complete, but allows some neurotransmitters to reach the collection device. If the inactivation system can be inhibited, measurement of release becomes more feasible. Unfortunately, specific inhibitors of GLU and ASP uptake (presumably the major inactivation system for these amino acids) have not yet been identified. Consequently, most studies of release from central-nervous-system synapses have been designed to optimize recovery of released neurotransmitters. Studies are often done *in vitro* using slices or synaptosomes, with release evoked by a rather nonspecific electrical or chemical stimulus. Release is commonly measured following uptake of a radioactive putative neurotransmitter which is assumed to mix with the endogenous neurotransmitter pool. Under such conditions, it has been reported that release of putative neurotransmitters can occur from neuronal and glial cell bodies (Minchin and Iversen, 1974; Sellstrom and

Hamberger, 1977). However, it seems that substances commonly thought to act as neurotransmitters, including neuropeptides, are released in a calcium-dependent manner, while non-neurotransmitters are not. Demonstration of release *in vitro* with chemical or electrical stimulation clearly does not offer the same kind of evidence that demonstration of release under normal conditions does, and such factors must be considered when analyzing data from release studies.

Lesion studies showing a decrease in GLU and ASP in the cochlear nucleus after loss of the auditory nerve prompted further studies to determine if the pool of GLU and ASP which was lost represented a neurotransmitter pool. Initially, we used radioactive GLU and ASP, preincubated with slices of the cochlear nucleus to allow uptake, and then measured release after depolarization with high concentrations of potassium. Although both GLU and ASP were taken up and concentrated in the slices, very little of the radioactivity was released in a calcium-dependent manner. Furthermore, the uptake of radioactive GLU and ASP into cochlear-nucleus slices actually increased with auditory-nerve lesion, an unexpected result if we postulate a major uptake system associated with terminals of the auditory nerve. A tentative explanation for these results is that an uptake system for GLU and ASP exists on glia in the cochlear nucleus and quantitatively overwhelms the uptake by nerve terminals. Based on the assumption that radioactive GLU and ASP were not taken up sufficiently into releasable pools, our second approach was to measure release of endogenous GLU and ASP from cochlear nucleus slices (Wenthold, 1979). Upon potassium depolarization, several amino acids were released, with the putative neurotransmitters GLU, ASP, GABA, and GLY being the major amino acids released. The release of these amino acids was also largely calcium-dependent. In order to address the question of whether or not any of the released amino acids originated from pools contained within the auditory nerve, release was measured in cochlear nucleus after degeneration of the auditory nerve. Calcium-dependent release of GLU was reduced 41% and that of ASP 26%, while release of GABA and GLY were unchanged. Alanine, generally not considered to be a neurotransmitter, also displayed a significant calcium-dependent release, which was decreased about 20% by auditory-nerve lesion.

Release of GLU and ASP has subsequently been studied in a number of laboratories, with all producing the same general conclusion, i.e., GLU and ASP appear to be released from auditory-nerve terminals (Table I). The most dramatic results were obtained using D-aspartate (D-ASP), a non-metabolized analog of GLU and ASP which appears to be taken up, packaged, and released by glutamatergic and aspartatergic terminals in the same manner as the actual neurotransmitter. Furthermore, some of the D-ASP which is taken up is retrogradely transported to the cell body. In the study referred to here (Potashner, 1983), the cochlear nucleus was prelabeled *in vitro* with radioactive D–ASP, and release was measured after electrical stimulation. It was found that loss of the auditory nerve reduced release of labeled D–ASP nearly 80% in the ventral cochlear nucleus and to a lesser extent in the dorsal cochlear nucleus. Interestingly, release of D–ASP was also significantly reduced by disarticulation of the middle-ear ossicles, which attenuates acoustic stimulation. It is possible that such disarticulation produced trauma to the cochlea, but even if cochlear trauma occurred, it is unlikely that substantial spiral ganglion cell loss would occur secondary to cochlear hair-cell loss in the time period used, namely, 2–13 days after surgery. It is still possible that damage occurred which directly affected the spiral ganglion cells, leading rapidly to their degeneration. However, the interesting alternative remains that a decrease in acoustic stimulation leads to a change in the capacity for neurotransmitter uptake, storage, or release, which is reflected in the diminished release of D–ASP. Certainly this possibility warrants further investigation.

TABLE 8-I
RELEASE OF GLUTAMATE AND ASPARTATE FROM THE COCHLEAR NUCLEUS

	Methods	*Results*
Wenthold (1979)	*In vitro* slice; endogenous amino acids; K^+ depolarization	GLU, ASP release decreased by cochlear ablation
Canzek and Reubi (1980)	*In vitro* slice; preload with ^{3}H-GLU or ^{3}H-GLN; K^+ depolarization	^{3}H-GLU, ^{3}H-ASP release decreased by cochlear ablation
Hansson *et al.* (1980)	*In vivo*; prelabel with ^{3}H-GLU or ^{3}H-ASP; stimulate with sound	Release of radioactivity increased with sound stimulation
Potashner (1983)	*In vitro* subdivisions; prelabel with ^{3}H-D-ASP; electric field stimulation	Release of radioactivity decreased with cochlear ablation
Jarlstedt *et al.* (1984)	*In vivo*; endogenous amino acids; stimulate with sound	GLU, ASP release increased with sound stimulation

Release of amino acids from the cochlear nucleus has also been studied *in vivo* by Jarlstedt and colleagues, (e.g., Jarlstedt *et al.*, 1984). The initial study (Hanson *et al.*, 1980) was done by infusing radioactive GLU or ASP into the cochlear nucleus with a push-pull cannula and measuring release in response to sound stimulation. A major criticism of this study is that the substance with which the released radioactivity was associated was not identified. Recently, the same group has studied release of endogenous amino acids *in vivo* using a probe containing a small loop of dialysis tubing (Jarlstedt *et al.*, 1984). It was reported that of the several amino acids assayed, the release of GLU and ASP were most affected by sound stimulation, with the release of both substances more than doubling. The release of other amino acids, including glycine, serine, and alanine were also increased, but to a lesser degree. The amino acids could have originated from any population of neurons or glia within the cochlear nucleus. This technique, however, may provide the capability of measuring release from defined regions of the cochlear nucleus. For the present results, in view of the *in vitro* studies, it is tempting to suggest that the released GLU and ASP originate from auditory-nerve terminals.

A question that one might expect the release studies to address is whether GLU or ASP is the more likely candidate for the auditory nerve neurotransmitter. As discussed elsewhere, the known receptors for GLU and ASP seem to accept either amino acid almost equally well. Therefore, the controlling factor determining if GLU or ASP is the neurotransmitter would be expected to be the release. The release studies to date, however, would suggest that both amino acids are released by the auditory nerve.

IV. BIOSYNTHESIS OF GLUTAMATE AND ASPARTATE IN THE COCHLEAR NUCLEUS

Because GLU and ASP are ubiquitous substances and apparently do not require dedicated synthetic and degradative pathways for their production as neurotransmitters, it has been difficult to determine which pathways are involved in production of neurotransmitter GLU and ASP, and how these steps are regulated. Several studies have suggested that glutamine (GLN) is the immediate precursor of neurotransmitter GLU and this conversion is catalyzed by the enzyme, glutaminase (EC 3.5.1.2, GLNase). Radioactive GLN is readily converted into releasable GLU in brain slices (Hamberger *et al.*, 1979), in synaptosome preparations (Bradford *et al.*, 1978), and in slice preparations after *in vivo* labeling (Ward *et al.*, 1983). Maintenance of the neurotransmitter pool of GLU appears to require GLN because both release and tissue levels of GLN are decreased in slices in media lacking GLN (Hamberger *et al.*, 1979). The synthesis of ASP has been studied less. The major enzyme involved in the metabolism of ASP is aspartate aminotransferase (EC 2.6.1.1, AATase). Other major enzymes associated with the metabolism of GLU and ASP are glutamate dehydrogenase (EC 1.4.1.3., GDH) and glutamine synthetase (EC 6.3.1.2., GS).

We were interested in determining which, if any, of these enzymes was involved in producing GLU and ASP in the auditory nerve. To address this question, we lesioned the auditory nerve and determined if any of these enzymes decreased in the cochlear nucleus (Wenthold, 1980). It was found that GDH remained unchanged, GS increased, and GLNase and AATase decreased. The increase in GS is consistent with a glial localization of this enzyme, as determined through immunocytochemistry (see Norenberg, 1983, for review), because upon degeneration of the auditory nerve, glial cells proliferate in the cochlear nucleus. The decreases in GLNase and AATase are modest (25–30% after 3 days), but are consistent with their ubiquity and their tendency to concentrate in auditory-nerve terminals.

These results provided insights as to how GLU and ASP might be synthesized in the cochlear nucleus, but alone did not provide any additional evidence that GLU and ASP were, in fact, neurotransmitters for the auditory nerve. It remained possible that all neurons, regardless of their neurotransmitter, contained high levels of AATase and GLNase in their presynaptic terminals. Analysis of different brain regions showed that the cochlear nucleus does not contain higher levels of these enzymes than other areas. However, the auditory nerve supplies a relatively small percentage of the total cochlear nucleus protein so that elevated levels of AATase and GLNase in the nerve terminals would not greatly influence the concentration for the whole nucleus. To address more specifically whether or not these enzymes were enriched in the auditory nerve, we compared concentrations in the axon region of the auditory nerve with those of the same region of several other nerves. Because almost all protein synthesis occurs in the cell body, proteins at the synapse are transported down the axon from the cell body. Amounts of proteins in the axon are less than those in the synaptic terminal, but should be related to the amounts in the terminal. We found that GLNase and AATase were 2–5 times higher in the auditory nerve than in the other nerves, while GDH was present at about the same level (Wenthold, 1980).

These results suggested that the auditory nerve has a greater capacity for the synthesis of GLU and ASP than many other neurons. They also suggested that other putative glutamatergic and aspartatergic neurons may also contain elevated levels of GLNase and AATase and that immunocytochemical localization of these enzymes might aid in the identification of such neurons. To investigate these possibilities more thoroughly, we produced antibodies against cytoplasmic AATase and obtained antibodies against phosphate-dependent GLNase (Curthoys *et al.*, 1976) for immunocytochemical localization of these enzymes. The immunocytochemistry verified the biochemical studies, showing that both AATase and GLNase are enriched in terminals of the auditory nerve (Fig. 1) (Altschuler *et al.*, 1981, 1984; Fex *et al.*, 1982; Wenthold and Altschuler, 1983). Therefore, these results provided the first direct evidence that auditory-nerve terminals contain the enzymes necessary to produce GLU and ASP and, thus, clearly satisfy one criterion for these amino acids being neurotransmitters of the auditory nerve.

Antibodies against AATase and GLNase have been used to study other putative glutamatergic and aspartatergic neurons. The studies, to date, indicate that putative glutamatergic and aspartatergic neurons are enriched in

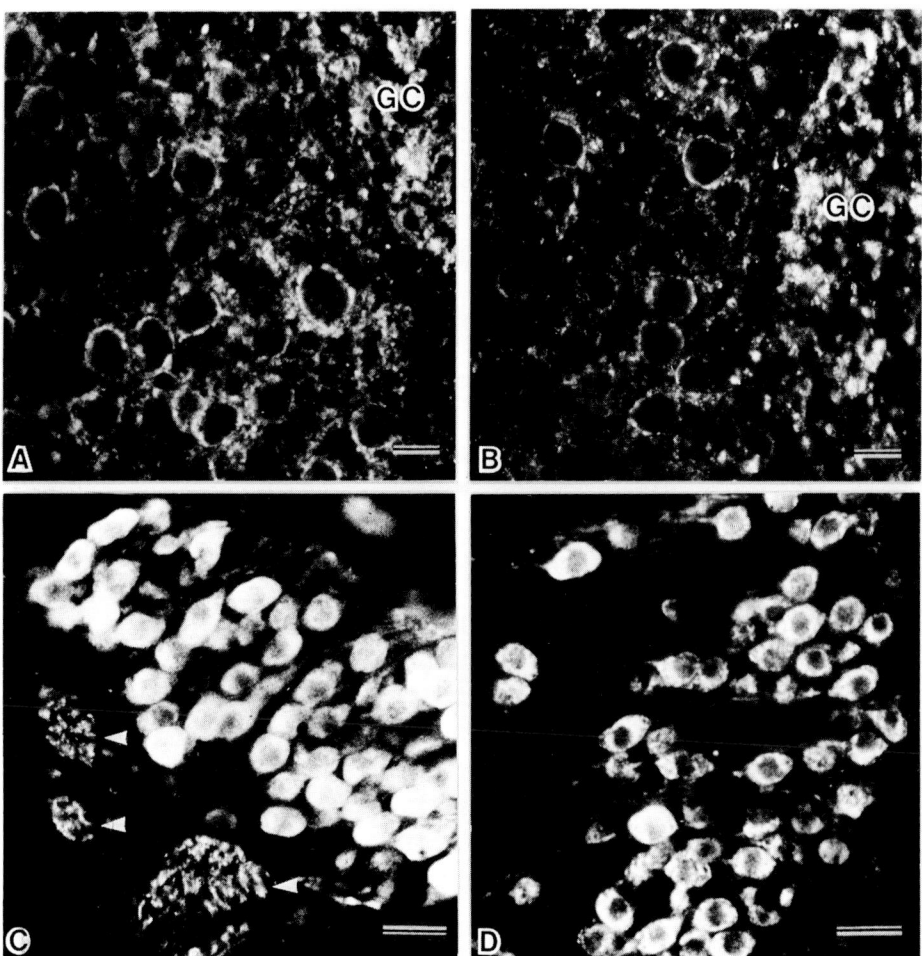

Figure 8-1. Fluorescence micrographs of the guinea-pig cochlear nucleus and spiral ganglion stained with antibodies against AATase (A, C) and GLNase (B, D). Immunofluorescent rings of AATase immunoreactivity (A) and GLNase immunoreactivity (B) around spherical cells of the anteroventral cochlear nucleus and labeling of granule cells (GC) are shown. The cell bodies of the auditory nerve, the spiral ganglion cells, are also intensely labeled with antibodies against AATase (C) or GLNase (D). In C, efferent fibers are also labeled (arrowheads). Primary antisera used at 1/1200 dilution. Bar = 25 μm (from Wenthold and Altschuler, 1983).

GLNase, AATase, or both, compared to other neurons (Wenthold and Altschuler, 1983). The variable distributions of these enzymes may indicate some, as yet unknown, relation concerning whether or not GLU or ASP is the neurotransmitter, or may suggest that different metabolic pathways are used in different populations of glutamatergic and aspartatergic neurons.

V. UPTAKE AND RETROGRADE TRANSPORT OF D-ASPARTATE IN THE AUDITORY NERVE

Many neurotransmitters are taken up into the terminals from which they are released by a specific high-affinity uptake system. This uptake was first observed for catecholamines (Axelrod, 1971), but is now known to be a property of most neurotransmitters, including GLU and ASP (Curtis and Johnston, 1974). Although the uptake system has been widely used in the study of many neurotransmitters, its use in the study of GLU and ASP has been limited because other structures, including glial cells, neuronal cell bodies, and perhaps even terminals of non-glutamatergic/aspartatergic neurons, can take up GLU and ASP (Schon and Kelly, 1974; Henn *et al.*, 1974; Balcar *et al.*, 1977). The function of the uptake system is believed to be twofold: it serves as an inactivation system by clearing the synaptic cleft of neurotransmitter and it serves a salvage function by recycling neurotransmitter.

Cuenod and his collaborators (Cuenod *et al.*, 1982) have observed that several putative neurotransmitters are not only taken up into terminals from which they are released, but are also retrogradely transported to the neuron's cell body (see Cuenod *et al.*, 1982 for review). While the L forms of GLU and ASP do not give retrograde labeling of putative glutamatergic/aspartatergic neurons, D-ASP does. It is suggested that the L forms of the amino acids are too rapidly metabolized, while D-ASP acts as a false transmitter and is taken up by the same transport system that transports the L forms of GLU and ASP. The necessary enzymes to rapidly metabolize D-ASP are not present in mammalian brain. The retrograde transport of D-ASP has been widely studied and appears to label most putative glutamatergic/aspartatergic neurons. A major concern with this technique is its selectivity. The retrograde transport step is likely to be nonspecific; both large and small exogenous molecules have been shown to be retrogradely transported in neurons when applied to the terminals in sufficient concentrations. Presumably, they are passively endocytosed or enter fibers and terminals damaged by the injection. Therefore, with ^{3}H-D-ASP, the selectivity would be provided by high-affinity uptake by the presynaptic terminals. Conceivably, a high concentration at the site of injection or damage from the injection could lead to nonspecific retrograde labeling. Therefore, it seems that control experiments are required to show that other D amino acids, such as D-leucine, are not retrogradely transported under the same conditions in which the D-ASP labeling was done.

After injection of D–ASP into the cochlear nucleus of the guinea pig or cat, both types of spiral ganglion cells were labeled (Oliver et al., 1983). At shorter survival times, there was heavy labeling of astrocytes and presynaptic terminals in the cochlear nucleus. Furthermore, the D–ASP which was taken up could be released *in vitro* with electrical stimulation. Controls showed that other pathways to the cochlear nucleus, for which the neurotransmitter is unknown, were not labeled after D–ASP injections into the cochlear nucleus. It is also cited as a positive control in these studies that D–ASP labeled granule cells and parallel fibers which comprise a putative glutamatergic system in the cerebellum. However, others have reported that this pathway is unlabeled by D–ASP (Cuenod et al., 1982). While such discrepancies may be due to the use of different techniques, they highlight the uncertainties that exist concerning the use of D–ASP as a marker for neurons using GLU or ASP as neurotransmitters.

VI. PHARMACOLOGICAL STUDIES ON THE AUDITORY-NERVE NEUROTRANSMITTER

The pharmacological properties of GLU and ASP have received considerable attention, yet receptors for these putative neurotransmitters remain poorly defined. At least three receptor types have been identified, based on the agonist they prefer: N-methyl-D-aspartate (NMDA) receptors, quisqualic acid (QA) receptors, and kainic acid (KA) receptors (McLennan, 1981; Watkins, 1981; Watkins et al., 1981; see also Martin, Chapter 11, this volume). Both GLU and ASP interact equally well with each receptor type and it is not known if they are the natural ligands of all, or any, of these receptors. Consequently, they are collectively referred to as excitatory amino-acid receptors, acknowledging the fact that GLU, ASP, or a related substance could be the endogenous ligand. The receptor types appear to be associated with different populations of putative glutamatergic/aspartatergic neurons, and this has been elegantly demonstrated in the hippocampus by means of autoradiographic localization (Monaghan et al., 1983).

The pharmacology of the auditory-nerve neurotransmitter has been studied iontophoretically by Martin and Adams (1979) and Martin (1980) in the cat, and by Caspary (Caspary and Harvey, 1978; 1979; Caspary et al., 1981) in the chinchilla, and *in vitro*, using bath application of drugs, by Nemeth et al. (1983) in the chicken. The results of these studies are consistent with GLU, ASP, or a related compound being the neurotransmitter of the auditory nerve. Interestingly, these results suggest the receptor to be of the NMDA type in cat and chinchilla and of the non-NMDA type in chicken. The interpretation of this finding awaits the elucidation of the nature of the multiple excitatory amino-acid receptors. It may suggest a different neurotransmitter is used in chicken or that the receptor molecule has undergone a very slight structural alteration in this species.

The auditory-nerve neurotransmitter has also been studied in the mouse cochlear nucleus *in vitro* (Oertel, 1984). In this study, it was reported that all cells tested in the ventral cochlear nucleus were insensitive to bath-applied GLU and ASP and that the synaptic responses to electrical stimulation of the auditory nerve were not blocked by bath application of D-α-aminoadipate. These findings are difficult to reconcile with the other pharmacological studies listed above. It is unlikely that all neurons in the anteroventral cochlear nucleus would be insensitive to GLU and ASP, since essentially all neurons in the central nervous system that have been tested respond to GLU and ASP (Watkins, 1978). Furthermore, kainic acid, a structural analog of GLU, destroys neurons in the cochlear nucleus, presumably through a mechanism involving a GLU-like receptor on the surface of these neurons. One possible explanation is that the bath-applied substances are not reaching the appropriate sites to exert their actions. The use of a non-metabolized analog of GLU in these studies, such as kainic acid, may be useful in determining if this explanation is tenable. The fact that D-α-aminoadipate does not block synaptic responses may suggest that an NMDA-type receptor is not present in the mouse cochlear nucleus, as may be the case in chicken. Unfortunately, antagonists for the other receptor classes were not tested, allowing no firm conclusions to be made.

VII. CONCLUSIONS

Several of the criteria necessary to demonstrate that a substance is a neurotransmitter have been met for GLU and ASP in the auditory nerve. Although each piece of evidence supporting this hypothesis could have alternative interpretations—and this must always be considered and must temper any conclusion we make—the fact that supporting evidence comes from many independent lines of research, including both presynaptic and postsynaptic data, argues strongly in favor of GLU and ASP being the auditory-nerve neurotransmitters. More conclusive support must await demonstration of selective presynaptic release and postsynaptic identity of action, which probably cannot be shown at this time, due to technical limitations. It should be stressed that the uncertainties surrounding the neurotransmitter status of GLU and ASP are not limited to the auditory nerve. In fact, the auditory nerve may be one of the most completely studied putative glutamatergic/aspartatergic systems. Therefore, significant advances in our knowledge of GLU and ASP in the auditory nerve will be due to, or result in, significant advances in the general field of GLU and ASP research. A very promising recent development, which awaits a thorough test, is the direct localization of GLU in tissue using a glutamate-specific antibody (Storm-Mathisen *et al.*, 1983). If it proves technically feasible to selectively localize GLU, this technique may resolve many uncertainties concerning GLU, including those involving the auditory nerve, and may rather easily identify neurons containing high levels of GLU. Furthermore, it may also be possible to distinguish between ASP-containing and GLU-containing neurons and, when coupled with antibodies against GLU- and ASP-metabolizing enzymes, this technique may give indications as to how these amino acids are produced and regulated in neurons.

A number of investigators have suggested that GLU and ASP are not themselves neurotransmitters, but rather are precursors or products of the true neurotransmitter. One popular notion is that the neurotransmitter is a small peptide made up largely, or exclusively, of GLU and ASP (Zaczek *et al.*, 1983). It seems very unlikely, however, that a peptide could meet the requirements for being the auditory-nerve neurotransmitter. Almost all known neuro-active peptides are synthesized in the cell body as part of a much longer molecule and transported to the presynaptic terminal. For the auditory nerve, not only would the cell need to expend a great deal of energy to produce the large precursor of the neuro-active peptide, but regulation of its production would be cumbersome and ineffective. When the store of peptide in the terminal was depleted, the cell body would need to respond and make more precursor, which would then be transported to the terminal. It is unlikely that this mechanism could assure the constant and sufficient supply of neurotransmitter needed by auditory-nerve terminals. Alternatively, it may be suggested that a postulated peptide might be synthesized enzymatically in the presynaptic terminals. Although this is not believed to be a standard route of neuro-active peptide production, it would not have the problems of supply and regulation that synthesis in the cell body involves. Furthermore, this would be consistent with GLU and ASP being concentrated in the presynaptic terminal, where they would be precursors. While such a mechanism of synthesis cannot be ruled out, many studies have been done on the metabolism of GLU and ASP in the central nervous system using radioactive amino acids, and such studies show that GLU and ASP are converted into the expected small-molecule products; there are no reports of the formation of peptides.

The research done on GLU and ASP in the auditory nerve has raised several interesting questions and provided some new approaches to characterize the auditory system. The discovery that the release of ^{3}H–D–ASP from cochlear nucleus slices is affected by disarticulation of middle ear ossicles (Potashner, 1983) suggests that this release may be a sensitive indicator of the status of the auditory nerve. Immunocytochemical localization of GLU-

and ASP-related enzymes may also be a useful parameter to study, for example, during development of the auditory nerve.

REFERENCES

Altschuler, R. A.; Neises, G. R., Harmison, G. G., Wenthold, R. J., and Fex, J.: Immunocytochemical localization of aspartate aminotransferase immunoreactivity in cochlear nucleus of the guinea pig. *Proc. Natl. Acad. Sci. U.S.A. 78:* 6553-6557, 1981.

Altschuler, R. A.; Wenthold, R. J., Schwartz, A. M., Haser, W. G., Curthoys, N. P., Parakkal, M., and Fex, J.: Immunocytochemical localization of glutaminase-like immunoreactivity in the auditory nerve. *Brain Res. 291:* 173-178, 1984.

Axelrod, J.: Noradrenaline; Fate and control of its biosynthesis. *Science 173:* 598-606, 1971.

Balcar, V. J.; Borg, J., and Mandel, P.: High affinity uptake of L-glutamate and L-aspartate by glial cells. *J. Neurochem. 28:* 87-93, 1977.

Bradford, H. F.; Ward, H. K., and Thomas, A. J.: Glutamate as a substrate for nerve endings. *J. Neurochem. 30:* 1453-1459, 1978.

Canzek, V., and Reubi, J. C.: The effect of cochlear nerve lesion on the release of glutamate, aspartate and GABA from cat cochlear nucleus. *Exp. Brain Res. 38:* 437-441, 1980.

Caspary, D. M., and Havey, D. C.: Effects of putative amino acid neurotransmitters on response patterns of neurons in the cochlear nucleus. *Neurosci. Abstr. 4:* 5, 1978.

Caspary, D. M., and Havey, D. C.: Glutamate and aspartate: Effects on threshold and response patterns in the cochlear nucleus. *Neurosci. Abstr. 5:* 17, 1979.

Caspary, D. M.; Havey, D. C., and Faingold, C. L.: Glutamate and aspartate: Alterations of thresholds and response patterns of auditory neurons. *Hearing Res. 4:* 325-333, 1981.

Cotman, C. W.; Foster, A., and Lanthorn, T.: An overview of glutamate as a neurotransmitter. In Di Chiara, G., and Gessa, G. L. (Eds.): *Glutamate as a Neurotransmitter.* New York, Raven Press, 1981, pp. 1-27.

Cuenod, M.; Bagnoli, P., Beaudet, A., Rustioni, A., Wiklund, L., and Streit, P.: Transmitter-specific retrograde labeling of neurons. In Chan-Palay, V., and Palay, S. L. (Eds.): *Cytochemical Methods in Neuroanatomy.* New York, A. R. Liss, 1982, pp. 17-44.

Curthoys, N. P.; Kuhlenschmidt, T., Godfrey, S. S., and Weiss, R. F.: Phosphate-dependent glutaminase from rat kidney. *Arch. Biochem. Biophys. 172:* 162-167, 1976.

Curtis, D. R., and Johnston, G. A. R.: Amino acid transmitters in the mammalian central nervous system. *Ergeb. Physiol. 69:* 97-188, 1974.

Fex, J.; Altschuler, R. A., Wenthold, R. J., and Parakkal, M. H.: Aspartate aminotransferase immunoreactivity in cochlea of guinea pig. *Hearing Res. 7:* 149-160, 1982.

Freeman, A. R.; Shank, R. P., Kephart, J., Dekin, M., and Wang, M.: A model for excitatory transmission at a glutamate synapse. In Di Chiara, G. and Gessa, G. L. (Eds.): *Glutamate as a Neurotransmitter.* New York, Raven Press, 1981, pp. 227-243.

Godfrey, D. A.; Carter, J. A., Berger, S. J., Lowry, O. H., and Matschinsky, F. M.: Quantitative histochemical mapping of candidate transmitter amino acids in cat cochlear nucleus. *J. Histochem. Cytochem. 25:* 417-431, 1977.

Godfrey, D. A.; Carter, J. A., Lowry, O. H., and Matschinsky, F. M.: Distribution of gamma-aminobutyric acid, glycine, glutamate and aspartate in the cochlear nucleus of the rat. *J. Histochem. Cytochem. 26:* 118-126, 1978.

Hamberger, A. C.; Chiang, C. H., Nylen, E. S., Scheff, S. W., and Cotman, C. W.: Glutamate as a CNS transmitter I. Evaluation of glucose and glutamine as precursors for the synthesis of preferentially released glutamate. *Brain Res. 168:* 513-530, 1979.

Hansson, E.; Jarlstedt, J., and Sellstrom, A.: Sound stimulated ^{14}C-glutamate release from the nucleus cochlearis. *Experientia 36:* 576-577, 1980.

Henn, F. A.; Goldstein, M. N., and Hamberger, A.: Uptake of the neurotransmitter candidate glutamate by glia. *Nature (Lond.) 249:* 663-664, 1974.

Jarlstedt, J.; Karlsson, B., and Hamberger, A.: *In vivo* studies of amino acid transmitters in the central auditory system of the guinea pig. In press, 1984.

Martin, M. R.: The effects of iontophoretically applied antagonists on auditory nerve and amino acid evoked excitation of anteroventral cochlear nucleus neurons. *Neuropharmacology 19:* 519-528, 1980.

Martin, M. R., and Adams, J. C.: Effects of DL-α-aminoadipate on synaptically and chemically evoked excitations of anteroventral cochlear nucleus neurons of the cat. *Neuroscience 4:* 1097-1105, 1979.

McLennan, H.: On the nature of the receptors for various excitatory amino acids in the mammalian central nervous system. In Di Chiara, G., and Gessa, G. L. (Eds.): *Glutamate as a Neurotransmitter.* New York, Raven Press, 1981, pp. 253-262.

Minchin, M. C. W., and Iversen, L. L.: Release of [^{3}H] gamma-aminobutyric acid from glial cells in rat dorsal root ganglia. *J. Neurochem. 23:* 533-540, 1974.

Monaghan, D. T.; Holets, V. R., Toy, D. W., and Cotman, C. W.: Anatomical distributions of four pharmacologically distinct ^{3}H-L-glutamate binding sites. *Nature 306:* 176-179, 1983.

Nemeth, E. F.; Jackson, H., and Parks, T. N.: Pharmacological evidence for synaptic transmission mediated by non-N-methyl-D-aspartate receptors in the avian cochlear nucleus. *Neurosci. Lett. 40:* 39-44, 1983.

Norenberg, M. D.: Immunohistochemistry of glutamine synthesis. In Hertz, L., Kvamme, E., McGeer, E. G., and Schousboe, A. (Eds.): *Glutamine, Glutamate and GABA in the Central Nervous System.* New York, Alan R. Liss, 1983, pp. 95-111.

Oertel, D.: Cells in the anteroventral cochlear nucleus are insensitive to L-glutamate and L-aspartate; excitatory synaptic responses are not blocked by D-α-aminoadipate. *Brain Res.* In press, 1984.

Oliver, D. L.; Potashner, S. J., Jones, D. R., and Morest, D. K.: Selective labeling of spiral ganglion and granule cells with D-aspartate in the auditory system of cat and guinea pig. *J. Neurosci. 3:* 455-472, 1983.

Potashner, S. J.: Uptake and release of D-aspartate in the guinea pig cochlear nucleus. *J. Neurosci. 41:* 1094-1101, 1983.

Schon, F., and Kelley, J. S.: Autoradiographic localization of [^{3}H]GABA and [^{3}H]glutamate over satellite glial cells. *Brain Res. 66:* 275-288, 1974.

Sellstrom, A., and Hamberger, A.: Potassium-stimulated gamma-aminobutyric acid release from neurons and glia. *Brain Res. 119:* 189-198, 1977.

Storm-Mathiesen, J.; Leknes, A. K., Bore, A. T., Vaaland, J. L., Edminson, P., Haug, F.-M. S., and Ottersen, O. P.: First visualization of glutamate and GABA in neurons by immunocytochemistry. *Nature 301:* 517-520, 1983.

Ward, H. K.; Thanki, C. M., and Bradford, H. F.: Glutamate and glucose as precursors of transmitter amino acids. *Ex vivo* studies. *J. Neurochem. 40:* 855, 1983.

Watkins, J. C.: Excitatory amino acids. In McGeer, E. G.; Olney, J. W., and McGeer, P. L. (Eds.): *Kainic Acid as a Tool in Neurobiology.* New York, Raven Press, 1978, pp. 37-69.

Watkins, J. C.: Pharmacology of excitatory amino acid receptors. In Roberts, P. J.; Storm-Mathiesen, J., and Johnstone, G. A. R. (Eds.): *Glutamate: Transmitter in the Central Nervous System.* Chichester, J. Wiley and Sons, 1981, pp. 1-24.

Watkins, J. C.; Davies, J., Evans, R. H., Francis, A. A., and Jones, A. W.: Pharmacology of receptors for excitatory amino acids. In Di Chiara, G., and Gess, G. L. (Eds.): *Glutamate as a Neurotransmitter.* New York, Raven Press, 1981, pp. 263-273.

Wenthold, R. J.: Glutamic acid and aspartic acid in subdivisions of the cochlear nucleus after auditory nerve lesion. *Brain Res. 143:* 544-548, 1978.

Wenthold, R. J.: Release of endogenous glutamic acid, aspartic acid and GABA from cochlear nucleus slices. *Brain Res. 162:* 338-343, 1979.

Wenthold, R. J.: Glutaminase and aspartate aminotransferase decrease in cochlear nucleus after lesion of the auditory nerve. *Brain Res. 190:* 293-297, 1980.

Wenthold, R. J., and Altschuler, R. A.: Immunocytochemistry of aspartate aminotransferase and glutaminase. In Hertz, L.; Kvamme, E., McGeer, E. G., and Schousboe, A. (Eds.): *Glutamine, Glutamate and GABA in the Central Nervous System.* New York, Alan R. Liss, Inc., 1983, pp. 33-50.

Wenthold, R. J., and Gulley, R. L.: Aspartic acid and glutamic acid levels in the cochlear nucleus after auditory nerve lesion. *Brain Res. 138:* 111-123, 1977.

Wenthold, R. J., and Gulley, R. L.: Glutamic acid and aspartic acid in the cochlear nucleus of the waltzing guinea pig. *Brain Res. 158:* 279-284, 1978.

Zaczek, R.; Koller, K., Cotter, R., Heller, D., and Coyle, J. T.: N-acetylaspartylglutamate: An endogenous peptide with high affinity for a brain "glutamate" receptor. *Proc. Natl. Acad. Sci. U.S.A. 80:* 1116-1119, 1983.

Chapter 9

IDENTIFICATION OF GLUTAMATERGIC AND ASPARTATERGIC PATHWAYS IN THE AUDITORY SYSTEM

STEVEN J. POTASHNER, D. KENT MOREST,
DOUGLAS L. OLIVER, AND DOYLE R. JONES

I. Introduction
II. Autoradiographic Studies
III. Biochemical Studies
IV. Conclusions
V. Functional Considerations
    A. The Ventral Cochlear Nucleus
    B. The Dorsal Cochlear Nucleus
    C. Ganglion Cells
    References

I. INTRODUCTION

The cochlear nucleus (CN) is the first site in the brain to receive information from the cochlea via the cochlear nerve and, thus, plays a key role in the processing of acoustic information. In the CN, specific kinds of synaptic organization have been correlated with characteristic neuronal discharge patterns evoked by acoustic stimuli (Morest *et al.*, 1973; Cant and Morest, 1979a, b; Tolbert *et al.*, 1982). For example, individual bushy cells in the anteroventral cochlear nucleus receive most of their excitatory input from a few, large cochlear nerve endings, which maintain a high degree of synaptic security. This synaptic organization preserves the signal transmitted by the cochlear afferents. In contrast, fusiform cells in the dorsal cochlear nucleus receive only a fraction of their input from the cochlear nerve and a large input from interneurons, such as granule cells. This kind of synaptic organization can produce significant transformations in the sensory signal. Where correlations can be established between specific kinds of synaptic organization and morphologic types of neurons, on the one hand, and electrical discharge patterns, on the other, there may be an opportunity to explain the mechanisms that generate the discharge patterns of these neurons to acoustic stimuli (Morest, 1975). Such explanations would depend in part on information about the transmitters used at particular synapses.

The transmitters in the CN are likely to include acetylcholine (Comis and Guth, 1974; Godfrey et al., 1977a, 1981, Chapter 10, this volume; Kimura et al., 1981), norepinephrine (Fuxe, 1965; Anden et al., 1966; Kromer and Moore, 1976), GABA, and glycine (Davies, 1975, 1977, Chapter 14, this volume; Fisher and Davies, 1976; Wenthold and Morest, 1976; Godfrey et al., 1977a, b; Wenthold, 1979; Canzek and Reubi, 1980). In addition, L-glutamate and/or L-aspartate (L-GLU/L-ASP) have been proposed as cochlear nerve transmitters (Godfrey et al., 1977b, 1978; Canzek and Reubi, 1980; Wenthold, 1981, Chapter 8, this volume; Martin, Chapter 11, this volume; Caspary, Chapter 12, this volume). However, most of the neurons which might use these compounds as transmitters remain to be identified. This should be possible with autoradiographic localization of transmitter-related markers (Iversen and Schon, 1973; Droz, 1975). For example, D-aspartate (D-ASP) may be a useful marker for L-GLU/L-ASP neurons (Oliver et al., 1981; Cuenod et al., 1982). L-GLU, L-ASP, and D-ASP are probably taken up by the same high-affinity transporter into glia and axonal endings of glutamatergic or aspartatergic neurons (Balcar and Johnston, 1972, 1973; Hokfelt and Ljungdahl, 1975; Davies and Johnston, 1976; Iversen, 1977; Johnson, 1978). After uptake, D-ASP is not significantly metabolized for several hours (Davies and Johnston, 1976; Takagaki, 1978). Uptake and retrograde transport of D-ASP have been reported in neurons which are presumed, on the basis of other evidence, to use L-GLU/L-ASP as a transmitter (Streit, 1980; Beaudet et al., 1981; Oliver et al., 1981). Finally, after uptake, D-ASP can be released in a Ca^{++}-dependent manner from neural pathways using L-GLU/L-ASP as transmitters (Malthe-Sorenssen et al., 1979, 1980; Beaudet et al., 1981; Potashner and Gerard, 1983). Thus, D-ASP probably gains access only to glutamatergic and aspartatergic neurons and some of it enters the presynaptic transmitter pools of these cells.

The aim of this study was to identify the types of neurons projecting to the CN and those intrinsic to the CN which could be labeled by D-ASP and which can take up and release D-ASP. We injected D-ASP into the CN and used autoradiographic techniques to follow its distribution *in situ* over a 48-h period. Shortly after injection, label was localized in glia and in certain axons. Later, the label was apparently transported retrogradely to the somata of granule cells in the CN and spiral ganglion cells in the cochlea. Companion biochemical studies indicated that the axon endings of these neurons probably mediate the high-affinity uptake and the release of D-ASP in the CN. These results are consistent with the proposal that the granule cells of the CN as well as the cochlear nerve fibers may use L-GLU/L-ASP as transmitters.[1]

II. AUTORADIOGRAPHIC STUDIES

In addition to intrinsic cells, the CN contains the synaptic endings of neurons in the spiral ganglion, the superior olivary complex, the nuclei of the lateral lemniscus, and the inferior colliculus (Fig. 1). An opportunity was provided for the axonal endings of these cells to take up the marker by injecting various concentrations of [^{3}H]-D-ASP (60 μM to 3 mM) into the CN of cats and guinea pigs. The injection site usually included much of the dorsal cochlear nucleus (DCN), and the dorsal region of the anteroventral cochlear nucleus (AVCN) and posteroventral cochlear nucleus (PVCN) (Fig. 1). Animals were sacrificed for autoradiography 15 min to 48 h after injection of the marker. Details of the methods have been reported elsewhere (Oliver et al., 1983).

Fifteen min after injection, label was localized mainly in the injection sites. The distribution of the labeling suggested the presence of D-ASP in glia and in axonal endings. Injection sites were characterized by the heavy labeling of the neuropil and the glial cell bodies (Fig. 2) with an abrupt transition to light or imperceptible labeling at the perimeter of the injection site. The large spherical somata of neurons in the ventral subdivisions of the CN appeared unlabeled or lightly labeled (Fig. 2). However, types of neurons known to receive cochlear afferents were usually surrounded by heavily labeled peri-somatic profiles resembling the large endings of cochlear nerve fibers (Fig. 2, asterisk). In addition, there appeared to be abundant labeling of smaller perisomatic structures, which could represent the smaller cochlear nerve endings or possibly the glial processes surrounding these endings.

Forty min after injection, label appeared both in and beyond the injection site over the fibers of the molecular layer in the DCN and the cochlear nerve root, providing further evidence of early uptake by axonal endings and suggesting the retrograde axonal transport of the marker. The molecular layer of the DCN consists predominantly of parallel fibers, the axons of the granule cells in the DCN and in the external granular layer of the ventral CN (Fig. 1B). The axonal endings of the parallel fibers are especially concentrated in the molecular layer (Lorente de No., 1981; Kane, 1974a; Mugnaini et al., 1980a, b). Label was prominent in the molecular layer within the injection site (Fig. 3). Rostral to the injection site, silver grains lay in a band over the fibers extending from the molecular layer into the external granular layer covering the AVCN and PVCN (Fig. 3, arrow). By contrast, 15 min after injection, the molecular layer of the DCN and the cochlear nerve root were scarcely labeled (Oliver et al., 1983). In the cochlear nerve root, at the lateral margin of the AVCN, well beyond the injection site, silver grains lay over the nerve fascicles, but not over the interfascicular glia (Fig. 4A), which were labeled only after 40 min (Fig. 4B).

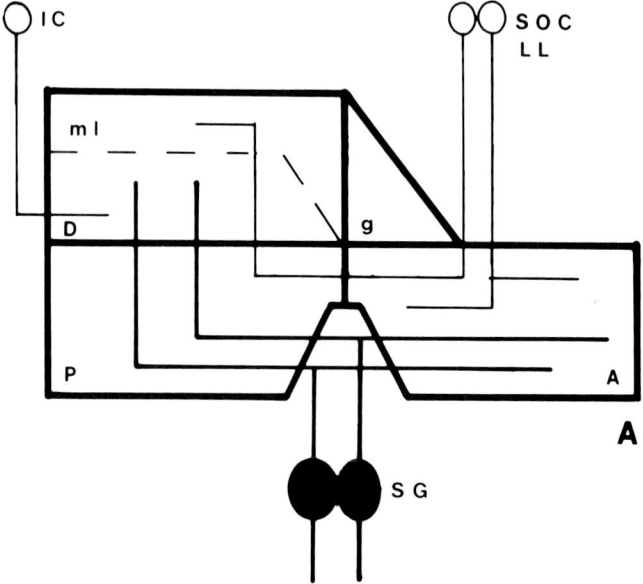

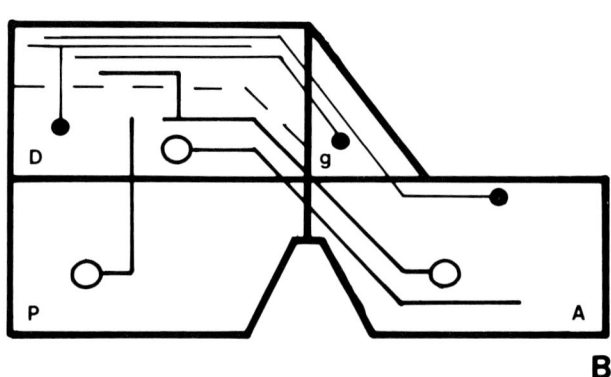

Figure 9-1. Schematic diagrams to show the extrinsic inputs (A) and some of the intrinsic connections (B) of the cochlear nucleus in a parasaggital view. Neurons labeled by D–ASP are indicated in black. Abbreviations: A, anteroventral cochlear nucleus; D, dorsal cochlear nucleus; g, external granular layer; IC, inferior colliculus; LL, nuclei of the lateral lemniscus; ml, molecular layer; P, posteroventral cochlear nucleus; SG, spiral ganglion; SOC, superior olivary complex. (Reprinted, with permission, from Oliver et al., 1983.)

Between 5 and 48 h after injection, label appeared in the granule cell bodies of the CN, but disappeared from the glia. In the same period, label appeared to be transported retrogradely to the somata of the types I and II

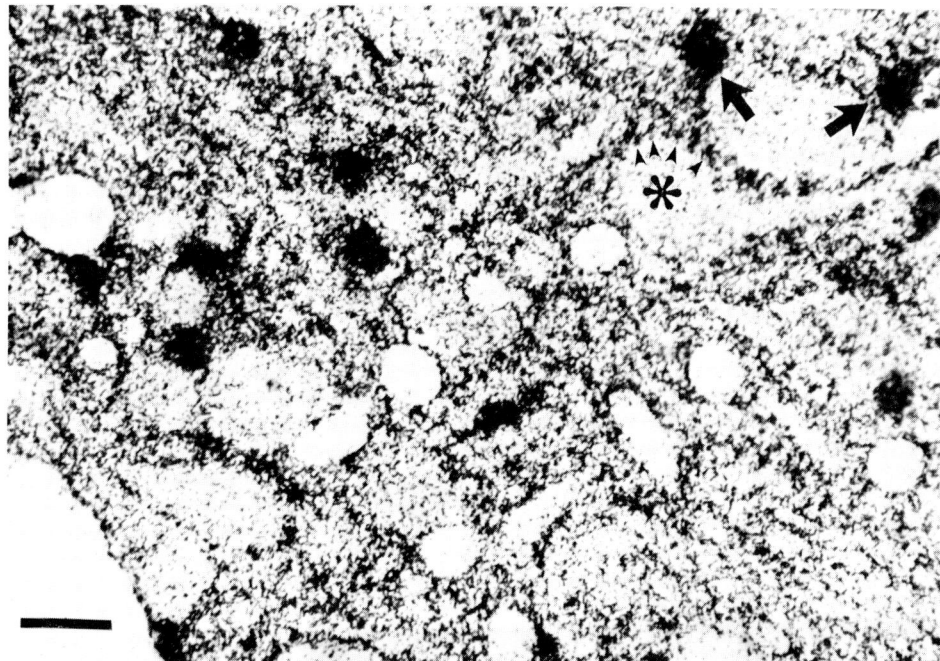

Figure 9-2. Labeling of the neuropil and glia in the PVCN of the guinea pig 15 min after injection of D-ASP into the cochlear nucleus. Label appears in astrocyte somata (arrows) and as clusters of grains (arrowheads) surrounding the somata of neurons, some of which are octopus cells (asterisk). Scale: 10 μm. (Modified from Oliver et al., 1983.)

spiral ganglion cells in the cochlea and to the organ of Corti (Figs. 5-8). At the longest survival time (48 h), virtually every spiral ganglion cell in regions of the cochlea innervating the injection site contained label (Fig. 7). In contrast, other neurons in the CN and other extrinsic neurons projecting to the CN remained unlabeled.

The temporal progression of labeling is consistent with the retrograde axonal transport of label from uptake sites in axonal endings in the CN to the somata of granule cells and spiral ganglion cells. Support for this conclusion is provided by a quantitative statistical analysis of counts of the autoradiographic silver grains in the cochlea (Jones et al., 1984). After injection of the marker into the CN, there appeared to be a wave of labeling which advanced from the CN to the cochlea via the cochlear nerve. With increasing survival time, there was an overall increase in the labeling density in the cochlear nerve and in the regions of the organ of Corti containing cochlear nerve processes and endings (Fig. 8, upper panel). Moreover, the differential labeling of basal and apical portions of the cochlear nerve suggests a proximo-distal transport of label (Fig. 8, lower panel). At early survival times, the relative labeling

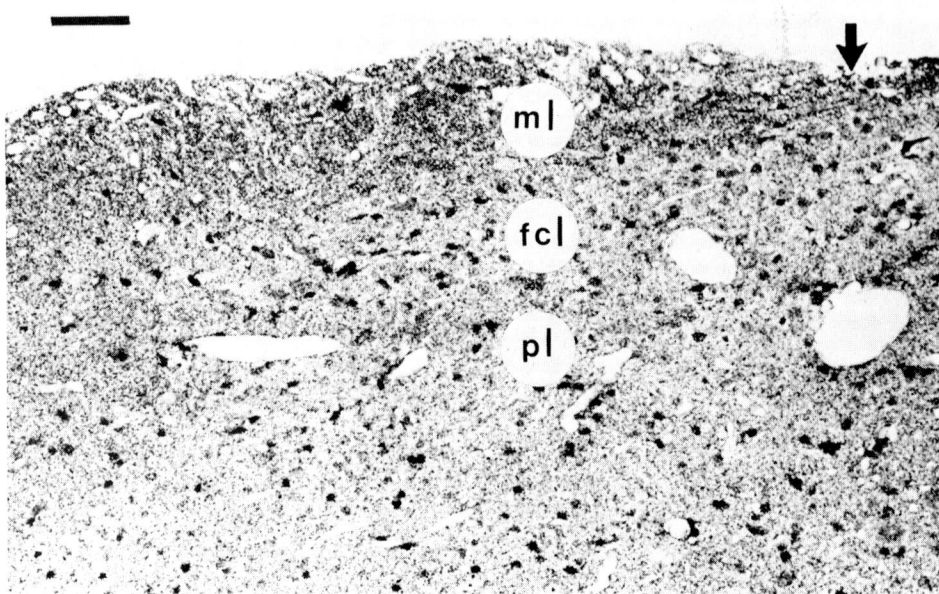

Figure 9-3. Labeling of the molecular layer of the DCN of the guinea pig 40 min after injection of D-ASP into the cochlear nucleus. Labeled astrocytes appear as small dark cells scattered evenly throughout the field. Label is concentrated in the neuropil of the molecular layer where labeled parallel fibers are evident (arrow). Abbreviations: ml, molecular layer; fcl, fusiform cell layer; pl, polymorphic layer. Scale: 100 μm. (Modified from Oliver et al., 1983.)

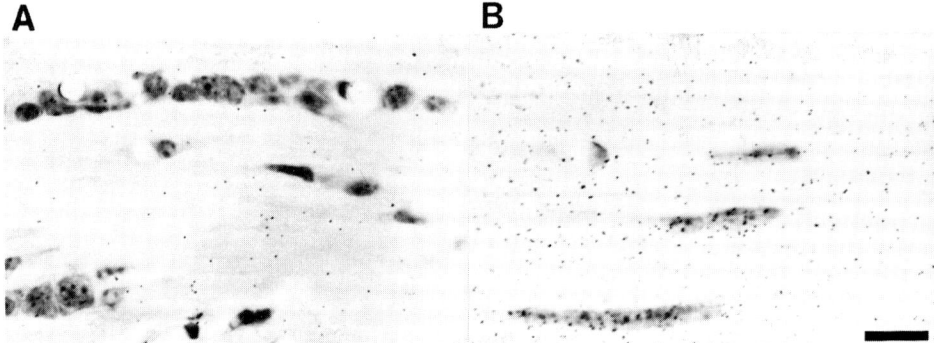

Figure 9-4. Labeling of the cochlear nerve root in the AVCN of the guinea pig 15 min (A) and 40 min (B) after injection of D-ASP into the cochlear nucleus. In panel A, label appears in the fiber fascicles of the cochlear nerve root, but not in the interfascicular glia. In B, the density of labeling in the cochlear nerve fibers has increased and label appears in the interfascicular glia. Scale: 20 μm. (Modified from Oliver et al., 1983.)

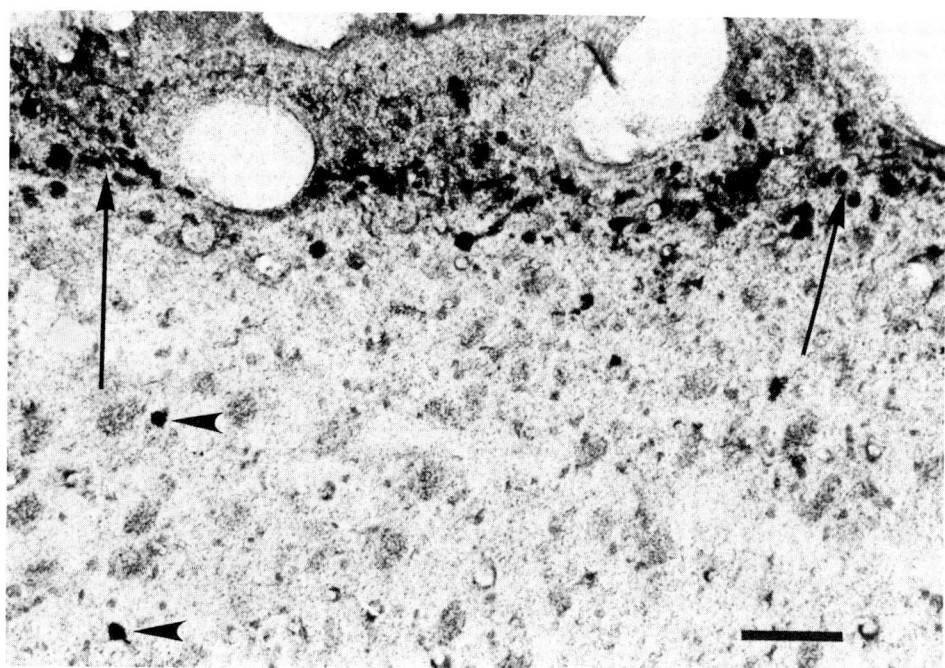

Figure 9-5. Labeled granule cells in the cat 48 h after injection of D–ASP into the DCN. Label appears in the somata of the granule cells in the external granular layer (arrows) and in several granule cells (arrowheads) in the underlying AVCN. Scale: 100 μm. (Modified from Oliver et al., 1983.)

density was high in parts of the nerve adjacent to the basal turn of the cochlea, proximal to the CN, compared to more distal areas of nerve near the apical turn of the cochlea. Later, as the overall density of labeling in the nerve increased, the labeling density of distal portions of the nerve became similar to that of proximal portions. The corresponding changes in silver grain counts are statistically significant.

III. BIOCHEMICAL STUDIES

Within the neuronal populations of the central nervous system, D–ASP is thought to be taken up and retrogradely transported only by glutamatergic and aspartatergic neurons (Oliver et al., 1981; Cuenod et al., 1982). Moreover, the marker appears to gain access to releasable transmitter pools in the synaptic endings of such neurons (Malthe-Sorenssen et al., 1979, 1980; Beaudet et al., 1981). Our autoradiographic studies suggest that the axons of the granule cells of the CN and of the spiral ganglion cells take up and retrogradely transport D–ASP and, therefore, may use L–GLU/L–ASP as a

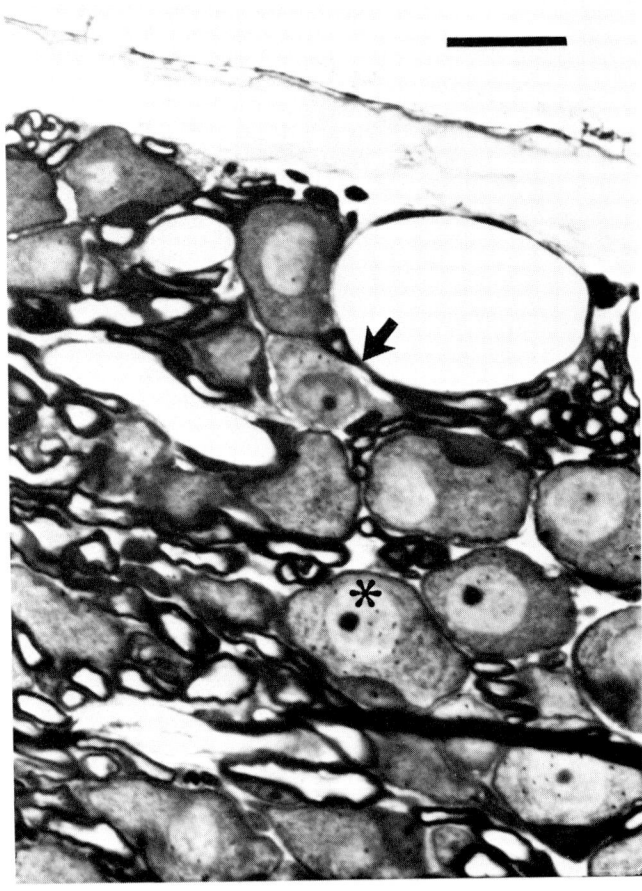

Figure 9-6. Labeled spiral ganglion cells in the cat cochlea 15 h after injection of D-ASP into the cochlear nucleus. Label appears in the large type I (asterisk) and the smaller type II (arrow) ganglion cells. Scale: 20 μm.

transmitter. Biochemical studies are consistent with this conclusion and suggest that D-ASP uptake and release in the CN is mediated by the axons of the granule and spiral ganglion cells. If the cochlear nerve endings accumulate D-ASP and if the marker gains access to transmitter pools in these endings, then lesions which selectively destroy the cochlear nerve should depress the high-affinity uptake and the release of D-ASP (Werman, 1966; Iversen, 1977; Orrego, 1979) at the projection sites of the nerve in the AVCN, PVCN, and DCN (Fig. 1). However, if the granule-cell endings also mediate these activities, they should contribute some of the uptake and release of D-ASP in the DCN (Fig. 1). Thus, after destruction of the cochlear nerve, D-ASP uptake and release should be

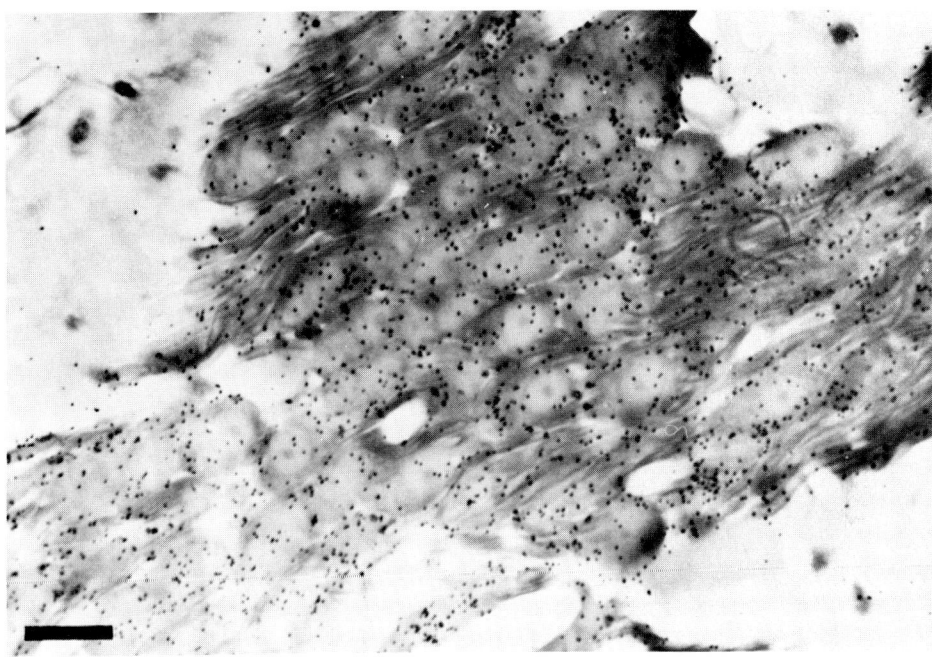

Figure 9-7. Label in the spiral ganglion in the cat cochlea 48 h after injection of D–ASP into the cochlear nucleus. Label appears in many fibers of the cochlear nerve and in every ganglion cell. Scale: 100 μm. (Reprinted, with permission, from Oliver et al., 1983.)

less depressed in the DCN than in the ventral subdivisions of the CN.

The uptake of D–ASP was measured by placing segments of tissue containing the AVCN, PVCN, or DCN of guinea pigs in an oxygenated Ringer solution, containing 0.1 μM [^{3}H]-D–ASP at 37° C. After 45 min, the tissue-associated radioactivity was determined, the radioactivity trapped in the extracellular spaces of the tissue was subtracted, and the resulting figure used to express the uptake as the tissue-to-medium ratio (T/M) of radioactivity. Release was measured by superfusing tissues, which had taken up [^{3}H]-D–ASP, with fresh Ringer containing no D–ASP. For a short period of time during the superfusion, electrical field stimulation was used to evoke the release of D–ASP. The radioactivity released to the superfusion fluid was determined and used to compute the amounts of D–ASP released spontaneously and in response to electrical stimulation. Details of the methods have been reported elsewhere (Oliver et al., 1983; Potashner, 1983).

D–ASP is taken up by cellular elements of the CN and gains access to releasable pools, which are presumed to be in synaptic endings. Segments of tissue containing the AVCN, PVCN, or DCN from unlesioned animals took up D–ASP from the bathing solution and achieved a higher tissue concentra-

150   *Auditory Biochemistry*

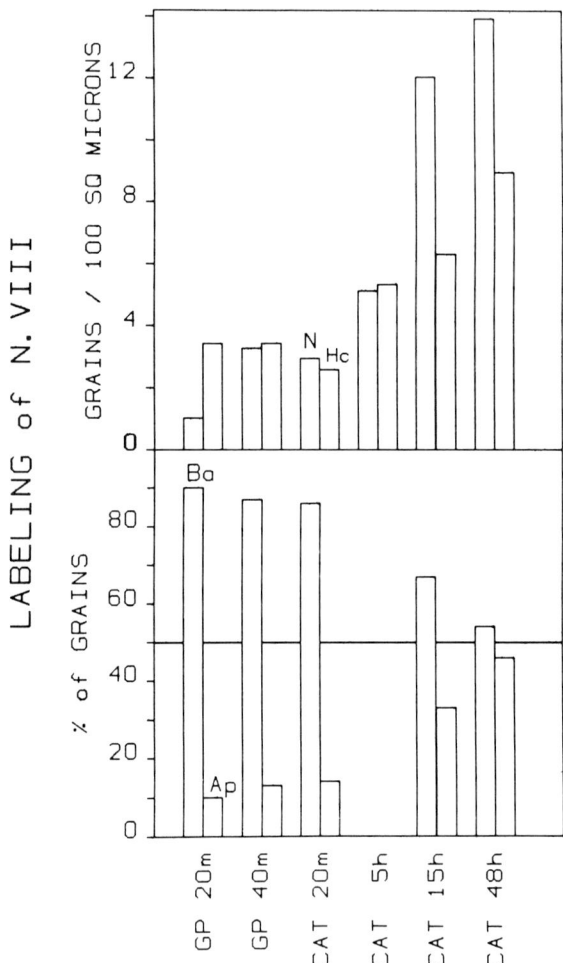

Figure 9-8. Progression of the labeling of the cochlear nerve (N.VIII) at several times after injection of D–ASP into the cochlear nucleus. In the upper panel, a time course is shown of the relative labeling density of part of the cochlear nerve in the cochlea (N) and of the neuronal tissue under the hair cells in the organ of Corti (Hc). The lower panel illustrates the relative distribution of label in portions of the cochlear nerve adjacent to the basal (Ba) and apical (Ap) turns of the cochlea at various times after injection of D–ASP into the cochlear nucleus. GP, guinea pig.

tion of the marker than that present in the medium (Fig. 9, controls). During the subsequent superfusion of some of the tissues, there was a small spontaneous loss of the marker to the medium (Fig. 10, control pre-stimulation release in the time-course plots). Electrical stimulation of the tissues for 4 min evoked a transient increase in the amount of D–ASP lost to the medium

(Fig. 10, controls in the time-course plots). The electrically-evoked release in each subdivision was quantitated and plotted in a histogram (Fig. 10). The electrically evoked release was suppressed by Ca^{++} deprivation (Fig. 10, LOW–Ca^{++} in histograms), suggesting that electrical stimulation probably evoked the release of D–ASP from synaptic endings (Rubin, 1974; Orrego, 1979).

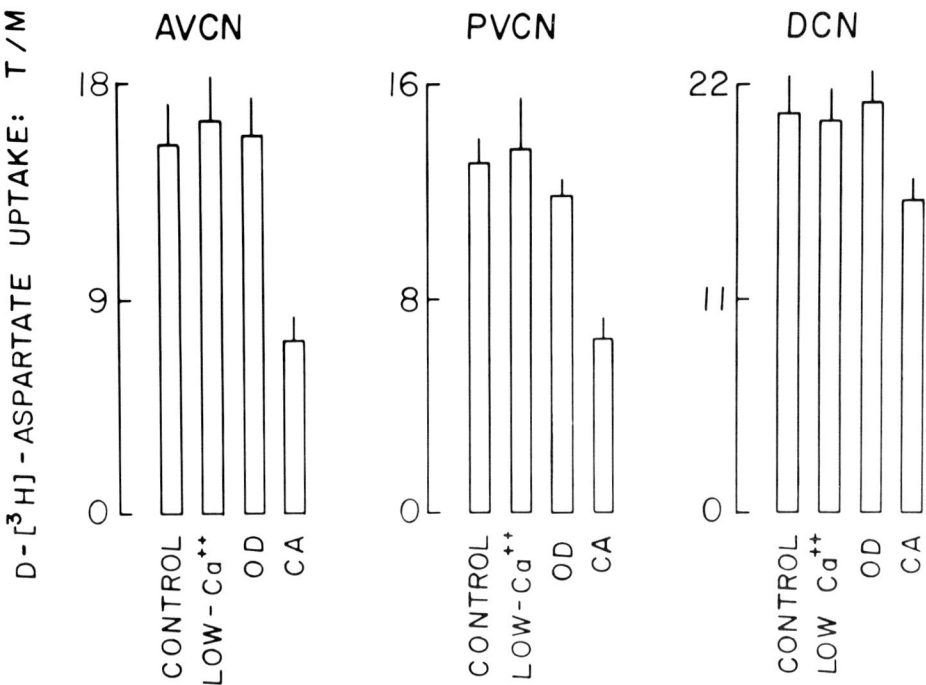

Figure 9-9. Uptake of D–ASP *in vitro* by subdivisions of the guinea pig cochlear nucleus. In control experiments, tissues were dissected from intact animals. In LOW-Ca^{++} experiments, the concentration of $CaCl_2$ in the medium was reduced 100 fold and 2 mM $MgCl_2$ was added. Uptake was measured in tissues taken from animals in which the auditory ossicles had been disarticulated (OD) or the cochleas ablated (CA). The data are means ± SEM. (Reprinted, with permission, from Potashner, 1983.)

Part of the uptake and release of D–ASP in the CN appears to be mediated by the cochlear nerve endings, because lesions which affect the nerve depress these activities in each subdivision of the CN. For example, in one group of guinea pigs, the middle ear ossicles were disarticulated bilaterally to attenuate acoustic stimulation of the cochleas and, thus, of the cochlear nerves (Fig. 1). Nine to thirteen days after surgery, the uptake of D–ASP appeared to be unaffected (Fig. 9, OD), but a modest depression of D–ASP release was apparent in each subdivision of the CN (Fig. 10, OD). In another

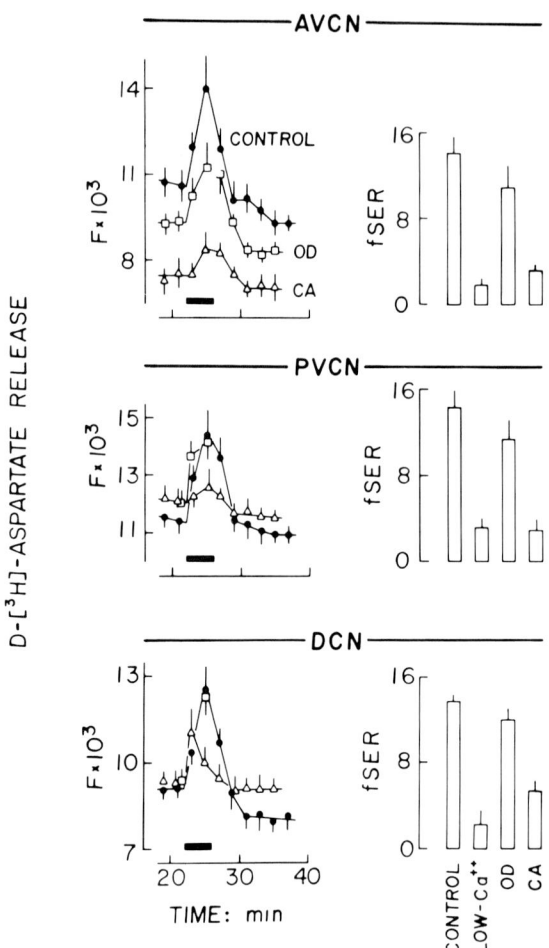

Figure 9-10. Release of D-ASP *in vitro* by subdivisions of the guinea-pig cochlear nucleus. The time course of the spontaneous and electrically evoked releases are illustrated by the plots on the left. The black horizontal bars denote the period of stimulation. The D-ASP released by the electrical stimulation is represented quantitatively in the histograms on the right. The data are means ± SEM. Computation of the data was described elsewhere (Potashner, 1983). Experimental conditions were as in Fig. 9. (Reprinted, with permission, from Potashner, 1983.)

group of guinea pigs, the cochleas were ablated bilaterally to initiate the degeneration of the cochlear nerve fibers (Fig. 1). Two to eight days after surgery, when the degeneration of the cochlear nerve endings in the CN should be well advanced (Gentchev and Sotello, 1973; Kane, 1974b; Wenthold and Gulley, 1977), the uptake and release of D-ASP was depressed in each division of the CN (Figs. 9 and 10, CA).

It is likely that the cochlear nerve endings mediate nearly all of the synaptic release of D–ASP in the ventral CN, as the loss of D–ASP release in the AVCN and PVCN after cochlear ablation was as extensive as the entire Ca^{++}-dependent portion of the electrically evoked release (Fig. 10). However, after the degeneration of cochlear nerve endings, elements remain in the DCN which mediate the uptake and release of D–ASP and, thus, may also use L–GLU/L–ASP as transmitters. This conclusion is supported by a comparison of the effects of cochlear ablation in the AVCN, PVCN, and DCN, which indicated that there was less inhibition of the uptake and release of D–ASP in the DCN than in the AVCN and PVCN. For example, after cochlear ablation, D–ASP uptake in the DCN was inhibited only by 22%, while that in the AVCN and PVCN was depressed by 50–55% (Fig. 9, CA); also, D–ASP release in the DCN was inhibited by 55–58%, while that in the AVCN and PVCN was depressed by 78–80% (Fig. 10, CA). However, some Ca^{++}-dependent release of D–ASP, presumably from synaptic endings, remained in the DCN after cochlear ablation (Fig. 10). The autoradiographic studies suggest that the residual uptake and release of D–ASP in the DCN may be mediated by the axons of the granule cells.

IV. CONCLUSIONS

The following can be concluded on the basis of the present findings. First, D–ASP injected into the CN is initially taken up by a high-affinity transporter into astrocytes and the synaptic endings of cochlear nerve fibers and granule cells. Second, the same endings that take up D–ASP release it during synaptic transmission. Third, D–ASP is subsequently transported retrogradely from the synaptic endings to the somata of granule cells in the CN and spiral ganglion cells in the cochlea. Finally, the granule cells and spiral ganglion cells may use L–GLU/L–ASP as transmitters.

The present results suggest that D–ASP uptake in the CN is mediated by a transporter located on astrocytes and certain synaptic endings. In the autoradiographic studies, the earliest labeling observed after injection of D–ASP into the CN appeared over astrocyte somata and over structures in the neuropil resembling axonal endings. This was most evident in the case of the large cochlear nerve endings or end-bulbs visible in the light microscope. The biochemical studies suggest that after uptake of D–ASP *in vitro*, the marker was present in synaptic endings. There was a marked, electrically evoked, Ca^{++}-dependent release of D–ASP from each division of the CN. Since the synaptic release of transmitters requires the depolarization of axons and the presence of Ca^{++} ions extracellularly (Rubin, 1974; Orrego, 1979), these findings imply that newly taken-up D–ASP gained access to transmitter pools in some synaptic endings.

The present results also suggest that the endings of cochlear nerve fibers and granule cells in the CN mediate the uptake and release of D–ASP. These activities were markedly depressed in each subdivision of the CN after lesions which affected the cochlear nerve. In the AVCN and PVCN, almost all of the Ca^{++}-dependent release of D–ASP appeared to be dependent on the integrity of the cochlear nerve endings. By contrast, in the DCN there was a persistence of D–ASP uptake and release after cochlear afferents degenerated. These activities might be mediated by any of the neuronal elements still innervating the DCN, including the axonal endings of central neurons belonging to descending auditory pathways (Elverland, 1977; Kane, 1977; Kane and Conlee, 1979), granule and other cells intrinsic to the CN which project to the DCN, and local neurons within the DCN (Brawer *et al.*, 1974; Kane, 1974b; Mugnaini *et al.*, 1980a, b). However, the labeling of the granule cells by D–ASP injected into the CN, together with the failure of the other types of neurons to accumulate the marker, implies that the axonal endings of granule cells mediate some of the uptake and release of D–ASP in the DCN.

The present study has demonstrated the selective labeling of granule and spiral ganglion cells after injections of D–ASP into the CN. The results are consistent with an early uptake of D–ASP by the axonal endings of granule cells and cochlear nerve fibers in the CN and the subsequent retrograde transport of the marker to the somata of these neurons. In the cochlear nerve root 15 min after injection of D–ASP, there was sparse labeling of fiber fascicles, while 40 min after injection the density of label had increased markedly. By 5–6 h, label began to appear in type I and II spiral ganglion cells of the cochlea and in their peripheral processes in the organ of Corti. Between 5 and 48 h, there was an overall proximo-distal increase in the labeling density in the cochlear nerve, the spiral ganglion, and the organ of Corti. In the molecular layer of the DCN and the external granular layer, the labeling pattern and its temporal progression are also consistent with retrograde transport of D–ASP by the parallel fibers of granule cells.

Evidence from other neural systems suggests that D–ASP can serve as an autoradiographic marker for neurons which probably use L–GLU/L–ASP as transmitters (see review by Cuenod *et al.*, 1982). Injection of D–ASP into the cerebellar cortex labels cerebellar granule cells (Oliver *et al.*, 1981) and cells in the inferior olive (Oliver *et al.*, 1981; Wiklund *et al.*, 1982), the source of climbing fibers innervating the cerebellar cortex. Independent biochemical evidence also suggests that these neurons may use L–GLU/L–ASP as transmitters (Young *et al.*, 1974; McBride *et al.*, 1976a, b, 1978; Nadi *et al.*, 1977; Roffler-Tarlov and Sidman, 1978; Sandoval and Cotman, 1978). Injection of D–ASP into the striatum labels a subpopulation of neurons in the cerebral cortex which project to the striatum (Streit, 1980; Le Vay and

Sherk, 1981) and are thought to use L-GLU as their transmitter (Divac *et al.*, 1977; McGeer *et al.*, 1977; Kim *et al.*, 1977; Fonnum *et al.*, 1981).

The present findings are consistent with the proposal that the spiral ganglion neurons and the granule cells use L-GLU/L-ASP as transmitters in the CN. Injection of D-ASP into the CN labeled types I and II spiral ganglion cells; this suggests that these neurons may use L-GLU/L-ASP as transmitters. Other evidence (see INTRODUCTION) suggests that D-ASP can serve as a biochemical marker for glutamatergic and/or aspartergic neurons, as it is taken up and released by the synaptic endings of these cells. In the present study, the uptake and release of D-ASP in the CN was dependent on the integrity of the cochlear nerve; this also suggests that these fibers may use L-GLU/L-ASP as transmitters. Corroboration comes from biochemical and pharmacological studies which provide strong evidence for L-GLU/L-ASP as the cochlear nerve transmitter (Wenthold, 1981, Chapter 8, this volume; Martin, Chapter 11, this volume; Caspary, Chapter 12, this volume).

The granule cells of the DCN and the external granular layer are local neurons whose axons form parallel fibers which make a large number of synapses within the DCN. Indeed, the granule cells in the guinea pig and the cat may well form as many synaptic contacts as the cochlear afferents. That the granule cells may also use L-GLU/L-ASP as transmitters is suggested by two observations. First, these cells become labeled after injections of D-ASP into the CN. Second, after destruction of the cochlear nerve, which is accompanied by a marked loss of D-ASP uptake and release in the AVCN and PVCN, there is some preservation of these activities in the DCN. This conclusion receives further support from evidence that after the destruction of the guinea pig cochlear nerve, the activities of enzymes that synthesize L-GLU and L-ASP and the concentrations of these amino acids were diminished slightly in the DCN, while relatively large deficits in these measures were observed in the ventral subdivisions of the CN (Wenthold, 1978, 1980).

V. FUNCTIONAL CONSIDERATIONS

A. THE VENTRAL COCHLEAR NUCLEUS

The autoradiographic labeling pattern with D-ASP suggests that the big cochlear nerve endings surrounding the large perikarya of bushy cells and octopus cells may use L-GLU/L-ASP as a transmitter. The fine structure of these endings in the cat and the guinea pig is characterized by large, clear spherical vesicles, a widened synaptic cleft, and a strongly asymmetric synaptic membrane complex—in other words, a so-called type I morphology (Cant and Morest, 1984). Combined electrophysiological and morphological findings indicate that these synaptic endings are exictatory (Pfeiffer, 1966; Morest, 1975; Bourk, 1976; Rhode et al., 1983a, b; Oertel, 1983). Thus we may postulate that the large endings of the cochlear nerve, including the end-bulbs of Held, make type I, excitatory synapses in the CN and use L-GLU/L-ASP as a transmitter.[2] Moreover, the collaterals of the large fibers and endings of the cochlear nerve that form smaller type I synapses might use the same transmitters. This hypothesis could be strengthened by localization of D-ASP in these endings with electron microscopic autoradiography.

If the large cochlear nerve synapses using L-GLU/L-ASP are excitatory, they no doubt account for some features of the electrical responses of neurons in the cochlear nucleus (see Cant and Morest, 1984). For example, the primary-like responses of the cochlear nerve fibers to acoustic stimulation are typically preserved by the spherical bushy cells in the rostral AVCN. This function presumably depends on a prepotent excitatory input from the large end-bulbs of the cochlear nerve, which provide for a high degree of synaptic security. Another example is the octopus cell in PVCN, which typically converts cochlear nerve input into responses which are most active at their onset and which remain relatively quiet for the rest of the stimulus. This pronounced onset activity and its tendency to occur consistently at a minimum latency may be a function of the spatial convergence of many different excitatory axons. This function could be explained in terms of the spatial arrangement of the many large type I endings from the cochlear nerve and the properties of chemical transmission at their synapses.

B. THE DORSAL COCHLEAR NUCLEUS

The autoradiographic labeling pattern with D-ASP suggests that the granule cells may function as excitatory interneurons using L-GLU/L-ASP as a transmitter.[2] A large number of granule cells must project to each fusiform cell in the DCN by way of parallel fibers in the molecular, polymorph, and fusiform cell layers (Kane, 1974a). Therefore, fusiform cell activity might be influenced via the granule cells. For example, granule cell axons, together with input from the cochlear nerve to both granule and fusiform cells, might form part of an excitatory feed-forward mechanism. In addition, descending pathways from more centrally located auditory nuclei (Cant and Morest, 1984) and fibers from the CN project heavily into the granule cell region and the DCN where they could influence the activity of granule cells and inhibitory neurons. These arrangements might contribute to the relatively high level of spontaneous activity and the complex electrical responses attributed to fusiform cells (Kane, 1974a, b; Evans and Nelson, 1973; Godfrey et al., 1975; Young and Brownell, 1976; Young, 1980) which are thought to depend on interactions with excitatory and inhibitory neurons.

There is a parallel between the morphology of the granule cells in the DCN and those in the cerebellar cortex. Also, both cell types are thought to use an acidic amino-acid transmitter. However, there is little support for a more detailed parallel between the other cell types and the intrinsic circuits in the DCN and those in the cerebellar cortex. For example, there is no convincing evidence that the DCN contains cell types analogous to cerebellar Purkinje cells or axonal endings analogous to those of the climbing fibers.

C. GANGLION CELLS

At least two mutually exclusive categories of spiral ganglion cells have been defined on the basis of their innervation patterns (Ginzberg and Morest, 1983, 1984). Ganglion cells with inner radial fibers innervate inner hair cells while those with outer spiral fibers innervate outer hair cells. The connections of each ganglion cell type with different subdivisions and with specific types of neurons in the CN is uncertain. These could be related to the different types of cochlear nerve endings, ranging from large end-bulbs to small boutons. Since several terminal types regularly occur on the same individual fibers, most types probably arise from the ganglion cell class innervating inner hair cells, because more than 90% of the ganglion cells belong to this category. A differential distribution of these terminal types on CN neurons might play a role in producing different kinds of electrical responses. Thus, the stellate cells in AVCN, often associated with a chopper response, receive small Type I endings but not large ones, whereas the bushy cells receive both sizes both sizes of terminals (see Cant and Morest, 1984) and are implicated in primary-like responses. Small endings with Type I morphology, including pleomorphic vesicles and nearly symmetric synaptic membrane complexes, were identified with cochlear nerve endings on octopus cells (Kane, 1973), and an inhibitory function suggested for them. However, if the ganglion cells forming these endings are labeled by D-ASP, as the present observations might seem to imply, then they also could use L-GLU/L-ASP as a transmitter in the CN.

There is a duality in the ganglion cell populations innervating inner and outer hair cells and in the distribution of large and small cochlear nerve fibers in the CN. There is indirect evidence that the small fibers from outer hair cells project more heavily to the DCN than to the ventral CN (Morest and Bohne, 1983). This duality in the cochlear nerve projections to the CN suggests the existence of inner and outer hair cell systems, perhaps analogous in some ways to the cone and rod systems of the visual pathway. Perhaps the inner hair cell system can provide a more discriminative, differentiating function than the outer hair cell system, or more resolution, in some sense not yet defined. The outer hair cell system might provide for more of an analytical or integrative function, that would include the cells with complex electrical responses, such as those found in the DCN (e.g., Young and Voigt, 1982).

These considerations might lead to the expectation that different transmitters would be used, either by inner and outer hair cells or by the ganglion cells connecting them with the CN. However, the available findings suggest that all of the spiral ganglion cells could use L-GLU/L-ASP as transmitters in the CN, regardless of type. It remains to be seen if other putative transmitters, e.g., certain peptides, might be localized in particular types of ganglion cells. Or perhaps more than one transmitter could be released at cochlear nerve endings, either at heterotypic synapses formed by the same axons or by different kinds of fibers.

ACKNOWLEDGMENTS

The authors are grateful to L. T. Andrus, S. Kwok, C. Sowers-Clift, and P. L. Tran for technical assistance. This work was supported by grants from the Deafness Research Foundation and the U.S. Public Health Service (R01 NS 19036, R01 NS 14347, F32 NS 00068).

NOTES

[1]Complete reports of the autoradiographic and the biochemical studies have appeared elsewhere (Oliver et al., 1983; Potashner, 1983; Jones et al., 1984).

[2]However, L-GLU has been known to produce an inhibitory effect as well as an excitatory one, depending on the properties of the postsynaptic membrane (see Nicoll and Alger, 1981).

REFERENCES

Anden, N. E.; Fuxe, K., and Lavesson, K.: Effect of large mesencephalic diencephalic lesions on the noradrenaline, dopamine and 5-HT neurons of the central nervous system. *Experientia.* 22: 842-843, 1966.

Balcar, V. J., and Johnston, G. A. R.: The structural specificity of the high affinity uptake of L-glutamate and L-aspartate by rat brain slices. *J. Neurochem.* 19: 2657-2666, 1972.

Balcar, V. J., and Johnston, G. A. R.: High affinity uptake of transmitters: Studies on the uptake of L-aspartate, GABA, L-gluatmate and glycine in cat spinal cord. *J. Neurochem.* 20: 529-539, 1973.

Beaudet, A.; Burkhalter, A., Reubi, J. C., and Cuenod, M.: Selective bidirectional transport of [^{3}H]-D-aspartate in the pigeon retinotectal pathway. *Neuroscience.* 6: 2021-2034, 1981.

Bourk, T. R.: *Electrical responses of neural units in the anteroventral cochlear nucleus of the cat.* Ph.D. Thesis. Boston, M. I. T., 1976.

Brawer, J. R.; Morest, D. K., and Kane, E. C.: The neuronal architecture of the cochlear nucleus of the cat. *J. Comp. Neurol.* 155: 251-299, 1974.

Cant, N. B., and Morest, D. K.: Organization of the neurons in the anterior division of the anteroventral cochlear nucleus of the cat. Light microscopic observations. *Neuroscience.* 4: 1909-1923, 1979a.

Cant, N. B., and Morest, D. K.: The bushy cells in the anteroventral cochlear nucleus of the cat. A study with the electron microscope. *Neuroscience.* 4: 1925-1945, 1979b.

Cant, N. B., and Morest, D. K.: The structural basis for stimulus coding in the cochlear nucleus of the cat. In Berlin, C. I. (Ed.): *Recent Developments in Hearing Science.* San Diego, College Hill Press, 1984.

Canzek, V., and Reubi, J. C.: The effect of cochlear nerve lesion on the release of glutamate, aspartate and GABA from cat cochlear nucleus *in vitro. Exp. Brain Res. 38:* 437-441, 1980.

Comis, S. D., and Guth, P. S.: The release of ACh from the cochlear nucleus upon stimulation of the crossed olivo-cochlear bundle. *Neuropharmacol. 13:* 633-641, 1974.

Cuenod, M.; Beaudet, A., Rustioni, A., Wiklund, L., and Streit, P.: Transmitter specific retrograde labeling of neurons. In Chan-Palay, V., and Palay, S. L. (Eds.): *Cytochemical Methods in Neuroanatomy,* New York, A. R. Liss, 1982, pp. 17-44.

Davies, W. E.: The distribution of GABA transminase-containing neurons in the cat cochlear nucleus. *Brain Res. 83:* 27-33, 1975.

Davies, W. E.: GABAergic innervation of the mammalian cochlear nucleus. *Collogues Inst. Natnl. Sante Rech. Med. 68:* 155-164, 1977.

Davies, L. P., and Johnston, G. A. R.: Uptake and release of D- and L-aspartate by rat brain slices. *J. Neurochem. 26:* 1007-1014, 1976.

Divac, I.; Fonnum, F., and Storm-Mathisen, J.: High affinity uptake of glutamate in terminals of corticostriatal axons. *Nature. 266:* 377-378, 1977.

Droz, B.: Autoradiography as a tool for visualizing neurons and neuronal processes. In Cowan, W. M., and Cuenod, M. (Eds.): *The Use of Axonal Transport for Studies of Neuronal Connectivity.* Amsterdam, Elsevier, 1975, pp. 128-154.

Elverland, H. H.: Descending connections between the superior olivary and cochlear nuclear complexes in the cat studied by autoradiographic and horseradish peroxidase methods. *Exp. Brain Res. 27:* 397-412, 1977.

Evans, E. F., and Nelson, P. G.: The responses of single neurons in the cochlear nucleus of the cat as a function of their location and the anesthetic state. *Exp. Brain Res. 17:* 402-427, 1973.

Fisher, S. K., and Davies, W. E.: GABA and its related enzymes in the lower auditory system of the guinea pig. *J. Neurochem. 27:* 1145-1155, 1976.

Fonnum, F.; Storm-Mathisen, J., and Divac, I.: Biochemical evidence for glutamate as neurotransmitter in corticostriatal and corticothalamic fibers in rat brain. *Neuroscience. 6:* 863-873, 1981.

Fuxe, K.: Evidence for the existence of monamine neurons in the CNS. IV: Distribution of monoamine nerve terminals in the CNS. *Acta Physiol. Scand. 64 (Suppl. 247):* 37-85, 1965.

Gentschev, T., and Sotelo, C.: Degenerative patterns in the ventral cochlear nucleus of the rat after primary deafferentation. An ultrastructural study. *Brain Res. 62:* 37-60, 1973.

Ginzberg, R. D., and Morest, D. K.: A study of cochlear innervation in the young cat with the Golgi method. *Hearing Res. 10:* 227-246, 1983.

Ginzberg, R. D., and Morest, D. K.: The fine structure of cochlear innervation in the cat. *Hearing Res.* (in press), 1984.

Godfrey, D. A.; Kiang, N. Y. S., and Norris, B. E.: Single unit activity in the dorsal cochlear nucleus of the cat. *J. Comp. Neurol. 162:* 269-284, 1975.

Godfrey, D. A.; Williams, A. D., and Matschinsky, F. M.: Quantitative histochemical mapping of enzymes of the cholinergic system in the cat cochlear nucleus. *J. Histochem. Cytochem.* 25: 397-416, 1977a.

Godfrey, D. A.; Carter, J. A., Berger, S. J., Lowry, O. H., and Matschinsky, F. M.: Quantitative histochemical mapping of candidate transmitter amino acids in cat cochlear nucleus. *J. Histochem Cytochem.* 25: 417-431, 1977b.

Godfrey, D. A.; Carter, J. A., Lowry, O. H., and Matschinsky, F. M.: Distribution of gamma-amino butyric acid, glycine, glutamate and aspartate in the cochlear nucleus of the rat. *J. Histochem. Cytochem.* 26: 118-126, 1978.

Godfrey, D. A.; Park, J. L., Rabe, J. R., Dunn, J. D., Smith, T. J., and Ross, C. D.: Quantitave evaluation of centrifugal cholinergic pathways to the rat cochlear nucleus. *Neurosci. Abstr.* 7: 56, 1981.

Hokfelt, T., and Ljungdahl, A.: Uptake mechanisms as a basis for histochemical identification and tracing of transmitter-specific neuron populations. In Cowan, W. M., and Cuenod, M. (Eds.): *The Use of Axonal Transport for Studies of Neuronal Connectivity.* Amsterdam, Elsevier, 1975, pp. 249-305.

Iversen, L. L.: Inactivation of neurotransmitters. In Cottrell, G. A., and Usherwood, P. N. R. (Eds.): *Synapses.* New York, Academic Press, 1977, pp. 137-153.

Iversen, L. L., and Schon, F.: The use of autoradiographic techniques for the identification and mapping of transmitter specific neurons in the CNS. In Mandell, A. (Ed.): *New Concepts in Neurotransmitter Regulation.* New York, Plenum Press, 1973, pp. 153-193.

Johnson, J. L.: The excitant amino acids, glutamic and aspartic acid, as transmitter candidates in the vertebrate CNS. *Prog. Neurobiol.* 10: 155-202, 1978.

Jones, D. R.; Morest, D. K., Oliver, D. L., and Potashner, S. J.: Transganglionic transport of D-aspartate from cochlear nucleus to cochlea—A quantitative autoradiographic study. (submitted), 1984.

Kane, E. C.: Octopus cells in the cochlear nucleus of the cat: Heterotypic synapses upon homeotypic neurons. *Int. J. Neurosci.* 5: 251-279, 1973.

Kane, E. C.: Synaptic organization in the dorsal cochlear nucleus of the cat: A light and electron microscopic study. *J. Comp. Neurol.* 155: 301-330, 1974a.

Kane, E. C.: Patterns of degeneration in the caudal cochlear nucleus of the cat after cochlear ablation. *Anat. Rec.* 179: 67-92, 1974b.

Kane, E. C.: Descending inputs to the cat dorsal cochlear nucleus: An electron microscopic study. *J. Neurocytol.* 6: 583-605, 1977.

Kane, E. C., and Conlee, J. W.: Descending inputs to the caudal cochlear nucleus of the cat. Degeneration and autoradiographic studies. *J. Comp. Neurol.* 187: 759-784, 1979.

Kim, J. S.; Hassler, R., Haug, P., and Paik, K. S.: Effect of frontal cortex ablation on striatal glutamic acid level in rat. *Brain Res.* 132: 370-374, 1977.

Kimura H.; McGeer, P. L., Peng, J. H., and McGeer, E. G.: The central cholinergic system studied by choline acetyl transferase immunohistochemistry in the cat. *J. Comp. Neurol.* 200: 151-201, 1981.

Kromer, L. F., and Moore, R. Y.: Cochlear nucleus innervation by central norepinephrine neurons in the rat. *Brain Res.* 118: 531-537, 1976.

LeVay, S., and Sherk, H.: The visual claustrum of the cat. I. Structure and connections. *J. Neurosci.* 1: 956-980, 1981.

Lorente de No, R.: *The Primary Acoustic Nuclei.* New York, Raven Press, 1981.

Malthe-Sorenssen, D.; Skrede, K. K., and Fonnum, F.: Ca^{++}-dependent release of D-[^{3}H]aspartate evoked by selective electrical stimulation of excitatory afferent fibers to hippocampal pyramidal cells *in vitro. Neuroscience.* 4: 1255-1263, 1979.

Malthe-Sorenssen, D.; Skrede, K. K., and Fonnum, F.: Release of D-[^{3}H]aspartate from the dorsolateral septum after electrical stimulation of the fimbria *in vitro*. *Neuroscience.* 5: 127-133, 1980.

McBride, W. J.; Aprison, M. H., and Kusano, K.: Contents of several amino acids in the cerebellum, brain stem, and cerebrum of the "staggerer", "weaver" and "nervous" neurologically mutant mice. *J. Neurochem.* 26: 867-870, 1976a.

McBride, W. J.; Nadi, N. S., Altman, J., and Aprison, M. H.: Effects of selective doses of x-irradiation on the levels of several amino acids in the cerebellum of the rat. *Neurochem. Res.* 1: 141-152, 1976b.

McBride, W. J.; Rea, M. A., and Nadi, N. S.: Effects of 3-acetylpyridine on the levels of several amino acids in different CNS regions of the rat. *Neurochem. Res.* 3: 793-801, 1978.

McGeer, P. L.; McGeer, E. G., Scherer, U., and Singh, K.: A glutamatergic corticostriatal path? *Brain Res.* 128: 369-373, 1977.

Morest, D. K.: Synaptic relationships of Golgi type II cells in the medial geniculate body of the cat. *J. Comp. Neurol.* 162: 157-194, 1975.

Morest, D. K., and Bohne, B. A.: Noise-induced degeneration in the brain and representation of inner and outer hair cells. *Hearing Res.* 9: 145-151, 1983.

Morest, D. K.; Kiang, N. Y. S., Kane, E. C., Guinan, J. J. Jr., and Godfrey, D. A.: Stimulus coding at caudal levels of the cat's auditory nervous system: II. Patterns of synaptic organization. In Moller, A. R. (Ed.): *Basic Mechanisms in Hearing.* New York, Academic Press, 1973, pp. 470-509.

Mugnaini, E.; Osen, K. K., Dahl, A., Freidrich, V. L. Jr., and Korte, G.: Fine structure of granule cells and related interneurons (termed Golgi cells) in the cochlear nuclear complex of cat, rat, and mouse. *J. Neurocytol.* 9: 536-570, 1980a.

Mugnaini, E.; Warr, W. B., and Osen, K. K.: Distribution and light microscopic features of granule cells in the cochlear nuclei of cat, rat, and mouse. *J. Comp. Neurol.* 191: 581-606, 1980b.

Nadi, N. S.; Kanter, D., McBride, W. J., and Aprison, M. H.: Effects of 3-acetylpyridine on several putative neurotransmitter amino acids in the cerebellum and medulla of the rat. *J. Neurochem.* 28: 661-662, 1977.

Nicoll, R. A., and Alger, B. E.: Synaptic excitation may activate a calcium dependent potassium conductance in hippocampal pyramidal cells. *Science.* 212: 957-959, 1981.

Oertel, D.: Synaptic responses and electrical properties of cells in brain slices of the mouse anteroventral cochlear nucleus. *J. Neurosci.* 3: 2043-2053, 1983.

Oliver, D. L.; Jones, D. R., Potashner, S. J., and Morest, D. K.: Evidence for selective uptake, release, and axonal transport of D-aspartate in the auditory system and cerebellum of cat and guinea pig. *Anat. Rec.* 199: 186A, 1981.

Oliver, D. L.; Potashner, S. J., Jones, D. R., and Morest, D. K.: Selective labeling of spiral ganglion and granule cells with D-aspartate in the auditory system of the cat and guinea pig. *J. Neurosci.* 3: 455-472, 1983.

Orrego, F.: Criteria for the identification of central neurotransmitters, and their application to studies with some nerve tissue preparations *in vitro*. *Neuroscience.* 4: 1037-1057, 1979.

Pfeiffer, R. R · Anteroventral cochlear nucleus: Wave forms of extracellularly recorded spike potentials. *Science.* 154: 667-668, 1966.

Potashner, S. J.: Uptake and release of D-aspartate in the guinea pig cochlear nucleus. *J. Neurochem.* 41: 1094-1101, 1983.

Potashner, S. J., and Gerard, D.: Kainate-enhanced release of D-[^{3}H]aspartate from cerebral cortex and striatum: Reversal by baclofen and pentobarbital. *J. Neurochem.* 40: 1548-1577, 1983.

Rhode, W. S.; Smith, P. H., and Oertel, D.: Physiological response properties of cells labeled intracellularly with horseradish peroxidase in cat dorsal cochlear nucleus. *J. Comp. Neurol. 213:* 426-447, 1983a.

Rhode, W. S.; Oertel, D., and Smith, P. H.: Physiological response properties of cells labeled intracellularly with horseradish peroxidase in cat ventral cochlear nucleus. *J. Comp. Neurol. 213:* 448-463, 1983b.

Roffler-Tarlov, S., and Sidman, R. L.: Concentrations of glutamic acid in cerebellar cortex and deep nuclei of normal mice and weaver, staggerer and nervous. *Brain Res. 172:* 269-283, 1978.

Rubin, R. P.: *Calcium and the Secretory Process.* New York, Plenum Press, 1974.

Sandoval, M. E., and Cotman, C. W.: Evaluation of glutamate as a neurotransmitter of cerebellar parallel fibers. *Neuroscience. 3:* 199-206, 1978.

Streit, P.: Selective retrograde labeling indicating the transmitter of neuronal pathways. *J. Comp. Neurol. 191:* 429-463, 1980.

Takagaki, G.: Sodium and potassium ions and accumulation of labelled D-aspartate and GABA in crude synaptosomal fraction from rat brain. *J. Neurochem. 30:* 47-56, 1978.

Tolbert, L. P.; Morest, D. K., and Yurgelun-Todd, D.: The neuronal architecture of the anteroventral cochlear nucleus of the cat in the region of the cochlear nerve root: Horseradish peroxidase labeling of identified cell types. *Neuroscience. 7:* 3031-3052, 1982.

Wenthold, R. J.: Glutamic and aspartic acid in subdivisions of the cochlear nucleus after auditory nerve lesion. *Brain Res. 143:* 544-548, 1978.

Wenthold, R. J.: Release of endogenous glutamic acid, aspartic acid, and GABA from cochlear nucleus slices. *Brain Res. 162:* 338-343, 1979.

Wenthold, R. J.: Glutamate and aspartate as transmitters for the auditory nerve. In Di Chiara, G., and Gessa, G. L. (Eds.): *Advances in Biochemical Psychopharmacology, Vol. 27: 'Glutamate as a Neurotransmitter'.* New York, Raven Press, 1981, pp. 69-78.

Wenthold, R. J.: Glutaminase and aspartate aminotransferase decrease in the cochlear nucleus after lesion of the auditory nerve. *Brain Res. 190:* 293-297, 1980.

Wenthold, R. J., and Gulley, R. L.: Aspartic and glutamic acid levels in the cochlear nucleus after auditory nerve lesion. *Brain Res. 138:* 111-123, 1977.

Wenthold, R. J., and Morest, D. K.: Transmitter related enzymes in the guinea pig cochlear nucleus. *Neurosci. Abstr. 2:* 28, 1976.

Werman, R.: Criteria for identification of a central nervous system transmitter. *Comp. Biochem. Physiol. 18:* 745-766, 1966.

Wiklund, L.; Toggenburger, G., and Cuenod, M.: Aspartate: Possible neurotransmitter in cerebellar climbing fibers. *Science 216:* 78-80, 1982.

Young, A. B.; Oster-Granite, M. L. Herndon, R. M., and Snyder, S. H.: Glutamic acid: Selective depletion by viral induced granule cell loss in hamster cerebellum. *Brain Res. 73:* 1-13, 1974.

Young, E. D.: Identification of response properties of ascending axons from dorsal cochlear nucleus. *Brain Res. 200:* 23-37, 1980.

Young, E. D., and Brownell, W. E.: Responses to tones and noise of single cells in dorsal cochlear nucleus of unanesthetized cats. *J. Neurophysiol. 39:* 282-300, 1976.

Young, E. D., and Voigt H. F.: Response properties of type II and type III units in dorsal cochlear nucleus. *Hearing Res. 6:* 153-169, 1982.

Chapter 10

CHOLINERGIC NEUROTRANSMISSION IN THE COCHLEAR NUCLEUS

Donald A. Godfrey, Jami L. Park, Jon D. Dunn, and C. David Ross

I. Introduction
II. Enzymes of Acetylcholine Metabolism: Visualization with Stains
III. Acetylcholine and Its Related Enzymes: Quantitative Studies
IV. Release of Acetylcholine
V. Receptors for Acetylcholine
VI. Role of Acetylcholine
VII. Conclusions
References

I. INTRODUCTION

Acetylcholine is one of the best established of the chemicals utilized by vertebrate neurons for intercellular communication, or neurotransmission, at synapses. Through many years of experimentation, it has become generally accepted that motoneurons innervating skeletal muscles, preganglionic autonomic neurons, and the postganglionic parasympathetic autonomic neurons are cholinergic, i.e., using acetylcholine as transmitter, and in the course of these experiments, the chemistry and pharmacology of acetylcholine have been extensively studied (Hebb, 1957, 1963; Gilman et al., 1980; MacIntosh, 1981; Cooper et al., 1982). More recently, transmitter functions for acetylcholine in interneurons of the corpus striatum and in basal forebrain neurons projecting to the cerebral cortex have been studied in relation to Parkinson's and Alzheimer's disease, respectively (Coyle et al., 1983; Lehmann and Langer, 1983). Within vertebrate sensory systems, acetylcholine as a neurotransmitter has, in general, been less well studied, perhaps because its role seems to be less prominent than in motor systems. Nevertheless, one of the better established vertebrate cholinergic neuronal pathways is that from the superior olive to the cochlea (olivocochlear) of the auditory system (Klinke, 1981; Guth and Melamed, 1982). A major advantage of studying transmitter chemistry in the initial brain structures of sensory systems is that the data can be correlated with physiological and behavioral data which are more definitive than for other parts of the brain. This is especially true for the auditory and visual systems.

The purpose of this chapter is to review the evidence concerning acetylcholine as a neurotransmitter in the first brain center of the auditory system, the cochlear nucleus. All projections from the cochlea to the brain, via the auditory division of the eighth cranial nerve, terminate at synapses within the cochlear nucleus. Here the coded information from the cochlea is re-coded in a variety of ways (Kiang et al., 1973) through synaptic convergence onto neurons, not only of the auditory nerve fibers but also of fibers from cochlear nucleus interneurons and fibers from neurons in other parts of the brain (Morest et al., 1973; Lorente de Nó 1981). The chemicals released as synaptic transmitters at any of these millions of synapses have only begun to be investigated. A simplifying assumption in the attempts to identify these chemicals is that all synapses of a particular neural pathway, such as the auditory nerve, would release the same neurotransmitter. This working assumption, based on what has previously been found for other neuronal systems, such as the skeletal motoneurons and parasympathetic neurons, may not be entirely true, but at least provides a first approximation around which studies can be organized. In approaching the neurotransmitter chemistry of a region like the cochlear nucleus where little is known, two major approaches can be followed: (1) try to identify the transmitter(s) of its most prominent synapses, in this case those of the auditory nerve, or (2) try to identify which synapses release a certain well-established transmitter, such as acetylcholine. This chapter follows the second approach.

Indicators of cholinergic neurotransmission include (1) the presence of acetylcholine and its related enzymes, choline acetyltransferase (ChAT) and acetylcholinesterase (AChE) (Fig. 1), (2) evidence for synaptic release of acetylcholine, (3) evidence for receptors with which acetylcholine interacts to produce effects on neuronal electrical activity, and which can be influenced by drugs affecting well-established cholinergic receptors. The following sections will present information relative to these various indicators.

II. ENZYMES OF ACETYLCHOLINE METABOLISM: VISUALIZATION WITH STAINS

Although AChE is not as definitive a marker for cholinergic neurons as ChAT (Fig. 1), methods for its measurement and visualization were established earlier, based on using acetyl thiocholine as substrate, which forms colored products related to the sulfur group of thiocholine (Koelle, 1963).

Rasmussen (1960, 1964, 1967) noted staining for AChE activity in two

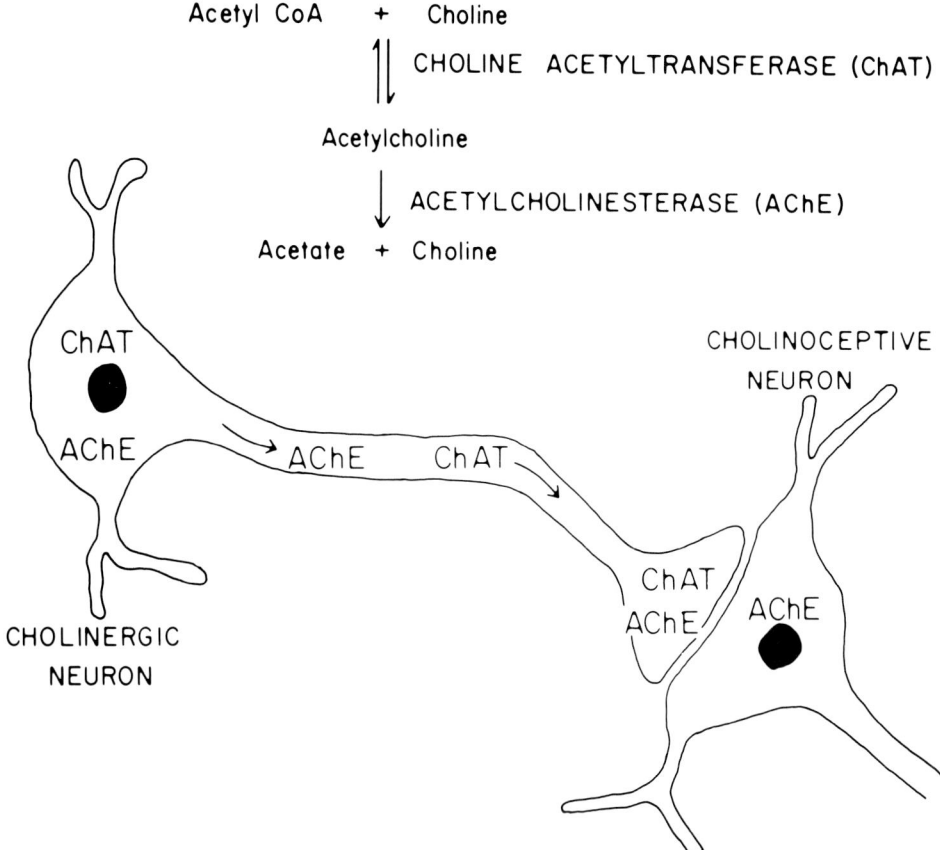

Figure 10-1. Neuronal localization of the enzymes of acetylcholine metabolism. Both the synthetic enzyme, choline acetyltransferase (ChAT), and the degradative enzyme, acetylcholinesterase (AChE), are located in the somata of cholinergic neurons, where they are made, and in the axons, wherein they are transported to the axonal terminals to regulate availability of acetylcholine for synaptic transmission. AChE is additionally present in noncholinergic neurons receiving cholinergic synapses, i.e., cholinoceptive neurons, where it is involved in termination of acetylcholine action at the postsynaptic membrane.

centrifugal pathways to the cat cochlear nucleus: (1) branches from the olivocochlear bundle and (2) a connection from the lateral superior olivary nucleus via the trapezoid body. He further mentioned that the latter connection was the more prominent. Stain was not seen in the auditory nerve fibers or fibers leaving the cochlear nucleus for the superior olivary complex. Subsequent workers have confirmed the staining of the olivocochlear branches and the lack of stain in the auditory nerve fibers and fibers projecting from the cochlear nucleus, but not the AChE-positive direct connection from the lateral superior olive (Osen and Roth, 1969; Brown and Howlett, 1972;

Moore and Osen, 1979; Osen and Mugnaini, 1980; Martin, 1981).

Within the cochlear nucleus itself, dark staining for AChE activity has been observed in the granular regions and in the superficial molecular and fusiform soma layers of the dorsal cochlear nucleus of the cat (Rasmussen, 1967; Osen and Roth, 1969), but only in the granular regions of the mouse (Martin, 1981) and rat (Osen and Mugnaini, 1980; Godfrey and Matschinsky, 1981), and nowhere in the human cochlear nucleus, which essentially lacks dorsal cochlear nucleus layering and granular regions (Moore and Osen, 1979). In an electron microscopic study of the anteroventral cochlear nucleus of cat and chinchilla, some non-auditory-nerve synaptic terminals, especially in the granular layer, stained for AChE activity (McDonald and Rasmussen, 1971), as well as some spindle-shaped neuronal somata. The staining pattern in these spindle-shaped somata resembles that for more well-established cholinergic neuronal somata elsewhere in the brain (Lewis and Shute, 1966).

Recently, stains for ChAT-like immunoreactivity have been applied to the brain (Kimura *et al.*, 1981; Armstrong *et al.*, 1983), but no definitive information about the cochlear nucleus has so far been obtained. For example, one report (Kimura *et al.*, 1981) shows "cholinoceptive" neurons in the cat cochlear nucleus, but the report's reliability is made questionable by the many "cholinergic" neurons shown in the lateral superior olivary nucleus, which has very low ChAT activity by direct quantitative assay (see Section III). In general, a close, but not complete, correspondence has been found in rat brain between neuronal somata which stain for ChAT-like immunoreactivity and those which stain "intensely" for AChE activity (Eckenstein and Sofroniew, 1983; Levey *et al.*, 1983; Satoh *et al.*, 1983).

III. ACETYLCHOLINE AND ITS RELATED ENZYMES: QUANTITATIVE STUDIES

Measurements of acetylcholine level as an indicator of the prominence of cholinergic mechanisms are limited because of its rapid changes in postmortem tissue (Kumar, 1973; Weintraub *et al.*, 1976), in contrast to the stability of ChAT and AChE activities (McGeer and McGeer, 1976; Godfrey and Matschinsky, 1981), its variability with the state of the animal (Modak *et al.*, 1976), and lack of an assay applicable to such small tissue samples as can be analyzed for ChAT (Fonnum, 1975) or AChE (McCaman *et al.*, 1968) activity. However, acetylcholine level is an important marker when studying the short-term dynamics of a cholinergic system. Comis and Davies (1969) measured the acetylcholine content of the cat cochlear nucleus, in the presence of hemicholinium to block acetylcholine synthesis, with and without electrical stimulation of the ipsilateral superior olive, and found more than an 80% decrease of acetylcholine content with olivary stimulation. (Acetylcholine released by the stimulation should be destroyed by AChE and thereby lost from the system.) This result is consistent with a cholinergic olivo-cochlear nucleus pathway, but could also fit with a non-cholinergic pathway activating cholinergic interneurons in the cochlear nucleus. Consistent with the AChE-staining results suggesting that the auditory nerve is not cholinergic, Comis and Davies (1969) also found that the acetylcholine content of the cochlear nucleus was not decreased during sound stimulation in cats with bilateral destruction of the superior olivary complex.

Probably the most convincing evidence that the auditory nerve contains few, if any, cholinergic fibers comes from quantitative measurements of ChAT activity. Although all cholinergic tracts examined contain high activity of this enzyme, the activity in the auditory nerve and its roots in the cochlear nucleus is extremely low in guinea pig (Wenthold and Morest, 1976; Fex and Wenthold, 1976), rat (Figs. 2, 3) (Godfrey and Matschinsky, 1981), cat (Fig. 3) (Godfrey et al., 1977), and chinchilla (unpublished observations). Furthermore, cochlear ablation had no significant effect on ChAT activity in the guinea pig cochlear nucleus (Wenthold and Morest, 1976).

Quantitative mapping of ChAT and AChE activities in the cochlear nucleus has provided several clues about the role of acetylcholine as a transmitter (Godfrey and Matschinsky, 1976, 1981; Godfrey et al., 1977). Although these procedures cannot attain the ultrastructural localization of staining methods, they can reach histological resolution (20 μm linear resolution), and the results are much less subject to personal impressions and interpretations. The overall activity of ChAT in the cochlear nucleus is consistent with a moderate role for acetylcholine (Fig. 2). Activities tend to be highest in parts of the anteroventral cochlear nucleus, including the overlying granular layer, and in the fusiform soma layer of the dorsal cochlear nucleus, but the distribution pattern may vary considerably across species (Fig. 3) (Godfrey et al., 1977; Godfrey and Matschinsky, 1981). The distribution of AChE activity within the cochlear nucleus agrees qualitatively with the staining pattern for AChE activity, including the very high activities in the molecular and fusiform soma layers of the cat. As in cerebellar cortex, these high AChE activities are out of proportion with much less impressive ChAT activities (Godfrey et al., 1977).

The derivations of ChAT and AChE activities in the cochlear nucleus have been investigated by mapping their distributions in animals with variously placed lesions. This approach is illustrated in Figs. 4–8 for the rat, wherein studies have progressed farthest. A major advantage of the quantitative mapping approach for such studies is that it provides objective measurements of the magnitudes of differential effects in cochlear nucleus subregions related to specific lesion locations (Fig. 7). Possible complications in interpre-

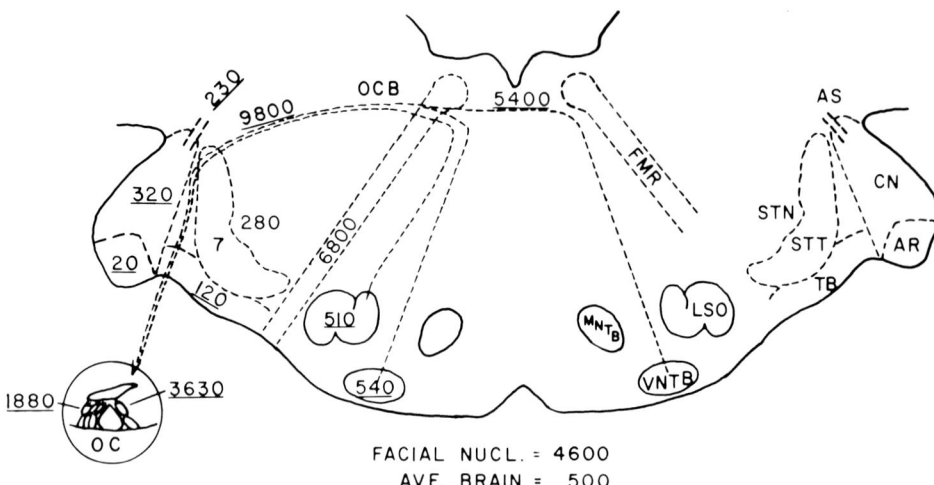

Figure 10-2. Schematic transverse section through the albino rat brainstem, showing average ChAT activities (μmol/kg dry wt/min) of auditory and nearby or related structures, based on previous publications (Godfrey and Matschinsky, 1981; Godfrey et al., 1983a, c, 1984). Activities for auditory structures are underlined. The value for the facial nucleus is included to represent a region of cholinergic somata, and the value for the facial motor root (FMR) represents a cholinergic tract. The value for the spinal trigeminal tract (STT) represents a non-cholinergic tract. The high activity of the olivocochlear bundle (OCB), actually including also centrifugal vestibular fibers, is included, both laterally and at the midline, as well as the activities of its nuclei of origin in the rat, the lateral superior olivary nucleus (LSO) and ventral nucleus of the trapezoid body (VNTB) (White and Warr, 1983), and of its major site of termination in the organ of Corti (OC). Activities for the three major tracts connecting to the cochlear nucleus (CN) are also shown: the major input connection via the auditory nerve root (AR) and the major output connections (which also contain some centrifugal inputs) via the trapezoid body (TB) and acoustic strias (AS). Other abbreviations are MNTB, medial nucleus of trapezoid body, and STN, spinal trigeminal nucleus. Not included in the picture is the average ChAT activity for the inferior colliculus: 105 μmol/kg dry wt/min.

tation of the results of such experiments from retrograde, transneuronal, or nonspecific effects have been discussed (Godfrey et al., 1983c).

The general results of the lesion studies, as summarized in Fig. 8, indicate that transection of virtually all centrifugal pathways to the cochlear nucleus, including both the olivocochlear bundle and trapezoid body components (columns 1-3), led to loss of 85-90% of ChAT activity in all cochlear nucleus subregions except granular regions, where 60-70% of the activity was lost. Cuts of the olivocochlear bundle sparing the trapezoid body (columns 4-7) led to consistent, but much less impressive losses of ChAT activity. When some of the trapezoid body was cut also (columns 8-9), there was a more impressive effect in some regions, especially at the rostrocaudal level of its interruption (Fig. 7). When the entire trapezoid body was cut laterally, there

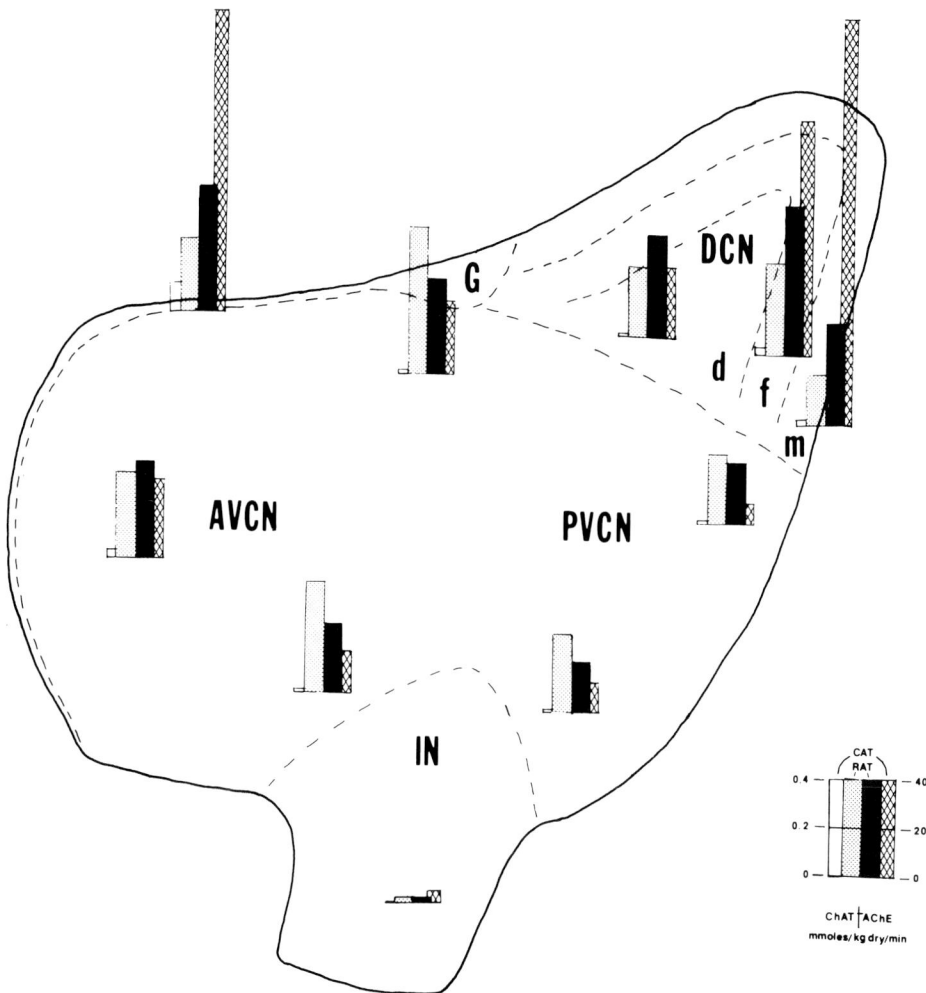

Figure 10-3. Average ChAT and AChE activities for subregions of rat and cat cochlear nucleus represented on a schematic side view of the cochlear nucleus. Note the generally much higher ChAT activities in the rat than in the cat, while AChE activities are more comparable, and the differences between rat and cat in the patterns of distribution of the enzyme activities. AVCN, anteroventral cochlear nucleus; DCN, dorsal cochlear nucleus, including molecular (m), fusiform soma (f), and deep (d) layers; G, granular region; IN, interstitial nucleus (auditory nerve root); PVCN, posteroventral cochlear nucleus.

was a prominent loss of ChAT activity even after 3 days (column 10), more impressive at 7 days (column 11). However, when the cut of the trapezoid body was farther medial, under the lateral superior olivary nucleus (column 12), the loss of ChAT activity in the cochlear nucleus was much less. A yet farther medial trapezoid body cut, near the medial nucleus of the trapezoid body (lesion 6 of Fig. 4), had little effect on ChAT activity in the cochlear nucleus (not shown). Destruction of the superior olivary complex (columns 13-16) had bilateral effects on ChAT activity in most cochlear nucleus regions, especially the dorsal cochlear nucleus. Any effect of cerebellar peduncle transection (lesion 7 of Fig. 4) was not obvious, and so is not shown in Fig. 8.

These findings suggest that ChAT activity in the rat cochlear nucleus is predominantly associated with two groups of centrifugal fibers (Fig. 9) which

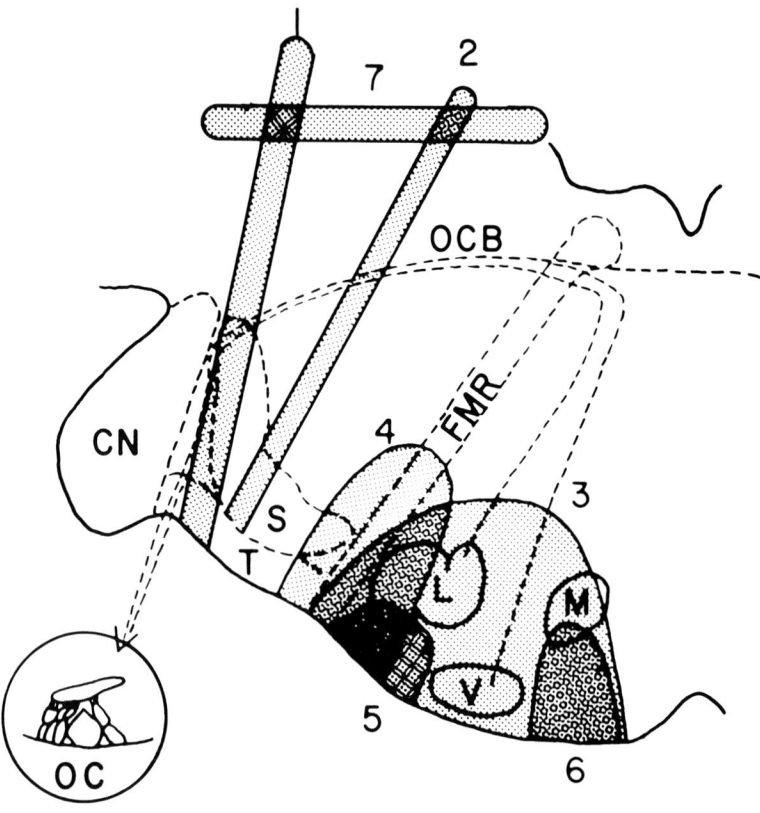

Figure 10-4. Schematic transverse half-section through the brain stem showing approximate locations of lesions whose effects upon ChAT activity in the rat cochlear nucleus were analyzed. CN, cochlear nucleus; FMR, facial motor root; L, lateral superior olivary nucleus; M, medial nucleus of trapezoid body; OC, organ of Corti; OCB, olivocochlear bundle; S, spinal trigeminal tract; T, trapezoid body; V, ventral nucleus of trapezoid body. Lesion 1 cuts virtually all centrifugal pathways to the cochlear nucleus; lesion 2 spares the trapezoid body and cerebellar peduncles; lesion 3 destroys the superior olivary complex; lesions 4-6 cut the trapezoid body at various mediolateral locations; lesion 7 cuts the cerebellar peduncles. All lesions are unilateral except 7, which includes both unilateral and bilateral cuts.

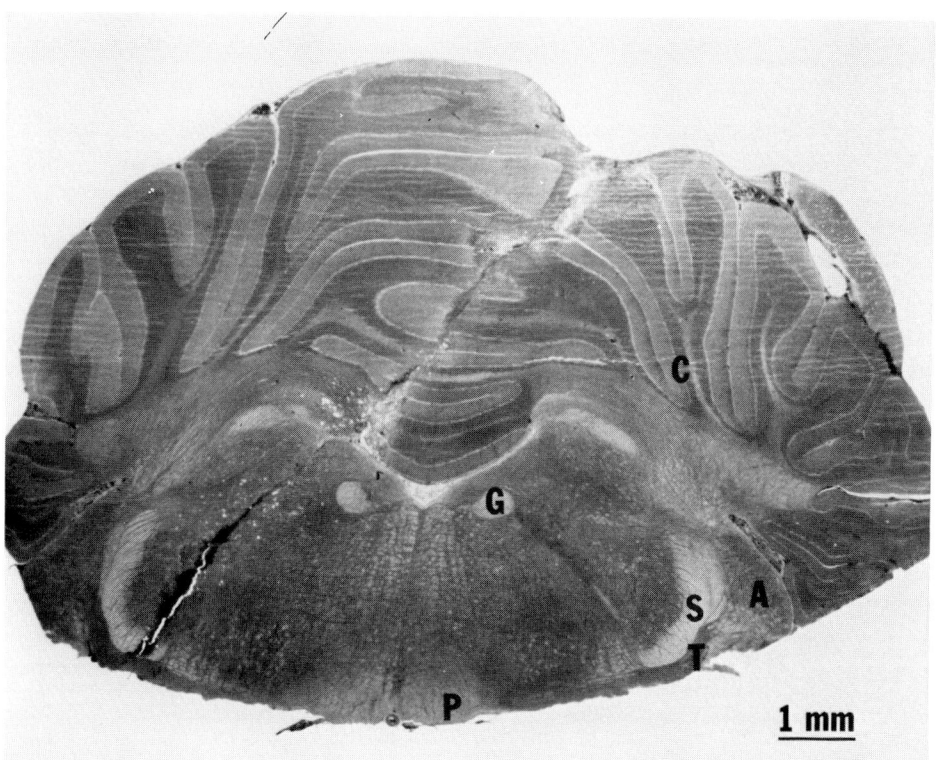

Figure 10-5. Unstained, freeze-dried, 20 μm-thick transverse section (section 93) through the cerebellum and brainstem of a rat (number 070982D) showing the path of a knife-cut lesion (type 2 of Fig. 4) interrupting the olivocochlear bundle and acoustic strias on one side, but sparing the trapezoid body. A, anteroventral cochlear nucleus; C, cerebellum; G, facial genu; P, pyramid; S, spinal trigeminal tract.

may be proposed on this basis to be cholinergic: branches from the olivocochlear bundle (White and Warr, 1983) may account for as much as 20% of the ChAT activity, but the major component (roughly 65%) of the activity seems related to fibers entering via the trapezoid body route. Such a conclusion is in line with Rasmussen's (1967) original suggestion, which has not been confirmed by the more recent studies of staining for AChE activity. In connection with Rasmussen's findings in the cat, preliminary chemical data for the cat appear similar to those for rat in that a large cut medial to the cochlear nucleus resulted in loss of most (present estimate 70%) of its ChAT activity.

The proposed cholinergic pathways would seem to emanate from, or else pass through, the vicinity of the superior olivary complex, about 55% from the ipsilateral side, 30% from the contralateral side (Fig. 9). There should be no argument about such an origin for collaterals from the olivocochlear

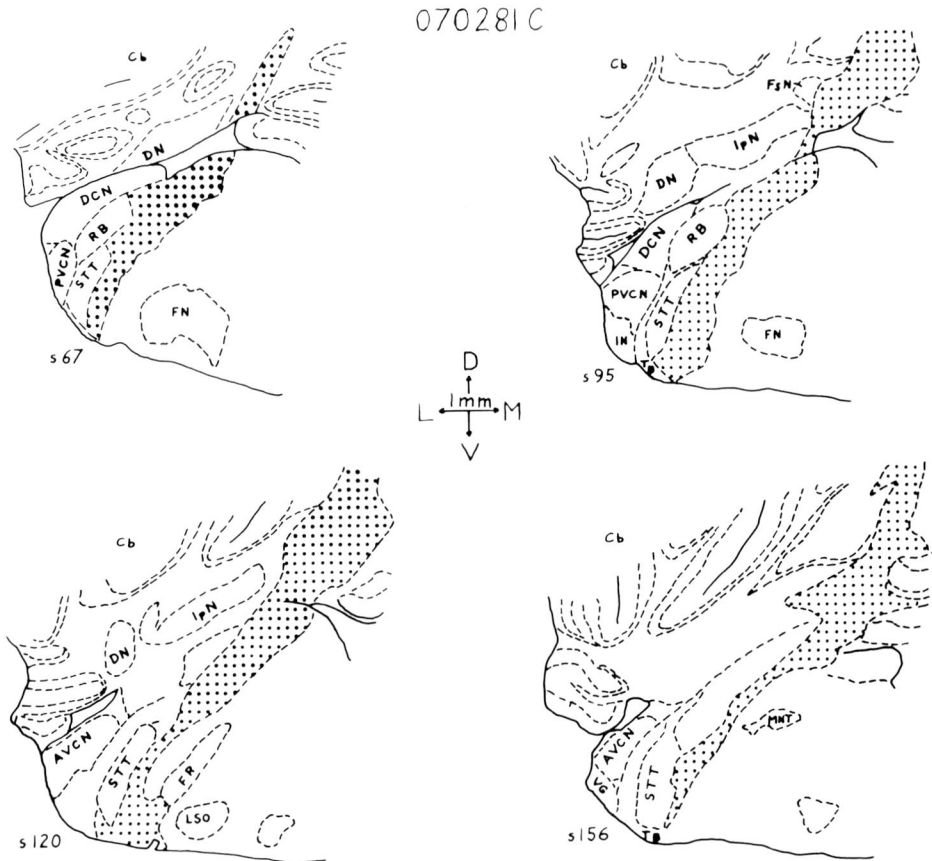

Figure 10-6. Tracings of four 20 μm-thick thionin-stained transverse sections through the brainstem and cerebellum of a rat (number 070281C) showing the location (stippled zone) of a lesion cutting the olivocochlear bundle, acoustic strias, and about half the rostrocaudal extent of the trapezoid body (trapezoid body present in sections 79-164, cut in 95-140). Sections are numbered consecutively from caudal to rostral. AVCN, anteroventral cochlear nucleus; Cb, cerebellum; DCN, dorsal cochlear nucleus; DN, dentate nucleus of cerebellum; FN, facial nucleus; FR, facial motor root; FsN, fastigial nucleus of cerebellum; IN, interstitial nucleus (auditory nerve root); IpN, interposed nucleus of cerebellum; LSO, lateral superior olivary nucleus; MNT, motor nucleus of trigeminal; RB, restiform body; STT, spinal trigeminal tract; TB, trapezoid body; VG, vestibular ganglion. Directional millimeter scale indicates dorsal (D), lateral (L), medial (M), and ventral (V) directions.

bundle to the cochlear nucleus (Adams, 1982; White and Warr, 1983), but the origin of fibers traveling by the trapezoid body route is more problematical. Rasmussen's suggestion of an origin from the lateral superior olivary nucleus in the cat does not fit well with very low ChAT activities measured in this

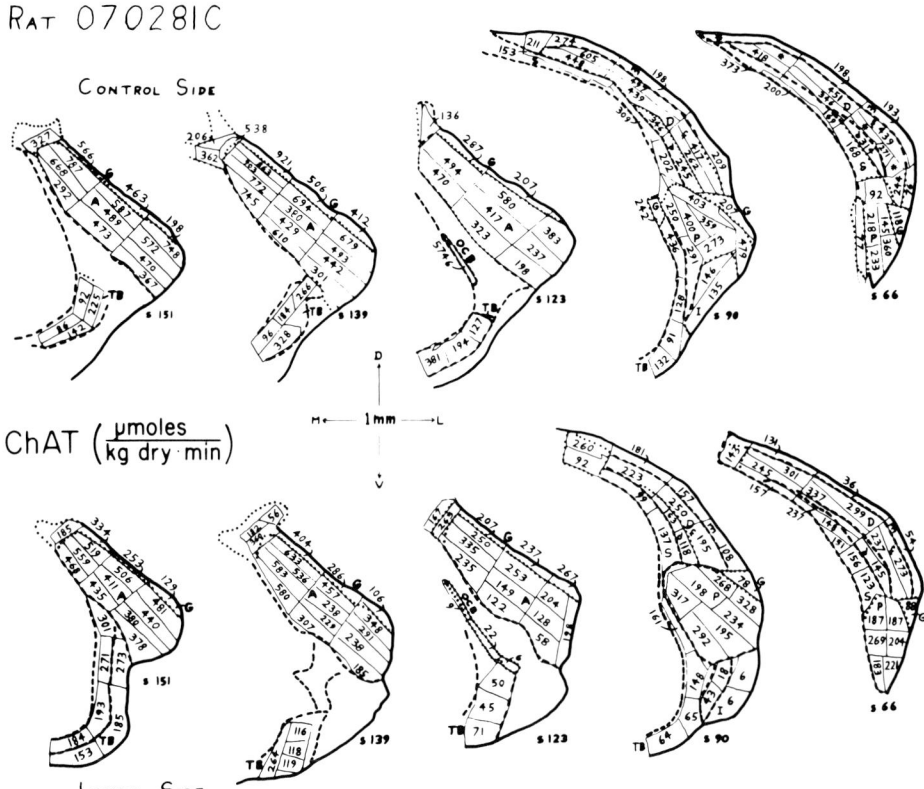

Figure 10-7. Distributions of ChAT activities in control and lesion-side cochlear nuclei of the rat with the lesion shown in Fig. 6. Four of the transverse sections are quite close to those traced for Fig. 6 (note section numbers). Thin lines are sample boundaries; thick lines are regional boundaries as seen in the freeze-dried sections themselves (solid and dashed) or adjacent thionin-stained or AChE-stained sections (dotted). Each matched pair of cochlear nucleus dissections was obtained from the same 20 μm-thick freeze-dried section (note same section number), and the activities measured in the same assay. Asterisks inside samples indicate that data were not obtained. A, anteroventral cochlear nucleus; D, dorsal cochlear nucleus, including molecular (m), fusiform soma (f), and deep (d) layers; G, granular region; I, interstitial nucleus (auditory nerve root); OCB, olivocochlear bundle (including also centrifugal vestibular fibers); P, posteroventral cochlear nucleus; S, acoustic strias; TB, trapezoid body. Note: (1) the dramatic loss of ChAT activity in the olivocochlear tract on the lesion side, (2) the loss of ChAT activity in the lesion-side trapezoid body in section 123, where it is cut by the lesion just medial to the location of samples (Fig. 6), (3) the greater reduction of ChAT activity in the lesion-side anteroventral cochlear nucleus in section 123 (50%), where the trapezoid body is cut, than in section 139 (37%), at the rostralmost extent of the trapezoid body transection, or section 151 (16%), rostal to the trapezoid body transection. Using a paired t-test for corresponding samples, lesion side activities could be shown to be significantly less ($p \leq 0.01$) than control side activities in large-cell portions of anteroventral cochlear nucleus of sections 123 and 139, granular region of sections 139 and 151 combined, and dorsal cochlear nucleus molecular and fusiform soma layers of sections 66 and 90 combined.

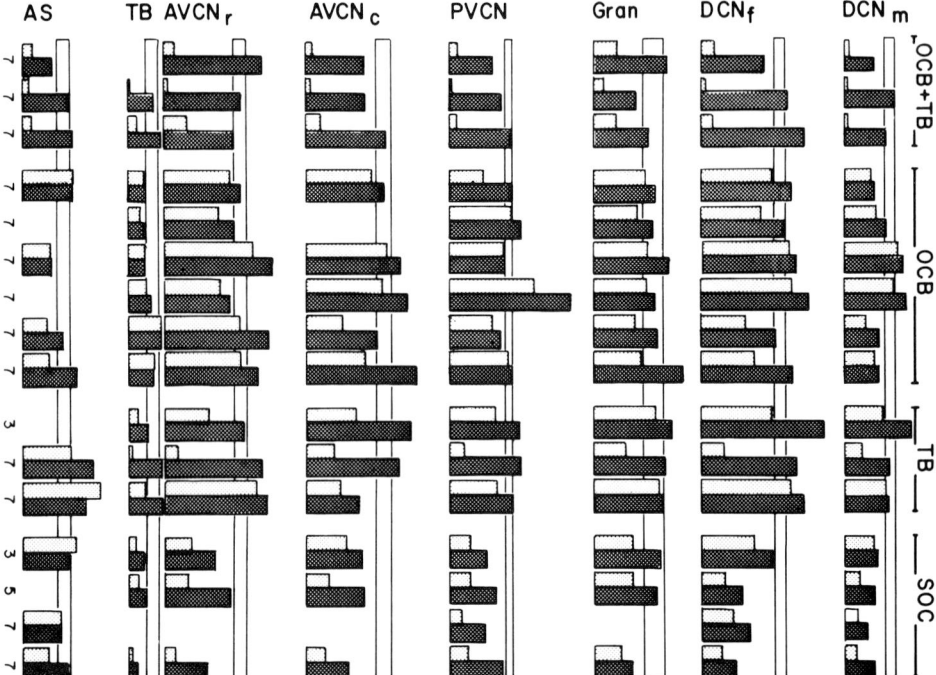

Figure 10-8. Summary of effects of lesions on ChAT activity in cochlear nucleus subregions and connecting tracts. Each row represents a particular region: DCN_m and DCN_f, molecular and fusiform soma layers of dorsal cochlear nucleus; Gran, granular region dorsolateral to large-cell region of anteroventral cochlear nucleus; PVCN, posteroventral cochlear nucleus (large-cell part); $AVCN_c$ and $AVCN_r$, caudal and rostral parts of anteroventral cochlear nucleus (large-cell parts); TB, trapezoid body; AS, acoustic strias. Each column of paired, lightly and darkly stippled bars represents a rat with a particular lesion. The darkly stippled bar gives the ChAT activity on the side contralateral to the lesion, while the lightly stippled bar gives the activity for the ipsilateral side. The long unshaded bar in each row gives the average ± standard error of control rat ChAT activity for each region. The first 3 columns represent lesions basically of type 1 of Fig. 4, cutting both olivocochlear bundle and trapezoid body (Godfrey et al., 1983c). The lesion of the first column spared the caudal 14% of the trapezoid body, while that of the third column spared the rostral 20% of the trapezoid body. The next 6 columns (4-9) represent lesions basically of type 2 of Fig. 4, cutting the olivocochlear bundle, but sparing trapezoid body. A section through the lesion of column 7 is shown in Fig. 5. Actually the lesion of column 8 cut the central half of the trapezoid body, sparing the caudal 19% and rostral 28% (Fig. 6), while that in column 9 cut a fourth of the trapezoid body, sparing the caudal 12% and the rostral 63%. The next 3 columns (10-12) represent lesions cutting the trapezoid body, but sparing the olivocochlear bundle, types 4 (10 and 11) and 5 (12) of Fig. 4. The lesion of column 11 damaged the medial part of the ventral cochlear nucleus and the lateral part of the lateral superior olivary nucleus, while that of column 10 did not. The last 4 columns (13-16) represent lesions destroying the superior olivary complex. That of column 15 also destroyed the rostral part of the cochlear nucleus. The number at the bottom of each column is the survival time, in days, between placement of the cut and sacrifice of the rat.

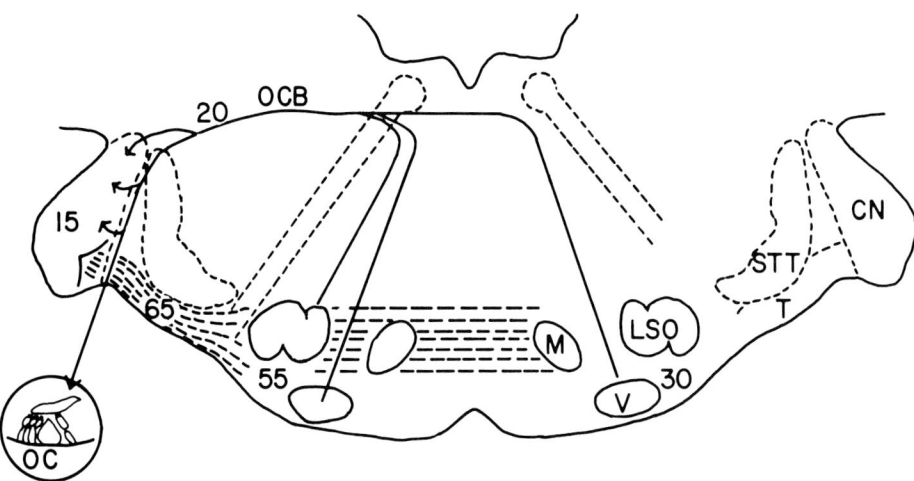

Figure 10-9. Schematic transverse section through the rat brainstem summarizing some conclusions of the lesion study concerning routes and origins of ChAT-containing centrifugal pathways to the cochlear nucleus. The percentages shown are estimates based on presently available data. Other pathways not shown might also make minor contributions. See text for description. CN, cochlear nucleus; LSO, lateral superior olivary nucleus; M, medial nucleus of trapezoid body; OC, organ of Corti; OCB, olivocochlear bundle; STT, spinal trigeminal tract; T, trapezoid body; V, ventral nucleus of trapezoid body.

nucleus (average 7 μmol/kg dry wt/min compared to 74 μmol/kg dry wt/min in dorsal hilus and 1,324 μmol/kg dry wt/min in the facial nucleus of the same cat). An origin from the AChE-positive cells adjacent to the lateral superior olivary nucleus does not seem to fit very well with their labeling by retrograde transport from the cochlea (Warr, 1975). Similarly in the rat, the AChE-positive somata in the lateral superior olivary nucleus, which do not stain as prominently as those of the ventral nucleus of the trapezoid body and have therefore not been noted by all workers (Brown and Howlett, 1972), seem related to the olivocochlear system (White and Warr, 1983). An attempt to destroy cholinergic neurons projecting from the superior olivary complex to the cochlear nucleus, without affecting fibers passing through, by injection of kainic acid (Masterton et al., 1979), was unsuccessful, as ChAT-containing neurons of the superior olive seem quite resistant to kainic acid (Godfrey et al., 1983b).

Although the origin of a ChAT-containing pathway to the cochlear nucleus via the trapezoid body remains thus unresolved, there are still numerous candidates among the many neurons projecting from the superior olivary complex to the cochlear nucleus (Elverland, 1977; Farley and Warr, 1981; Adams, 1983; Spangler et al., 1983). The immunohistochemical staining method for ChAT, together with retrograde labeling methods, may hold the most promise for resolving this issue. Some possibilities to consider include: the pathway might exist in the rat, but not in the cat; the somata of origin might not stain intensely for AChE activity (Levey et al., 1983; Eckenstein and Sofroniew, 1983); some olivocochlear neurons might have collaterals from near their somata that follow a trapezoid body route to the cochlear nucleus; or some of the axons of neurons innervating the stapedius muscle of the middle ear might send collaterals through the trapezoid body to the cochlear nucleus before exiting the brain in the facial nerve. Many somata of these stapedius motoneurons are located close to the superior olivary complex (Fullerton et al., 1983; Shaw and Baker, 1983), stain for AChE activity (Fullerton et al., 1983), and would be expected, like other motoneurons, to be cholinergic.

IV. RELEASE OF ACETYLCHOLINE

Only one study has measured release of acetylcholine from the cochlear nucleus during stimulation of centrifugal pathways (Comis and Guth, 1974). A small increase in acetylcholine release during stimulation was detected from the dorsal cochlear nucleus, but decreased release from the ventral cochlear nucleus. These disappointing results, compared to those obtained for the cochlea (Norris and Guth, 1974), may relate to the stimulation being applied only to the crossed olivocochlear bundle. If the connections in the cat are like those in the rat (Figs. 8, 9), then the crossed olivocochlear bundle would contribute only a minor fraction of the cholinergic synapses in the cochlear nucleus, and this contralateral connection might be most prominent in the dorsal cochlear nucleus. Stimulation within the superior olivary complex, as in the experiments of Comis and Davies (1969), might give a more clear-cut result. A second problem, especially related to the failure to demonstrate release in the ventral cochlear nucleus, may have resulted from difficulties with the bioassay used for acetylcholine.

V. RECEPTORS FOR ACETYLCHOLINE

In the peripheral nervous system, each major type of cholinergic synapse studied—skeletal neuromuscular junction, autonomic ganglion, and postganglionic parasympathetic—has a different type of cholinergic receptor (nicotinic types II and I, and muscarinic, respectively), based on their responses to various cholinergic agonists and antagonists (Gilman et al., 1980). It is not surprising, then, that cholinergic receptors of the central nervous system may have some unique characteristics differing from those of peripheral receptors (Morley et al., 1979; Barnard and Dolly, 1982). Nevertheless, as a first approximation in studying central cholinergic receptors, nicotinic and muscarinic types have been sought using chemicals verified primarily in the peripheral nervous system. Two approaches to studying cholinergic receptors in the central nervous system have been by quantitative measurement or visualization of binding sites and by studying the effects of microapplication to neurons of cholinergic agonists and antagonists.

Binding studies using α-bungarotoxin, which blocks the nicotinic acetylcholine receptor at the neuromuscular junction, have reported moderate amounts of binding (Morley et al., 1977, Chapter 13, this volume) through out the rat cochlear nucleus (Hunt and Schmidt, 1978) and some binding in most parts of the mouse cochlear nucleus (Arimatsu et al., 1981). A study of the binding of 3-quinuclidinylbenzilate (QNB), a muscarinic antagonist, has indicated relatively large amounts of binding in superficial parts of the rat cochlear nucleus (Wamsley et al., 1981). Whipple and Drescher (1984) reported that the density of muscarinic receptors, measured by QNB binding, was 1.6 times higher in the dorsal cochlear nucleus of the guinea pig than in the ventral cochlear nucleus.

Microapplication of acetylcholine while extracellularly recording discharges of single neurons in the ventral cochlear nucleus has rather consistently resulted in either increased or unchanged rate of discharge in cat (Comis and Whitfield, 1968; Comis, 1970; Martin and Adams, 1979) and chinchilla (Caspary et al., 1983). Two groups reported excitation in about one-fourth of the tested neurons (Martin and Adams, 1979; Caspary et al., 1983), while the other reported sensitivity in "a high percentage" of neurons (Comis and Whitfield, 1968). In the dorsal cochlear nucleus of the chinchilla, the predominant effect of applied acetylcholine, where there was any, was decreased firing rate of neurons (about 40% of tested neurons) (Caspary et al., 1983). No significant effects on the patterns of neuronal discharge were noted (Caspary et al., 1983). These fairly modest effects of applied acetylcholine are consistent with the moderate activities of ChAT in most parts of the cochlear nucleus (Section III). The decrease in neuronal discharge rate found with application of acetylcholine in the dorsal cochlear nucleus might correlate with the decrease in neuronal activity found with electrical stimulation of the crossed olivocochlear bundle fibers (Starr and Wernick, 1968), but the decreased neuronal activity in the ventral cochlear nucleus during the olivocochlear stimulation does not fit with the usually excitatory influence of acetylcholine there. This may relate to indirect effects of olivocochlear bundle stimulation mediated through its effect on the cochlea, especially since, following cochlear destruction, the effect of olivocochlear stimulation on neuronal spontaneous activity in the cochlear nucleus was predominantly excitatory rather than inhibitory (Starr and Wernick, 1968). Supporting this suggestion is the finding by Comis (1970) that blockage of the olivocochlear effect on the cochlea by application of strychnine to the cochlear round window sometimes abolished the inhibition of cochlear nucleus neuronal activity by electrical stimulation in parts of the superior olivary complex. Comis and Whitfield (1968) correlated excitation of anteroventral cochlear nucleus neurons by acetylcholine with excitation by electrical stimulation in the vicinity of the lateral superior olivary nucleus. Furthermore, activation by olivary stimulation could be blocked by microapplication to neurons of atropine (a muscarinic blocker), or of dihydro-β-erythroidine or gallamine (neuro-muscular nicotinic blockers). The possibility of muscarinic receptor involvement is supported by the reports of much slower onset and longer duration of action of acetylcholine as compared to excitatory (Martin and Adams, 1979) or inhibitory (Caspary et al., 1983) amino acids. However, Caspary et al (1983) found no consistent antagonism of the acetylcholine effects by atropine, and irreversible excitation of both dorsal and ventral cochlear nucleus neurons by the neuromuscular nicotinic blocker d-tubocurarine. Thus, the available pharmacological findings leave the nature of cholinergic receptors in the cochlear nucleus generally uncharacterized.

VI. ROLE OF ACETYLCHOLINE

The results presented in the earlier sections suggest that acetylcholine has a modest role in the processing of auditory information throughout the cochlear nucleus and that this primarily involves feedback modulation of the ascending pathway.

A small portion of the proposed cholinergic influence would seem to involve a side-branch of the feedback system affecting the organ of Corti in the cochlea. Whether the major (trapezoid body) portion is solely related to the cochlear nucleus, or is also part of a larger system remains to be determined. The behavioral effect of such a feedback influence might be expected to be fairly subtle, and has been investigated in cats by examining the effect of applying atropine onto the cochlear nucleus at its lateral surface (Pickles and Comis, 1973). The assumption, based on the previous experiments of this group (Comis and Whitfield, 1968), was that this should block cholinergic synapses in the cochlear nucleus, although one would technically expect only muscarinic synapses to be blocked. A consistent effect was found upon the cats' abilities to discriminate signals in noise (Pickles and Comis, 1973), which was later conceptualized as a role for cholinergic centrifugal pathways in determining critical bandwidth (Pickles, 1976).

VII. CONCLUSIONS

From the results presented it is obvious that many more experiments will be necessary before the significance of cholinergic synapses in the cochlear nucleus can be fully understood. The clarification of the across-species differences in distributions of ChAT and AChE activities, identification of the various neurons contributing to cholinergic synapses in the cochlear nucleus, need of obtaining more definitive data on acetylcholine release, need to characterize the exact nature of the cholinergic receptors, and the need for further clarification of the behavioral effects of the cholinergic pathways on hearing are only a few of the remaining questions, whose answers will require cooperative efforts of investigators with varieties of expertise.

One might question the value of pursuing the investigation of a neurotransmitter with a relatively modest role in the cochlear nucleus, compared to that of the auditory nerve transmitter. However, when exploring a virtually unknown area, the opportunity to begin with an aspect for which the tools of solution are to a considerable extent already available, should not be overlooked. Understanding the role of the cholinergic system in the cochlear nucleus can provide a solid foundation from which to explore more difficult aspects of its transmitter chemistry.

ACKNOWLEDGMENTS

Supported by Oral Roberts University intramural funds and NIH grant NS 17176. We are grateful to Katrina Beranek for technical assistance and production of figures, to Kelly Ferrell for typing, and to the Oral Roberts University Photography and Word Processing Centers for preparation of the manuscript.

REFERENCES

Adams, J. C.: Collaterals of labyrinthine efferent axons. *Soc. Neurosci. Abstr. 8:* 149, 1982.

Adams, J. C.: Cytology of periolivary cells and the organization of their projections in the cat. *J. Comp. Neurol. 215:* 275–289, 1983.

Arimatsu, Y.; Seto, A., and Amano, T.: An atlas of α-bungarotoxin binding sites and structures containing acetylcholinesterase in the mouse central nervous system. *J. Comp. Neurol. 198:* 603–631, 1981.

Armstrong, D. M.; Saper, C. B., Level, A. I., Wainer, B. H., and Terry, R. D.: Distribution of cholinergic neurons in rat brain: demonstrated by the immunocytochemical localization of choline acetyltransferase. *J. Comp. Neurol. 216:* 53–68, 1983.

Barnard, E. A., and Dolly, J. O.: Peripheral and central nicotinic ACh receptors—how similar are they? *Trends Neurosci. 5:* 325–327, 1982.

Brown, J. C., and Howlett, B.: The olivocochlear tract in the rat and its bearing on the homologies of some constituent cell groups of the mammalian superior olivary complex: a thiocholine study. *Acta Anat. 83:* 505–526, 1972.

Caspary, D. M.; Havey, D. C., and Faingold, C. L.: Effects of acetylcholine on cochlear nucleus neurons. *Exp. Neurol. 82:* 491–498, 1983.

Comis, S. D.: Centrifugal inhibitory processes affecting neuroens in the cat cochlear nucleus. *J. Physiol. 210:* 751-760, 1970.

Comis, S. D., and Davies, W. E.: Acetylcholine as a transmitter in the cat auditory system. *J. Neurochem. 16:* 423-429, 1969.

Comis, S. D. and Guth, P. S.: The release of acetylcholine from the cochlear nucleus upon stimulation of the crossed olivo-cochlear bundle. *Neuropharmacol. 13:* 633-641, 1974.

Comis, S. D., and Whitfield, I. C.: Influence of centrifugal pathways on unit activity in the cochlear nucleus. *J. Neurophysiol. 31:* 62-68, 1968.

Cooper, J. R.; Bloom, F. E., and Roth, R. H.: *The Biochemical Basis of Neuropharmacology.* New York, Oxford, 1982.

Coyle, J. T.; Price, D. L., and DeLong, M. R.: Alzheimer's disease: a disorder of cortical cholinergic innervation. *Science 219:* 1184-1190, 1983.

Eckenstein, F., and Sofroniew, V.: Identification of central cholinergic neurons containing both choline acetyltransferase and acetylcholinesterase and of central neurons containing only acetylcholinesterase. *J. Neurosci. 3:* 2286-2291, 1983.

Elverland, H. H.: Descending connections between the superior olivary and cochlear nuclear complexes in the cat studied by autoradiographic and horseradish peroxidase methods. *Exp. Brain Res. 27:* 397-412, 1977.

Farley, G. R., and Warr, W. B.: Some recurrent projections of the superior olive to anteroventral and dorsal cochlear nuclei in cat. *Soc. Neurosci. Abstr. 7:* 56, 1981.

Fex, J., and Wenthold, R. J.: Choline acetyltransferase, glutamate decarboxylase and tyrosine hydroxylase in the cochlea and cochlear nucleus of the guinea pig. *Brain Res. 109:* 575-585, 1976.

Fonnum, F.: Radiochemical assays for choline acetyltransferase and acetylcholinesterase. In Marks, N., and Rodnght, R. (Eds.): *Research Methods in Neurochemistry, Vol. 3.* New York, Plenum, 1975, pp. 253-275.

Fullerton, B. C.; Joseph, M. P., Guinan, J. J., Jr., and Norris, B. E.: Stepedius motoneurons in the cat. *Soc. Neurosci. Abstr. 9:* 1085, 1983.

Gilman, A. G.; Goodman, L. S., and Gilman, A.: *The Pharmacological Basis of Therapeutics.* New York, MacMillan, 1980.

Godfrey, D. A., and Matschinsky, F. M.: Approach to three-dimensional mapping of quantitative histochemical measurements applied to studies of the cochlear nucleus. *J. Histochem. Cytochem. 24:* 697-712, 1976.

Godfrey, D. A., and Matschinsky, F. M.: Quantitative distribution of choline acetyltransferase and acetylcholinesterase activities in the rat cochlear nucleus. *J. Histochem. Cytochem. 29:* 720-730, 1981.

Godfrey, D. A.; Park, J. L., Dunn, J. D., and Ross, C. D.: Choline acetyltransferase activity in the cochlea of the rat. *Physiologist 26:* A-104, 1983a.

Godfrey, D. A.; Park, J. L., Dunn, J. D., and Ross, C. D.: Effects of kainic acid on the superior olivary complex of the rat. *Soc. Neurosci. Abstr. 9:* 211, 1983b.

Godfrey, D. A.; Park, J. L., Rabe, J. R., Dunn, J. D., and Ross, C. D.: Effects of large brain stem lesions on the cholinergic system in the rat cochlear nucleus. *Hearing Res. 11:* 133-156, 1983c.

Godfrey, D. A.; Park, J. L., and Ross, C. D.: Choline acetyltransferase and acetylcholinesterase in centrifugal labyrinthine bundles of rats. *Hearing Res.* In press, 1984.

Godfrey, D. A.; Williams, A. D., and Matschinsky, F. M.: Quantitative histochemical mapping of enzymes of the cholinergic system in cat cochlear nucleus. *J. Histochem. Cytochem. 25:* 397-416, 1977.

Guth, P. S., and Melamed, B.: Neurotransmission in the auditory system: a primer for

pharmacologists. *Annu. Rev. Pharmacol. Toxicol.* 22: 383-412, 1982.
Hebb, C. O.: Biochemical evidence for the neural function of acetylcholine. *Physiol. Rev.* 37: 196-220, 1957.
Hebb, C. O.: Formation, storage, and liberation of acetylcholine. In Eichler, O., and Farah, A. (Eds.): *Handbuch Der Experimentellen Pharmakologie, Vol. V., Cholinesterases and Anticholinesterase Agents.* Berlin, Springer-Verlag, 1963, pp. 55-88.
Hunt, S., and Schmidt, J.: Some observations on the binding patterns of α-bungarotoxin in the central nervous system of the rat. *Brain Res.* 157: 213-232, 1978.
Kiang, N. Y. S.; Morest, D. K., Godfrey, D. A., Guinan, J. J., Jr., and Kane, E. C.: Stimulus coding at caudal levels of the cat's auditory nervous system: I. Response characteristics of single units. In Moller, A. R. (Ed.): *Basic Mechanisms in Hearing.* New York, Academic Press, 1973, pp. 455-478.
Kimura, H.; McGeer, P. L., Pen, J. H., and McGeer, E. G.: The central cholinergic system studied by choline acetyltransferase immunohistochemistry in the cat. *J. Comp. Neurol.* 200: 151-201, 1981.
Klinke, R.: Neurotransmitters in the cochlea and the cochlear nucleus. *Acta Otolaryng.* 91: 541-554, 1981.
Koelle, G. B.: Cytological distributions and physiological functions of cholinesterases. In Eichler, O., and Farah, A. (Eds.): *Handbuch Der Experimentellen Pharmakologie, Vol. V., Cholinesterases and Anticholinesterase Agents.* Berlin, Springer-Verlag, 1963, pp. 187-298.
Kumar, S.: Estimation of rat brain acetylcholine I. Effect of time lapse between killing and freezing of the brain. *Experientia* 29: 284, 1973.
Lehmann, J., and Langer, S. Z.: The striatal cholinergic interneuron: synaptic target of dopaminergic terminals? *Neuroscience* 10: 1105-1120, 1983.
Levey, A. I.; Wainer, B. H., Mufson, E. J., and Mesulam, M.-M.: Colocalization of acetylcholinesterase and choline acetyltransferase in the rat cerebrum. *Neuroscience* 9: 9-22, 1983.
Lewis, P. R., and Shute, C. C. D.: The distribution of cholinesterase in cholinergic neurons demonstrated with the electron microscope. *J. Cell Sci.* 1: 381-390, 1966.
Lorente de Nó, R.: *The Primary Acoustic Nuclei.* New York, Raven Press 1981.
MacIntosh, F. C.: Acetylcholine. In Siegel, G. J.; Albers, R. W., Agranoff, B. W., and Katzman, R. (Eds.): *Basic Neurochemistry.* Boston, Little, Brown and Company, 1981, pp. 183-204.
Martin, M. R.: Acetylcholinesterase-positive fibers and cell bodies in the cochlear nuclei of normal and reeler mutant mice. *J. Comp. Neurol.* 197: 153-167, 1981.
Martin, M. R., and Adams, J. C.: Effects of DL-α-aminoadipate on synaptically and chemically evoked excitation of anteroventral cochlear nucleus neurons of the cat. *Neuroscience* 4: 1097-1105, 1979.
Masterton, R. B.; Glendenning, K. K., and Hutson, K. A.: Preservation of trapezoid body fibers after biochemical ablation of superior olives with kainic acid. *Brain Res.* 173: 150-159, 1979.
McCaman, M. W.; Tomey, L. R., and McCaman, R. E.: Radiometric assay of acetylcholinesterase activity in submicrogram amounts of tissue. *Life Sci.* 7: 233-244, 1968.
McDonald, D. M., and Rasmussen, G. L.: Ultrastructural characteristics of synaptic endings in the cochlear nucleus having acetylcholinesterase activity. *Brain Res.* 28: 1-18, 1971.
McGeer, P. L., and McGeer, E. G.: Enzymes associated with the metabolism of catecholamines, acetylcholine and GABA in human controls and patients with Parkinson's disease and Huntington's chorea. *J. Neurochem.* 26: 65-76, 1976.
Modak, A. T.; Weintraub, S. E., McCoy, T. H., and Stavinoha, W. B.: Use of 300-msec microwave irradiation for enzyme inactivation: a study of effects of sodium pentobarbital

on acetylcholine concentration in mouse brain regions. *J. Pharmacol. Exp. Ther.* 197: 245–252, 1976.

Moore, J. K., and Osen, K. K.: The cochlear nuclei in man. *Am. J. Anat.* 154: 393–418, 1979.

Morest, D. K.; Kiang, N. Y. S., Kane, E. C., Guinan, J. J., Jr., and Godfrey, D. A.: Stimulus coding at caudal levels of the cat's auditory nervous system: II. Patterns of synaptic organization. In Moller, A. R. (Ed.): *Basic Mechanisms in Hearing.* New York, Academic Press, 1973, pp. 479–504.

Morley, B. J.; Kemp, G. E., and Salvaterra, P.: α-Bungarotoxin binding sites in the CNS. *Life Sci.* 24: 859–872, 1979.

Morley, B. J.; Lorden, J. F., Brown, G. B., Kemp, G. E., and Bradley, R. J.: Regional distribution of nicotinic acetylcholine receptor in rat brain. *Brain Res.* 134: 161–166, 1977.

Norris, C. H., and Guth, P. S.: The release of acetylcholine (ACH) by the crossed olivo-cochlear bundle (COCB). *Acta Otolaryng.* 77: 318–326, 1974.

Osen, K. K., and Mugnaini, E.: Acetylcholinesterase (AChE)-positive afferents to the cochlear nuclei in the rat. *Neurosci. Lett. Suppl.* 5: S 148, 1980.

Osen, K. K., and Roth, K.: Histochemical localization of cholinesterases in the cochlear nuclei of the cat, with notes on the origin of acetylcholinesterase-positive afferents and the superior olive. *Brain Res.* 16: 165–185, 1969.

Pickles, J. O.: Role of centrifugal pathways to cochlear nucleus in determination of critical bandwidth. *J. Neurophysiol.* 39: 394–400, 1976.

Pickles, J. O., and Comis, S. D.: Role of centrifugal pathways to cochlear nucleus in detection of signals in noise. *J. Neurophysiol.* 36: 1131–1137, 1973.

Rasmussen, G. L.: Anatomic relationships of the ascending and descending auditory systems. In Fields, W. S., and Alford, B. R. (Eds.): *Neurological Aspects of Auditory and Vestibular Disorders.* Springfield, Illinois, Charles C. Thomas, 1964, pp. 5–19.

Rasmussen, G. L.: Efferent connections of the cochlear nucleus. In Graham, A. B. (Ed.): *Sensorineural Hearing Processes and Disorders.* Boston, Little, Brown, & Co., 1967, pp. 61–75.

Rasmussen, G. L.: Efferent fibers of the cochlear nerve and cochlear nucleus. In Rasmussen, G. L., and Windle, E. F. (Eds.): *Neural Mechanisms of the Auditory and Vestibular Systems.* Springfield, Illinois, Charles C. Thomas, 1960, pp. 105–115.

Satoh, K.; Armstrong, D. M., and Fibiger, H. C.: A comparison of the distribution of central cholinergic neurons as demonstrated by acetylcholinesterase pharmacohistochemistry and choline acetyltransferase immunohistochemistry. *Brain Res. Bull.* 11: 693–720, 1983.

Shaw, M. D., and Baker, R.: The locations of stapedius and tensor tympani motoneurons in the cat. *J. Comp. Neurol.* 216: 10–19, 1983.

Spangler, K. S.; Henkel, C. K., and Cant, N.: Organization of descending projections from the superior olivary complex to the cochlear nuclei in the cat. *Soc. Neurosci. Abstr.* 9: 497, 1983.

Starr, A., and Wernick, J. S.: Olivocochlear bundle stimulation: effects on spontaneous and tone-evoked activities of single units in cat cochlear nucleus. *J. Neurophysiol.* 31: 549–564, 1968.

Wamsley, J. K.; Lewis, M. S., Young, W. S. III, and Kuhar, M. J.: Autoradiographic localization of muscarinic cholinergic receptors in rat brainstem. *J. Neurosci.* 1: 176–191, 1981.

Warr, W. B.: Olivocochlear and vestibular efferent neurons of the feline brain stem: their location, morphology and number determined by retrograde axonal transport and acetylcholinesterase histochemistry. *J. Comp. Neurol.* 161: 159–182, 1975.

Weintraub, S. T.; Modak, A. T., and Stavinoha, W. B.: Acetylcholine: postmortem increases in rat brain regions. *Brain Res.* 105: 179–183, 1976.

Wenthold, R. J., and Morest, D. K.: Transmitter related enzymes in the guinea pig cochlear nucleus. *Soc. Neurosci. Abstr. 2:* 28, 1976.

Whipple, M. R., and Drescher, D. G.: Muscarinic receptors in the cochlear nucleus and auditory nerve of the guinea pig. *J. Neurochem. 43:* 192–198, 1984.

White, J. S., and Warr, W. B.: The dual origins of olivocochlear neurons in the albino rat. *J. Comp. Neurol. 219:* 203–214, 1983.

Chapter 11

THE PHARMACOLOGY OF AMINO ACID RECEPTORS AND SYNAPTIC TRANSMISSION IN THE COCHLEAR NUCLEUS

MICHAEL R. MARTIN

I. Introduction
II. Methods
III. Excitatory Amino Acids
    A. Historical Background
        1. Structure-Activity Studies
        2. The Search for Antagonists
        3. Current Status
    B. The Pharmacology of Auditory Nerve Transmission
        1. *In Vivo* Studies in Mammals
        2. *In Vitro* Studies in Chicken
IV. Inhibitory Amino Acids
    A. Historical Background
    B. The Mammalian Cochlear Nucleus
V. Summary and Conclusions
References

I. INTRODUCTION

The purpose of this review is not to summarize all of our knowledge of cochlear nucleus pharmacology, but to focus on the types of amino acid receptors that are present and their possible function in synaptic transmission and integration of sensory information. The reasons for special interest in amino acids are that the excitatory amino acids glutamate (GLU) and aspartate (ASP) and the inhibitory amino acids γ-aminobutyrate (GABA) and glycine (GLY) are present in the cochlear nucleus (Godfrey *et al.*, 1977, 1978; Wenthold, 1978), are released in a calcium-dependent manner (Wenthold, 1979; Canzek and Reubi, 1980; Hansson *et al.*, 1980), and uptake mechanisms are present (Oliver *et al.*, 1981; Schwartz, 1981; Jones *et al.*, 1982). In addition, Whitfield and Comis (1966) demonstrated that

GLU excites and GABA depresses cochlear nucleus units. However, the question remains, are these phenomena nonspecific and related only to general cellular metabolism, or are they associated with synaptic transmission?

Over the years, a number of investigators have endeavored to develop criteria that can be helpful in establishing a transmitter role for a compound and differentiating between synaptic and nonsynaptic transmitter-like actions (see also Chapter 6, this volume). Two of these criteria are:

1. The action of the substance on the postsynaptic neuron should be identical to that of the natural transmitter.
2. This effect should be antagonized by drugs which specifically obstruct the postsynaptic action of the natural transmitter.

In both situations the substance must interact with the same postsynaptic receptor as the natural transmitter. The ability to determine whether this is the case is dependent on the number of different types of receptors present on the postsynaptic neuron that the substance activates, the specificity of action of antagonists, and the method of drug application. A discussion of these points will form the basis of this review.

II. METHODS

There are at least four possible routes for drug administration in *in vivo* preparations, intravenous (or intra-arterial) injection, topical application, pressure ejection from micropipettes, and microiontophoresis. There are advantages and disadvantages to each. Intravenous injections have the advantage that the drug may reach all receptors on the neuron equally. Thus, receptors on distal dendrites will be affected to the same degree as those more proximal. The disadvantage is that only strychnine (STR) and bicuculline (BIC), the inhibitory amino-acid antagonists, are capable of crossing the blood-brain barrier. Topical application onto the nucleus circumvents this problem, but the drug concentration will vary through the depth of the nucleus, and mechanical disturbances make single unit studies impractical. Pressure ejection and microiontophoresis from microelectrodes avoid mechanical disturbance problems, but drugs applied in this manner are restricted to acting only on receptors near the microelectrode tip, thus missing postsynaptic receptor sites located some distance from the soma. Of the two methods, microiontophoresis has been preferred because the necessary equipment requirements have been more readily met.

There are several components to the design of microiontophoresis protocols, each of which relates to the two criteria listed above for identifying transmitter substances. The first is to establish that the agonist actions are dependent on specific amino-acid receptors. To do this, two substances are alternately applied and their effects on neuronal activity compared. For example, on spinal Renshaw cells, excitatory amino acid-induced responses are compared to acetylcholine-induced responses. Equi-effective doses are applied until consistently repeatable controls for each drug are obtained. Then, either a cholinergic or an excitatory amino acid antagonist is applied during the alternating agonist applications and the differential response is observed. Finally, the antagonist application is terminated, and agonist-induced responses are allowed to return to control levels. This type of experiment establishes two things: first, the agonist behaves like the natural transmitter, either exciting or depressing postsynaptic activity, and second, the agonist-induced response is receptor-specific, as demonstrated by the differential effect of the antagonists.

The demonstration that a drug excites or depresses neuronal activity and is selectively sensitive to antagonists does not necessarily permit the conclusion that the associated receptors play a role in synaptic transmission. The receptors may well be extrasynaptic or related to unknown synapses or synapses not under investigation. To determine if a defined receptor is associated with a given synapse requires application of antagonists, whose receptor specificity has been determined against a series of agonists, during a period of synaptically-evoked activity. Ideally, all this information is gathered on the same units. In this way, the action of the agonist can be compared with that of the natural transmitter, the specificity of the antagonist defined, and the sensitivity of the natural synaptic transmitter determined, with the result that receptors with known characteristics can be correlated with the synaptic receptor.

Since most studies on receptor differentiation and synaptic transmission in the *in vitro* cochlear nucleus have been conducted on units in the anteroventral cochlear-nucleus (AVCN) division, it is important to review the physiology and anatomy of this region. Two types of units can be recognized physiologically: prepotential and non-prepotential. Prepotential units have a characteristic waveform associated with unit action potentials (Pfeiffer, 1966; Bourk, 1976). A "prepotential" precedes each action potential by approximately 0.5 ms; this prepotential has been shown to be presynaptic. Cells that receive large calyceal endings on their somata, i.e., the large spherical or bushy cells of the anterior AVCN with their tightly compacted dendritic tree (Brawer and Morest, 1975; Tolbert and Morest, 1978), have been associated with these prepotential waveforms. In other words, cells with auditory nerve synapses concentrated on their somata correspond to units with prepotentials. The remaining cells of the AVCN have long, branching dendrites (Brawer and Morest, 1975; Tolbert and Morest, 1978). Non-prepotential units, then, correspond to cells with auditory nerve terminals scattered along these long dendrites.

III. EXCITATORY AMINO ACIDS

A. HISTORICAL BACKGROUND

1. Structure-Activity Studies

During the course of an investigation of the effects of various central-nervous-system (CNS) metabolites (Curtis et al., 1960), it became obvious that a unique class of amino acids represented by GLU and ASP, distinct from compounds such as acetylcholine or norepinephrine, produced strong excitations of spinal neurons when applied microiontophoretically. However, the apparent absence of an enzymatic mechanism for removing the applied amino acids led these investigators to conclude that these substances were not excitatory transmitters. Later studies (Curtis and Watkins, 1963) tended to confirm this view since, unlike acetylcholine and norepinephrine, GLU and ASP indiscriminately excited all cells. Of the compounds tested, N-methyl-D-aspartate (NMDA) was the most potent. The significance of these findings remained obscure for a number of years. Nevertheless, structure-activity studies continued.

Initially, it was believed that GLU and ASP were equipotent. Duggan (1974) re-evaluated the relative potencies of these two amino acids and concluded that cat spinal interneurons, which were excited by dorsal root stimulation with a short latency, were slightly more sensitive to GLU than ASP, whereas ASP was the more effective of the two on Renshaw cells, which are believed not to receive monosynaptic primary afferent inputs. This was the first indication that there might be multiple receptors for excitatory amino acids. At about this time, the amino acid kainate (KA) was tested for the first time (Johnston et al., 1974) and found to be even more potent than NMDA. When the potency of KA and NMDA were compared on a similar group of spinal interneurons and Renshaw cells (McCulloch et al., 1974), the differential sensitivity was even more striking than that for GLU and ASP, giving rise to the notion of selective receptor agonists. The following year Biscoe et al. (1975) reported a third potent amino acid, quisqualate (QU).

2. The Search for Antagonists

Following the initial discovery of the action of GLU and ASP, the search for antagonists capable of distinguishing between excitatory amino acids and other excitants, such as acetylcholine, was unsuccessful. During the early 1970's a series of compounds were suggested to be antagonists, but generally their potency and selectivity were poor. However, L-glutamate diethylester

(GDEE) (Haldemann et al., 1972) and 1-hydroxy-3-aminopyrrolid-2-one (HA-966) (Davies and Watkins, 1973) have proven useful. Both of these compounds reduced responses induced by GLU and ASP. Unfortunately, both compounds were weak antagonists of acetylcholine-induced responses as well.

Surprisingly, the magnesium ion (Mg^{++}) proved to be the first substance that could be used to distinguish clearly between the actions of various excitatory amino acids. Following the suggestion (McCulloch et al., 1974) that KA and NMDA activated different receptors, a study of the differential sensitivity of these two agonists by Davies and Watkins (1977) showed that KA-induced responses were insensitive to Mg^{++}, and NMDA-induced responses were abolished by Mg^{++}. ASP-induced responses also proved to be more sensitive than those induced by GLU. However, Mg^{++} was also a weak acetylcholine antagonist.

A breakthrough in the search for selective antagonists occurred with the discovery of the properties of D-α-aminoadipate (DAA) (Biscoe et al., 1977, 1978). Not only did DAA permit differentiation between NMDA- and KA-induced responses and ASP- and GLU-induced responses, as did Mg^{++}, but it also had no effect on acetylcholine-induced responses. Furthermore, DAA was a selective antagonist of dorsal root- compared to ventral root-evoked excitation of Renshaw cells. This was the first evidence that an excitatory amino acid had a transmitter role at an identified mammaliam CNS synapse.

3. Current Status

A series of compounds that have been identified as selective excitatory amino acid antagonists (McLennan et al., 1981; Watkins, 1981a, b) are shown in Table I. Through a series of studies, a concept has evolved that there are at least three types of receptors in vertebrates corresponding to the structural analogs of ASP and GLU; these are KA, NMDA, and QU. The concept suggests that each of the agonists interacts with a specific receptor, and that each receptor can be differentiated based on its sensitivity to a series of antagonists. These include 2, 3-cis-piperidine dicarboxylate (PDA) (KA, NMDA, and QU receptors), γ-D-glutamylglycine (γDGG) (KA and NMDA receptors), GDEE (QU receptors), and DAA, along with D-α-aminosuberate, HA966, Mg^{++}, and 2-amino-5-phosphonovalerate (2APV) (NMDA receptors). GLU and ASP appear to be mixed agonists, with the ability to interact with each of the three receptor types, and having variable sensitivity to all of the antagonists.

TABLE 11-I
EXCITATORY AMINO ACID ANTAGONIST AND AGONIST INTERACTIONS IN
THE VERTEBRATE CENTRAL NERVOUS SYSTEM

Agonists	Antagonists			
	DAA, DAS, 2APV, HA966, Mg^{++}	GDEE	γDGG	PDA
Kainate	N.E.	N.E.	↓↓↓	↓↓↓
N-Methyl-D-aspartate	↓↓↓	N.E.	↓↓↓	↓↓↓
Quisqualate	N.E.	↓↓↓	↓	↓↓↓
Glutamate	↓↓	↓↓	↓↓↓	↓↓↓
Aspartate	↓↓↓	↓	↓↓↓	↓↓↓

Increase in number of arrows indicates increasingly effective antagonist action against agonist-induced responses. N.E., no effect; DAA = D-α-aminoadipate; DAS = D-α-aminosuberate; 2APV = 2-amino-5-phosphonovalerate; HA966 = 1-hydroxy-3-aminopyrrolid-2-one; Mg^{++} = magnesium ions; GDEE = L-glutamate diethylester; γDGG = γ-D-glutamylglycine; PDA = 2,3,-cis-piperidine dicarboxylate.

B. THE PHARMACOLOGY OF AUDITORY NERVE TRANSMISSION

1. *In Vivo* Studies in Mammals

In the first detailed study on excitatory amino acid receptor pharmacology in the cochlear nucleus, Martin and Adams (1979) demonstrated that prepotential and non-prepotential units do not respond in the same way to excitatory amino acids. Like most other neurons in the CNS, non-prepotential units are readily excited by GLU, ASP, NMDA, and KA. In contrast, prepotential units rarely respond to these agonists. Usually, they do not respond at all or, after an initial increase in firing rate, they appear to become overdepolarized and stop firing. The response of spinal motorneurons is similar (Zieglgänsberger and Puil, 1973). The depolarization-block mechanism is not clear; it is believed to result from a large localized increase in the conductance of the soma membrane leading to shunting and sodium inactivation.

When these first studies on the response of cochlear nucleus neurons to excitatory amino acids and antagonists were conducted, only a few agonists and antagonists were available. In studying DLAA (the racemic mixture of α-aminoadipate), Martin and Adams (1979) tested the specificity of the antagonist by comparing the differential sensitivities of responses induced by a series of excitatory compounds, including GLU, ASP, NMDA, KA, and acetylcholine (ACh), on non-prepotential units. As has been found in other CNS regions, DLAA blocked NMDA, reduced ASP more than GLU, and KA-

or ACh-evoked responses were not affected. The sensitivity of tone-evoked excitation to DLAA was then tested. DLAA was a potent antagonist of tone-evoked excitation of prepotential units. However, DLAA had very little effect on non-prepotential tone-evoked activity. Low stimulus-intensity tones were used for non-prepotential units, which probably led to the different results for the two types of units as found later (Martin, 1980).

A more complete list of antagonists were used in a later study by Martin (1980). The specificities of the actions of the antagonists, DAA, HA-966, Mg^{++}, and GDEE, were tested on excitatory amino acid-induced excitations of cochlear nucleus units. GDEE had no effect on NMDA-evoked responses and reduced responses to GLU more than ASP. DAA, HA-966, and Mg^{++} were effective antagonists of NMDA-evoked responses, and reduced ASP more than GLU-evoked responses on non-prepotential units. However, GDEE had no effect on tone-evoked activity. DAA, HA-966, and Mg^{++} also consistently reduced tone-evoked responses of both non-prepotential and prepotential units. Stimuli used for non-prepotential units were 20 and 30 dB above threshold in the later study (Martin, 1980), compared to 0 to 20 dB for the earlier study (Martin and Adams, 1979). By increasing the stimulus intensity, there is an increased probability of DAA-sensitive auditory nerve terminals being active near the microiontophoretic electrode. Caspary *et al.* (1981, Chapter 12, this volume) have reported data from the posteroventral cochlear nucleus (PVCN) corroborating these findings. PVCN units responded to microiontophoretic applications of GLU, ASP, and NMDA. DLAA also reduced responses to these three excitatory amino acids as well as to tone-evoked excitation.

Evidence, some of which is circumstantial, suggests that all three of the currently-known, excitatory amino-acid receptor types — KA, QU, and NMDA — are present on mammalian AVCN units. Since AVCN units are sensitive to KA (Martin and Adams, 1979), GDEE reduces GLU-evoked responses more than ASP-evoked responses (and has no effect on NMDA-evoked responses), and DAA (and DLAA, HA-966, and Mg^{++}) block NMDA-evoked responses (Martin, 1980). A summary of agonist and antagonist interactions on AVCN non-prepotential units is given in Table II. Since NMDA receptor antagonists also block tone-evoked excitations, it is probable that the NMDA receptor is the postsynaptic receptor at the mammalian auditory nerve-cochlear nucleus synapse.

2. *In Vitro* Studies in Chicken

Recently, Nemeth *et al.* (1983) reported that the auditory nerve-evoked field potential in the nucleus magnocellularis in an isolated whole brain-

TABLE 11-II
EXCITATORY AMINO ACID ANTAGONIST EFFECTS ON
AGONIST AND AUDITORY NERVE TRANSMITTER-INDUCED
RESPONSES IN THE MAMMALIAN COCHLEAR NUCLEUS

Agonist	Antagonist DAA, DAS, HA966, Mg^{++}	GDEE
Kainate	N.E.	N.T.
N-Methyl-D-aspartate	↓↓↓	N.E.
Glutamate	↓↓	↓↓↓
Aspartate	↓↓↓	↓↓
Auditory nerve transmitter	↓↓	N.E.

Symbols and abbreviations as in Table 11-I. N.T. = not tested.

stem preparation of the chicken was sensitive to certain excitatory amino acid antagonists. Specifically, bath-applied PDA and γDGG depressed orthodromically-evoked activity. APV, DAA, and GDEE had no effects. However, the specificities of the antagonists were not tested against agonist-induced responses. Using a tissue slice preparation of the chicken nucleus magnocellularis, Martin (1985) has confirmed these results, and, in addition, noted that KA, NMDA, and QU all produced dose-dependent alterations of antidromic nucleus magnocellularis field potentials (Table III). The responses induced by all three agonists were sensitive to PDA. DAA and GDEE reversibly blocked only the actions of NMDA and QU, respectively. The conclusion that can be drawn from these two studies is that the transmitter in the chicken, like the mammal, appears to be an excitatory amino acid. However, the postsynaptic receptor in the chicken appears to be of the KA type.

IV. INHIBITORY AMINO ACIDS

A. HISTORICAL BACKGROUND

The evolution of the concept of separate GABA-preferring and GLY-preferring receptors in the CNS is not as circuitous as that of separate receptors for excitatory amino acids. An examination of the process of evolution of this concept, does, however, reinforce the importance of specific antagonists and appropriate experimental design to neuropharmacology. Reviews of the early literature can be found in Curtis (1965), Curtis and Watkins (1965), Curtis and Crawford (1969), and Curtis and Johnston

TABLE 11-III
EXCITATORY AMINO ACID ANTAGONIST EFFECTS ON
AGONIST AND AUDITORY NERVE TRANSMITTER-INDUCED
RESPONSES IN THE CHICKEN NUCLEUS MAGNOCELLULARIS

Agonist	Antagonist		
	DAA	GDEE	PDA
Kainate	N.E.	N.E.	↓↓↓
N-Methyl-D-aspartate	↓↓↓	N.E.	↓↓↓
Quisqualate	N.E.	↓↓↓	↓↓↓
Auditory nerve transmitter	N.E.	N.E.	↓↓↓

Symbols and abbreviations as in Table 11-I.

(1974). This story is much like a lock and key; most all of the pieces were present from the start, it only remained to put the correct key into the proper lock for the story to unfold.

The inhibitory actions of a series of monocarboxylic amino acids, including GABA, GLY, β-alanine (β-ALA), and taurine (TAU) were known from the earliest microiontophoretic studies (Curtis and Watkins, 1960, 1961). Of these compounds, special interest was given to GABA because of its apparent potency and as a compound unique to the CNS. Because of the similarity of the structure of GABA to the other monocarboxylic amino acids, it was assumed that they would all behave in a similar manner. As a result, a transmitter role for these amino acids was doubted for many years, since the only known antagonist of postsynaptic inhibition, strychnine (STR) (Bradley et al., 1953; Eccles et al., 1954), had no effect on GABA-induced inhibitions. Picrotoxin, which had been reported to be an effective GABA antagonist in crayfish (cf Takeuchi and Takeuchi, 1969), had variable and inconsistent effects in vertebrate preparations. It was a number of years before attention was turned to the other naturally-occurring inhibitory amino acids, and strychnine was shown to be a selective and reversible antagonist of GLY-induced inhibitions (Curtis et al., 1968a, b), thereby establishing GLY as a major transmitter in the spinal cord (cf Werman et al., 1967, 1968), and demonstrating the presence of at least two inhibitory amino-acid receptor types: GABA-preferring and GLY-preferring.

The understanding of the actual role of GABA in synaptic transmission was still hampered by the absence of effective antagonists. In 1970, McLennan showed that the convulsant bicuculline (BIC) was an effective GABA antagonist in the crayfish. Soon BIC was also shown to be an effective antagonist of both GABA-induced and certain synaptically-evoked inhibitions in the vertebrate CNS as well (Curtis et al., 1971a, b). Today STR and BIC are considered to be the pharmacological tools of choice in differentiating GABA-preferring and GLY-preferring receptors.

Table IV contains lists of currently known agonist-antagonist interactions in two regions of the CNS, compared with the cochlear nucleus. Based on this information, two types of inhibitory GABA receptors are known: one in spinal and one in cortical regions. Only one type of GLY receptor is known, and this receptor appears to play a role in synaptic transmission, perhaps only in spinal and brainstem structures. As with the extrasynaptic excitatory amino acid receptors, the role of GLY receptors in cortical regions is unknown.

TABLE 11-IV
INHIBITORY AMINO ACID ANTAGONIST AND AGONIST INTERACTIONS
IN THREE REGIONS OF THE CENTRAL NERVOUS SYSTEM

	Antagonists					
	Cochlear Nucleus		Spinal Cord		Cerebral Cortex	
Agonists	STR	BIC	STR	BIC	STR	BIC
Glycine	↓	N.E.	↓	N.E.	↓	N.E.
Taurine	↓	N.E.	↓	N.E.	↓	↓
β-Alanine	↓	N.E.	↓	N.E.	↓	↓
Muscimol	N.E.	↓	N.E.	↓	N.E.	↓
GABA	N.E.	↓	N.E.	↓	N.E.	↓

Arrows indicate antagonist action on agonist-induced responses. N.E. = no effect; STR = strychnine; BIC = bicuculline.

B. THE MAMMALIAN COCHLEAR NUCLEUS

The agonist-antagonist receptor interactions for inhibitory amino acids have been examined on AVCN units by Martin et al. (1982). The purpose of this microiontophoretic study was to determine the types of inhibitory amino-acid receptors present in this region and whether they correspond to the receptors found in spinal or cortical regions. The results of this study showed that responses induced by GLY, ALA, and TAU in the cochlear nucleus are STR-sensitive and BIC-insensitive. This is the same pattern seen in the spinal cord. GABA and muscimol-induced responses were also like those found in the spinal cord; they were BIC-sensitive and STR-insensitive. Thus, there are at least two inhibitory amino acid receptor types corresponding to those found in other brainstem and spinal cord structures.

The role of these receptors in synaptic transmission in the cochlear nucleus is unknown. Electrical or single tone-evoked activity (Whitfield and Allanson,

1958; Whitfield and Comis, 1966; Pirsig *et al.*, 1968; Caspary *et al.*, 1979; Watanabe, 1979; Martin *et al.*, 1982; Martin and Dickson, 1983; Martin and Penix, 1983), two-tone interaction (Whitfield and Comis, 1966; Watanabe, 1979; Martin *et al.*, 1982; Martin and Dickson, 1983), and lateral inhibition (Martin and Dickson, 1983) in the cochlear nucleus were all STR- and BIC-insensitive. In addition, inhibition of cochlear nucleus units by lateral superior olive stimulation is also insensitive to STR (Comis, 1970). BIC has not been tested on this inhibition. Intravenous STR does, however, reduce contralateral tone-evoked depression of field potentials in the cochlear nucleus following ipsilateral electrical stimulation of the cochlea (Pirsig *et al.*, 1968). Because the STR was given intravenously, the site of antagonism is not known and may well have been the lateral superior olive (Moore and Caspary, 1983) and not the cochlear nucleus.

V. SUMMARY AND CONCLUSIONS

From this review, it can be seen that at least five amino acid receptors are present in the cochlear nucleus. Some of the receptors play a role in synaptic transmission, for example, the NMDA-receptor in mammals and the KA-receptor in chicken. Future studies, using intracellular recording and patch-clamp techniques coupled with pharmacological techniques, should provide new insights into how these receptors function in the transduction process and ultimately in the sensory-integration process. In addition, agonist and antagonist tools, that are now available, will enable us to identify and characterize inhibitory amino-acid synapses within the cochlear nucleus. It is just a matter of time and research effort before the proper combination of experimental design and pharmacological manipulation is found.

ACKNOWLEDGMENTS

The author wishes to thank Drs. R. Altschuler, J. Fex, and E. V. Porter for reviewing, and M. L. Adams for typing the manuscript.

REFERENCES

Biscoe, T. J.; Davies, J., Dray, A., Evans, R. H., Francis, A. A., Martin, M. R., and Watkins, J. C.: D-α-aminoadipate as a selective antagonist of amino acid-induced and synaptic excitation of mammalian spinal neurons. *Nature. (London)* 270: 743–745, 1977.

Biscoe, T. J.; Davies, J., Dray, A., Evans, R. H., Martin, M. R., and Watkins, J. C.: D-α-aminoadipate, α, ϵ-diaminopimelic acid and HA-966 as antagonists of amino acid-induced and synaptic excitation of mammalian spinal neurons *in vitro*. *Brain Res.* 148: 543–548, 1978.

Biscoe, T. J.; Evans, R. H., Headley, P. M., Martin, M., and Watkins, J. C.: Domoic and quisqualic acids as potent amino acid excitants of frog and rat spinal neurons. *Nature. (London)* 255: 166-167, 1975.

Bourk, T. R.: *Electrical responses of neural units in the anteroventral cochlear nucleus of the cat.* Ph.D. Thesis. Boston, M. I. T., 1976.

Bradley, K.; Easton, D. M., and Eccles, J. C.: An investigation of primary or direct inhibition. *J. Physiol. (London)* 122: 474-488, 1953.

Brawer, J. R., and Morest, D. K.: Relations between auditory nerve endings and cell types in the cat's anteroventral cochlear nucleus seen with the Golgi method and Nomarski optics. *J. Comp. Neur.* 160: 491-506, 1975.

Canzek, V., and Reubi, J. C.: The effect of cochlear nerve lesion on the release of glutamate, aspartate and GABA from cat cochlear nucleus. *Exp. Brain Res.* 38: 437-441, 1980.

Caspary, D. M.; Havey, D. C., and Faingold, C. L.: Effects of microiontophoretically applied glycine and GABA on neuronal response patterns in the cochlear nuclei. *Brain Res.* 172: 179-185, 1979.

Caspary, D. M.; Havey, D. C., and Faingold, C. L.: Glutamate and aspartate: alteration of thresholds and response patterns of auditory neurons. *Hearing Res.* 4: 325-333, 1981.

Comis, S. D.: Centrifugal inhibitory processes affecting neurones in the cat cochlear nucleus. *J. Physiol. (London)* 210: 751-760, 1970.

Curtis, D. R.: The actions of amino acids upon mammalian neurones. In Curtis, D. R., and McIntyre, A. K. (Eds.): *Studies in Physiology, Presented to J. C. Eccles.* Berlin, Springer-Verlag, 1965, pp. 34-42.

Curtis, D. R., and Crawford, J. M.: Central synaptic transmission — microelectrophoretic studies. *Annu. Rev. Pharmacol.* 9: 209-240, 1969.

Curtis, D. R.; Duggan, A. N., Felix, D., and Johnston, G. A. R.: Bicuculline, an antagonist of GABA and synaptic inhibition in the spinal cord of cat. *Brain Res.* 32: 69-96, 1971a.

Curtis, D. R.; Duggan, A. W., Felix, D., Johnston, G. A. R., and McLennan, H.: Antagonism between bicuculline and GABA in the cat brain. *Brain Res.* 33: 57-73, 1971b.

Curtis, D. R.; Hösli, L., and Johnston, G. A. R.: A pharmacological study of the depression of spinal neurones by glycine and related amino acids. *Exp. Brain Res.* 6: 1-18, 1968a.

Curtis, D. R.; Hösli, L., Johnston, G. A. R., and Johnston, I. H.: The hyperpolarization of spinal motoneurones by glycine and related amino acids. *Exp. Brain Res.* 5: 235-258, 1968b.

Curtis, D. R., and Johnston, G. A. R.: Amino acid transmitters in the mammalian central nervous system. *Ergeb. Physiol.* 69: 97-188, 1974.

Curtis, D. R.; Phillis, J. W., and Watkins, J. C.: The chemical excitation of spinal neurones by certain acidic amino acids. *J. Physiol. (London)* 150: 656-682, 1960.

Curtis, D. R., and Watkins, J. C.: The excitation and depression of spinal neurones by structurally related amino acids. *J. Neurochem.* 6: 117-141, 1960.

Curtis, D. R., and Watkins, J. C.: Analogues of glutamic and γ-amino-n-butyric acids having potent actions on mammalian neurones. *Nature (London)* 191: 1010-1011, 1961.

Curtis, D. R., and Watkins, J. C.: Acidic amino acids with strong excitatory actions on mammalian neurones. *J. Physiol. (London)* 166: 1-14, 1963.

Curtis, D. R., and Watkins, J. C.: The pharmacology of amino acids related to gamma-aminobutyric acid. *Pharmacol. Rev.* 17: 347-392, 1965.

Davies, J., and Watkins, J. C.: Microelectrophoretic studies on the depressant action of HA966 on chemically and synaptically excited neurones in the cat cerebral cortex and cuneate nucleus. *Brain Res.* 59: 311-322, 1973.

Davies, J., and Watkins, J. C.: Effect of magnesium ions on the responses of spinal neurones to excitatory amino acids and acetylcholine. *Brain Res.* 130: 364-368, 1977.

Duggan, A. W.: The differential sensitivity of L-glutamate and L-aspartate of spinal interneurones and Renshaw cells. *Exp. Brain Res. 19:* 522–528, 1974.

Eccles, J. C.; Fatt, P., and Kiketsu, K.: Cholinergic and inhibitory synapses in a pathway from motor axon collaterals to motoneurons. *J. Physiol. (London) 126:* 524–562.

Godfrey, D. A.; Carter, J. A., Berger, S. J., Lowry, O. H., and Matschinsky, F. M.: Quantitative histochemical mapping of candidate transmitter amino acids in cat cochlear nucleus. *J. Histochem. Cytochem. 25:* 417–431, 1977.

Godfrey, D. A.; Carter, J. A., Lowry, O. H., and Matschinsky, F. M.: Distribution of gamma-aminobutyric acid, glycine, glutamate and aspartate in the cochlear nucleus of the rat. *J. Histochem. Cytochem. 26:* 118–126, 1978.

Haldeman, S.; Huffman, R. D., Marshall, K. C., and McLennan, H.: The antagonism of the glutamate induced and synaptic excitation of thalamic neurones. *Brain Res. 39:* 419–425, 1972.

Hansson, E.; Jarlstedt, J., and Sellström, Å.: Sound-stimulated ^{14}C-glutamate release from the nucleus cochlearis. *Experientia 36:* 576–577, 1980.

Johnston, G. A. R.; Curtis, D. R., Davies, J., and McCulloch, R. M.: Spinal interneurone excitation by conformationally restricted analogues of L-glutamic acid. *Nature (Lond.) 248:* 804–805, 1974.

Jones, D. R.; Oliver, D. L., Potashner, S. J., and Morest, D. K.: Retrograde axonal transport of D-aspartate from cochlear nucleus to Type II spiral ganglion cells in the cat. *Assoc. Res. Otolaryngol. Abstr. 5:* 90, 1982.

Martin, M. R.: The effects of iontophoretically-applied antagonists on auditory nerve and amino acid-evoked excitation of anteroventral cochlear nucleus neurons. *Neuropharmacology 19:* 519–528, 1980.

Martin, M. R.: Excitatory amino acid pharmacology and auditory nerve transmission in the nucleus magnocellularis of the chicken. *Soc. Neurosci. Abstr. 10,* 1984. In press.

Martin, M. R., and Adams, J. C.: Effects of DL-α-aminoadipate on synaptically and chemically-evoked excitation of anteroventral cochlear nucleus neurons of the cat. *Neuroscience 4:* 1091–1105, 1979.

Martin, M. R., and Dickson, J. W.: Lateral inhibition in the anteroventral cochlear nucleus of the cat: a microiontophoretic study. *Hearing Res. 9:* 35–41, 1983.

Martin, M. R.; Dickson, J. W., and Fex, J.: Bicuculline, strychnine and depressant amino acid responses in the anteroventral cochlear nucleus of the cat. *Neuropharmacology 21:* 201–207, 1982.

Martin, M. R., and Penix, L. P.: Comparison of the effects of bicuculline and strychnine on brain stem auditory evoked potentials in the cat. *Brit. J. Pharmacol. 78:* 75–77, 1983.

McCulloch, R. M.; Johnston, G. A. R., Game, C. J. A., and Curtis, D. R.: The differential sensitivity of spinal interneurones and Renshaw cells to kainate and N-methyl-D-aspartate. *Exp. Brain Res. 21:* 515–518, 1974.

McLennan, H.: Bicuculline and inhibition of crayfish stretch receptor neurones. *Nature (London) 228:* 674–675, 1970.

McLennan, H.: On the nature of the receptors for various excitatory amino acids in the mammalian central nervous system. In DiChiara, G., and Gessa, G. L. (Eds.): *Glutamate as a Neurotransmitter.* New York, Raven Press, 1981, pp. 253–262.

Moore, M. J., and Caspary, D. M.: Strychnine blocks binaural inhibition in lateral superior olivary neurons. *J. Neurosci. 3:* 237–242, 1983.

Nemeth, E. F.; Jackson, H., and Parks, T. N.: Pharmacological evidence for synaptic transmission by non-N-methyl-D-aspartate receptors in the Avian cochlear nucleus. *Neurosci. Lett. 40:* 39–44, 1983.

Oliver, D. L.; Jones, D. R., Potashner, S. J., and Morest, D. K.: Evidence for selective uptake, release and axonal transport of D-aspartate in the auditory system and cerebellum of cat and guinea pig. *Anat. Rec. 199:* 186A, 1981.

Pfeiffer, R. R.: Anteroventral cochlear nucleus: wave forms of extracellularly recorded spike potentials. *Science. 154:* 667–668, 1966.

Pirsig, W.; Pfalz, R., and Sadanaga, M.: Postsynaptic auditory crossed efferent inhibition in the ventral cochlear nucleus and its blocking by strychnine nitrate (guinea pig). *Kumamoto Med. J.* 21: 75–82, 1968.

Schwartz, I. R.: The differential distribution of label following uptake of ^{3}H-labeled amino acids in the dorsal cochlear nucleus of the cat. *Exp. Neurol. 73:* 601–617, 1981.

Takeuchi, A., and Takeuchi, N.: A study of the action of picrotoxin on the inhibitory neuromuscular junction of the crayfish. *J. Physiol. (London) 205:* 377–391, 1969.

Tolbert, L. P., and Morest, D. K.: Patterns of synaptic organization in the cochlear nucleus of the cat. *Soc. Neurosci. Abstr. 4:* 11, 1978.

Watanabe, T.: Funneling mechanism in hearing. *Hearing Res. 1:* 111–119, 1979.

Watkins, J. C.: Pharmacology of excitatory amino acid transmitters. *Adv. Biochem. Psychopharmacol.* 29: 205–212, 1981a.

Watkins, J. C.: Pharmacology of excitatory amino acid receptors. In Roberts, P. J.; Storm-Mathiesen, J., and Johnston, G. A. R. (Eds.): *Glutamate: Transmitter in the Central Nervous System.* Chichester, J. Wiley and Sons, 1981b, pp. 1–24.

Wenthold, R. J.: Glutamic acid and aspartic acid in subdivisions of the cochlear nucleus after auditory nerve lesion. *Brain Res. 143:* 544–548, 1978.

Wenthold, R. J.: Release of endogenous glutamic acid, aspartic acid and GABA from cochlear nucleus slices. *Brain Res. 162:* 338–343, 1979.

Werman, R.; Davidoff, R. A., and Aprison, M. H.: Evidence for glycine as the principal transmitter mediating postsynaptic inhibition in the spinal cord of the cat. *J. Gen. Physiol. 50:* 1093–1094, 1967.

Werman, R.; Davidoff, R. A., and Aprison, M. H.: Inhibitory action of glycine on spinal neurones in the cat. *J. Neurophysiol. 31.* 81–95, 1968.

Whitfield, I. C., and Allanson, J. T.: A study of the effect of some neurally active drugs on inhibition in the auditory pathway. *Arch. Ital. Biol. 96:* 29–37, 1958.

Whitfield, I. C., and Comis, S. D.: *The Role of Inhibition in Information Transfer. The Interaction of Centrifugal and Centripetal Stimulation on Neurones of the Cochlear Nucleus.* Final Report, (Part 2), Grant 63-115. United States Air Force, Air Force European Office of Aerospace Research, 1966.

Zieglgänsberger, W., and Puil, E. A.: Actions of glutamic acid on spinal neurones. *Exp. Brain Res. 17:* 35–49, 1973.

Chapter 12

THE EFFECTS OF INHIBITORY AND EXCITATORY AMINO-ACID NEUROTRANSMITTERS ON THE RESPONSE PROPERTIES OF BRAINSTEM AUDITORY NEURONS

DONALD M. CASPARY, LEONARD P. RYBAK, AND CARL L. FAINGOLD

I. Introduction
II. Brief Review of Methodology
III. Is an Excitatory Amino Acid the Transmitter at Auditory Nerve Synapses in the Cochlear Nucleus?
    A. Effects of Excitatory Amino Acids and Specific Agonists on Tone-Evoked Activity
    B. Effects of Excitatory Amino-Acid Antagonists on Tone-Evoked Activity
    C. Effect of Baclofen on the Response Properties of Cochlear-Nucleus Neurons
    D. Summary of the Effects of Excitatory Amino Acids and Related Substances on the Response Properties of Neurons in the Cochlear Nucleus
IV. Glycine and GABA May Mediate Inhibitory Responses Seen for Neurons of the Dorsal Cochlear Nucleus
    A. Effects of Glycine on Neurons in the Dorsal Cochlear Nucleus
V. Glycine-Mediated Binaural Inhibition in the Lateral Superior Olivary Nucleus
    A. Effects of Glycine and Strychnine on Neurons of the Lateral Superior Olive
VI. Conclusion
    References

I. INTRODUCTION

Recent data suggest that excitatory and inhibitory amino acids may be involved in processing acoustic information in brainstem auditory structures (see Chapters 8, 9, and 11, this volume). The present chapter will describe our efforts to identify specific roles for neurotransmitter substances in the generation of known physiologic responses of neurons in both the cochlear nucleus (CN) and superior olivary complex (SOC). Previous chapters have presented data which strongly implicate an excitatory amino acid as the neurotransmitter at auditory-nerve synapses in the CN (Chapters 8, 9, and 11), while the inhibitory amino-acid neurotransmitters, GABA and glycine, are implicated in the CN and in SOC (see Chapters 14 and 15).

Assuming the identity of a given substance as a neurotransmitter at a particular synapse within the auditory system, specific questions can be asked (Fig. 1): 1. What are the roles of the excitatory and inhibitory amino-acid neurotransmitters in the generation of the area-specific response patterns of CN neurons? 2. What transmitter(s) mediate(s) binaural inhibition in the lateral superior olivary nucleus?

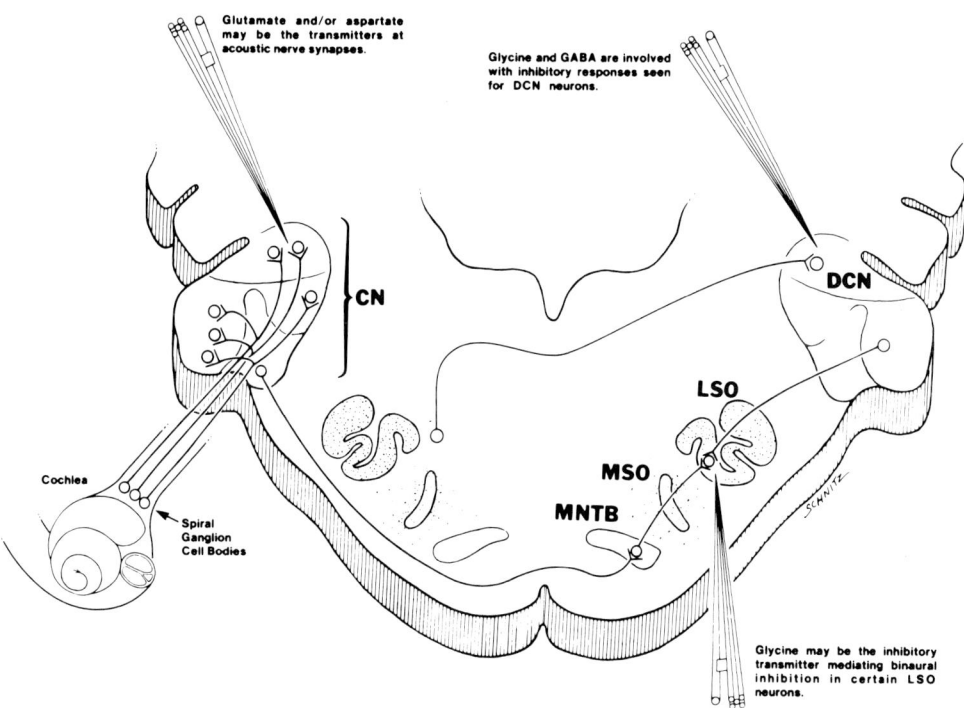

Figure 12-1. Schematic representation of the three principal sites examined in the present study. Localization of recorded neurons was based on histological track reconstruction, some extracellular marking, and correlation with known response patterns. (Adapted from Moore and Caspary, 1983.)

The possible role of the excitatory amino acids as transmitters at auditory-nerve synapses can be tested by iontophoretically applying these substances, in the presence or absence of agents known to interact at synapses involving excitant amino-acid neurotransmitters (i.e., receptor antagonists, inhibitors of release, etc.), onto individual CN neurons and observing their effects on: (1) response thresholds, (2) rate-intensity functions, (3) discharge rates, and (4) response patterns. The findings presented in this chapter support the hypothesis that an excitatory amino acid or similar substance is the neurotransmitter at auditory-nerve synapses in the CN.

The binaurally-responsive neurons in the SOC, the second major brainstem auditory structure, respond to interaural differences in intensity with varying degrees of inhibition, which can also be iontophoretically examined. Neurons in the lateral superior olivary nucleus (LSO) receive input from both ears and are thought to code for the localization of sound in space (Masterton et al., 1967; Goldberg and Brown, 1969; Erulkar, 1972; Tsuchitani, 1978; Caird and Klinke, 1984). These neurons typically show an excitatory response with ipsilateral acoustic stimulation and a profound reduction of response with binaural acoustic stimulation (Guinan et al., 1972a, b; Tsuchitani, 1978; Caird and Klinke, 1984). Data will be presented in this chapter which suggest that glycine mediates binaural inhibition in the LSO.

II. BRIEF REVIEW OF METHODOLOGY

Chinchillas (*Chinchilla laniger*) weighing 300–600 g, were anesthetized with sodium pentobarbital (40 mg/kg), followed by ketamine-HCl (40 mg/kg) with supplements of sodium pentobarbital to maintain anesthesia. Because of the effect of pentobarbital on neuronal response patterns (Young and Brownell, 1976; Ritz and Brownell, 1982) and on amino-acid neurotransmitter release (Collins et al., 1981; Potashner and Lake, 1981), some animals were anesthetized and maintained with ketamine alone, with wound edges infiltrated with lidocaine.

Details of the surgical and stereotaxic approaches for both CN (Caspary, 1972; Caspary et al., 1979) and SOC (Moore and Caspary, 1983) have been previously described. Extracellular recordings were obtained with a glass micropipet (tip diameter 1.0–1.5 μm) glued alongside a 5-barrel micropipet (Glass Company of America; outer tip diameter 8–10 μm) so that the single-barrel tip protruded 3–7 μm beyond the tip of the 5-barrel microelectrode (Havey and Caspary, 1980). Four barrels of the 5-barrel microelectrode were filled with the substances to be tested (Table I). A balancing or summating channel was employed to alleviate current effects; this fifth barrel and the recording micropipet were filled with either 2.0 M potassium acetate, fast green FCF in 2.0 M sodium chloride (Woolf, 1981), or horseradish peroxidase (HRP) (Spencer et al., 1978). Tuning curves and characteristic frequency (CF) were determined for all neurons encountered. Post-stimulus time histograms (PSTH's) and interspike interval histograms (ISIH's), as well as continuous tracing of spike counts, were obtained for all neurons before, during, and subsequent to iontophoretic application of drugs. Stimulus conditions included either 50 or 100 presentations of coherent tone bursts (50 or 100 ms duration, 5 ms rise-fall) presented at the rate of 2/s or 4/s (Caspary et al., 1981; Moore and Caspary, 1983; Caspary et al., 1984).

TABLE 12-I
IONTOPHORETIC DRUGS AND ACTIONS

Agent	Concentration	pH	Source	Proposed Action
L-α-Glutamic acid, monsodium salt	0.5 M	7.0	Sigma	Excitatory-neurotransmitter quisqualate-receptor agonist
L-Glutamic acid diethylester HCl	0.2 M	3.5	Sigma	Excitatory amino-acid quisqualate-receptor antagonist
L-Aspartic acid	0.5 M	8.0	Sigma	Excitatory-neurotransmitter NMDA-receptor agonist
N-Methyl-D-aspartate	0.01 M	7.0	Sigma	Excitatory-neurotransmitter NMDA-receptor agonist
D-α-Aminoadipate	50 mM	7.0	Sigma	Excitatory amino-acid NMDA-receptor antagonist
DL-2-Amino-5-phosphonovaleric acid (APV)	50 mM	7.0	Cambridge	Excitatory amino-acid NMDA-receptor antagonist
γ-Amino-N-butyric acid	0.5 M	3.5-4.0	Sigma	Inhibitory-neurotransmitter agonist
Bicuculline methiodide	0.005 M	3.0	Sigma	GABA-receptor antagonist
Glycine	0.5 M	3.5-4.0	Sigma	Inhibitory-neurotransmitter agonist
Strychnine HCl	0.01 M	3.0	Sigma	Glycine-receptor antagonist

Classification of the PSTH response pattern was determined at 25–35 dB above threshold at CF, and each test substance was ejected continuously for the duration of a histogram (25 s). Drug application was continued through a series of histograms if spike counts (rate meter plots) indicated a continuing change from the initial control. When no further alteration of response was noted, drug application was terminated and a series of control histograms was collected until response patterns and spike counts returned to pre-drug levels. These procedures were repeated at successive intensities and at several drug dosages if time permitted.

III. IS AN EXCITATORY AMINO ACID THE TRANSMITTER AT AUDITORY NERVE SYNAPSES IN THE COCHLEAR NUCLEUS?

An excitatory amino acid has been proposed as the neurotransmitter at auditory-nerve synapses in the CN (see Chapters 8, 9, and 11, this volume; for review, see Klinke, 1981; Guth and Melamed, 1982). The distribution of glutamate and aspartate in different regions of the CN has been mapped microhistochemically, and parallels, to some extent, the distribution of auditory-nerve synapses (Godfrey et al., 1977, 1978). Radioactive aspartate and glutamate, when injected in the cochlea, label different structures in the CN (Kane, 1979). D-aspartate, when injected into the CN, labels auditory-nerve endings (Oliver et al., 1983), while uptake and release of D-aspartate are altered by cochlear ablation and disarticulation (Potashner, 1983). Sectioning of the auditory nerve or crushing of the cochlea reduces the level of aspartate and glutamate in the CN in proportion to the regional distribution of auditory-nerve synapses (Wenthold, 1979; Thalmann et al., 1980). Results from immunocytochemical studies suggest that aspartate aminotransferase, the enzyme required for conversion of oxaloacetate to aspartate, and glutaminase, an enzyme involved in glutamate metabolism, are similarly distributed and are reduced following sectioning of auditory nerve (Wenthold, 1980; Altschuler et al., 1981; Wenthold and Altschuler, 1983). Synaptic release of glutamate and aspartate from the CN has been demonstrated both *in vivo* and *in vitro* (Wenthold, 1979; Canzek and Reubi, 1980; Hansson et al., 1980). The studies reviewed above and the work presented in Chapters 8, 9, 11 by Wenthold, Potasher, and Martin, respectively, suggest that an excitatory amino acid or similar substance satisfies many of the criteria which are necessary to prove that a substance is a neurotransmitter. Additional criteria, which can be tested using the technique of microiontophoresis, include the ability of a substance to mimic the action of tone-evoked, synaptically-released transmitter (in this case) and the ability of pharmacologic agents, which interact with the synaptically-released transmitter, to interact with the exogenously-applied substance in an identical manner.

A. EFFECTS OF EXCITATORY AMINO ACIDS AND SPECIFIC AGONISTS ON TONE-EVOKED ACTIVITY

Table I lists the substances which were iontophoretically applied onto dorsal cochlear nucleus (DCN) and posteroventral cochlear nucleus (PVCN) neurons, and their proposed actions from studies on other systems (Watkins et al., 1981; Fonnum, 1984). We have looked at the excitatory amino acids themselves, glutamate and aspartate, as well as the excitatory-amino-acid, aspartate-preferring, receptor agonist, N-methyl-D-aspartate (NMDA). PSTH's at CF were used to characterize response patterns of CN neurons. "Pauser build-up" and "build-up" types of response patterns in the DCN, as well as "on" and "chopper" response types from the postero-ventral CN (PVCN), were the predominant, area-specific response types examined. Over 500 CN neurons have been examined using iontophoresis, of which 317 were tested with the excitatory amino acids and/or one of their receptor agonists. Iontophoretic application of glutamate, aspartate, or NMDA resulted in 5-15 dB decreases in threshold, dramatically-shifted rate-intensity functions (upward and to the left), and, for certain restricted response types, altered PSTH response patterns (Caspary and Havey, 1978; Caspary et al., 1981).

A representative PVCN neuron (Fig. 2) displayed an increasingly tonic, more "primary-like", response with increasing doses of aspartate. Application of aspartate lowered the threshold (10 dB) and clearly shifted the rate-intensity function, upward and to the left (Fig. 3B). The large parallel upward shift produced by 200 nA of aspartate most likely represents an induction of spontaneous activity, which is also reflected as a break in the dose-response curve (Fig. 3A). The break point in Fig. 3A could represent the point of aspartate-induced depolarization where the discharge threshold was exceeded in both the tone-evoked and non-tone-evoked conditions, and could indicate that this point is close to the reversal potential of the tone-evoked EPSP (unpublished intracellular results). Typically, neurons in the CN displayed similar responses to both glutamate and aspartate. The neuron in Figs. 2, 3A, and 3B responded to glutamate in a manner similar to that displayed for aspartate. Passage of current alone (200 nA) did not elicit any change in the response pattern or threshold of this neuron.

Iontophoretic application of excitatory amino-acid neurotransmitters also induced changes in response patterns, thresholds, and rate-intensity functions of DCN neurons (Fig. 4). The classifying control histograms for this neuron (Fig. 4) display the "build-up" type of response pattern described for fusiform cells of the DCN (Caspary, 1972; Evans and Nelson, 1973; Rhode et al., 1983). Rate-intensity functions of DCN neurons typically shifted upward with the application of excitatory amino acids or their agonists, and changes in response pattern often involved a filling-in of the "build-up"

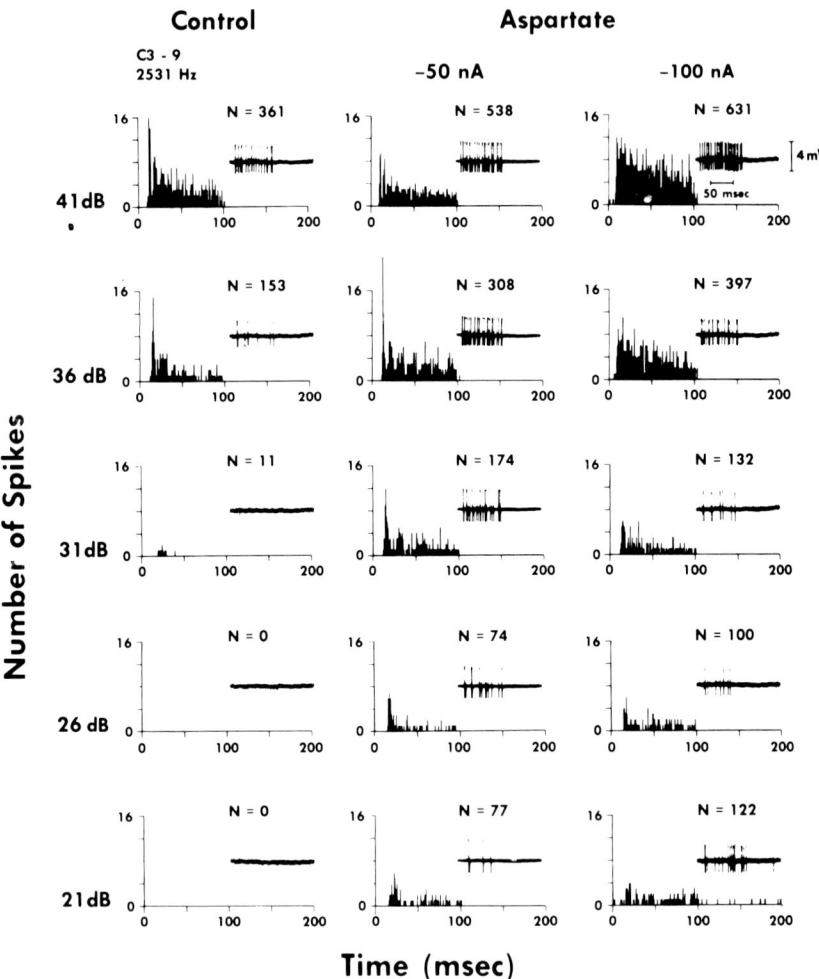

Figure 12-2. Post-stimulus time histograms showing the typical effects of excitatory amino acids on responses of a unit from the posteroventral cochlear nucleus of the chinchilla. Stimuli were 100 ms tone bursts (2,531 Hz) presented at 2/s. Responses were summed over 50 trials. Stimulus intensities, in dB SPL, are shown at left. Discharge rates are in total number of spikes. Comparison of the control responses (shown in the left column) to those obtained during iontophoretic application of 50 nA and 100 nA of aspartate (shown in the middle and right columns, respectively) reveals that application of aspartate not only increased discharge rate in a dose-dependent manner, but also lowered CF threshold by 10 dB, and resulted in an increasingly tonic, more "primary-like" discharge pattern. Five sweeps of raw data are shown in the inset of each histogram. Rise-fall times of 5 ms were utilized for all stimuli. PSTH bin widths are 200 μs for this and subsequent PSTH's; "N" refers to the total number of spikes shown in the histogram.

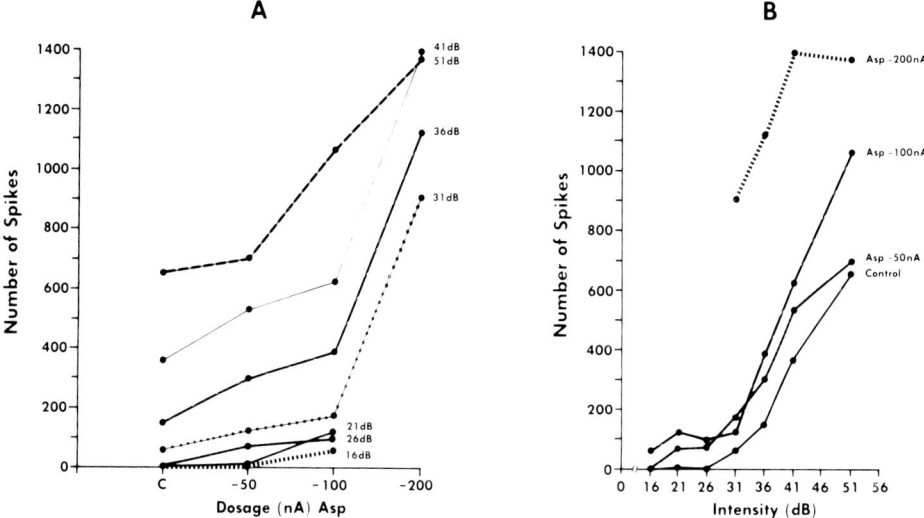

Figure 12-3. A. Dose-response curves for the same unit shown in Fig. 2. Iontophoretic application of aspartate produced a dose-dependent increase in discharge rate at each stimulus intensity studied. Application of 200 nA of aspartate resulted in a dramatic increase in firing rate across intensity. B. The same data, displayed as rate-intensity functions, resulted in parallel curves at dose levels below 100 nA, while 200 nA of aspartate caused a shift in the function.

portion of the PSTH at lower intensities, with the induction of an initial "pauser" peak at higher intensities (Fig. 4). Effects on both DCN and PVCN neurons tended to be greatest near threshold, and the "build-up," "pauser build-up," and "on"-response types displayed a resistance to induction of spontaneous activity by the exogenously-applied excitants.

B. EFFECTS OF EXCITATORY AMINO-ACID ANTAGONISTS ON TONE-EVOKED ACTIVITY

Specific excitatory amino-acid receptor types have been discussed by Martin in Chapter 11 of this volume. We have examined the receptor antagonists D-α-aminoadipate (DAA) and DL-2-amino-5-phosphonovaleric acid (APV). In some cases, the glutamate-preferring receptor agonist, quisqualate, and its antagonist, glutamic acid diethylester (GDEE), were used. Although GDEE rarely blocked tone-evoked or spontaneous activity, GDEE was effective in blocking the excitation caused by iontophoretic application of glutamate. On the other hand, the NMDA-receptor antagonist, DAA (or in some earlier experiments, the racemic mixture DLAA) was effective in blocking tone-evoked activity as well as the effects of exogenously-

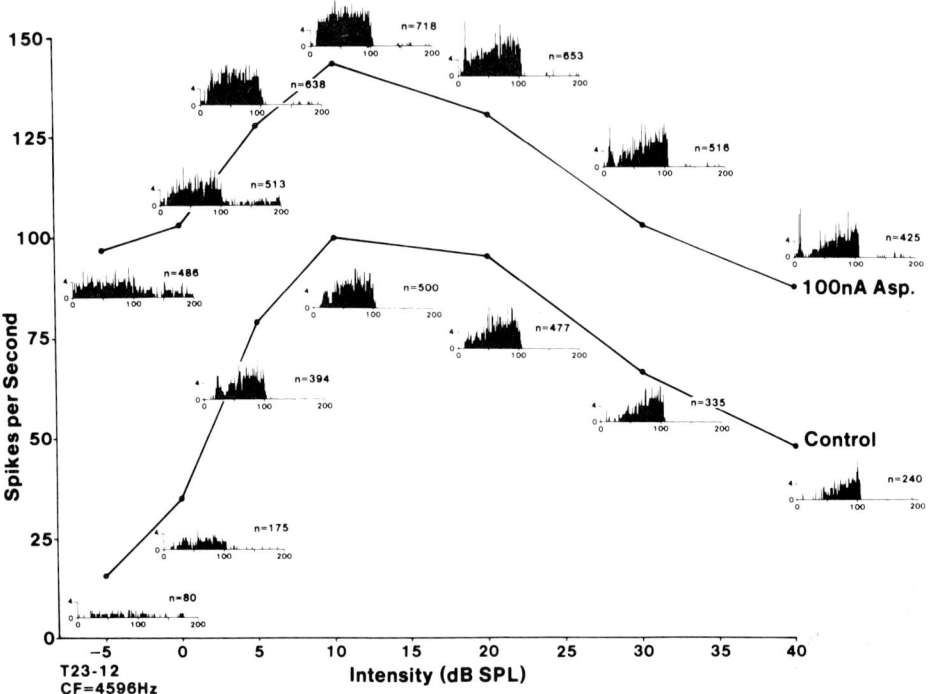

Figure 12-4. The effects of iontophoretic application of 100 nA of aspartate onto a "build-up" neuron from dorsal cochlear nucleus are shown on a rate-intensity function along with the corresponding post-stimulus time histograms. Stimuli were 100-ms tone bursts at CF (4,596 Hz) presented at 2/s and were summed over 50 trials. Discharge rates in the rate-intensity function are in spikes/s. Note that aspartate application caused an overall increase in discharge rate, as well as an increase in spontaneous activity below threshold and did not alter the nonmonotonic rate-intensity function. At higher intensities, aspartate application resulted in a slight alteration in the response pattern, as illustrated by the initial "pauser" peak appearing in the histograms, while at lower intensities, some filling-in of the initial "build-up" and thus, flattening of the response, was observed.

applied NMDA or aspartate in both DCN and PVCN neurons. In later experiments, we tested APV, a more recently-developed NMDA receptor antagonist, and observed it to be effective in suppressing tone-evoked responses. Iontophoretic application of DLAA onto a PVCN "on-L" neuron (Godfrey et al., 1975) altered both tone-evoked synaptic activity and the effects of exogenously-applied aspartate (Fig. 5). The control rate-intensity function of this neuron was distinctly non-monotonic and its PSTH response pattern became increasingly "phasic" with increasing intensity. Iontophoretic application of aspartate altered both the rate-intensity function and the response pattern of this neuron (top trace and histograms, Fig. 5). Tone-evoked synaptic activity was blocked by 100 nA of DLAA (bottom trace, Fig. 5), shifting the rate-intensity function downward. Simultaneous application of aspartate and DLAA blocked both the tone-evoked excitation and the excitation produced by iontophoretic application of aspartate. A more typical but somewhat less dramatic effect was seen for another PVCN neuron (Fig. 6). DAA shifted the rate-intensity function, at suprathreshold intensities, downward and to the right (15–20%). The agonist NMDA shifted the curve upward and to the left, with modest changes in response patterns. This effect was blocked by the simultaneous application of DAA and NMDA. Thus, excitatory amino-acid antagonists alter the tone-evoked response properties of CN neurons, a finding which partially satisfies the criterion of pharmacological identity needed to establish a substance as a neurotransmitter at a specific synapse.

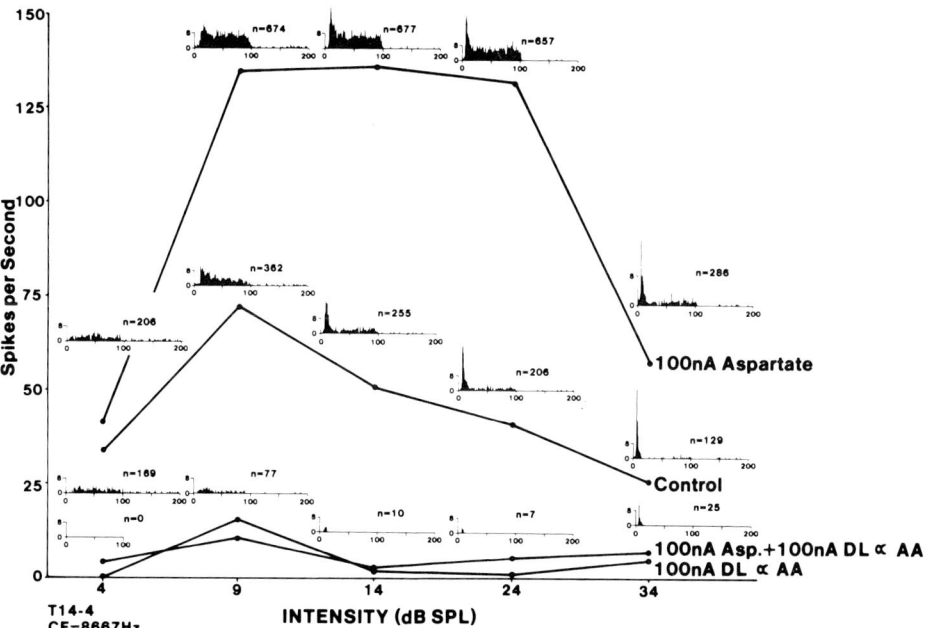

Figure 12-5. Rate-intensity curves, obtained from an "on" unit located in posteroventral cochlear nucleus, are displayed with accompanying poststimulus time histograms. Stimuli were 100-ms tone bursts at CF (8,667 Hz) presented at 2/s and were summed over 50 trials. Discharge rate is in spikes/s in the rate-intensity function. Iontophoretic application of aspartate caused an increase in discharge rate and altered the response pattern; however, it had no effect on the non-monotonic nature of the rate-intensity function. Application of the excitatory amino acid antagonist DL-α-aminoadipate (DLAA) greatly suppressed the firing rate at all stimulus intensities tested. Simultaneous application of aspartate and DLAA blocked tone-evoked activity as well as the effects of exogenously-applied aspartate. (Note that no histograms are displayed with the latter curve.)

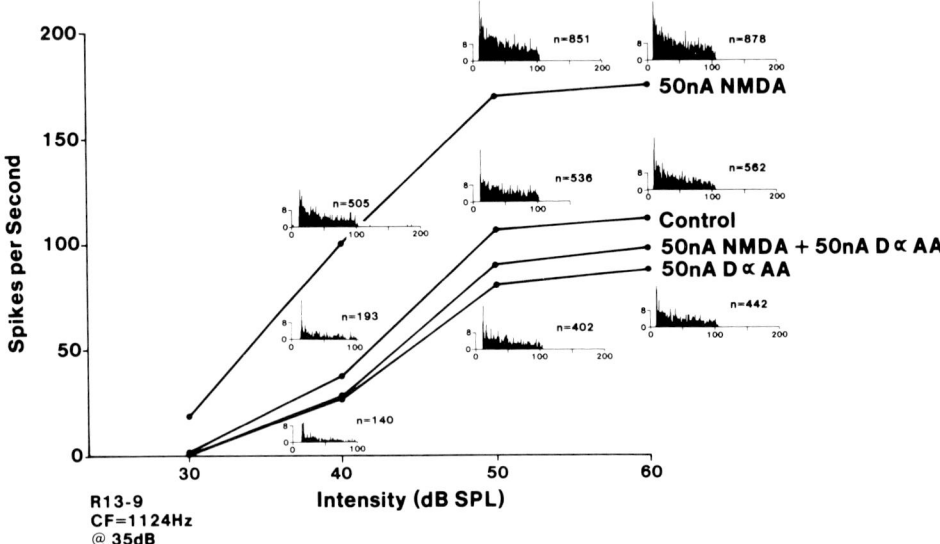

Figure 12-6. Effects of iontophoretic application of N-methyl-D-aspartate (NMDA) and the proposed NMDA-receptor antagonist, D-α-aminoadipate (DAA), on a rate-intensity function obtained from a neuron located in posteroventral cochlear nucleus. Discharge rate is in spikes/s. The corresponding post-stimulus time histograms represent the neuron's response to 100-ms tone bursts at CF (1,124 Hz) presented at the rate of 2/s. Responses were summed over 50 trials. NMDA application resulted in an upward shift of the rate-intensity function while DAA application shifted the curve down. When applied simultaneously with NMDA, DAA appeared to block both the synaptically-released transmitter and the exogenously-applied agonist. (Note that histograms for the NMDA + DAA curve are not shown.)

C. EFFECT OF BACLOFEN ON THE RESPONSE PROPERTIES OF COCHLEAR-NUCLEUS NEURONS

Release of excitatory amino acids is reportedly reduced by baclofen (Potashner, 1978, 1979), a lipophilic GABA-mimetic used to treat the spasticity of multiple sclerosis and spinal cord injury (Pinto DeSilva et al., 1972). Baclofen also has potential as both an analgesic and an anticonvulsant (Cutting and Jordan, 1980; Sawynok and LaBella, 1982; Terrence et al., 1983), and, although considered to be a safe pharmacologic agent, it can produce auditory and visual hallucinations upon abrupt withdrawal (Lees et al., 1977; Stein, 1977). Baclofen has a close structural resemblance to GABA, but increasing evidence suggests that the levorotatory form of baclofen may function presynaptically to inhibit the release of excitatory amino-acid neurotransmitters (Bowery, 1982). Baclofen inhibits the *in vitro* synaptic release of glutamate and aspartate in a calcium-dependent fashion (Potashner,

1978, 1979), and iontophoretic, bath, or systemic application of baclofen can reduce excitation in systems where an excitatory amino acid is the suspected neurotransmitter (Cain and Simmonds, 1982; Olpe et al., 1982). Far-field, auditory-evoked response may be altered by systemic baclofen (Martin, 1982). Excitatory postsynaptic potentials in spinal motor neurons are greatly depressed by small systemic doses of baclofen (Pierau and Zimmerman, 1973; Davidoff and Sears, 1974; Davies, 1981; Collins et al., 1982), and several studies show that baclofen reduces primary afferent depolarization in the spinal cord, both *in vivo* and *in vitro* (see Bowery, 1982; Caspary et al., 1984, for review). Baclofen effects appear to be bicuculline-insensitive, a finding which, along with data from a number of binding studies, lends support to the theory that baclofen may be acting presynaptically through a bicuculline-insensitive $GABA_B$ receptor to inhibit the release of excitatory amino-acid neurotransmitters. Differential binding studies by Bowery (1982) and others suggests that the mechanism of baclofen's action may be a selective ligand for a presynaptic $GABA_B$ receptor and that this binding is bicuculline-insensitive (see Bowery, 1982; Caspary et al., 1984, for review).

Iontophoretic application of (-)baclofen was used as a tool to test further the hypothesis that an excitatory amino acid may be the transmitter at auditory nerve synapses. The action of (-)baclofen was compared to that of iontophoretically-applied GABA and bicuculline, as well as to that of DAA and aspartate. Figure 7 displays the effects of iontophoretic application of (-)baclofen on the tone-evoked responses of seven neurons representing all subdivisions of the CN. (-)Baclofen's greatest effect on tone-evoked activity was usually near threshold, although greater than 50% suppression was observed for many neurons even at 40 dB above CF threshold (neurons R88-3, R93-13, R-92-10, Fig. 7). Neurons with restricted discharge characteristics, i.e., "on" or "build-up" type neurons, displayed total or near-total suppression of tone-evoked responses throughout the intensity ranges examined. The responses of a typical "build-up" neuron from the DCN were nearly completely suppressed with (-)baclofen (Fig. 8). (-)Baclofen consistently suppressed the tone-evoked response, which resulted in a dramatic downward shift of the rate-intensity functions. Optimum (-)baclofen doses were between 0 and 40 nA, both for DCN and VCN neurons, and depended, in part, on the distance between the recording and drug barrels (Caspary et al., 1984).

The effect of (-)baclofen differed from the effects of GABA with respect to both time course and magnitude of response (Caspary et al., 1984). GABA desensitization ("fade"; see Krnjevic, 1981, for review) commonly occurred, while the response to (-)baclofen application was never found to fade. The difference in time course and magnitude of effect between iontophoretically-applied GABA and (-)baclofen is clearly seen when one studies the rate-meter plot for a neuron responding to 25-dB-SPL tone-burst stimulation (Fig. 9).

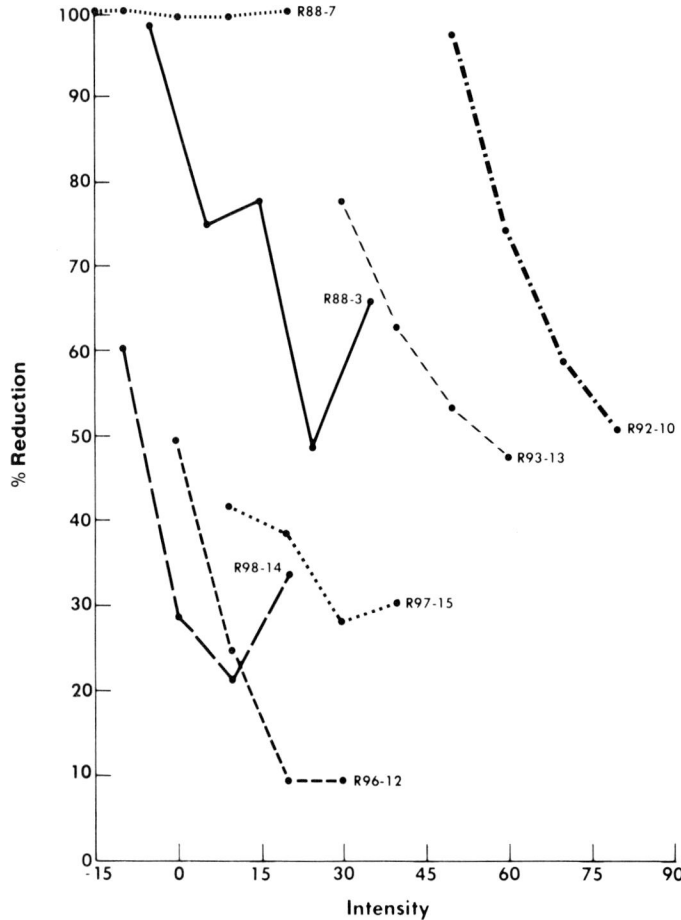

Figure 12-7. Effects of iontophoretically-applied (-)baclofen on the tone-evoked responses of seven neurons representing all subdivisions of the cochlear nucleus are displayed as percent reduction of control vs. stimulus intensity. All tone bursts were presented at CF. Note that while a significant reduction in response was frequently observed at high intensities, (-)baclofen's greatest effect was at lower intensities. (From Caspary et al., 1984.)

Bicuculline never antagonized the (-)baclofen response and was only partially effective in blocking GABA response. The firing of these same CN neurons was also suppressed by DAA, while aspartate or NMDA application produced excitation in those neurons tested. This (-)baclofen hypothesis is presented in Fig. 10. Iontophoretically-applied (-)baclofen (Fig. 10B) may act presynaptically to inhibit the tone-evoked release of the excitant transmitter

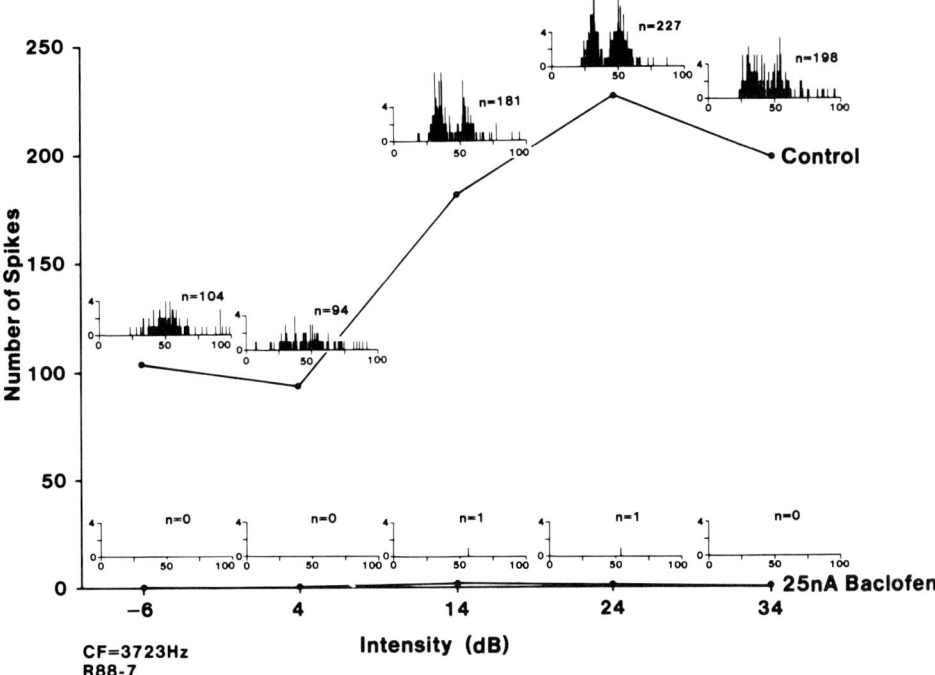

Figure 12-8. Rate-intensity functions with accompanying post-stimulus time histograms illustrate the effects of 25 nA of iontophoretically-applied (-)baclofen on a unit from dorsal cochlear nucleus. Stimuli were 50-ms tone bursts at CF (3,723 Hz) presented at the rate of 4/s. Responses were summed over 100 trials. Discharge rate is in total number of spikes. Near total suppression of the tone-evoked response was observed for all intensities examined.

at auditory-nerve synapses. These findings make several assumptions, including: (1) that an excitatory amino acid may be a transmitter at acoustic nerve synapses in the CN, and (2) that presynaptic $GABA_B$ receptors do indeed exist. These data lend further support, albeit circumstantially, to the idea that an excitatory amino acid is released at the auditory nerve synapses in the CN.

D. SUMMARY OF THE EFFECTS OF EXCITATORY AMINO ACIDS AND RELATED SUBSTANCES ON THE RESPONSE PROPERTIES OF NEURONS IN THE COCHLEAR NUCLEUS

Table II summarizes our findings that the excitatory amino acids, glutamate and aspartate, as well as their agonists, are capable of: (1) increasing discharge rates, (2) altering response patterns of "on", "pauser build-up", "build-up," and "chopper" neurons, (3) lowering CF thresholds, and

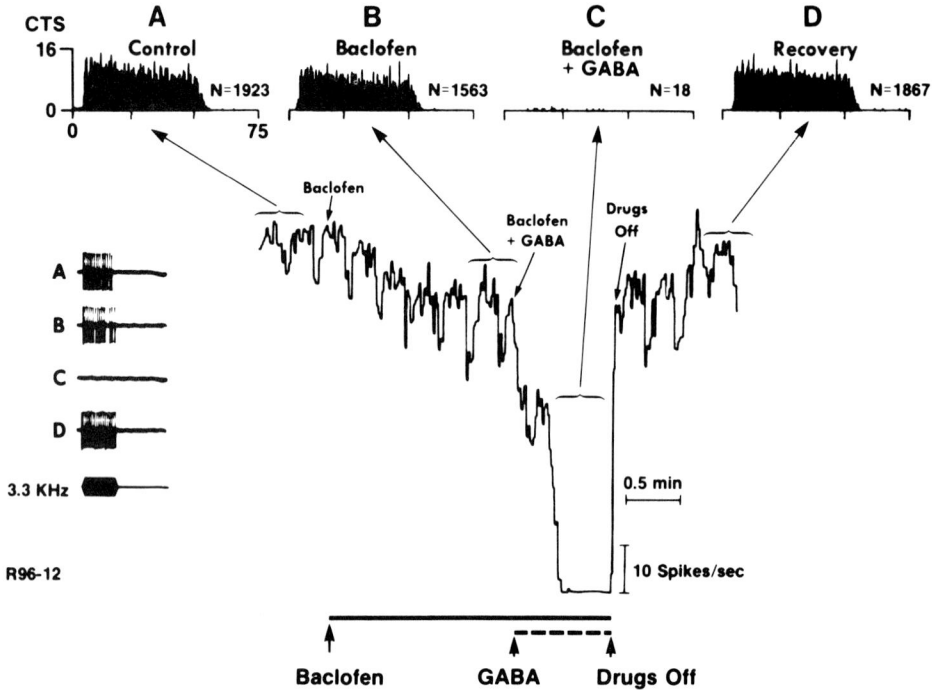

Figure 12-9. A rate-meter plot with representative post-stimulus time histograms and spike train segments display the effects of 50 nA of iontophoretically-applied (-)baclofen and 60 nA of GABA on a unit from the ventral cochlear nucleus. Stimuli were 50-ms tone bursts at CF (3.3 kHz) at 25 dB SPL, and were presented at the rate of 4/s. Responses were summed over 100 trials. The control condition (A) is followed by 3 min of continuous (-)baclofen application. Baclofen suppressed the response with little change in the "primary-like" response pattern (B). Simultaneous application of (-)baclofen and GABA resulted in an abrupt suppression of firing (C), illustrating the difference in time course and magnitude of effect of these two substances. When (-)baclofen and GABA application were discontinued (arrow), the unit recovered rapidly from GABA, followed by a slower recovery from (-)baclofen (D). (From Caspary et al., 1984.)

(4) shifting rate-intensity functions upward and to the left (Table II). On the other hand, substances thought to antagonize excitatory amino-acid neurotransmitters and their agonists, especially those thought to be acting on NMDA receptors, appear to provide partial blockade of the effects of iontophoretically-applied excitatory amino acids and their agonists, and, in some cases, block tone-evoked synaptic activity as well. These antagonists tend to (1) decrease discharge rates, (2) raise thresholds, (3) shift rate-intensity functions downward and to the right, and (4) rarely alter response patterns (Table II). Finally, baclofen, a substance thought to act presynaptically to inhibit the release of excitatory amino-acid neurotransmitters, tends to

Tone Evoked Transmitter Release

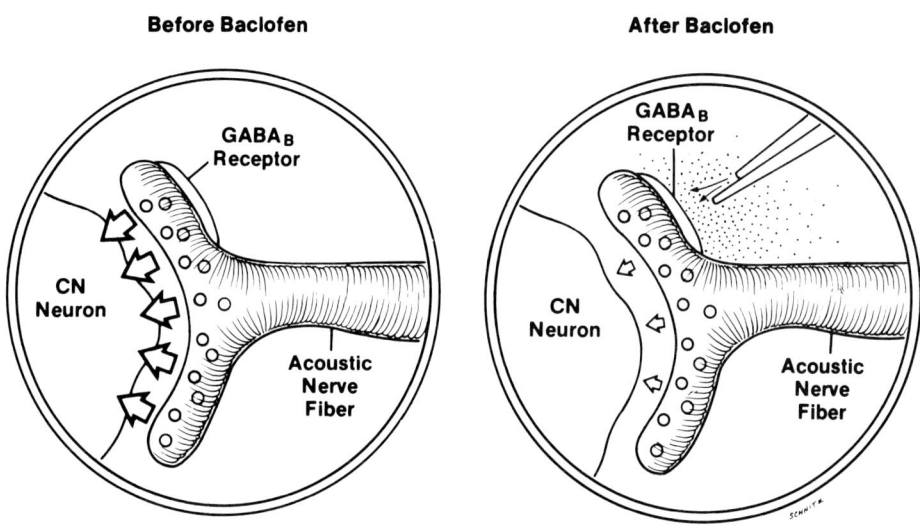

Figure 12-10. A drawing illustrating the hypothesis of (-)baclofen's proposed effect on excitatory amino-acid release at auditory-nerve synapses. Synaptic release of transmitter (arrows) onto a cochlear nucleus neuron (illustration on left) is reduced after iontophoretic application of (-)baclofen. (-)Baclofen is thought to act presynaptically through a putative $GABA_B$ receptor, which inhibits transmitter release.

(1) decrease discharge rates, (2) raise thresholds, (3) shift rate-intensity functions downward and to the right, and (4) rarely alter response patterns (Table II).

TABLE 12-II
EFFECT ON TONE-EVOKED ACTIVITY

Properties Affected	AGONISTS			ANTAGONISTS		OTHER
	Glutamate	Aspartate	NMDA	DAA	APV	Baclofen
Discharge rate	Increased	Increased	Increased	Decreased	Decreased	Decreased
Threshold	Decreased	Decreased	Decreased	Increased	Increased	Increased
Response pattern	Altered	Altered	Altered	Rarely altered	Rarely altered	Rarely altered
Rate-intensity function	Up and to the left	Up and to the left	Up and to the left	Down and to the right	Down and to the right	Down and to the right

The effects of these substances on response properties of CN neurons strongly support a role for excitatory amino acids or similar substances (Zaczek et al., 1982) in excitatory neurotransmission in the CN. The action of the frequently-utilized antispasticity agent, baclofen, on acoustically-evoked CN responses suggests a possible future role for this agent in the treatment of certain acoustically-related pathological conditions.

IV. GLYCINE AND GABA MAY MEDIATE INHIBITORY RESPONSES SEEN FOR NEURONS OF THE DORSAL COCHLEAR NUCLEUS

The morphology of the DCN has been extensively described, using both light and electron microscopy (for review, see Brugge and Geisler, 1978; Tsuchitani, 1978; Osen and Mugnaini, 1981; Rhode et al., 1983; Aitken et al., 1984). DCN neurons display complex inhibitory properties which include: (1) inhibitory side-bands or inhibitory areas found within the excitatory regions, (2) non-monotonic rate-intensity functions, and (3) "build-up" or "pauser build-up" types of PSTH response patterns at CF (Caspary, 1972; Evans and Nelson, 1973; Britt and Starr, 1976; Young and Brownell, 1976; Voigt and Young, 1980; Young and Voigt, 1982; Rhode et al., 1983). These physiologic response properties, as well as the finding of a large number of non-primary types of synaptic endings described for this area (Kane, 1974; Kane and Finn, 1977; Kane, 1978; Mugnaini et al., 1980), suggest that several potent inhibitory mechanisms are operative in the DCN. Intracellular studies (Caspary, 1972; Britt and Starr, 1976; Rhode et al., 1983) suggest the presence of inhibitory postsynaptic potentials, and a number of histochemical and cytochemical studies strongly implicate GABA and glycine as inhibitory neurotransmitters of this area (for review, see Chapters 14 and 15, this volume; Fisher and Davies, 1976; Klinke, 1981; Guth and Melamed, 1982). Godfrey and colleagues (1976, 1977), using techniques developed by Lowry (1953), mapped the regional CN distribution of GABA and glycine, and found that the highest concentrations occurred in the DCN, both for the cat and the white rat. These studies indicated that GABA and glycine were present in low concentrations in the auditory nerve, while concentrations 40 times greater were found within the granular regions of the DCN. Strychnine, a specific receptor antagonist for glycine, has been reported to bind to cells in the DCN with little or no binding in VCN (Zarbin et al., 1981). Iontophoretic application of glycine raised thresholds and altered the "pauser build-up" and "build-up" response patterns recorded in DCN neurons (Caspary et al., 1979).

A. EFFECTS OF GLYCINE ON NEURONS IN THE DORSAL COCHLEAR NUCLEUS

Iontophoretically-applied glycine suppressed tone-evoked activity and altered the contour of "build-up" and "pauser build-up" responses of DCN neurons (Fig. 11). This inhibitory amino acid was effective at very low currents (e.g., simply turning off the retaining current, Fig. 11), displaying a greater than 40% reduction in discharge rate at 35 dB SPL. Glycine altered the rate-intensity function, shifting it downward in parallel to the control curve at the lower doses, while shifting it to the right and altering its contour with application of 50 nA (Fig. 12).

Further systematic studies of the effect of iontophoretic application of GABA and glycine are needed to attempt to delineate: (1) whether different cell types and/or response types within the DCN are selectively responsive to either GABA or glycine and to their specific antagonists bicuculline and strychnine, and (2) which aspects of the different inhibitory-response properties are mediated by which inhibitory neurotransmitter.

V. GLYCINE-MEDIATED BINAURAL INHIBITION IN THE LATERAL SUPERIOR OLIVARY NUCLEUS

The SOC is the first structure in the ascending auditory pathway to receive bilateral input and, thus, neurons in this structure are thought to code for localization of sound in space (Guinan *et al.*, 1972a; Brugge and Geisler, 1978; Tsuchitani, 1978; Jenkins and Masterton, 1982; Caird and Klinke, 1984). Binaural neurons in the lateral superior olive (LSO) receive their predominant contralateral input from the ventral cochlear nucleus (VCN) of the opposite side, via "secure" synapses in the ipsilateral medial nucleus of the trapezoid body (MNTB), and receive their major ipsilateral projection directly from the homolateral CN (Fig. 1). Neurons in the LSO classically exhibit an excitatory response with ipsilateral acoustic stimulation and a profound reduction of response with binaural acoustic stimuatlion. (Fig. 13). Data suggesting that glycine may be the inhibitory transmitter of the MNTB come from ultrastructural studies showing Type II synaptic endings on the somata of the principal cells of the LSO (Cant, 1983; White, 1983), endings which have flattened vesicles and symmetrical junctions in the rat (White, 1983) and cat (Cant, 1983). The endings appear to be of the kind identified as glycinergic in spinal-cord interneurons (Bodian, 1966; Hokfelt and Ljundakai, 1971). This type of ending is also present on the somata and proximal dendrites of the principal cells of the chinchilla LSO (Fig. 14). The two endings in Fig. 14 resemble those described as glycinergic (Bodian, 1966; Hokfelt and Ljundakai, 1971) in spinal-cord interneurons,

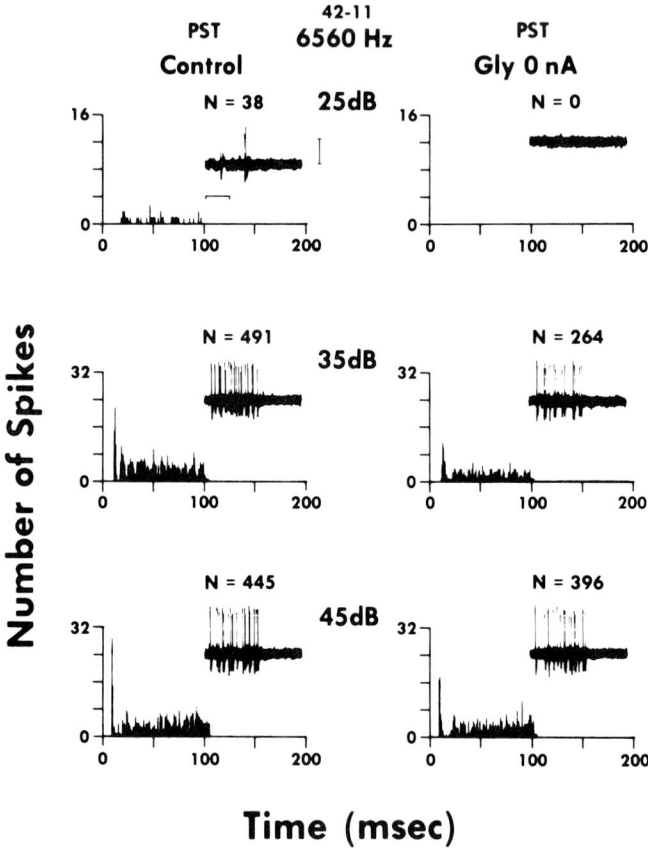

Figure 12-11. The effects of glycine on post-stimulus time histograms obtained from a unit from the deep region of dorsal cochlear nucleus are shown. Stimuli were 100-ms tone bursts at CF (6,560 Hz), presented at 4/s, with responses being summed over 50 trials. Control histograms are shown in the left column and the effects of iontophoretic application of glycine are shown in the right column at three stimulus intensities. Simply turning off the retaining current reduced the discharge rate of this neuron at each intensity tested. Glycine effects were greatest near threshold and for the 35-dB SPL condition, with both pattern and rate affected. Five sweeps of raw data are shown in the inset of each histogram (vertical bar = 1.5 mv; horizontal bar = 0.25 msec).

where considerable support exists for glycine as the inhibitory neurotransmitter (Young and Macdonald, 1983). It has also been reported that tritiated glycine may be taken up preferentially by this type of ending in the SOC (Schwartz, 1982, Chapter 15, this volume). The glycine-receptor antagonist, strychnine, displays intense binding to LSO neurons, but not to the neurons in the medial superior olivary nucleus (Zarbin et al., 1981). Moore and Caspary (1983) reported that glycine, iontophoretically-applied during ipsi-

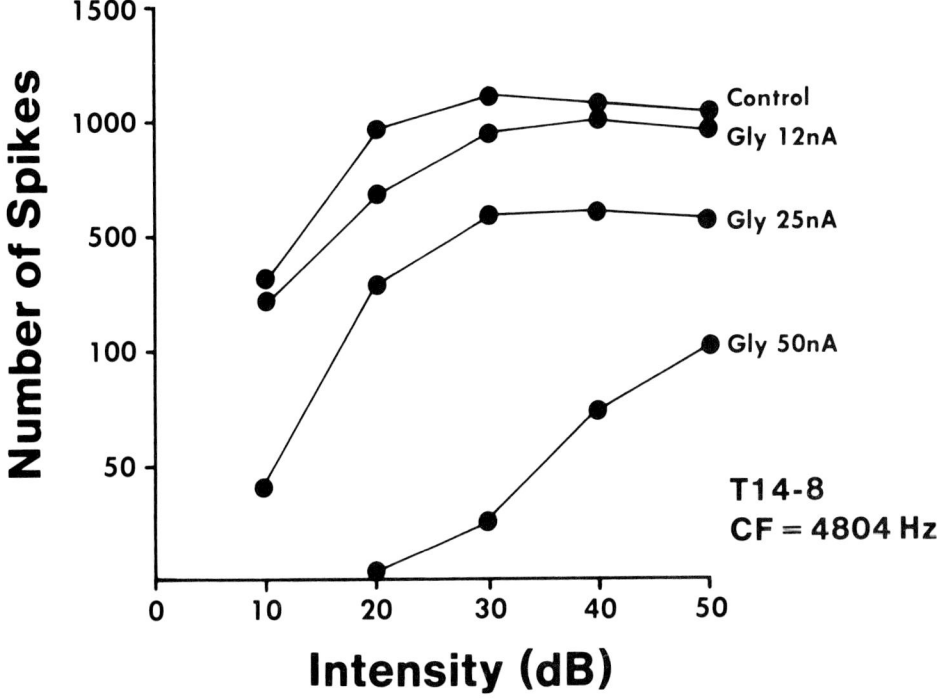

Figure 12-12. Rate-intensity functions display the effects of iontophoretic application of glycine ejected at three current levels onto a "chopper" neuron from deep dorsal cochlear nucleus. Discharge rate is in number of spikes. The rate-intensity functions shifted to the right with increasing dose levels of glycine. At 50 nA of glycine, the function is no longer parallel to the control curve.

lateral acoustic stimulation, mimics the effect of binaural stimulation, while iontophoretically-applied strychnine, an alkaloid with an affinity for the glycine receptor three times that of glycine (Davidson, 1976), blocks the inhibitory effect of binaural acoustic stimulation (see Fig. 15).

A. EFFECTS OF GLYCINE AND STRYCHNINE ON NEURONS OF THE LATERAL SUPERIOR OLIVE

Chinchilla LSO neurons displayed "on"-type phasic responses to ipsilateral stimulation (Fig. 15A), while binaural stimulation resulted in a total suppression of the response (Fig. 15B). Iontophoretic application of glycine during ipsilateral stimulation appeared to mimic the binaural condition (Fig. 15C), while application of strychnine blocked the binaural inhibition (Fig. 15E), resulting in a response pattern which resembled the ipsilateral control condition. Strychnine also reversibly blocked the effect of exogenously-applied glycine

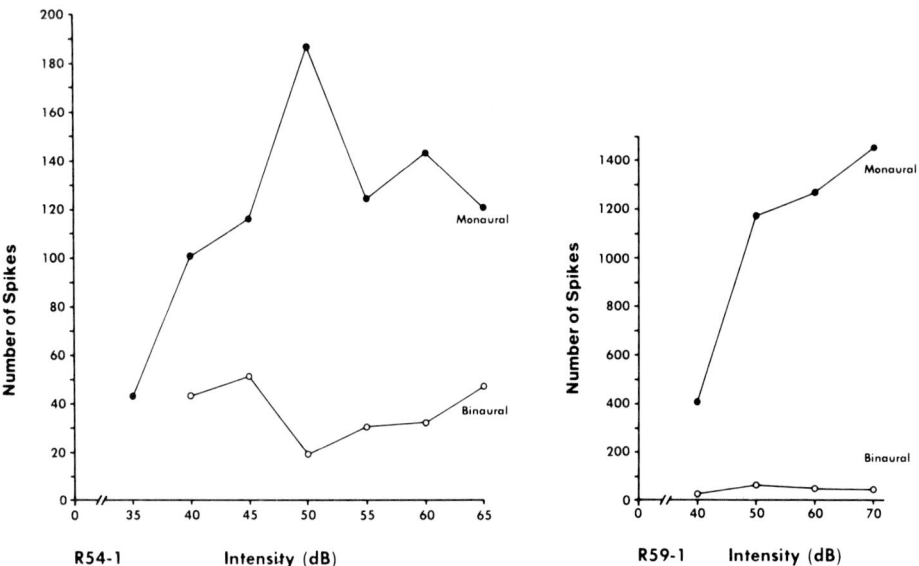

Figure 12-13. Rate-intensity curves for two representative neurons from the lateral superior olive are displayed. Firing rate is in total number of spikes. These units show the typical excitatory response to monaural (ipsilateral) acoustic stimulation and suppression of the response with binaural acoustic stimulation.

(Fig. 15F), after which the neuron recovered to the pre-drug condition (Fig. 15G, 15H). GABA usually had no effect on the ipsilateral response (Fig. 15L), further suggesting the specificity of the glycine effect. Similar effects were seen across intensities (Fig. 16), with binaural stimulation during strychnine application resembling the ipsilateral control, whereas the binaural control condition (bottom trace) resulted in total suppression of the response. Thus, glycine applied during ipsilateral stimulation appears to mimic the effect of binaural acoustic stimulation, while strychnine application appears to antagonize or block the effect of binaural inhibition, which, along with studies described above, implicate glycine as the transmitter mediating this response. These findings further suggest that glycine may be a transmitter mediating binaural inhibition in certain LSO neurons and that the projection to the LSO from the MNTB may be glycinergic.

VI. CONCLUSION

In order to understand certain neurologically-based communication abnormalities, such as receptive aphasia, neural presbycusis, auditory perceptual dysfunction, and, perhaps, sensory-induced seizures, we must begin to examine the neurochemical aspects of central sensory processing. It is con-

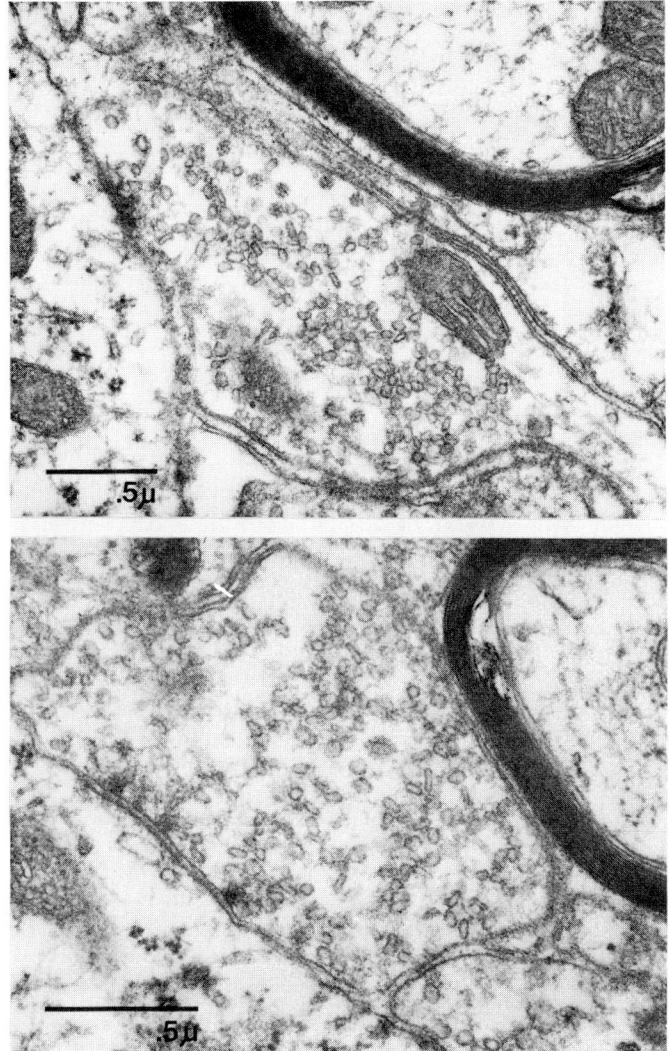

Figure 12-14. Two examples of synaptic endings on the somata of principal cells of the lateral superior olivary nucleus. These endings have small, flattened vesicles and symmetrical synaptic contacts, and most likely represent the glycinergic inhibitory input from the medial nucleus of the trapezoid body (see text).

ceivable that certain communicative disorders may involve neurochemical fault(s). The studies reported here represent attempts to characterize the neurotransmitters which mediate tone-evoked synaptic transmission at specific sites within the central auditory pathway. Putative neurotransmitters and their respective antagonists were applied in tests of mimicry (identity of

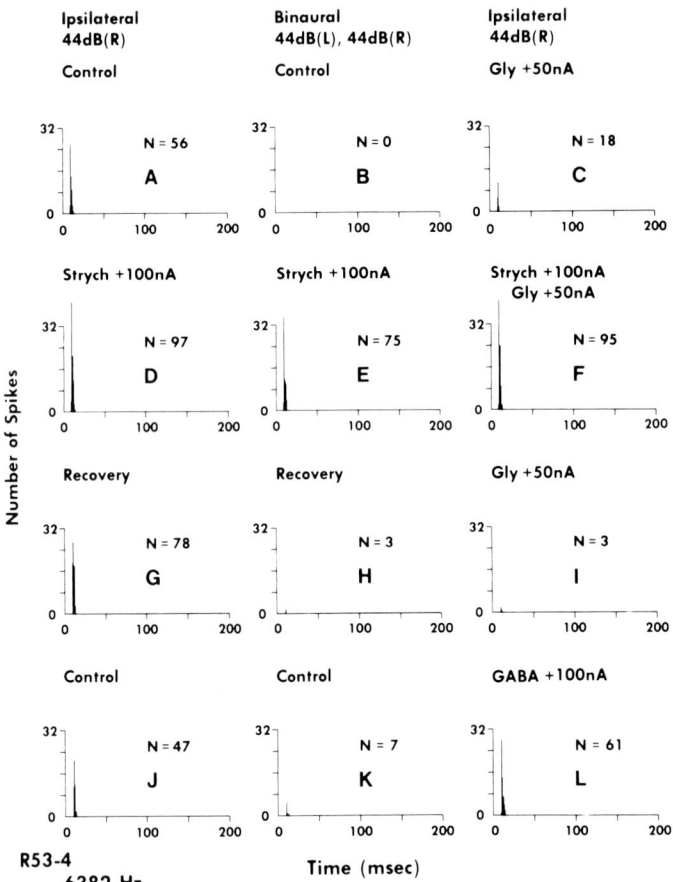

Figure 12-15. Post-stimulus time histograms obtained from a neuron located in the lateral superior olive. Stimuli were 100-ms tone bursts at CF (6,382 Hz) at 44 dB SPL, presented at 2/s. Responses were summed over 50 trials. Discharge rate is in number of spikes. Responses to ipsilateral stimulation are shown in the left and right columns, while responses to binaural stimulation are shown in the middle column. This unit displayed an "on"-type response to ipsilateral stimulation (A), which was totally suppressed with binaural acoustic stimulation (B). Iontophoretic application of glycine during ipsilateral stimulation resulted in suppression of the response (C), mimicking the binaural condition. Strychnine, a specific glycine antagonist, when applied during ipsilateral stimulation, enhanced the response to ipsilateral stimulation (D). Strychnine application during binaural stimulation blocked the binaural inhibition (E). Simultaneous application of glycine and strychnine during ipsilateral stimulation (F) resulted in reversible blocking of the effects of the exogenously-applied glycine. Responses returned to pre-drug levels during both ipsilateral (G) and binaural (G) stimulation. Glycine application was repeated (I), after which the neuron resumed its control discharge rate observed during both ipsilateral (J) and binaural (K) stimulation. GABA application appeared to have no effect on this neuron (L). (From Moore and Caspary, 1983.)

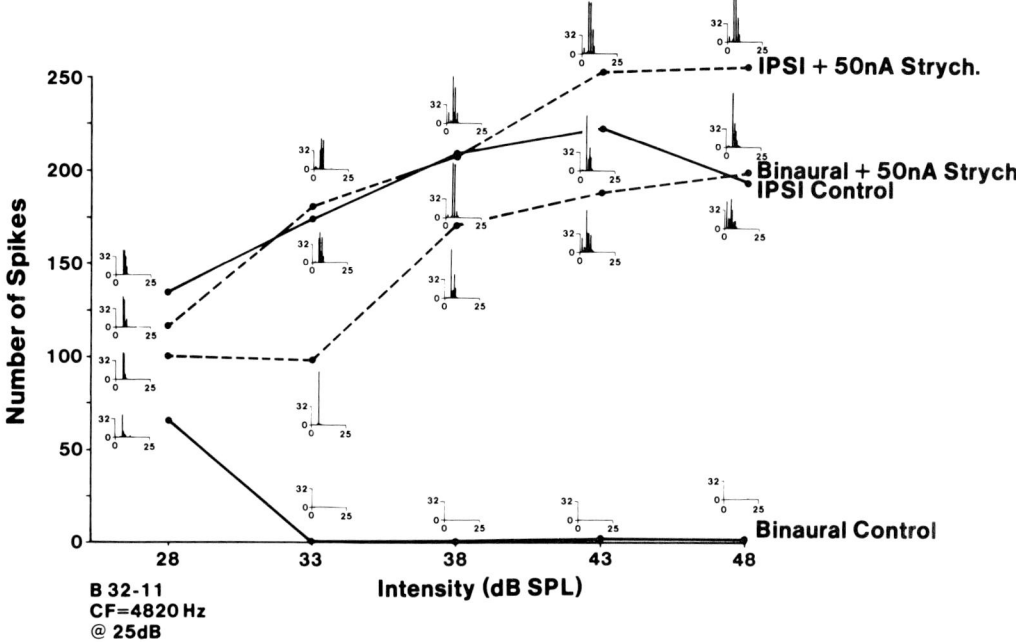

Figure 12-16. Rate-intensity curves and accompanying post-stimulus time histograms for a neuron from the lateral superior olive. Stimuli were 50-ms tone bursts at CF (4,820 Hz), delivered at the rate of 4/s. Responses were summed over 100 trials. Discharge rate is in number of spikes. Iontophoretic application of 50 nA of strychnine during ipsilateral stimulation increased the firing rate of the unit. Binaural inhibition (bottom curve) was blocked by strychnine application.

action) and antagonism (pharmacological identity) of the synaptically-released compound. Although we have examined neurotransmitter actions on several important parameters, such as alteration of threshold, spontaneous activity, response patterns, and rate-intensity functions, as measures of sensory processing, the effects of neurotransmitters on many other coding features have not yet been investigated. We have utilized established, auditory-physiologic paradigms in the context of the known synaptic morphology in an attempt to interpret how certain amino-acid neurotransmitters may be involved in the processing of the acoustic message. These initial steps will, it is hoped, lead to characterization of the role of neurotransmitters in generating the synaptic events involved in the coding of acoustic information, with the eventual goal of specific pharmacologic intervention in the treatment of communicative disorders.

ACKNOWLEDGMENTS

Ms. D. C. Havey and Dr. M. J. Moore made substantial contributions to segments of the studies presented in this chapter. The authors are indebted to Ms. K. Berry for her competent technical assistance and help with the manuscript. Mr. G. Schnitz and W. Andrea of the Department of Biomedical Communications were invaluable in the preparation of the illustrations. We also thank Ms. T. Kissel for typing the manuscript. These studies were supported by NIH grant NS 15640, the Deafness Research Foundation, the Pearson Family Foundation, and funds from the Central Research Committee of Southern Illinois University School of Medicine.

REFERENCES

Aitken, L. M.; Irvine, D. R. F., and Webster, W. R.: Central neural mechanisms of hearing. In Geiger, S. R.; Darian-Smith, I., Brookhart, J. M., and Mountcastle, V. B. (Eds.): *Handbook of Physiology – the Nervous System III*. Baltimore, Williams and Wilkins, 1984, pp. 675-737.

Altschuler, R. A.; Neises, G. R., Harmison, G. G., Wenthold, R. J., and Fex, J.: Immunocytochemical localization of aspartate aminotransferase immunoreactivity in cochlear nucleus of the guinea pig. *Proc. Natl. Acad. Sci. U.S.A. 78:* 6553-6557, 1981.

Bodian, D.: Synaptic types of spinal motoneurons: An electron microscopic study. *Bull. Hopkins Hosp. 119:* 16-45, 1966.

Bowery, N. G.: Baclofen: 10 years on. *Trends Pharmacol. Sci. 3:* 400-403, 1982.

Britt, R., and Starr, A.: Synaptic events and discharge patterns of cochlear nucleus. *J. Neurophysiol. 39:* 162-178, 1976.

Brugge, J. F., and Geisler, C. D.: Auditory mechanisms of the lower brainstem. *Annu. Rev. Neurosci. 1:* 63-94, 1978.

Cain, C. R., and Simmonds, M. A.: Effects of baclofen on the olfactory cortex slice preparation. *Neuropharmacology 21:* 371-373, 1984.

Caird, D., and Klinke, R.: Processing of binaural stimuli by cat superior olivary complex neurons. *Exp. Brain Res. 52:* 385-399, 1984.

Cant, N. B.: The fine structure of the lateral superior olivary nucleus of the cat. *Soc. Neurosci. Abstr. 225:* 5, 1983.

Canzek, V., and Reubi, J. C.: The effect of cochlear nerve lesion on the release of glutamate, aspartate, and GABA from cat cochlear nucleus in vitro. *Exp. Brain Res. 38:* 437-441, 1980.

Caspary, D. M.: Classification of subpopulations of neurons in the cochlear nuclei of the kangaroo rat. *Exp. Neurol. 37:* 131-151, 1972.

Caspary, D. M., and Havey, D. C.: Effects of putative amino acid neurotransmitters on response patterns of neurons in the cochlear nuclei. *Soc. Neurosci. Abstr. 4:* 5, 1978.

Caspary, D. M.; Havey, D. C., and Faingold, C. L.: Effects of microiontophoretically applied glycine and GABA on neuronal response patterns in the cochlear nuclei. *Brain Res. 172:* 179-185, 1979.

Caspary, D. M.; Havey, D. C., and Faingold, C. L.: Glutamate and aspartate: Alteration of thresholds and response patterns of auditory neurons. *Hearing Res. 4:* 325-333, 1981.

Caspary, D. M.; Rybak, L. P., and Faingold, C. L.: Baclofen reduces tone evoked activity of cochlear nucleus neurons. *Hearing Res.* 1984. In press.

Collins, G. G. S.; Anson, J., and Kelly, E. P.: Baclofen: Effects on evoked field potentials and amino acid neurotransmitter release in the rat olfactory cortex slice. *Brain Res. 238:* 371-383, 1982.

Collins, G. G. S.; Anson, J., and Probett, G.: Patterns of endogenous amino acid release from slices of rat and guinea-pig olfactory cortex. *Brain Res. 209:* 231-234, 1981.

Cutting, D. A., and Jordan, C. C.: Baclofen as a potential analgesic agent. *Scot. Med. J. 25:* S17-S22, 1980.

Davidoff, R. A., and Sears, E. S.: The effects of Lioresal on synaptic activity in the isolated spinal cord. *Neurology 24:* 957-963, 1974.

Davidson, N.: *Neurotransmitter Amino Acids. Chapter 3. Glycine.* New York, Academic Press, 1976, pp. 39-56.

Davies, J.: Selective depression of synaptic excitation in cat spinal neurones by baclofen: An iontophoretic study. *Brit. J. Pharmacol. 72:* 373-384, 1981.

Erulkar, S. D.: Comparative aspects of spatial localization of sound. *Physiol. Rev. 52:* 237-360, 1972.

Evans, E. F., and Nelson, P. G.: On the functional relationship between the dorsal and ventral divisions of the cochlear nucleus of the cat. *Exp. Brain Res. 17:* 428-442, 1973.

Fisher, S. K., and Davies, W. E.: GABA and its related enzymes in the lower auditory system of the guinea pig. *J. Neurochem. 27:* 1145-1155, 1976.

Fonnum, F.: Glutamate: A neurotransmitter in mammalian brain. *J. Neurochem. 42:* 1-11, 1984.

Godfrey, D. A.; Carter, J. A., Berger, S. J., Lowry, O. H., and Matschinsky, F. M.: Quantitative histochemical mapping of candidate transmitter amino acids in cat cochlear nucleus. *J. Histochem. Cytochem. 25:* 417-431, 1977.

Godfrey, D. A.; Carter, J. A., Lowry, O. H., and Matschinsky, F. M.: Distribution of gamma-aminobutyric acid, glycine, glutamate and aspartate in the cochlear nucleus of the rat. *J. Histochem. Cytochem. 26(2):* 118-126, 1978.

Godfrey, D. A.; Kiang, N. Y. S., and Norris, B. E.: Single unit activity in the posterior ventral cochlear nucleus of the cat. *J. Comp. Neurol. 162:* 247-268, 1975.

Godfrey, D. A., and Matschinsky, F. M.: Approach to three-dimensional mapping of quantitative histochemical measurements applied to studies of the cochlear nucleus. *J. Histochem. Cytochem. 24:* 697-712, 1976.

Goldberg, J. M., and Brown, P. B.: Response of binaural neurons of dog superior olivary complex to dichotic tonal stimuli: some physiological mechanisms of sound localization. *J. Neurophysiol. 32:* 613-636, 1969.

Guinan, J. J.; Guinan, S. S., and Norris, B. E.: Single auditory units in the superior olivary complex. I. Responses to sounds and classifications based on physiological properties. *Int. J. Neurosci. 4:* 101-120, 1972a.

Guinan, J. J.; Norris, B. E., and Guinan, S. S.: Single auditory units in the superior olivary complex. II. Locations of unit categories and tonotopic organization. *Int. J. Neurosci. 4:* 147-166, 1972b.

Guth, P. S., and Melamed, B.: Neurotransmission in the auditory system: A primer for pharmacologists. *Ann. Rev. Pharmacol. Toxicol. 22:* 383-412, 1982.

Hansson, E.; Jarlstedt, J., and Sellstrom, A.: Sound-stimulated ^{14}C-glutamate release from the nucleus cochlearis. *Experientia 36:* 576-577, 1980.

Havey, D. C., and Caspary, D. M.: A simple technique for constructing "piggy-back" multibarrel microelectrodes. *Electroenceph. Clin. Neurophysiol. 48:* 249-251, 1980.

Hokfelt, T., and Ljundakai, A.: Light and electron microscopic autoradiography on spinal cord slices after incubation with labeled glycine. *Brain Res. 32:* 189–194, 1971.

Jenkins, W. M., and Masterton, R. B.: Sound localization: Effects of unilateral lesions in central auditory system. *J. Neurophysiol. 47:* 987–1016, 1982.

Kane, E. C.: Synaptic organization in the dorsal cochelar nucleus of the cat: A light and electron microscopic study. *J. Comp. Neurol. 155:* 301–329, 1974.

Kane, E. S.: Central transport and distribution of labelled glutamic and aspartic acids to the cochlear nucleus in cats: an autoradiographic study. *Neuroscience 4:* 729–745, 1979.

Kane, E. S.: Primary afferents and the cochlear nucleus. In Naunton, R. F., and Fernandez, C. (Eds.): *Evoked Electrical Activity in the Auditory Nervous System.* New York, Academic Press, 1978, pp. 337–352.

Kane, E. S., and Finn, R. C.: Descending and intrinsic inputs to dorsal cochlear nucleus of cats: A horseradish peroxidase study. *Neuroscience 2:* 897–912, 1977.

Klinke, R.: Neurotransmitters in the cochlea and the cochlear nucleus. *Acta Otolaryng. 91:* 541–554, 1981.

Krnjevic, K.: Desensitization of GABA receptors. In Costa, E.; Di Chiara, G., and Gessa, G. L. (Eds.): *GABA and Benzodiazepine Receptors.* New York, Raven Press, 1981, pp. 111–120.

Lees, A. J.; Clarke, C. R. A., and Harrison, M. J.: Hallucinations after withdrawal of baclofen. *Lancet 8016:* 858, 1977.

Lowry, O. H.: The quantitative histochemistry of the brain: Histological sampling. *J. Histochem. Cytochem. 1:* 420–428, 1953.

Martin, M. R.: Baclofen and the brain stem auditory evoked potential. *Exp. Neurol. 76:* 675–680, 1982.

Masterton, B.; Jane, J. A., and Diamond, I. T.: Role of brainstem auditory structures in sound localization. I. Trapezoid body, superior olive, and lateral lemniscus. *J. Neurophysiol. 30:* 341–359, 1967.

Moore, M. J., and Caspary, D. M.: Strychnine blocks binaural inhibition in lateral superior olivary neurons. *J. Neurosci. 3:* 237–242, 1983.

Mugnaini, E.; Osen, K. K., Dahl, A., Friedrich, V. L., Jr., and Korte, G.: Fine structure of granule cells and related interneurons (termed Golgi cells) in the cochlear nuclear complex of cat, rat and mouse. *J. Neurocytol. 9:* 537–570, 1980.

Oliver, D. L.; Potashner, S. J., Jones, D. R., and Morest, D. K.: Selective labeling of spiral ganglion and granule cells with D-aspartate in the auditory system of cat. *J. Neurosci. 3:* 455–472, 1983.

Olpe, H. R.; Baudry, M., Fagni, L., and Lynch, G.: The blocking action of baclofen on excitatory transmission in the rat hippocampal slice. *J. Neurosci. 2:* 698–703, 1982.

Osen, K. K., and Mugnaini, E.: Neuronal circuits in the dorsal cochlear nucleus. In Syka, J., and Aitken, L. (Eds.): *Neuronal Mechanisms of Hearing.* New York, Plenum Press, 1981, pp. 119–125.

Pierau, F., and Zimmerman, P.: Action of GABA-derivative on postsynaptic potentials and membrane properties of cats' spinal motoneurons. *Brain Res. 54:* 376–380, 1973.

Pinto DeSilva, O.; Polikar, M., and Loustalot, P.: A review of clinical trials with Lioresal. In Birkmayer, W. (Ed.): *Spasticity—A Topical Survey.* Vienna, Hans Huber Press, 1972, pp. 192–207.

Potashner, S. J.: Baclofen: effects on amino acid release. *Can. J. Physiol. Pharmacol. 56:* 150–154, 1978.

Potashner, S. J.: Baclofen: effects on amino acid release and metabolism in slices of guinea pig cerebral cortex. *J. Neurochem. 32:* 103–109, 1979.

Potashner, S. J.: Uptake and release of D-aspartate in the guinea pig cochlear nucleus. *J. Neurochem. 41:* 1094-1101, 1983.

Potashner, S. J., and Lake, N.: Action of baclofen and pentobarbital on amino acid release. In Di Chiara, G., and Gessa, G. L. (Eds.): *Glutamate as a Neurotransmitter.* New York, Raven Press, 1981, pp. 139-145.

Rhode, W. S.; Smith, P. H., and Oertel, D.: Physiological response properties of cells labeled intracellularly with horseradish peroxidase in cat dorsal cochlear nucleus. *J. Comp. Neurol. 213:* 427-447, 1983.

Ritz, L. A., and Brownell, W. E.: Single unit analysis of the posteroventral cochlear nucleus of the decerebrate cat. *Neuroscience 7:* 1995-2010, 1982.

Sawynok, J., and LaBella, F. S.: On the involvement of GABA in the analgesia produced by baclofen, muscimol and morphine. *Neuropharmacol. 2:* 397-403, 1982.

Schwartz, I. R.: Differential tritiated amino acid labeling of synaptic terminals in the cat medial superior olivary complex. *Assoc. Res. Otolaryngol. Abstr. 5:* 21, 1982.

Spencer, H. J.; Lynch, G., and Jones, R. K.: The use of somatofugal transport of horseradish peroxidase for tract tracing and cell labeling. In Robertson, R. T. (Ed.): *Methods in Physiological Psychology, Vol. II.* New York, Academic Press, 1978, pp. 291-316.

Stein, R.: Hallucinations after sudden withdrawal of baclofen. *Lancet,* pp. 44-45, 1977.

Terrence, C. F.; Potter, D. M., and Fromm, G. H.: Is baclofen an analgesic? *Clin. Neuropharmacol. 6(3):* 241-245, 1983.

Thalmann, R.; Comegys, T. H., Thalmann, I, and Webster, D. B.: Distribution of aspartate and glutamate in cochlear nucleus (CN) following destruction of organ of Corti (OC). *J. Acoust. Soc. Am. 67:* S77, 1980.

Tsuchitani, C.: Lower auditory brainstem structures of the cat. In Naunton, R. F., and Fernandez, C. (Eds.): *Evoked Electrical Activity in the Auditory Nervous System.* New York, Academic Press, 1978,

Voigt, H. F., and Young, E. D.: Evidence of inhibitory interactions between neurons in dorsal cochlear nucleus. *J. Neurophysiol: 44:* 76-96, 1980.

Watkins, J. C.; Davies, J., Evans, R. H., Francis, A. A., and Jones, A. W.: Pharmacology of receptors for excitatory amino acids. In Di Chiara, G., and Gessa, G. L. (Eds.): *Glutamate as a Neurotransmitter.* New York, Raven Press, 1981, pp. 263-273.

Wenthold, R. J.: Glutaminase and aspartate aminotransferase decrease in the cochlear nucleus after lesion of the auditory nerve. *Brain Res. 190:* 293-297, 1980.

Wenthold, R. J.: Release of endogenous glutamic acid, aspartic acid and GABA from cochlear nucleus slices. *Brain Res. 162:* 338-343, 1979.

Wenthold, R., and Altschuler, R.: Immunocytochemistry of aspartate aminotransferase and glutaminase. In Hertz, L.; Kvamme, E., McGeer, E., and Schousboe, A. (Eds.): *Glutamine, Glutamate, and GABA in the Central Nervous System. Neurology and Neurobiology.* New York, Alan R. Liss, Inc., 1983, pp. 33-50.

White, J. S.: Fine structure of the lateral superior olivary nucleus in the albino rat. *Soc. Neurosci. Abstr. 225:* 4, 1983.

Woolf, N. K.: Precise extracellular marking in the auditory nerve with high impedance micropipettes. *Hearing Res. 4:* 121-125, 1981.

Young, A. B., and Macdonald, R. L.: Glycine as a spinal cord neurotransmitter. In Davidoff, R. A. (Ed.): *Handbook of the Spinal Cord.* New York, Marcel Dekker, 1983, pp. 1-43.

Young, E. D., and Brownell, W. E.: Responses to tones and noise of single cells in dorsal cochlear nucleus of unanesthetized cats. *J. Neurophysiol. 39:* 282-300, 1976.

Young, E. D., and Voigt, F.: Response properties of type II and type III units in dorsal cochlear nucleus. *Hearing Res. 6:* 153-169, 1982.

Zaczek, R.; Koller, K., and Coyle, J. T.: N-acetyl-aspartyl-glutamate: An endogenous peptide with agonist properties at a glutamate receptor in brain. *Soc. Neurosci. Abstr. 8:* 403, 1982.

Zarbin, M. A.; Wamsley, J. K., and Kuhar, M. J.: Glycine receptor: Light microscopic autoradiographic localization with [^{3}H]strychnine. *J. Neurosci. 1:* 532–547, 1981.

Chapter 13

PUTATIVE NEUROTRANSMITTER RECEPTORS IN THE CENTRAL AUDITORY SYSTEM

BARBARA J. MORLEY, GLENN R. FARLEY, AND ERIC JAVEL

I. Introduction
II. Overview of Cholinergic Mechanisms
III. Cholinergic Receptors
IV. Evidence for Cholinergic Receptors in the Auditory System
V. The Physiological Characterization of Cholinergic Receptors in the Inferior Colliculus
VI. Conclusions
References

I. INTRODUCTION

A major area of emphasis in the effort to understand the biochemistry of the auditory system is the identification of putative neurotransmitters. From the many neurochemicals now known to be present in neural tissues, we have just begun to identify compounds with a function relevant to neurotransmission. An important criterion for establishing a neurochemical as a neurotransmitter is the demonstration of the presence of an endogenous receptor that mediates a physiological response.

Although the necessity for demonstrating the presence and physiological relevance of a receptor is clear, documenting neurotransmitter receptor function has proven difficult. In the case of many putative neurotransmitters, acceptable ligands for receptor binding assays are not available. Also, there are few, if any, known antagonists for many putative neurotransmitters.

One of the few neurotransmitters that has been amenable to intensive investigation is acetylcholine (ACh). ACh is present in most auditory structures, including the cochlea, the cochlear nucleus, the inferior colliculus, the superior olivary complex, the medial geniculate nucleus, and the nuclei of the lateral lemniscus. Our research has been directed at providing evidence that functional ACh receptors (AChR's) exist in auditory structures, with an emphasis on the nicotinic AChR's (nAChR's) present in the inferior colliculus (IC).

Before discussing the role of AChR's in the IC, we will provide an overview of some contemporary views of cholinergic mechanisms and receptors. These discussions should help to clarify the rationale we have used to investigate auditory AChR's.

II. OVERVIEW OF CHOLINERGIC MECHANISMS

ACh was the first endogenous neurotransmitter to be identified. Because of its presence in several neural structures, it remains one of the most important neurochemicals and is a major focal point in contemporary neurobiological research. ACh is present in high concentrations in brain areas related to the processing of sensory information in most species, suggesting that sensory systems have retained many common neurochemical pathways throughout evolution. The chemical structure of ACh is identical in all tissues, but it subserves several diverse and complex functions. This demonstrates the important fact that the neural response to ACh depends upon the characteristics of the *receptor* sites.

The most important function of ACh in the brain is presumed to be as a mediator of electrophysiological activity, i.e., as a neurotransmitter or neuroregulator. Evidence for a neurotransmitter role for ACh has been *inferred* from the presence of ACh and neurochemicals associated with ACh, such as choline acetyltransferase (ChAT) and acetylcholinesterase (AChE). ChAT is the specific enzyme responsible for the synthesis of ACh. Because ChAT is not known to have another function, it is considered to be a specific "marker" for cholinergic neurons. Its presence in perikarya of neuronal groups is currently assumed to be unequivocal evidence for the production of ACh within the cell, and the presumed function of such ACh is as a neurotransmitter or neuroregulator. ACh and ChAT activity should also be found in the axon terminals of these cells. Stimulation of the terminals should result in the release of ACh. Pre-synaptically, these cells should also demonstrate high-affinity choline uptake. Post-synaptic receptor sites should be found that are capable of interacting with ACh.

AChE is the enzyme that catalyzes the hydrolysis of ACh. The current model of how a cholinergic neuron works is that AChE is present *postsynaptically*, where its catalytic action is believed to be the primary mechanism for terminating the action of ACh. *Presynaptic* AChE is found in nerve-muscle connections, in cholinergic neurons in the sympathetic and parasympathetic nervous systems, in the axons of the olivocochlear bundle, and, presumably, in several areas of the central nervous system (CNS). In the CNS, AChE is contained in neurons that apparently do not contain ChAT (e.g., Eckenstein and Sofroniew, 1983) and are presumably neither cholinergic nor cholinoceptive. Pre-synaptic AChE may function as a trophic factor or may represent a phylogenetic "remnant" of a previously cholinergic or cholinoceptive neuron. However, AChE has also been co-localized with peptides (e.g., Morley *et al.*, 1984) in at least some neuronal groups that are not cholinergic. Although the major mode of terminating the action of ACh is presumed to be via AChE, it should be noted that the inactivation of ACh can also occur by diffusion and/or by interaction with the receptor. Certain primitive synapses known to use ACh as the neurotransmitter do not contain cholinesterases (see Michelson and Zeimal, 1973, for a discussion of these issues). For this reason, AChE is not utilized as a reliable marker for cholinergic activity.

Although the appearance of ChAT and ACh is usually taken as good evidence for the presence of ACh as a neurotransmitter, ACh may also have nonspecific effects and local hormonal actions, or may directly alter ion permeability. Because of the difficulties in demonstrating a neurotransmitter function for ACh simply from its presence, it is extremely important to provide evidence for endogenous receptor sites that can specifically interact with ACh.

III. CHOLINERGIC RECEPTORS

Cholinergic receptors are often classified as nicotinic or muscarinic. This classification was defined at the turn of the century, primarily by researchers demonstrating that certain tissues respond differently to nicotine and muscarine.

Nicotinic sites are often considered to be excitatory and muscarinic sites have often been found to be inhibitory. The nicotinic action is often thought to be rapid, while muscarinic responses may be slow. In the CNS, including the auditory nuclei, most of the responses to ACh have been found to be slow and are therefore assumed to be muscarinic. The original classification of nicotinic and muscarinic receptors, however, may be fortuitous and may not apply to many central and peripheral structures, including those of the auditory system.

The understanding of the nAChR has been developed largely from research on the nerve-muscle synapse of skeletal muscle. This receptor has been well studied, using biochemical, physiological, and pharmacological techniques. The nAChR is activated by nicotine and other *agonists*, most notably carbamylcholine chloride (carbachol). Substances that block the binding of ACh to the receptor site are called nicotinic *antagonists*. These include d-tubocurarine (curare) and several neurotoxins, such as α-bungarotoxin (BuTX) and α-cobra toxin.

Much important information about receptors has been learned from studying the peripheral nerve-muscle synapse. These studies have shown us that a receptor may consist of more than one subunit, that agonists and antagonists may bind to separate subunits of the receptor, and that an ion channel may be an integral part of a receptor. While the ACh binding site on neural tissues may be considered to be a "recognition site," complex relations between the binding of antagonists on various subunits and physiological blocking activity also exist. This has made direct comparisons between *in vivo* and *in vitro* experiments difficult in some tissues. It should also be noted that both biochemical and physiological studies employing agonists must be interpreted cautiously, since agonists change the conformational state of a receptor and thus alter the pharmacological properties of the receptor.

There are other pharmacological agents that affect nicotinic responses at neural synpases, such as the synapses of the spinal Renshaw cell and the sympathetic ganglion. Nicotine and curare have effects at these synapses that are similar to those found at the neuromuscular synapse. However, cytisine is a potent agonist at ganglionic synapses, but not at synapses of skeletal muscle. Additionally, cholinergic responses at ganglionic synapses are blocked by substances which only weakly antagonize skeletal muscle nAChR's, such as hexamethonium, a bisquaternary ammonium salt, and mecamylamine, a secondary amine. Since the physiological *actions* of these pharmacological agents are not known, we will refer to them as *ganglionic agonists* and *ganglionic blockers*, respectively.

Receptors activated by ACh and muscarine have been identified in target organs of the postganglionic parasympathetic nervous system, such as smooth muscle and heart, and in mammalian brain. At these receptor sites, cholinergic responses typically have a slow onset, are inhibitory, and are blocked by atropine, but not by nicotine. Muscarinic responses may be mediated by a "second messenger," such as cyclic GMP, and this may account for the slow response onset. Muscarinic receptors can be studied in biochemical receptor binding experiments using the antagonists atropine, quinuclidinyl benzilate (QNB), and propylbenzilylcholine mustard (PrBCM). In biochemical experiments, the binding of QNB or PrBCM is blocked by atropine, but not by nicotinic agents, confirming that QNB and PrBCM are specific to muscarinic sites. Both QNB and PrBCM, however, are preferable to atropine as muscarinic antagonists, because they are nearly irreversible and bind with little nonspecific binding.

All of the nicotinic and muscarinic substances described above have been utilized in physiological and pharmacological studies of cholinergic receptors. However, only a few are appropriate for biochemical studies. Ligands that have been radiolabeled and are useful for studying cholinergic receptor sites include ACh, nicotine, BuTX, curare, QNB, atropine, and PrBCM.

In dealing with these ligands, it is important to recognize that the entity being demonstrated is a "binding site." Only with the proper isolation, and biochemical and pharmacological characterization, can these binding sites be proven to be actual neurotransmitter-receptor sites.

IV. EVIDENCE FOR CHOLINERGIC RECEPTORS IN THE AUDITORY SYSTEM

Several studies have indicated that ACh, ChAT, and AChE are present throughout the auditory system (Koelle, 1954; Shute and Lewis, 1967; Palkovits and Jacobowitz, 1974; Cheney et al., 1975; Kobayashi et al., 1975; Kuhar, 1976; Saelens and Simke, 1976; Godfrey et al., 1977; Hoover et al., 1978; Vizi and Palkovits, 1978; Kimura et al., 1981). Evidence that ACh is functionally relevant to auditory processing comes from biochemical experiments demonstrating the binding of cholinergic receptor ligands in auditory nuclei (Morley et al., 1977; Segal et al., 1978; Rotter et al., 1979; Arimatsu et al., 1981; Morley and Kemp, 1981; Wamsley et al., 1981) and from experiments demonstrating that the application of ACh and cholinergic agents affect physiological activity (Curtis and Koizumi, 1961; Tebecis, 1970; Watanabe and Simada, 1973; Caspary et al., 1983; Farley et al., 1983).

In receptor binding studies, the presence of AChR's in auditory nuclei has been inferred from the specific binding of four cholinergic ligands: QNB, PrBCM, ACh, and BuTX.

The localization of mAChR's has been reported in autoradiographic studies using QNB and PrBCM. Muscarinic receptors have been localized in the nuclei of the lateral lemniscus, both the dorsal and ventral cochlear nuclei (Whipple and Drescher, 1984), and the IC, using QNB (Wamsley et al., 1981). These same areas also show high density of PrBCM binding (Rotter et al., 1981). Generally, there is a good correlation between the distributions of these two muscarinic-receptor ligands in the auditory nuclei. However, Wamsley et al., (1981) suggest that the binding of QNB is restricted to certain subdivisions of the cochlear nucleus, while PrBCM is more uniformly distributed.

In our studies, we have measured the concentration of BuTX in a biochemical assay in several areas of the central auditory system. These data are summarized in Table I. The presence of BuTX binding in these auditory nuclei has been confirmed by autoradiography in several laboratories (Hunt and Schmidt, 1978; Segal et al., 1978; Arimatsu et al., 1981). A high density of BuTX binding (2.9 pmol/g wet wt) is found in the IC. In biochemical assays of micro-dissected areas throughout the IC, we have found the binding to be distributed fairly uniformly (Morley, unpublished observations). However, autoradiographs of rat-brain IC show a higher concentration in the superficial lamina. The binding of BuTX to IC is saturable, of high-affinity, and has a cholinergic pharmacology, suggesting that it represents true receptor sites.

TABLE 13-I
CONCENTRATIONS OF BuTX BINDING IN AUDITORY NUCLEI

Area	pmol/g wet wt	pmol/mg protein
Cochlear Nucleus	1.14 ± 0.69	0.044 ± 0.015
Inferior Colliculus	2.92 ± 1.15	0.069 ± 0.016
Medial Geniculate Nucleus	1.48 ± 0.12	0.048 ± 0.012
Superior Olivary Complex	0.67 ± 0.04	0.030 ± 0.005
Auditory Cortex	2.64	0.033

BuTX binds to IC tissue with a single association rate constant of 2.6×10^7 min^{-1}. The upper limit for the dissociation constant is approximately 2×10^{-10} M and the half-life of the toxin-receptor complex is 5–8 h. This indicates that the association of BuTX with the IC receptor is slower than the binding of BuTX to skeletal-muscle AChR, and that the binding to the IC receptor is reversible.

The binding of BuTX to IC tissue clearly has a nicotinic cholinergic pharmacology (Fig. 1). Binding is inhibited by low concentrations of curare, nicotine, carbachol, and ACh. The ganglionic blocker, mecamylamine, and the muscarinic antagonist, atropine, produce negligible inhibition of binding.

The concentration of another type of nAChR in the IC has also been demonstrated, using tritiated ACh (with muscarinic sites saturated with atropine) (Schwartz et al., 1982). The concentration of this nAChR is 2.4 pmol/g wet wt. Although the concentration is similar to that found using BuTX, it is likely that these ligands are measuring different sites. The pharmacology of ACh binding is nearly opposite to that found with BuTX binding. The most potent inhibitor of this nicotinic ACh binding site is the ganglionic agonist, cytisine. Curare and BuTX negligibly affect the binding of this ACh site. Thus, the biochemical receptor assays suggest the presence of two nicotinic binding sites and at least one muscarinic receptor site in auditory nuclei.

Most auditory nuclei contain binding sites for both nicotinic and muscarinic ligands. Perhaps one exception is the superior olivary complex (SOC), which has a moderate concentration of uniformly-distributed BuTX sites (Hunt and Schmidt, 1978), but has not been reported to have muscarinic receptor sites.

Of particular interest is the IC, which contains high concentrations of *both* nicotinic and muscarinic receptors. The data from the biochemical binding studies (see Table II) suggest that there are receptors throughout the IC for

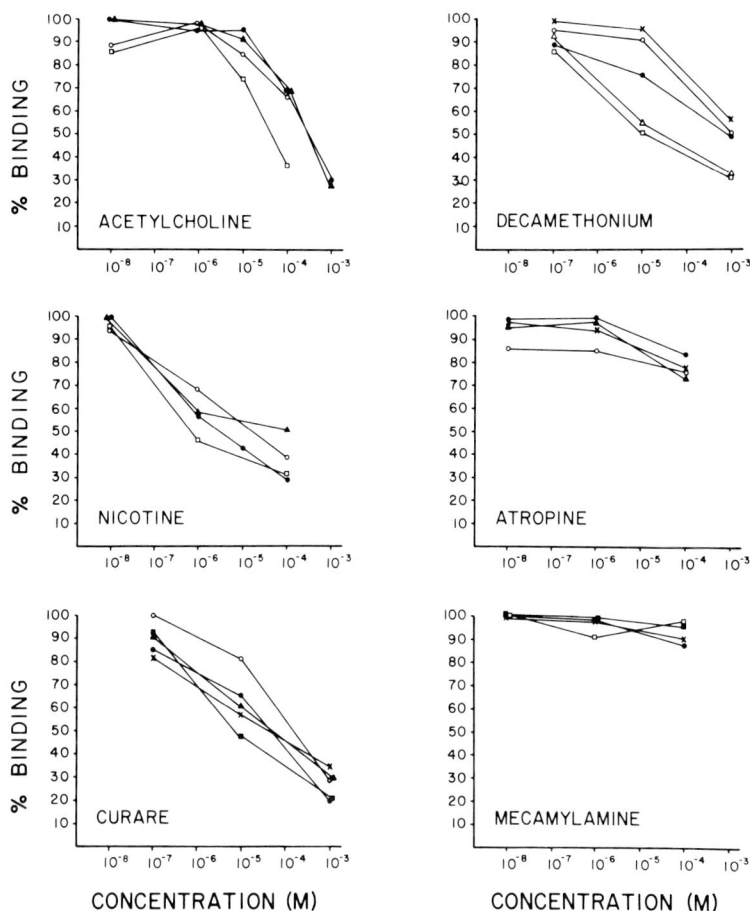

Figure 13-1. The pharmacological profile of BuTX binding to the IC (■—■) is compared to several other brain areas: Hippocampus (•—•), hypothalamus (o—o), raphe (□–□), superior olivary complex (X—X), reticular formation (▲—▲), and cortex (△—△). The inhibition experiments were carried out by incubating the tissue sample, 1.5×10^{-9} M BuTX, and various concentrations of the drug at room temperature for 60 min and assaying for toxin binding with a CM-50 column as previously described (Morley et al., 1977).

ACh, indicating that the presence of ACh and other cholinergic "markers" represents one or more endogenous cholinergic pathways terminating in the IC. The presence of two populations of nAChR's is suggested by the high concentrations of both agonist (nicotinic-ACh) binding and antagonist (BuTX) binding. At least one muscarinic-receptor type is suggested by the binding of the antagonists QNB and PrBCM. From the autoradiographic data, it appears that the concentrations of both nicotinic and muscarinic receptors may be higher in the external layers of the IC (see Table II).

TABLE 13-II
EVIDENCE FOR CHOLINERGIC RECEPTORS IN THE INFERIOR COLLICULUS

Receptor Ligand	Concentration	Distribution	Technique	Reference
Nicotinic:				
^{125}I-BuTX	2.9 pmol/g	Ubiquitous	Biochemical assay	Morley et al. (1977)
^{125}I-BuTX	High	Highest in superficial lamina	Autoradiography	Segal et al. (1978); Hunt and Schmidt (1978); Arimatsu et al. (1981)
^{3}H-ACh	2.4 pmol/g		Biochemical assay	Schwartz et al. (1982)
Muscarinic:				
^{3}H-QNB	High	External layer	Autoradiography	Wamsley et al. (1981)
^{3}H-PrBCM	Moderate	Not reported	Autoradiography	Rotter et al. (1979)

The biochemical receptor-binding studies are suggestive of an endogenous role for ACh in auditory processing. However, as we have previously mentioned, it is necessary to provide evidence of the biological significance of receptor binding sites. We have attempted to do this by investigating electrophysiological activity in the IC while iontophoretically applying cholinergic agents.

V. THE PHYSIOLOGICAL CHARACTERIZATION OF CHOLINERGIC RECEPTORS IN THE INFERIOR COLLICULUS

In our physiological studies, we have attempted to characterize cholinergic receptors *in vivo* and to relate our physiological observations with *in vitro* receptor-binding measurements. In these studies, we have investigated the effects of cholinergic agents on single-unit activity in the IC of rats (see Table III for a list of the agents utilized). We have characterized units according to their responsiveness to agonists (ACh, nicotine, carbachol). In some cases, we have also attempted to assess the effects of antagonists on agonist-induced physiological activity.

In these studies, a six-barreled "piggy-back" microelectrode is advanced into the IC while a probe stimulus of wide-band noise is delivered until responses of a single neuron are isolated. During an experiment, on-line analyses of post–stimulus-time histograms (PSTH's), response areas, response-intensity functions, tuning curves, and spike counts are collected. When a unit is isolated, its characteristic frequency (CF) is determined, either by means of a tuning curve or a response area. A response-intensity function is then taken at CF to determine threshold and maximum-response levels.

TABLE 13-III
LIST OF AGENTS USED

Agent	Concentration	pH	Source	Presumed Action
Carbamylcholine chloride (Carbachol)	0.5 M	6.3	Sigma	Cholinergic agonist
Acetylcholine chloride (ACh)	0.5 M	6.2	Sigma	Cholinergic agonist
Nicotine tartrate (Nic)	0.25 M	4.0	Pfaltz and Bauer	Nicotinic agonist
Atropine methyl nitrate (Atropine)	0.5 M	4.5	Sigma	Muscarinic antagonist
Scopolamine hydrobromide (Scopolamine)	0.5 M	4.0	Sigma	Muscarinic antagonist
d-Tubocurarine chloride (Curare)	0.05 M	4.0	Sigma	Neuromuscular nicotinic antagonist
Mecamylamine hydrochloride (Mecamylamine)	0.5 M	4.0	Merck[1]	Nicotinic ganglionic antagonist
Dihydro-beta-erythroidine hydrobromide (DHE)	0.1 M	4.0	Merck[1]	Nicotinic antagonist
BuTX	10 mg/ml	None	Miami Serpentarium[2]	Irreversible neuromuscular nicotinic blocker (reversible in CNS)

[1] Mecamylamine and DHE were obtained as a gift from Merck, Sharp & Dohme Research Laboratories, Rahway, NJ.
[2] Crude *Bungarus multicinctus* venom, obtained from the Miami Serpentarium, was purified with CM-50 and CM-25 gel chromatography. The resulting purified fraction was dissolved in 0.05 M NaCl.

A CF tone-burst at an intensity producing half-maximal driven responses is chosen as a test probe in order to assess both increases and decreases in activity. Typically, the data are analyzed in terms of the changes in response rates and shapes of the PSTH's.

In our studies of the cholinergic properties of the IC, we have primarily utilized carbachol as the agonist. Carbachol typically produces the same effects as ACh (see Table IV), but carbachol is more potent because it is resistant to degradation by AChE. We have also compared the effects of nicotine with carbachol. Nicotine is typically excitatory, and neurons that are excited by nicotine are also typically excited by carbachol, suggesting nAChR's exist on many cells in the IC. Both carbachol and ACh produce increases and decreases in tone-evoked responses, supporting the idea that inhibitory AChR's are also present in the IC (see Fig. 2). These findings are consistent with the results of biochemical receptor-binding assays that suggest that AChR's with nicotinic and muscarinic properties are localized in the IC.

TABLE 13-IV
COMPARISON OF EFFECTS OF AGONISTS

	Carbachol Effect		
	Excited	No effect	Inhibited
ACh Effect			
Excited	8 (37%)	2 (9%)	1 (4%)
No effect	0 (0%)	1 (4%)	1 (4%)
Inhibited	1 (4%)	2 (4%)	6 (27%)

This table shows data for nine cells representing the possible outcome of comparing the actions of two agents on a given unit. The cell entries are the number of units (and the corresponding percentage of the total tested) for which the comparison provided a particular result (e.g., excitation with agent 1 and inhibition with agent 2). The total number of units tested with both agents was 22.

On the assumption that ACh and carbachol activate both nAChR's and mAChR's, we have attempted to classify our recording units on the basis of the effects of pharmacological agents with reported blocking or antagonist properties at other neuronal cholinergic synapses. The substances we have used as blocking agents or antagonists are listed in Table III, and the results are summarized in Table V. In our studies, mecamylamine decreased the sound-evoked response rate of half of the units excited by carbachol, suggesting that mecamylamine is at least partially effective in blocking an endogenous cholinergic, excitatory (perhaps nicotinic) response. A typical PSTH recorded in the presence of mecamylamine is shown in Fig. 3. Neither of the muscarinic antagonists tested was consistently able to reverse the inhibitory effects of carbachol.

When responses were recorded in the presence of the classic nicotinic antagonist, curare, we observed a large increase in sound-evoked discharge rate (Fig. 3). This excitation was observed both in neurons that responded to carbachol and those that were unaffected by carbachol. We found no evidence that curare blocked or antagonized the agonist effects. There are previous reports of this paradoxical effect of curare in neural tissues (Curtis and Crawford, 1969). This action is poorly understood, but may represent an interaction of curare with ion channels. That is, it has been shown that curare blocks ion channels, but not ACh recognition sites, at certain parasympathetic ganglionic synapses (Ascher et al., 1979). Although curare has been shown to be an agonist at immature neuromuscular synapses (Trautmann, 1982), we believe that the excitatory effect we observe, following the iontophoresis of curare in the IC, is not a specific action at cholinergic receptor sites, because this action is not restricted to units that are responsive to carbachol.

The effect of BuTX was unlike that of curare. BuTX was without effect on

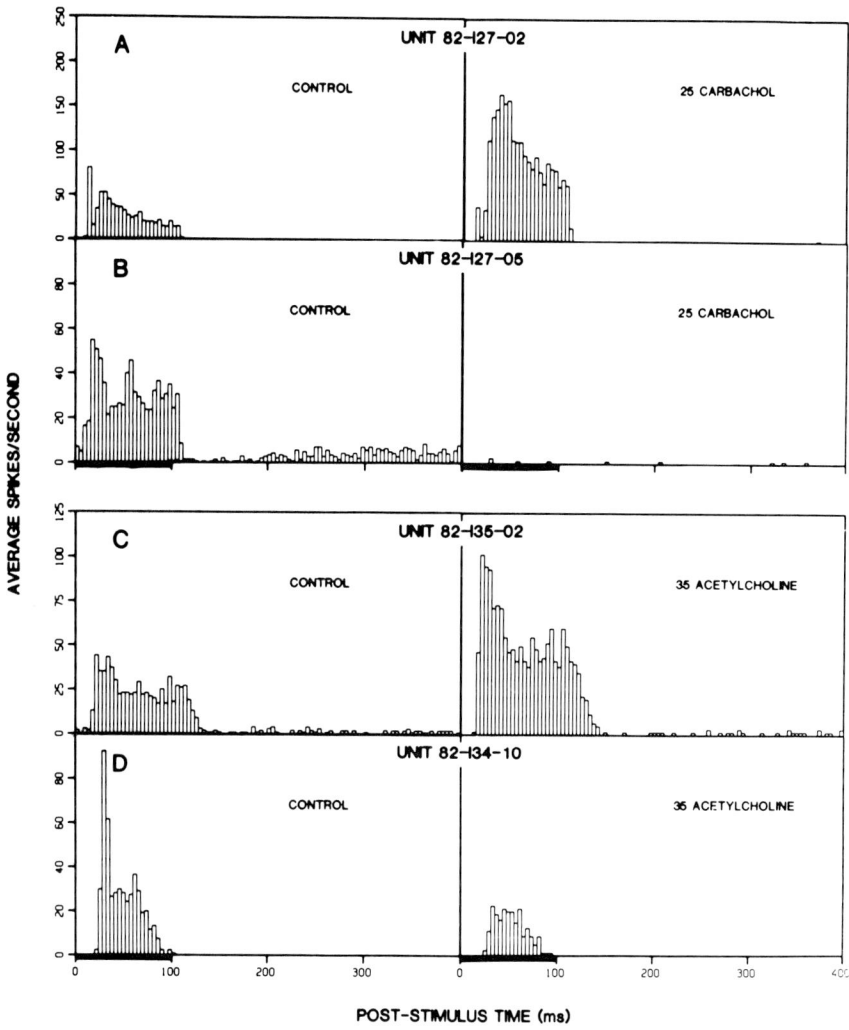

Figure 13-2. Effect of agonists on sound-evoked responses of cells in the rat IC. PSTH's for control and drug conditions in different units show that both carbachol (A, B) and ACh (C, D) can be both excitatory and inhibitory on IC neurons. Note that PSTH shapes are relatively similar when comparing control versus drug conditions, with the exception of panel D, which is an example of the small percentage of units encountered whose PSTH shapes were radically changed. Numbers adjacent to agonist names indicate ejection current in nanoamperes (nA).

the majority of units. In a small number of cases, however, BuTX prevented the inhibitory effects of carbachol (Fig. 4). This effect was not specific to inhibitory responses; we have also found that BuTX blocks

TABLE 13-V
AGENT EFFECTS WHEN DELIVERED ALONE

Agent	Excited	No Effect	Inhibited	Number Tested
Carbachol	48 (52%)	18 (19%)	27 (29%)	93
ACh	11 (50%)	2 (9%)	9 (41%)	22
Nicotine	7 (70%)	2 (20%)	1 (10%)	10
Mecamylamine	3 (13%)	5 (22%)	15 (65%)	23
Scopolamine	3 (14%)	7 (33%)	11 (53%)	21
DHE	6 (30%)	11 (55%)	3 (15%)	20
Curare	43 (96%)	2 (4%)	0 (0%)	45
Atropine	30 (83%)	5 (14%)	1 (3%)	36

The data above are presented in a format similar to that of Table I. The total comparisons made were 270.

excitatory, carbachol-induced responses (Morley et al., 1983).

The absence of a strong physiological blocking effect by BuTX has been reported by others and is interpreted to suggest a complex relation between ACh and antagonist binding sites at these receptors (Morley and Kemp, 1981, Morley et al., 1983). Fex and Adams (1978) have reported reversible blocking of the effects of olivocochlear neurons in the cochlea with BuTX. We have not been able to detect specific binding for BuTX in biochemical assays of the cochlea (Morley, unpublished observations), suggesting that this ligand may be capable of blocking cholinergic responses nonspecifically or at muscarinic sites. The presence of mAChR's in the cochlea has been documented (James et al., 1983).

Atropine methyl nitrate (AMN) is a muscarinic antagonist. We also observed apparent non-specific excitation produced by this form of atropine. There are no other data suggesting that atropine has a nonspecific effect in neural tissue. However, atropine binds to two separate sites in the brain, and only one of these is likely to be an mAChR (Farrow and O'Brien, 1973). Although AMN is a potent muscarinic antagonist, it has some ganglionic-blocking properties if used in high concentrations. We believe that this characteristic of AMN does not account for the paradoxical finding (i.e., the non-specific excitation), because its excitatory action was not restricted to neurons affected by carbachol, and excitation by atropine was found even at low "dose" (current intensity) levels.

One pharmacological agent of interest, which we have not tested, is strychnine. Strychnine has long been known to inhibit presumed cholinergic neurotransmission in the cochlea (see Guth et al., 1976, for a discussion). These data have long been difficult to interpret. Although strychnine has

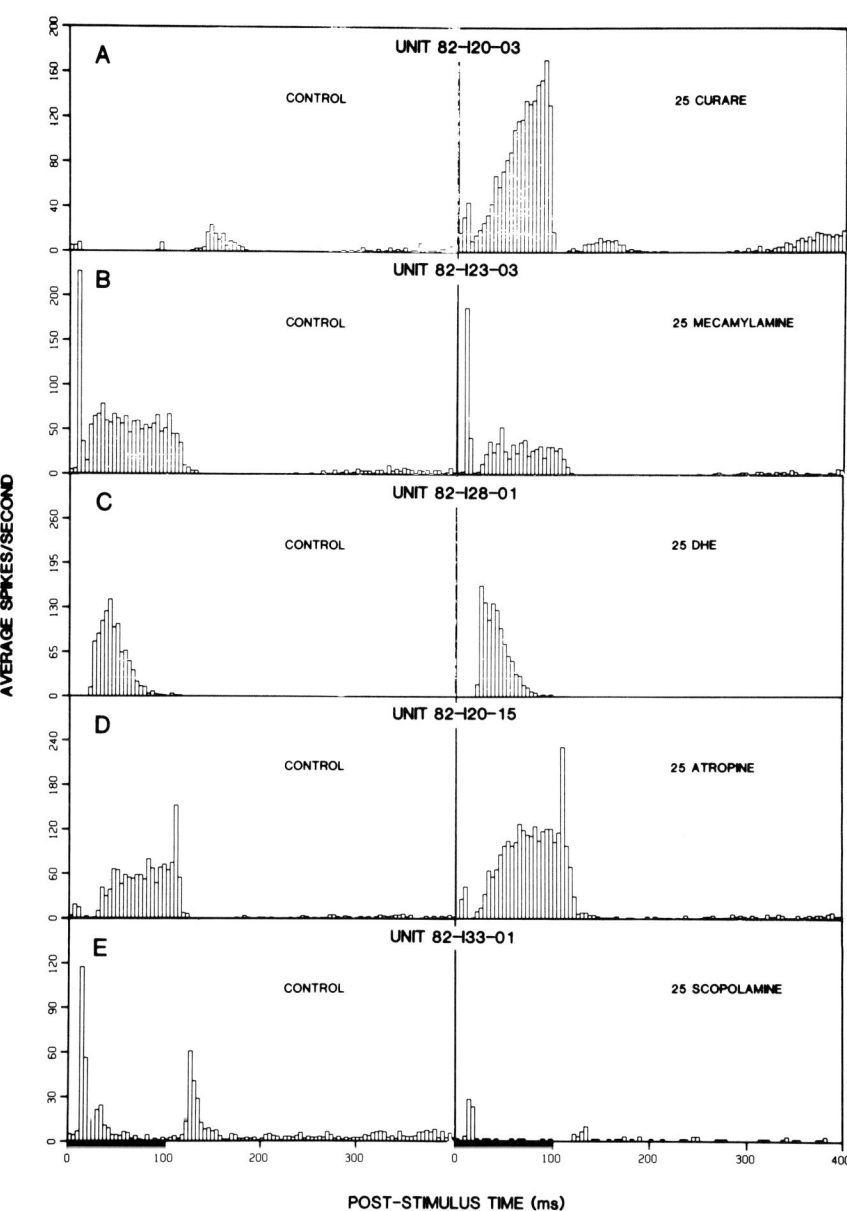

Figure 13-3. Effects of antagonists' PSTH's from different units show representative changes from control conditions, induced by five antagonists when delivered alone. Except for panel A, PSTH shapes are relatively similar when comparing control versus drug conditions. Numbers adjacent to agonist names indicate ejection current, in nA.

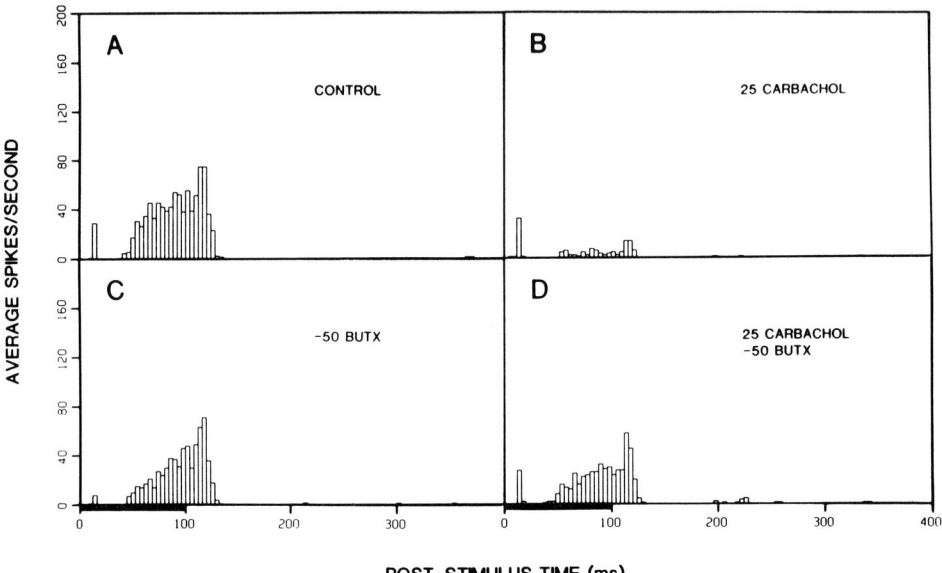

Figure 13-4. Antagonism of carbachol by BuTX. For this unit, carbachol was strongly inhibitory (B) relative to the control condition (A). BuTX (C) was mildly inhibitory relative to control when delivered alone, but diminished the inhibitory effect of carbachol when delivered concurrently (D). Note also the qualitative similarity between PSTH shapes for the different conditions. Numbers adjacent to agonist names indicate ejection current in nA.

typically not been considered to be a specific cholinergic antagonist, there is good evidence that strychnine can block ACh at certain synapses and can also inhibit the biochemical binding of BuTX to neural tissues (Ono and Salvaterra, 1981). This is interesting, in light of the observations that radiolabeled strychnine binds in high concentration to several areas of the auditory system (Zarbin et al., 1981).

In summary, the physiological data support the hypothesis that the binding of putative, cholinergic-receptor ligands to sites in the IC represents endogenous AChR's. We obtained both increases and decreases in response rates following the iontophoretic application of cholinergic agonists. We interpret these data to indicate that both nicotinic and muscarinic receptors exist in the IC. The effects of at least some antagonists were found to be consistent with this distinction.

In these studies, the principal effects we observed following the iontophoretic application of cholinergic agents occurred in the presence of strong, sound-evoked activity. In addition, the onsets of these responses were relatively slow. We interpret these results to suggest that the effect of the endogenous cholinergic input is to *regulate ongoing auditory processing* (Farley et al., 1983).

We have not yet been successful in correlating *in vitro* binding kinetics with the effects of iontophoretically-applied cholinergic agents. While these data may suggest the possibility that cholinergic receptors in the IC have mixed nicotinic-muscarinic properties, the biochemical, receptor-binding-inhibition experiments clearly indicate separate populations of receptor sites. These data emphasize the difficulties in studying brain receptors *in vivo* and suggest that many factors may significantly alter the responsiveness of a receptor to iontophoretically-applied pharmacological agents. These facts may include the conformational state of a receptor, ongoing physiological activity, the proximity of a receptor to the synapse, its proximity to other neurotransmitter receptors, and the possibility that pharmacological agents act as multiple neural sites. In addition, these data indicate the necessity of utilizing information from several disciplines in studying neurotransmitter receptors. While biochemical binding assays may be more helpful for identifying and characterizing receptors, a biological role for a binding site must also be confirmed. Correlating biochemical and physiological data, although difficult, will be necessary for a complete understanding of the role of endogenous neurochemicals in neurotransmission.

VI. CONCLUSIONS

1. ChAT and ACh are present in the IC of mammals, indicating that ACh is probably an endogenous neurotransmitter or neuroregulator in the IC.

2. On the basis of biochemical receptor-binding assays, there are probably three kinds of cholinergic receptors in the IC. One of these appears to have a pharmacology similar to that of the skeletal-muscle nicotinic receptor and one of these appears to have a pharmacology similar to that of the nicotinic ganglionic receptor. A third class appears to have a muscarinic pharmacology.

3. Based on our physiological experiments, units can either increase or decrease their response in the presence of ACh and cholinergic agonists. Excitatory responses can be further differentiated on the basis of antagonism by mecamylamine. These data confirm that some receptors in the IC have a ganglion-like pharmacology and some have a muscarinic pharmacology. Nicotinic antagonists produce mixed effects that are inconsistent with their action at the skeletal-muscle nAChR; therefore, it was not possible to characterize a third class of receptors, based on antagonist activity.

4. To date, we have not obtained perfect agreement between the results of

biochemical-inhibition experiments using cholinergic agents in receptor binding assays and the *in vivo* analysis of the effects of cholinergic agents. Part of the reason for this lack of agreement may lie in the difficulty in obtaining receptors *in vitro* that are in the same conformational state as exists *in vivo*. It is also possible that receptors with different pharmacological properties exist on the same neurons in the same IC, and act simultaneously to control neuronal activity.

REFERENCES

Arimatsu, Y.; Seto, A., and Amano, T.: An atlas of α-bungarotoxin binding sites and structures containing acetylcholinesterase in the mouse central nervous system. *J. Comp. Neurol.* 198: 603–631, 1981.

Ascher, P.; Lowy, W. A., and Rang, H. P.: Studies on the mechanism of action of acetylcholine antagonists of rat parasympathetic ganglion cells. *J. Physiol. (Lond.)* 295: 139–170, 1979.

Caspary, D. M.; Havey, D. C., and Faingold, C. L.: Effects of acetylcholine on cochlear nucleus neurons. *Exp. Neurol.* 82: 491–498, 1983.

Cheney, D. L.; LeFevre, H. F., and Racagni, G.: Choline acetyltransferase activity and mass fragmentographic measurement of acetylcholine in specific nulcei and tracts of rat brain. *Neuropharmacol.* 14: 801–809, 1975.

Curtis, D. R., and Crawford, J. M.: Central synaptic transmission—microelectrode studies. *Annu. Rev. Pharmacol.* 9: 209–240, 1969.

Curtis, D. R., and Koizumi, K.: Chemical transmitter substances in brain stem of cat. *J. Neurophysiol.* 24: 80–90, 1961.

Eckenstein, F., and Sofroniew, M. V.: Identification of central cholinergic neurons containing both choline actyltransferase and acetylcholinesterase and of central neurons containing only acetylcholinesterase. *J. Neurosci.* 3: 2286–2291, 1983.

Farley, G. R.; Morley, B. J.; Javel, E., and Gorga, M. P.: Single-unit responses to cholinergic agents in the rat inferior colliculus. *Hearing Res.* 11: 73–91, 1983.

Farrow, J. T., and O'Brien, R. D.: Binding of atropine and muscarine to rat brain fraction and its relation to the acetylcholine receptor. *Mol. Pharmacol.* 9: 33–40, 1973.

Fex, J., and Adams, J. C.: α-Bungarotoxin blocks reversibly cholinergic inhibition in the cochlea. *Brain Res.* 159: 440–444, 1978.

Godfrey, D. A.; Williams, A. D., and Matschinsky, F. M.: Quantitative distribution of choline acetyltransferase and acetylcholinesterase activities in the rat cochlear nucleus. *J. Histochem. Cytochem.* 29: 720–730, 1977.

Guth, P. S.; Norris, C. H., and Bobbin, R. P.: The pharmacology of transmission in the peripheral auditory system. *Pharmacol. Rev.* 28: 95–125, 1976.

Hoover, D. B.; Muth, E. A., and Jacobowitz, D. M.: A mapping of the distribution of acetylcholine, choline acetyltransferase and acetylcholinesterase in discrete areas of rat brain. *Brain Res.* 153: 295–306, 1978.

Hunt, S., and Schmidt, J.: Some observations on the binding patterns of α-bungarotoxin in the central nervous system of the rat. *Brain Res.* 157: 213–232, 1978.

James, W. M.; Cheatham, M. A., and Klein, W. L.: Muscarinic acetylcholine receptor binding in the guinea pig cochlea. *Hearing Res.* 9: 113–121, 1983.

Kimura, H.; McGeer, P. L., Peng, J. H., and McGeer, E. G.: The central cholinergic system studied by choline acetyltransferase immunohistochemistry in the cat. *J. Comp. Neurol.* 200: 151-201, 1981.

Kobayashi, R. M.; Brownstein, M., Saavedra, J. M., and Palkovits, M.: Choline acetyltransferase content in discrete regions of the rat brain stem. *J. Neurochem.* 24: 637-640, 1975.

Koelle, G. B.: The histocehmical localization of cholinesterases in the central nervous system of the rat. *J. Comp. Neurol.* 100: 211-218, 1954.

Kuhar, M. J.: The anatomy of cholinergic neurons. In Goldberg, A. M., and Hanin, I. (Eds.): *Biology of Cholinergic Neruons.* New York, Raven Press, 1976, pp. 3-27.

Michelson, M. J., and Zeimal, E. V.: *Acetylcholine.* New York, Pergamon Press, 1973.

Morley, B. J.; Farley, G. R., and Javal, E.: Nicotinic acetylcholine receptors in mammalian brain. *Trends Pharmacol. Sci.* 4: 225-227, 1983.

Morley, B. J., and Kemp, G. E.: Characterization of a putative nicotinic acetylcholine receptor in mammalian brain. *Brain Res. Rev.* 3: 81-104, 1981.

Morley, B. J.; Lorden, J. F., Brown, G. B., Kemp, G. E., and Bradley, R. J.: Regional distribution of nicotinic acetylcholine receptors in rat brain. *Brain Res.* 134: 161-166, 1977.

Morley, B. J.; Spangler, K., and Javel, E.: An atlas of somatostatincontaining cells and fibers in the developing and adult cat brain, 1984. Submitted for publication.

Ono, J. K., and Salvaterra, P. M.: Snake α-toxin effects on cholinergic and non-cholinergic responses of Aplysia Californica neurons. *J. Neurosci.* 1: 259-270, 1981.

Palkovits, M., and Jacobowitz, D. M.: Topographic atlas of catecholamines and acetylcholinesterase-containing neurons in the rat brain. II. Hindbrain (mesencephalon, rhombencephalon). *J. Comp. Neurol.* 157: 29-42, 1974.

Rotter, A.; Birdsall, N. J. M., Field, P. M., and Raisman, G.: Muscarinic receptors in the central nervous system of the rat. II. Distribution of binding of [^{3}H] propylbenzilylcholine mustard in the midbrain and hindbrain. *Brain Res. Rev.* 1: 167-183, 1979.

Saelens, J. K., and Simke, J. P.: Acetylcholine and choline concentrations of various biological tissues. In Goldberg, A. M., and Hanin, I. (Eds.): *Biology of Cholinergic Function,* New York, Raven Press, 1976, pp. 661-682.

Schwartz, R. D.; McGee, R., and Kellar, K. J.: Nicotinic cholinergic receptors labeled by [^{3}H] acetycholine in rat brain. *Mol. Pharmacol.* 22 56-62, 1982.

Segal, M.; Dudai, Y., Amsterdam, A.: Distribution of an α-bungarotoxinbinding cholinergic nicotinic receptor in rat brain. *Brain Res.* 148: 105-119, 1978.

Shute, C. D., and Lewis, P. R. The ascending cholinergic reticular system: neocortical, olfactory and subcortical projections. *Brain* 40: 494-521, 1967.

Tebecis, A. K.: Properties of cholinoceptive neurons in the medial geniculate nucleus. *Brit. J. Pharmacol.* 38: 117-137, 1970.

Trautmann. A.: Curare can open and block ionic channels associated with cholinergic receptors. *Nature (Lond.)* 298, 272-275, 1982.

Vizi, S., and Palkovits, M.: Acetylcholine content in different regions of the rat brain. *Brain Res. Bull.* 3: 93-96, 1978.

Wamsley, J. K.; Lewis, M. S.; Young, W. S. III, and Kuhar, M. J.: Autoradiographic localization of muscarinic cholinergic receptors in rat brainstem. *J. Neurosci.* 1: 176-191, 1981.

Watanabe, T., and Simada, S.: Pharmacological properties of cats' collicular auditory neurons. *Jap. J. Physiol.* 23: 291-308, 1973.

Whipple, M. R., and Drescher, D. G.: Muscarinic receptors in the cochlear nucleus and auditory nerve of the guinea pig. *J. Neurochem.* 43: 192-198, 1984.

Zarbin, M. A.; Wamsley, J. K., and Kuhar, M J.: Glycine receptor: light microscopic autoradiographic localization with [^{3}H] strychnine. *J. Neurosci.* 1: 532-547, 1981.

Chapter 14

THE NATURE OF NEUROTRANSMITTERS IN THE MAMMALIAN LOWER AUDITORY SYSTEM

W. Ewart Davies and Christopher Owen

I. Introduction
II. Cochlea and Auditory Nerve
   A. Excitation
   B. Inhibition
III. Cochlear Nucleus
   A. Excitation
   B. The Role of Acetylcholine
   C. Inhibition
IV. Inferior Colliculus
V. Audiogenic-Seizure-Susceptible Mice
IV. Summary and Conclusions
   References

I. INTRODUCTION

Research into the nature and function of neurotransmitters of the auditory system is quite sparse, compared to corresponding studies on the visual system. This is despite the fact that in many ways the auditory system is more amenable to investigation than the visual system. Whilst the first-, second-, and third-order sensory cells in vision are closely packed within the various layers of the retina, the equivalent cells in audition are segregated in the organ of Corti, spiral ganglion, and cochlear nucleus, respectively. This should make research into the nature of transmitters and their communication function easier, even though the end organ is less accessible than the retina.

Despite the advantages of the auditory system for neurotransmitter research, the identity of the primary afferent transmitter effective between the hair cells and the spiral ganglion cell dendrites steadfastly remains unidentified, and until recently (Wenthold, 1981; Potashner, 1983), little data were available as to the nature of the neurotransmitter of the spiral ganglion cell representation in the cochlear nucleus.

Some of the earlier studies showed that the conventional neurotransmitters, acetylcholine (Rossi, 1961) and the catecholamines (Fex et al., 1965) had only minor functions, if any, in the cochlea, eighth nerve, and cochlear nucleus. Similarly, GABA, undoubtedly the most well-authenticated inhibitory transmitter, appears to have little or no function in the cochlea (Tachibana & Kuriyama, 1974) although it probably has major functions at higher levels (Fex and Wenthold, 1976; Davies, 1977).

With the advent of more sensitive analytical methods, further progress has been made recently into the involvement of lesser known putative transmitters, both of a positive and a negative nature (Davies, 1982; Drescher et al., 1983; Hoffman et al., 1983).

II. COCHLEA AND AUDITORY NERVE

A. EXCITATION

There seems to be little doubt that the stimulus transfer from the hair cells to the dendrites of the spiral ganglion cells is mediated by chemical neurotransmission. The anatomical structure of the afferent synapse, first described by Smith and Sjöstrand (1961) in the guinea pig, has been confirmed in many other species, e.g., the monkey (Engstrom and Ades, 1973), rabbit (Borg et al., 1974), and cat (Dunn and Morest, 1975). The picture emerging is clearly that the afferent endings, particularly in the inner hair cells, contain synaptic vesicles and would be expected to be chemically mediated. The evidence for the outer hair cells' afferent endings, however, is less clear (Gulley and Reese, 1977), but there also are likely to be chemically mediated, although possibly secreting a different transmitter than the inner hair cells. The physiological picture emerging from the studies of very many workers, e.g., Ishi et al. (1971) and Furukawa and Matsuura (Chapter 2, this volume), also suggests clearly that the transmission is chemically mediated.

The nature of the primary afferent transmitter(s) steadfastly remains unidentified. The mass of published work has only served to eliminate most of the orthodox candidates. Acetylcholine, probably the premier central-nervous-system excitatory transmitter, undoubtedly has no function in the afferent system (Guth et al., 1981). Similarly, GABA (Fex and Wenthold, 1976), the catecholamines (Klinke and Evans, 1977), and serotonin (Klinke and Oertel, 1977) have to be eliminated as possibilities. The excitatory amino acids, glutamate and aspartate, pose different problems. Whilst the application of glutamate and aspartate to the cochlea produces excitation that is to be expected of the afferent transmitter (Bobbin and Thompson,

1978), their ubiquitous occurrence in all tissues and their indiscriminate ability to excite neural tissue devalues the data as being indicative of their identity as the afferent transmitter. Elegant studies on the nature of substances released into the cochlear perilymph during sound stimulation (Sewell et al., 1978; Drescher et al., 1983, Chapter 4, this volume) argue against glutamate or aspartate being the afferent transmitter. Drescher et al. (1983) showed an increase in the release of a GABA-like substance during stimulation, with little or no effect on the rate of release of glutamate and aspartate. The similar studies of Sewell et al. (1978) showed again little change in the release rate of the two amino acids, but showed the release of a hitherto unidentified substance, the "auditory nerve-activating substance."

Many other miscellaneous transmitter candidates have also been eliminated as possible transmitters in the cochlea (Guth et al., 1981). It would therefore appear that the afferent transmitter at the hair cell is likely to be a substance or substances not yet identified as having orthodox transmitter activity in other systems.

B. INHIBITION

Whilst the cochlea represents the source of many inhibitory phenomena, the only documented one to involve a neurotransmitter is the olivo-cochlear bundle (OCB). The bilaterally-represented OCB, first described by Rasmussen (1942), arises in the olivary nuclei and terminates primarily, but far from exclusively, on the outer hair cells. It is now fairly well established (for review see Klinke, 1981) that the inhibitory OCB uses acetylcholine as its transmitter; whether it does so exclusively is rather doubtful, particularly as a single cochlea receives two inputs from the OCB (the crossed and uncrossed) and has at least two kinds of endings, some innervating the hair cells and some innervating the afferent nerve endings (Iurato, 1974). Very recently, further doubt has been cast on the picture. James et al. (1983) showed that the bulk of ligand receptors binding QNB (a specific muscarinic ligand) are found in the spiral ganglion and eighth nerve and not in the sensory part of the organ. Furthermore, Arnold and Wang (1983) have shown that in the monkey, spiral ganglion cells have vesicle-containing nerve endings which they postulated to arise from the efferent system. Consequently, it is more than feasible that further efferent transmitters still remain to be identified. Such a possibility is enhanced by the study of Hoffman et al. (1983) who showed the existence of met-enkephalin in the cochlea. Further, met-enkephalin appears to exist in efferent fibers of the organ of Corti (Fex and Altschuler, 1981; Chapter 1, this volume) and is co-localized with acetylcholin esterase in OCB cells in the lateral superior olivary complex (Altschuler et al., 1983). Although Davies (1982) was unable to detect any Substance P in the guinea pig organ of Corti, other polypeptides may yet prove to be present. The involvement of other known inhibitory substances appears to be minimal. GABA is present in only very small quantities in the cochlea (Fex and Wenthold, 1976) and the catecholamines likewise appear to have no function there (Densert, 1974; Bobbin and Thompson, 1978).

III. COCHLEAR NUCLEUS

A. EXCITATION

The leading contenders for the role of neurotransmitter of eighth nerve endings terminating in the cochlear nucleus are undoubtedly the amino acids glutamate and aspartate. The evidence for aspartate has become quite convincing over the past few years (Bird et al., 1978; Wenthold, 1978; Martin and Adams, 1979). Aspartic acid not only exists in high quantities in the eighth nerve and the predominant area of representation of cochlear nerve fibers in the cochlear nucleus, but its concentration decreases dramatically in the cochlear nucleus after eighth-nerve lesions. The cochlear nucleus can also effectively take up and release D-aspartic acid (Potashner, 1983; Potashner et al., Chapter 9, this volume).

B. THE ROLE OF ACETYLCHOLINE

The evidence for acetylcholine having no function in the eighth-nerve stimulation of the cochlear nucleus is most convincing, and the role of acetylcholine in the cochlear nucleus is exclusively as a mediator of feedback pathways from higher centers. Some neurophysiological studies (Comis and Davies, 1969), and numerous neurochemical studies (Godfrey and Matschinsky, 1981; Godfrey et al., Chapter 10, this volume) have made it clear that the cholinergic fibers terminating in the cochlear nucleus are centrifugal fibers and not centripetal ones, although it is still a possibility that there is intrinsic cholinergic activity within this nucleus.

C. INHIBITION

The cochlear nucleus provides an abundance of inhibitory phenomena (Whitfield, 1967). The origins of such phenomena may be centrifugal, centripetal, or reflected in neuronal activity by mechanical, electrical, or chemical interactions in the organ of Corti. The role of GABA in the cochlear nucleus has been the subject of both neurophysiological (Whitfield and Comis, 1966) and neurochemical (Fisher and Davies, 1976) studies. Whilst the extensive studies of Fisher and Davies (1976) and of Davies (1981) indicate that most of the GABAergic activity is intrinsic and not mediated by long-axoned centrifugal or centripetal fibers, the elegant immunological studies of Moore and Moore (1984) could be interpreted as indicating otherwise. Their cochlear nucleus mapping of glutamic acid decarboxylase (GAD), an excellent marker of GABAergic activity, showed abundant GAD-containing nerve endings but very few GAD-containing cell bodies. It is possible that, relatively speaking, the concentration of GAD in terminals is much higher than in the cell bodies giving rise to them, and would, therefore, not be so clearly indicated by the anti-GAD serum.

The cytoarchitecture of the cochlear nucleus, however, is very complex (Osen, 1969) and, in all probability, both intrinsic and extrinsic GABAergic activity exists with different representation in the dorsal and ventral nuclei.

Of the other established neurotransmitters, the catecholamines, particularly noradrenaline, appear to be involved both in the dorsal and ventral cochlear nuclei. Two centrifugal pathways appear to be involved (Kromer and Moore, 1976), although the origin of such fibers is uncertain. Similarly, met-enkephalin-immunohistochemically-positive cells have been found in both the ventral and dorsal cochlear nucleus (Altschuler, 1979), but again, the functional significance of such cells is unknown.

IV. INFERIOR COLLICULUS

Innervation of the inferior colliculus is extremely complex. The bulk of its centripetal input arises from the contralateral cochlear nucleus (Osen, 1972) and the ipsilateral lemniscal nuclei. Whilst a vast literature is available (Whitfield, 1967) on the neuronal responses of collicular neurons to sound stimulation, very little has been published on the nature of its transmitters. A prominent neurochemical feature of the inferior colliculus is its remarkably high concentrations of GAD and GABA, suggesting a premier role of GABA in information processing at the collicular level. However, Fisher and Davies (1976) and Davies (1981) were unable to detect any changes in the collicular GAD levels with lesions placed centrally to the cochlear nucleus. This would suggest that the bulk of the inhibitory processes in the inferior colliculus, which is probably GABAergic in nature, either arises in the olivary and lemniscal nuclei, or is centrifugal in nature, arising from the medial geniculate. The data of Moore and Moore (1984) fit this picture reasonably well, as the bulk of GAD in their studies appears in the nerve endings of the inferior colliculus, although they found some cell bodies staining in the capsular region.

Other than for GABA, there seem to be very little data available on neurochemical profiles of transmitter-like substances in the inferior colliculi.

Other than distributional studies on the amino acids glutamate and aspartate, there is no suggestion as to the mediator of excitation in the inferior colliculus, and there are no suggestions that either of those amino acids is of particular importance in this nucleus.

V. AUDIOGENIC-SEIZURE SUSCEPTIBLE MICE

Alterations in the free amino-acid content of affected parts of the brain have been noted in many epileptic models. Amongst the most common of these are decreases in levels of GABA and glutamate, along with a rise in those of glycine, although the latter change is always secondary to the seizure occurrence, and never causal. In addition, decreased taurine levels have been observed in many experimental epilepsies (Emson, 1978).

A convenient model for the study of epilepsy is the DBA/2J mouse, which has a high, genetically-determined susceptibility to sound-induced (audiogenic) seizures which varies with age, the seizure incidence showing a reduction or absence at ages earlier and later than between 21–29 days of age. The basis of this age-related seizure susceptibility is unknown, although several factors have been implicated, amongst them defects in amino-acid neurotransmission. For example, drugs that increase GABA levels reduce both severity of and susceptibility to audiogenic seizures, whilst GABA, glycine and taurine have been shown to exert a dose-dependent, anti-convulsant effect in audiogenic seizure-prone rats, with taurine the most potent (for review, see Jobe and Laird, 1981).

Whilst whole brain levels of GABA, for instance, have been measured in DBA/2J mice and found not to differ significantly from those in controls (Sykes and Horton, 1982), no regional studies have been carried out in areas likely to be involved in the etiology of audiogenic-seizure susceptibility. It is worth noting that variations in levels of amino acids between susceptible and resistant animals have been noted in other species and in other seizure-prone strains of mice (for review, see Jobe and Laird, 1981).

Audiogenic seizures are thought to be primarily a form of brainstem epilepsy. Work from lesion studies has suggested that the primary pathway of audiogenic seizures is via subcortical pathways that predominantly involve the central nucleus of the inferior colliculus and the deep superior colliculus. Bilateral lesions of the inferior colliculus will block audiogenic seizures in mice, whilst direct electrical stimulations of this area will mimic a sound-induced seizure. The cerebellum is also thought to be involved, there being evidence that it has actions upon auditory receptors and/or afferent cochlear fibers. Lesions in this area increase both the incidence and severity of audiogenic seizures (Seyfried, 1982).

We have measured the levels of five free amino acids—glutamate, aspartate, glycine, taurine, and GABA—in the inferior colliculi, the cochlear nuclei, the cerebellum, the primary auditory cortex, and also in the organ of Corti. The determinations were performed at ages during (20–25 and 29 days) and after (30–36 and 45–50 days) the period of maximal susceptibility to audiogenic seizures. Corresponding measurements were carried out on age-matched BALB/c mice, which are resistant to audiogenic seizures at these ages.

Amino acids were extracted from the regions examined by a standard acetone extraction procedure (Osborne, 1973) and the sample combined with o-phthalaldehyde (OPA) and analyzed by high-pressure liquid chromatography (HPLC) using a method similar to that of Lindroth and Mopper (1979). Primary amines form highly fluorescent adducts in basic aqueous solutions with OPA and, used in conjunction with HPLC, can be used to detect amino acid levels at the picomole level. However, with the HPLC system employed (column : μ Bondapak C_{18} [Lichosorb RP 18 : 5 μM]; mobile phase: 43:57 methanol : phosphate pH 6.8), taurine cochromatographs with a number of contaminants, so, in order to quantify taurine levels, it was necessary to pass the sample through an ion-exchange column (2.0-cm AG 1-X8, 100/200 mesh in Cl^- form over AG 50W-X8, 200/400 mesh in H^+ form), which selectively removes all amino acids, etc., except taurine from the extract (Larsen et al., 1980).

The results obtained in this study show that there are no significant differences in the concentrations of glutamate, aspartate, glycine, taurine, and GABA between the two strains of mice in any region examined (Figs. 1–3). However, in the four brain regions examined, glutamate, aspartate, glycine, and GABA all exhibit marked increases in levels with age; these observations were noted in mice and other species by several workers (Agrawal et al., 1968), this change being characteristic of amino-acid neurotransmitters during development. Of particular interest is the rise of aspartate levels in the cochlear nucleus, where aspartate is a likely candidate for the role of the secondary auditory transmitter (Wenthold, 1981). The levels of aspartate in the mice aged 45–50 days are over three times those in mice aged 20–25 days. This marked increase could be construed as further evidence supporting the theory that aspartate may be a transmitter in the cochlear nucleus, though it will be noted that aspartate exhibits an equally marked increase in the cerebellum where it has no known transmitter function.

There were also no significant differences in the levels of taurine between the two strains in any brain region at any age (Fig. 3). However, in contrast to the other amino acids, taurine exhibited an age-related decrease in concentration as the animal matured, as opposed to the increase shown by other amino acids measured. This contrary behavior of taurine has been noted by others (Agrawal et al., 1968), but the reason for the decrease is unclear. It has been suggested that as the animal matures, increased efficiency of

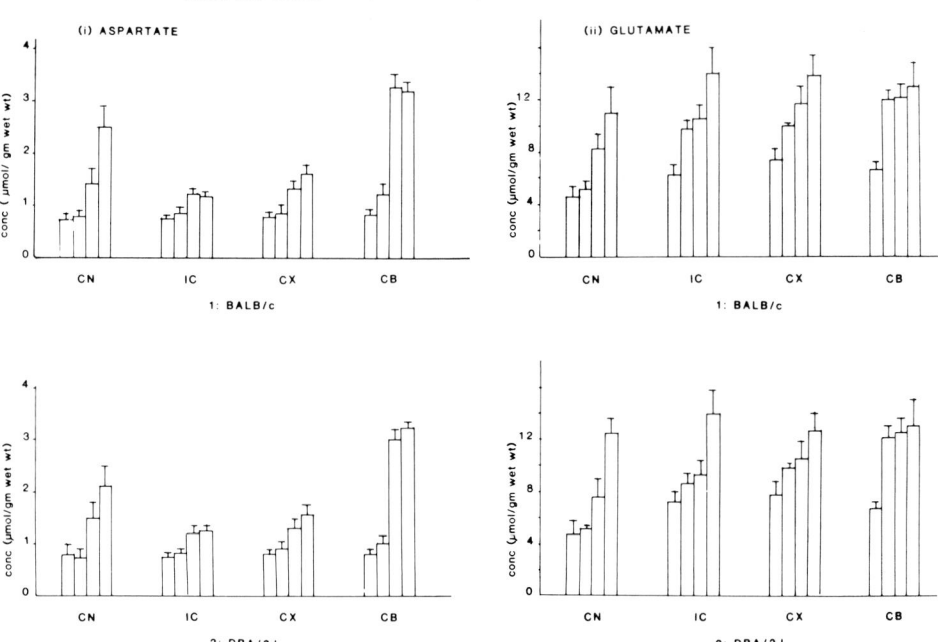

Figure 14-1. Levels of aspartic and glutamic acid during the age periods 20–25 days, 29 days, 30–36 days, and 45–50 days (left to right). Values are the means of 3 to 8 determinations ± standard error of the mean (SEM). CN, cochlear nucleus; IC, inferior colliculus; CX, primary auditory cortex; CB, cerebellum.

uptake systems or catabolizing enzymes may be responsible, but in any event, the functional significance of this fall is unclear.

In the organ of Corti, there was again no significant difference in the levels of any of the amino acids at any age between the two strains, and neither was there a significant age-related increase or decrease in the level of any amino acid measured (Fig. 4). This could be construed as evidence to suggest that none of the amino acids measured is a likely candidate for the transmitter at the hair cell—spiral ganglion cell synapse. Levels found in this area for taurine and GABA appear similar to those reported earlier, whilst those of glutamate are lower (Contreras and Bachelard, 1979; Guth et al., 1981).

In conclusion, it appears that variations in the levels of glutamate, aspartate, glycine, taurine, and GABA are unlikely to be related to the etiology of audiogenic seizures. If the inferior colliculi and/or cerebellum are the sites of seizure susceptibility, gross deficits in the levels of these five amino acids seem not to be involved. However, these results do not preclude deficits in binding, release, or uptake of these compounds being involved in this form of epilepsy.

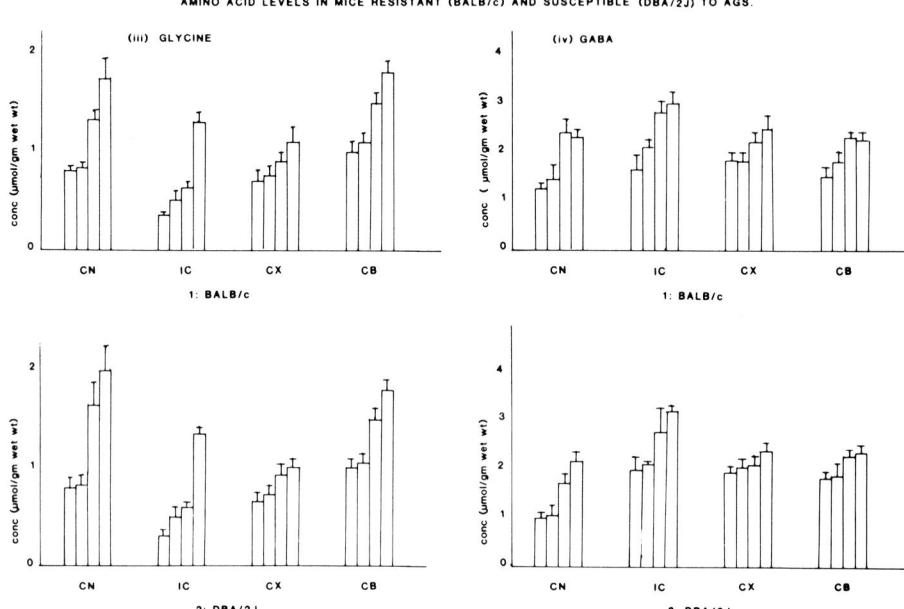

Figure 14-2. Levels of glycine and GABA during the age periods 20–25 days, 29 days, 30–36 days, and 45–50 days (left to right). Values are the means of 3 to 8 determinations ± SEM. CN, cochlear nucleus; IC, inferior colliculus; CX, primary auditory cortex; CB, cerebellum.

VI. SUMMARY AND CONCLUSIONS

Figure 5 summarizes our knowledge of neurotransmitters involved in the lower auditory system. It is obvious that there are still many gaps in our knowledge, and furthermore, that there are still transmitters hitherto unidentified and probably undiscovered. With the advent of immunohistochemistry and the recently indicated involvement of the gastrointestinal tract polypeptides in brain function, it would appear that the highest potential for new transmitters may well turn out to be in the area of polypeptide investigations.

Whilst the amino acids still present good possibilities, it remains a major problem to distinguish between metabolic functions and neurotransmitter functions of certain amino acids. Although the experiments described here on audiogenic-seizure susceptible mice appear to have shed no light on amino-acid neurotransmitter function in the auditory system, the use of genetic mouse variants may still be a good potential source of investigation.

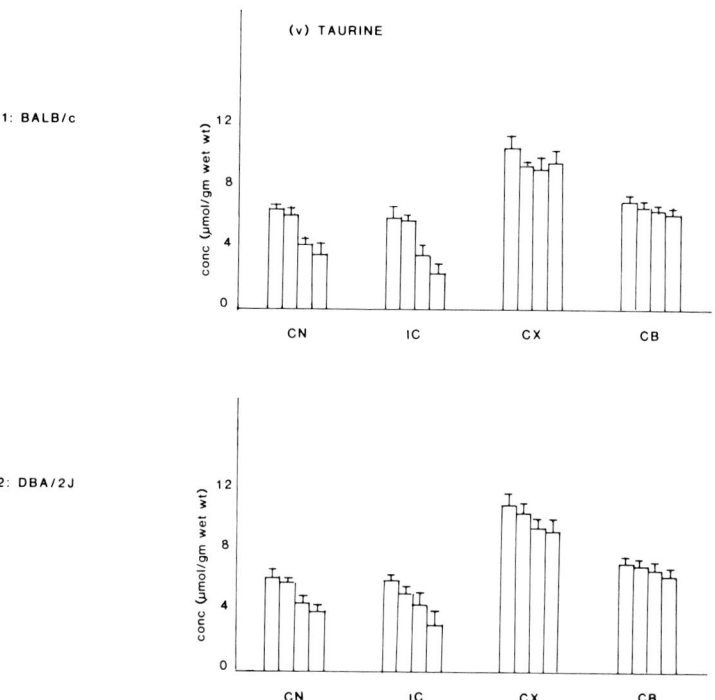

Figure 14-3. Levels of taurine during the age periods 20–25 days, 29 days, 30–36 days, and 45–50 days (left to right). Values are the means of 3 to 8 determinations ± SEM. CN, cochlear nucleus; IC, inferior colliculus; CX, primary auditory cortex; CB, cerebellum.

ACKNOWLEDGMENT

We gratefully acknowledge The Colt Foundation for financial assistance.

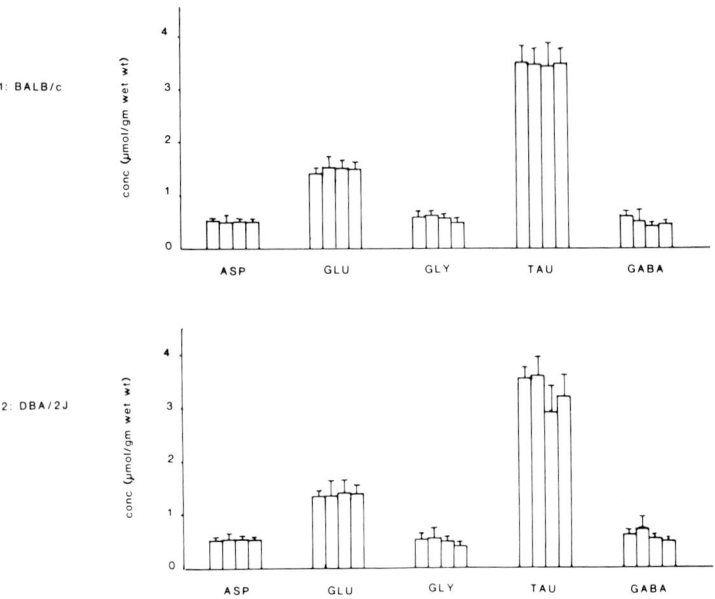

Figure 14-4. Levels of the amino acids aspartate (ASP), glutamate (GLU), glycine (GLY), taurine (TAU), and GABA during the age periods 20–25 days, 29 days, 30–36 days, and 45–50 days (left to right). Values are the means of 3 to 8 determinations ± SEM.

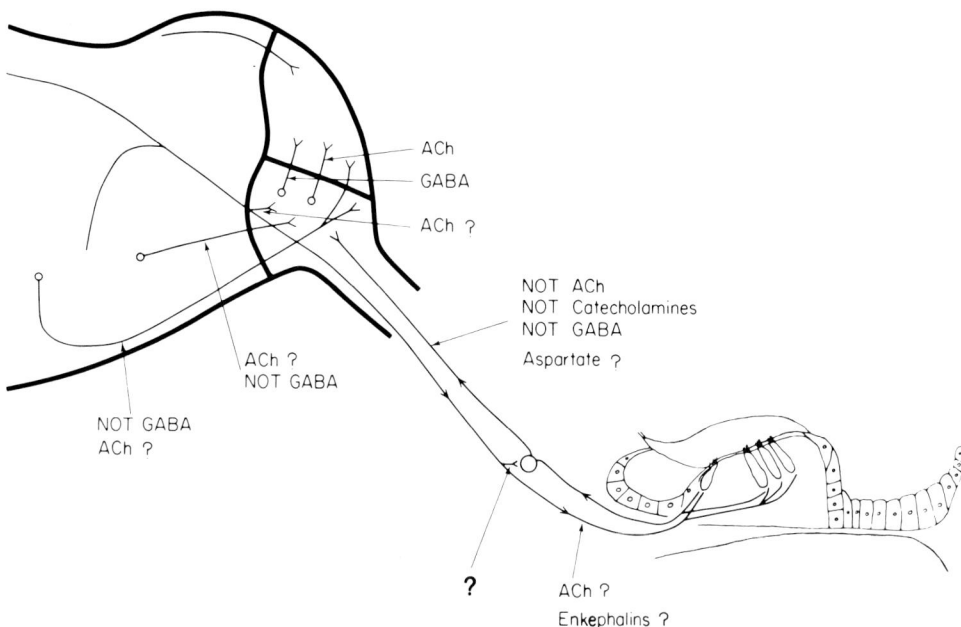

Figure 14-5. Summary of the involvement of orthodox neurotransmitters in the cochlea and cochlear nucleus.

REFERENCES

Agrawal, H. C.; Davis, J. M., and Himwich, W. A.: Developmental changes in mouse brain: weight, water content and free amino acids. *J. Neurochem. 15:* 917-923, 1968.

Altschuler, R. A.: Met-enkephalin positivity in the small cells of the deep cochlear nucleus and posteroventral nucleus of the rat. *Neurosci. Abstr. 9:* 15, 1979.

Altschuler, R. A.; Parakkal, M. H., and Fex, J.: Localization of enkephaline-like immunoreactivity in acetylcholinesterase-positive cells in the guinea pig lateral superior olivary complex that project to the cochlea. *Neuroscience 9:* 621-630, 1983.

Arnold, W., and Wang, J. B.: The spiral ganglion of the rhesus monkey. *Laryngol. Rhinol. Otol. (Stuttgart) 62:* 371-377, 1983.

Bird, S. J.; Gulley, R. L., Wenthold, R. J., and Fex, J.: Kainic acid injections result in degeneration of cochlear nucleus cells innervated by the auditory nerve. *Science 202:* 1087-1089, 1978.

Bobbin, R. P., and Thompson, M. H.: Effects of putative transmitters on afferent cochlear transmission. *Ann. Otol. Rhinol. Laryngol. 87:* 185-190, 1978.

Borg, E.; Densert, O., and Flock, A.: Synaptic vesicles in the cochlea. *Acta Otolaryngol. 78:* 321-322, 1974.

Comis, S. D., and Davies, W. E.: Acetylcholine as a transmitter in the cat auditory system. *J. Neurochem. 16:* 423-429, 1969.

Contreras, N. E. I. R., and Bachelard, H. S.: Some neurochemical studies on auditory regions of mouse brain. *Exp. Brain Res. 36:* 573-584, 1979.

Davies, W. E.: Gabaergic innervation of the mammalian cochlear nucleus. In Portmann, M., and Aran, J. M. (Eds.): *Inner Ear Biology. Vol. 68* Paris, INSERM, 1977, pp. 155-164.

Davies, W. E.: The absence of substance P from guinea pig auditory structures. *Arch. Otorhinolaryngol. 234:* 135-137, 1982.

Densert, O.: Adrenergic innervation in the rabbit cochlea. *Acta Otolaryngol. 78:* 345-356, 1974.

Drescher, M. J.; Drescher, D. D., and Medina, J. E.: Effect of sound stimulation at several levels on concentrations of primary amines, including neurotransmitter candidates, in perilymph of the guinea pig inner ear. *J. Neurochem. 41:* 309-320, 1983.

Dunn, R. A., and Morest, D. K.: Receptor synapses without synaptic ribbons in the cochlea of the cat. *Proc. Natl. Acad. Sci. U.S.A.*, 1975.

Emson, P. C.: Biochemical and metabolic changes in epilepsy. In Barbeua, A., and Huxtable, R. J. (Eds.): *Taurine and Neurological Disorders.* New York, Raven Press, 1978, pp. 319-338.

Engstrom, H., and Ades, H. W.: The ultrastructure of the organ of Corti. In Freidmann, E. (Ed.): *The Ultrastructure of the Sensory Organs.* New York, Elsevier, 1973, pp. 83-152.

Fex, J., and Altschuler, R. A.: Enkephalin-like immunoreactivity of olivocochlear nerve fibers in cochlea of guinea pig and cat. *Proc. Natl. Acad. Sci. U.S.A. 78:* 1255-1259, 1981.

Fex, J.; Fuxe, K., and Lennerstrand, G.: Absence of monamines in olivo-cochlear fibres in cat. *Acta Physiol. Scand. 64:* 259-262, 1965.

Fex, J., and Wenthold, R. J.: Cholineacetyltransferase, glutamate decarboxylase and tyrosine hydroxylase in the cochlea and cochlear nucleus of the guinea pig. *Brain Res. 109:* 575-585, 1976.

Fisher, S. K., and Davies, W. E.: GABA and its related enzymes in the lower auditory system of the guinea pig. *J. Neurochem. 27:* 1145-1155, 1976.

Godfrey, D. A., and Matchinsky, F. M.: Quantitative distribution of cholineacetyltransferase and acetylcholinesterase activities in the rat cochlear nucleus. *J. Histochem. Cytochem. 29:* 720-730, 1981.

Gulley, R. L., and Reese, T. S.: Freeze fracture studies on the synapses in the organ of Corti. *J. Comp. Neurol. 171:* 517-544, 1977.

Guth, P.; Sewell, W. F., and Tachibana, M: The pharmacology of the cochlear afferents and cochlear nucleus. In Brown, R. D., and Daigneault, E. A. (Eds.) *Pharmacology of Hearing.* New York, Wiley, 1981, pp. 99-136.

Hoffman, D. W.; Altschuler, R. A., and Fex, J.: High performance liquid chromatographic identification of enkephalin-like peptides in the cochlea. *Hearing Res. 9:* 71-78, 1983.

Ishii, Y.; Matsuura, S., and Furukawa, T.: Quantal nature of transmission at the synapse between the hair cells and VIII nerve fibers. *Jpn. J. Physiol. 21:* 79-89, 1971.

Iurato, S.: Efferent innervation of the cochlea. In Keidel, W. D., and Neff, W. D. (Eds.): *Handbook of Sensory Physiology. Vol. VI.* Berlin, Springer, 1974, 261-282.

James, W. M.; Cheatham, M. A., and Klein, W. L.: Muscarinic acetylcholine receptor binding in the guinea pig cochlea. *Hearing Res. 9:* 113-121, 1983.

Jobe, P. C., and Laird, H. E.: Neurotransmitter abnormalities as determinants of seizure susceptibility and intensity in the genetic models of epilepsy. *Biochem. Psychopharm. 30:* 3137-3144, 1981.

Klinke, R.: Neurotransmitters in the cochlea and the cochlear nucleus. *Acta Otolaryngol. 91:* 541-554, 1981.

Klinke, R., and Evans, E. F.: Evidence that catecholamines are not the afferent transmitter in the cochlea. *Exp. Brain Res. 28:* 315-324, 1977.

Klinke, R., and Oertel, W.: Evidence that 5HT is not the afferent transmitter in the cochlea. *Exp. Brain Res. 30:* 141-143, 1977.

Kromer, L. F., and Moore, R. Y.: Cochlear nucleus innervation by central norephinephrine neurons in the rat. *Brain Res. 118:* 531-537, 1976.

Larsen, B. R.; Gross, D. S., and Chang, S. Y.: A rapid method for taurine quantitation using high performance liquid chromatography. *J. Chromatogr. Sci. 18:* 233-236, 1980.

Lindroth, P., and Mopper, K.: High performance liquid chromatographic determination of subpicomolar amounts of amino acids by precolumn fluorescence demonstration with O-phthaldialdhyde. *Anal. Chem. 51:* 1667-1674, 1979.

Martin, M. R., and Adams, J. C.: The effects of DL-a-aminoadipate on synaptically and chemically evoked excitation of anteroventral cochlear nucleus neurons of the cat. *Neuroscience 4:* 1097-1105, 1979.

Moore, J. K., and Moore, R. Y.: Localisation of GAD immunoreactivity in brain stem auditory nuclei. *Assoc. Res. Otolaryngol. Abstr. 7:* 28, 1984.

Osborne, N. N.: The analysis of amines and amino acids in micro-quantities of tissue. *Prog. Neurobiol. 1:* 299-322, 1973.

Osen, K. K.: Cytoarchitecture of the cochlear nuclei in the cat. *J. Comp. Neurol. 136:* 453-484, 1969.

Osen, K. K.: Projection of the cochlear nuclei on the inferior colliculus in the cat. *J. Comp. Neurol. 144:* 355-371, 1972.

Potashner, S. J.: Uptake and release of D-aspartate in the guinea pig cochlear nucleus. *J. Neurochem. 41:* 1094-1101, 1983.

Rasmussen, G. L.: An efferent cochlear bundle. *Anat. Rec. 82:* 441, 1942.

Rossi, G.: L'acetylcholinesterase au cours du development de l'orielle interne du cobaye. *Acta Otolaryng. Suppl. 170:* 1-91, 1961.

Sewell, W. F.; Norris, C. H., Tachibana, M., and Guth, P. S.: Detection of an auditory nerve-activating substance. *Science 202:* 910-912, 1978.

Seyfried, T. N.: Convulsive disorders. In Foster, H. L.; Small, J. D., and Fox, J. G. (Eds.): *The Mouse in Biochemical Research. Experimental Biology and Oncology, Vol. IV.* New York, Academic Press, 1982, pp. 97-124.

Smith, C. A., and Sjöstrand, F. S.: Structure of the nerve endings on the external hair cells of the guinea pig as studied by serial sections. *J. Ultrastruct. Res. 5:* 523-556, 1961.

Sykes, C. C., and Horton, R. W.: Cerebral glutamic acid decarboxylase activity and γ-aminobutyric acid concentrations in mice susceptible or resistant to audiogenic seizures. *J. Neurochem. 39:* 1489-1491, 1982.

Tachibana, M., and Kuriyama, K.: Gamma aminobutyric acid in the lower auditory pathway of the guinea pig. *Brain Res. 69:* 370-374, 1974.

Wenthold, R. J.: Glutamic acid and aspartic acid in subdivisions of the cochlear nucleus after auditory nerve lesions. *Brain Res. 143:* 544-548, 1978.

Wenthold, R. J.: Glutamate and aspartate as neurotransmitters for the auditory nerve. In Di Chiara, G., and Gersa, G. L. (Eds.): *Advances in Biochemical Psychopharmacology. Vol. 27: Glutamate as a Neurotransmitter.* New York, Raven Press, 1981, pp. 69-78

Whitfield, I. C.: *The Auditory Pathway.* London, Arnold, 1967.

Whitfield, I. C., and Comis, S. D.: The interaction of centrifugal and centripetal stimulation on neurones of the cochlear nucleus. Final Reports Part II. AF EOAR 63-115, 1966.

Chapter 15

AUTORADIOGRAPHIC STUDIES OF AMINO ACID LABELING OF NEURAL ELEMENTS IN THE AUDITORY BRAINSTEM

Ilsa R. Schwartz

I. Introduction
II. Methods
III. The Dorsal Cochlear Nucleus
IV. The Anterior Ventral Cochlear Nucleus
V. The Medial Superior Olivary Nucleus
VI. The Lateral Superior Olivary Nucleus
VII. Summary and Conclusions
    References

I. INTRODUCTION

Identification of chemical properties with individual synaptic populations can only be obtained with morphologic methods. Immunocytochemical procedures offer one approach, but the number of specific antibodies to appropriate marker enzymes or other proteins, and indeed their specificity, is presently limited. By comparison, virtually all of the putative neurotransmitters are available in radiolabeled form and have been shown to be useful markers in the central nervous system. Thus, the combined use of radioautographic morphologic analysis, with control and manipulation of conditions affecting high-affinity uptake systems or other factors affecting labeling, provides a fruitful way to establish the relations between a number of auditory synaptic-terminal populations and chemical and metabolic properties related to their synaptic interactions. This information is essential for understanding complex events in the auditory system.

The ability of neural elements to accumulate selectively radiolabeled compounds can be used to probe both the anatomy of auditory brainstem neurons and their chemical properties. Once a compound has been shown to be taken up by a specific population of neural elements, the ability to detect the compounds autoradiographically, at the light-microscopic level, allows the distribution of these elements to be mapped with relative ease over a wide portion of the brainstem. The ability to study these neural elements in the same specimens at the electron microscopic level allows the details of their ultrastructure, as well as their relations to neurons and other synaptic populations, to be analyzed.

The survival of the uptake properties of tissues in *in-vitro* preparations of fresh brain slices allows the parameters affecting the uptake and release of labeled compounds to be studied in controlled chemical environments. Using slices, individual differences between non-inbred animals can be controlled by screening adjacent slices from a single nuclear area with several different probes or conditions. By working with large slices, the properties of several different auditory areas can be directly compared.

Among the putative neurotransmitters, the compounds which are identified as selective markers can be used to provide insights into the transmitter-related properties of the marked neural elements. In searching for potential selective markers for different populations of auditory neural elements, transmitter-related compounds are good candidates because: (1) different neurons are most likely to differ in terms of their transmitters, and (2) whatever the transmitter, mechanisms must exist for rapidly clearing it from the synaptic cleft, either through inactivation by chemical alteration, by reuptake into the presynaptic element, or by uptake into the postsynaptic element or nearby glia. An accumulating body of data implicates a number of amino acids as putative neurotransmitters in the auditory system. Because all of these amino acids and many of their precursors and breakdown products are readily available in radiolabeled form, our search for selective markers was initiated among the putative-neurotransmitter amino acids and related compounds.

Despite a considerable body of information about the number, origin, and types of axons, as well as classes of synaptic endings found on individual auditory neurons, especially in the cochlear nucleus (CN) and superior olivary complex (SOC), a significant percentage of terminals cannot be unambiguously assigned to a specific class on purely morphologic grounds. As many as 30–50% of terminals in the medial superior olivary nucleus (MSO) are unassignable (Schwartz, 1980). Further, synaptic terminals of several types of interneurons, if they cannot be morphologically distinguished from one another, also cannot be experimentally separated, because their locally distributed axons cannot be selectively lesioned or labeled by standard anterograde or retrograde transport techniques. The identification of markers for specific synaptic-terminal populations in the SOC and CN will allow us to describe the number, distribution, and interrelations of these terminals on specific auditory neurons.

II. METHODS

The procedures used in preparing tissues for incubation and in processing them for light microscopic (LM) autoradiography and electron microscopic (EM) autoradiography have been described previously (Schwartz, 1980, 1982a; Schwartz and Bok, 1979). In brief, the brains of anesthetized cats are rapidly chilled by ventricular perfusions of cold salt solutions. The tissue is kept cold, dissected, and mechanically chopped or sliced by hand into slices approximately 200 μM thick. The slices are incubated in oxygenated salt solutions containing micromolar amounts of labeled compounds, usually for a period of 20 min at room temperature. The slices are then rinsed briefly to remove the unbound labeled compounds, fixed, and prepared for autoradiography. During the incubations, adjacent slices of the same tissue can be treated under different conditions, or with different substrates, or inhibitors to provide important controls. Table I lists the compounds examined in each auditory structure and the number of animals in which each compound was studied. Listing of other variables examined, including the amino acids compared in each animal, are available in Schwartz (1984a, b, c).

TABLE 15-I
COMPOUNDS TESTED IN THE SUPERIOR OLIVARY COMPLEX AND COCHLEAR NUCLEUS

Compound	Number of Animals	
	CN	SOC
GABA	13 (5)	11 (5)
GLY	17 (4)	16 (8)
L-ASP	7 (4)	5 (2)
D-ASP	4 (0)	4 (1)
GLU	13 (4)	11 (4)
ALA	3 (2)	3 (2)
TAU	5 (2)	5 (3)
GLN	3 (0)	3 (1)
PRO	3 (0)	3 (0)
ARG	1 (0)	1 (0)
ORN	1 (0)	1 (0)
LEU	0 (0)	1 (0)
Total Animals:	26 (8)	23 (9)

The numbers of animals examined by light microscopy are given without parentheses; the numbers of those examined by electron microscopy are parenthesized. Abbreviations: GABA = g-aminobutyric acid; GLY = glycine; L-ASP = L-aspartic acid; D-ASP = D-aspartic acid; GLU = L-glutamic acid; ALA = alanine; TAU = taurine; GLN = glutamine; PRO = proline; ARG = arginine; ORN = ornithine; LEU = leucine.

III. THE DORSAL COCHLEAR NUCLEUS

Studies of the differential labeling with tritiated amino acids were begun in the dorsal cochlear nucleus (DCN), where the laminar organization of the tissue could be used to help in establishing the validity of any labeling patterns obtained. Figure 1 illustrates some of the neural elements found in the molecular, fusiform-cell, and deep layers of the DCN. The molecular layer contains primarily large numbers of small unmyelinated fibers. These fibers include the axons of granule cells (the parallel fibers) as well as axons of small neurons in both the molecular layer and deeper layers (Lorente de No, 1933, 1979, 1981). The parallel fibers predominate in the inner molecular layer, while the axons of small, non-granule cells found in the molecular layer are concentrated in the outer molecular layer immediately beneath the ependymal cells (Moore, 1984; personal communication). The arborization of the primary cochlear afferents is confined to the deep layer and the basal dendrites of the fusiform cells (Lorente de No, 1933; see review, 1981; Cohen et al., 1972). A number of descending projections to the DCN have also been identified (Kane, 1976a, b, 1977; Elverland, 1977; Kane and Finn, 1977; Kane and Conlee, 1979; Farley and Warr, 1981; Conlee and Kane, 1982) as well as projections from the AVCN to the DCN (Adams and Warr, 1976; Jones and Casseday, 1979; Mugnaini et al., 1980; Adams, 1983).

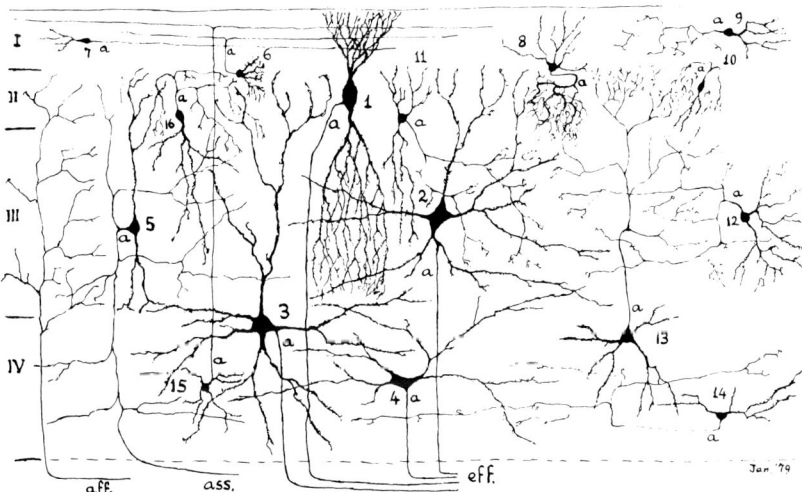

Figure 15-1. This figure illustrates several cell types which contribute axons to the molecular layer, as well as the restricted pattern of the distribution of the primary afferents to the level of the fusiform cell bodies and below. (From Lorente de No, 1981).

After incubation of DCN slices with a variety of amino acids (GABA, GLU, GLY, L–ASP, D–ASP, ALA, TAU, GLN, ARG, ORN), three different populations of synaptic terminals could be distinguished on the basis of differences in the localization of label, preferential accumulation of label from GABA, GLY, and GLU incubations, and ultrastructural features.

Figures 2–4 compare the localization of label in comparable slices through the fusiform and molecular layers after incubations with GABA, GLU, and GLY, respectively. In the slice incubated with GABA (Fig. 2) the label is primarily in the outer molecular layer. In the slice incubated with GLU (Fig. 3), the label is concentrated in the inner molecular layer. After incubation with GLY (Fig. 4), most of the label is evenly distributed over all the layers. When this material was examined electron-microscopically, we confirmed that the vast majority of clusters of silver grains seen with LM autoradiography were found over synaptic terminals. The terminals labeled by GABA, GLU, and GLY differed not only in their distribution, as determined by light microscopy, but also in their ultrastructural characteristics. After GABA incubations, the labeled elements were thin, unmyelinated axons and their associated boutons in the outer molecular layer (Fig. 5). After GLU incubations, the labeled endings were seen to arise from smalldiameter, unmyelinated axons, although the labeled terminals were slightly larger than those labeled by GABA incubations, and were found deeper in the molecular layer (Fig. 6). Competition experiments have shown that the GLU labeling pattern in the DCN can be inhibited by an excess of unlabeled GLU or ASP, but not by GABA. Thus, the GABA- and GLU-labeled populations are truly distinct. After GLY incubations, not only are the labeled terminals found distributed fairly evenly throughout the molecular, fusiform, and deep layers, but the majority of them are larger than those labeled by either GABA or GLU, and are usually found synapsing on cell bodies and dendrites (Fig. 7).

None of the localization patterns observed in the DCN corresponds to the distribution of the primary cochlear afferents. The GABA localization pattern in the cat outer molecular layer is consistent with the GAD localization studies of Moore (1984) in the guinea pig, where a population of GAD-positive small cells in the molecular layer and a region of GAD-positive terminals immediately beneath the ependymal layer were demonstrated. It would appear that the endings which are GAD-positive may also have a high-affinity uptake system for GABA. Thus, the GABA localization pattern could mark the distribution of the axons of a small cell in the molecular layer. Alternatively, the GABA localization pattern might represent a subpopulation of the granule cell axons, possibly those arising from cells in the deeper layers, or axons of some other, presently-unidentified cell type, either within the CN or central to it.

The distribution of GLU-labeled terminals deep in the molecular layer is

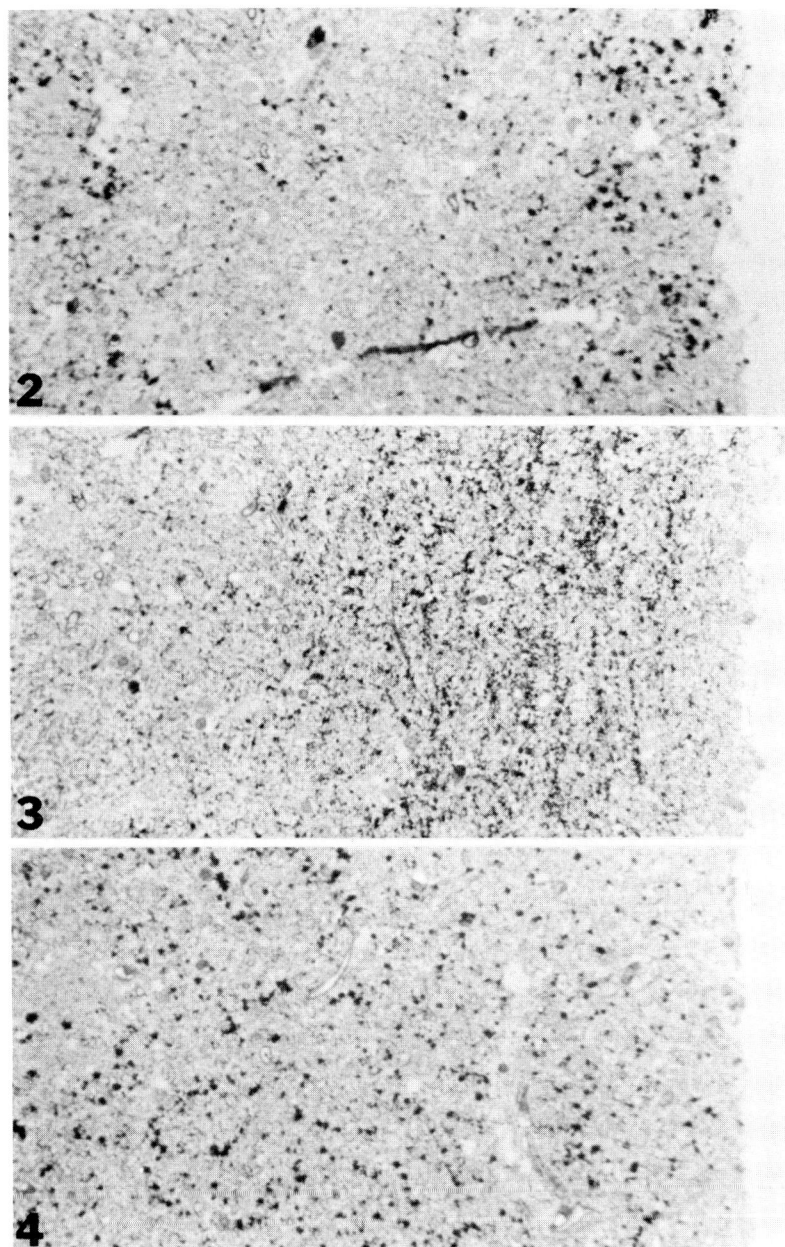

Figures 15-2–15-4. Light-microscopic autoradiographs of comparable DCN sections incubated, respectively, with GABA, GLU, and GLY and exposed for 28 or 29 days. Fig. 2 is from cat IS254, and Figs. 3 and 4 are from cat X430. × 495. In Fig. 2, the clumps of silver grains are concentrated in the outer molecular layer, although a few are found in the inner molecular layer, fusiform cell layer, and deep DCN. In Fig. 3, the clusters of silver grains in the inner molecular layer seem to be arranged in chains parallel to the surface of the DCN. In Fig. 4, note the relative absence of label over blood vessels (arrowheads), compared with the density of silver grains over elements in the neuropil of the molecular, fusiform-cell, and deep layers of the DCN. (From Schwartz, 1981).

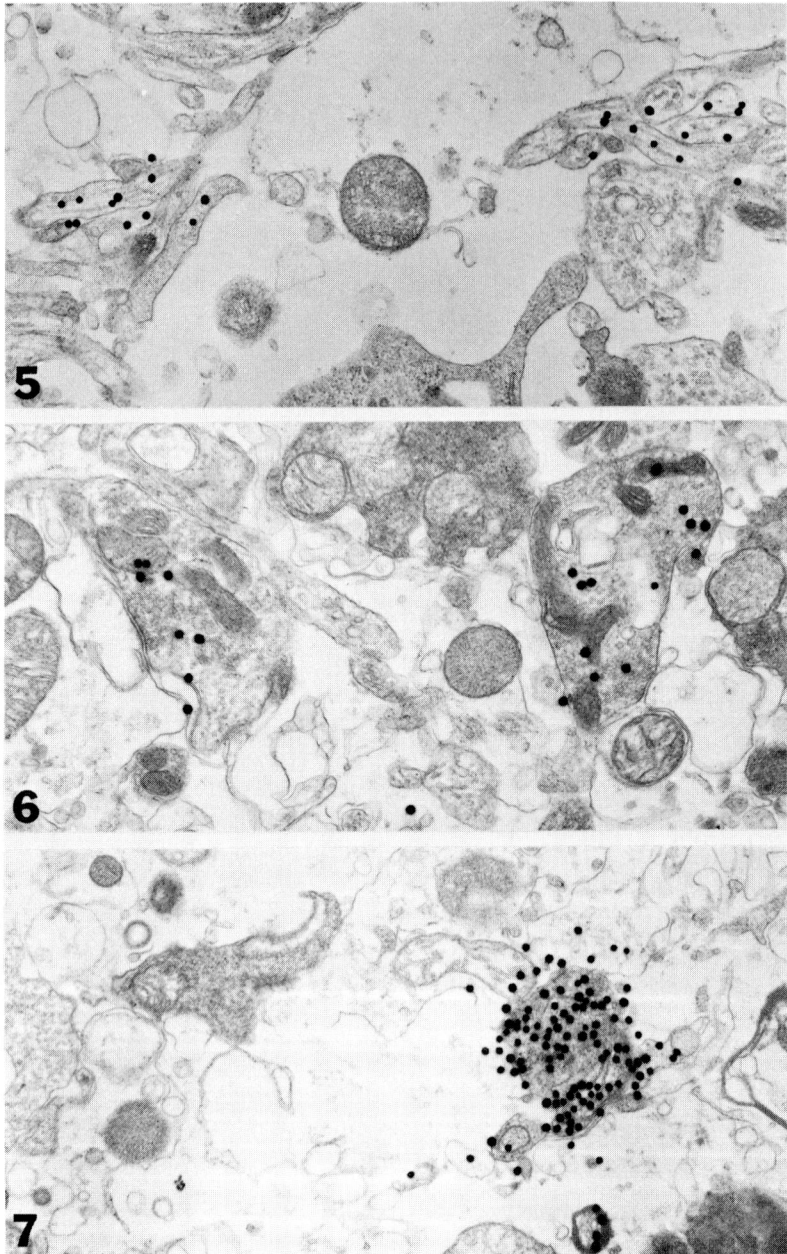

Figures 15-5–15-7. Electron-microscopic autoradiographs of synaptic terminals labeled after incubation with GABA, GLU, and GLY, respectively. × 14,100. In Fig. 5 (cat IS254, 21-day exposure), the localization of silver grains is over longitudinally-sectioned parallel fibers in the outer molecular layer. In Fig. 6 (cat IS245, 25-day exposure), localization of silver grains is shown over two synaptic terminals in the outer molecular layer. In Fig. 7 (cat IS250, 25-day exposure), there is a heavy concentration of silver grains over a synaptic terminal contacting a dendrite in the deep DCN. (From Schwartz, 1981).

most consistent with the distribution of the granule-cell axons (parallel fibers), although other populations might also contribute to the observed patterns of label localization. Parallel fibers in the cerebellum appear to be glutamatergic (Hudson et al., 1976; Sandoval and Cotman, 1978; Levi and Gallo, 1981).

After incubations with TAU, label was found over some cells in the molecular layer. After ASP, little uptake was seen, although one experiment suggested pattern differences related to length of incubation, with higher label concentration beneath the ependymal layer after 1–2 min incubations. Preliminary results from recent experiments have shown heavy labeling of the DCN molecular layer after incubations with D–ASP. The D–ASP distribution of labeling appears to cover the areas labeled after both GABA and GLU incubations.

Although a growing body of data suggests that ASP or GLU or some structurally-related compound may be a transmitter at the auditory-nerve afferent synapses in the cochlear nucleus (Godfrey et al., 1977; Martin, 1980; Wenthold, 1980, 1981; Guth et al., 1981; Guth and Melamed, 1982), our findings indicate the absence of a high-affinity uptake system for these compounds at, or close to, these endings. While synthesis might be the major source of the transmitter, the absence of a major uptake of precursors or breakdown products suggests that any conclusions about a transmitter role for GLU and ASP at these endings should be viewed with caution.

IV. THE ANTERIOR VENTRAL COCHLEAR NUCLEUS

The distinctive relation between the cochlear afferents and the large spherical cells in the anterior ventral cochlear nucleus (AVCN) provides another area where the underlying anatomy can simplify the interpretation of labeling patterns. Information about the distribution of non-cochlear endings is also available (Cant and Morest, 1979a, b; Cant, 1981). When comparable slices of the AVCN are incubated with a variety of amino acids (GABA, GLY, GLU, ASP, ALA, TAU, ARG, ORN, GLN), distinctive labeling patterns are observed only with GABA and GLY (Figs. 8–11) (Schwartz, 1983b).

After GABA and GLY incubations, labeled terminals are prominent on spherical cells. The frequency and pattern of distribution is the same with both GABA and GLY. EM autoradiography has shown that the labeled terminals do not have the ultrastructural characteristics associated with primary cochlear afferents (Fig. 12). As in the DCN, the absence of high-affinity uptake systems for ASP, GLU, and related compounds in or around the cochlear afferent terminals is striking, although some endings in the molecular layer are labeled following ASP or GLU incubations. The validity

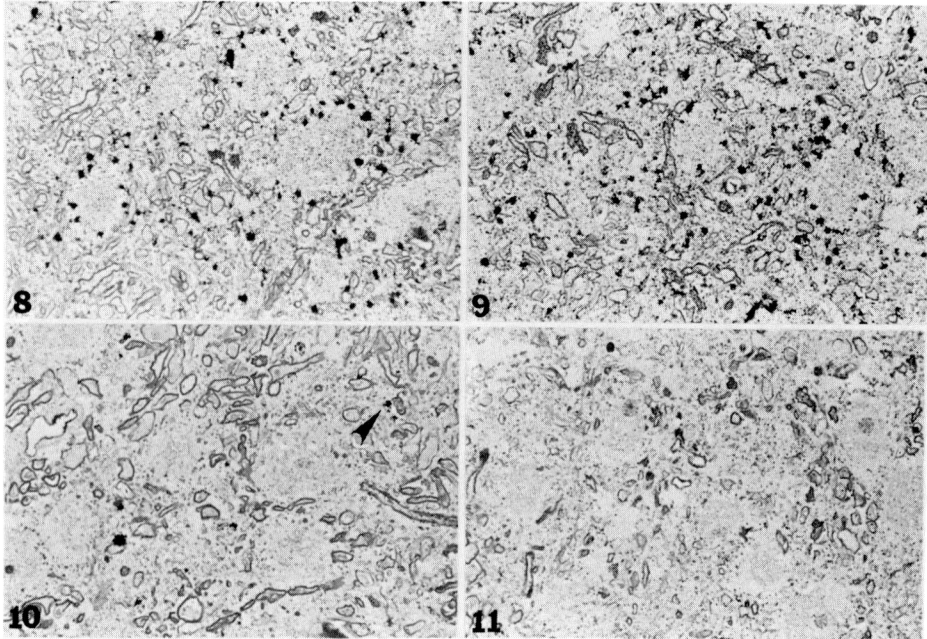

Figures 15-8–15-11. Localization patterns of silver grains observed in slices of the AVCN from cat IS250, incubated, respectively, with GLY (Fig. 8), GABA (Fig. 9), GLU (Fig. 10), and L-ASP (Fig. 11). The size and number of labeled terminals around spherical cells is comparable in the GABA- and GLY-incubated sections. A single labeled terminal (arrow) is seen in the GLU-incubated slice. No labeled terminal is present on any of the cells in the L-ASP incubated slice. × 76.

of our observations is supported by Oertel (1984). In a mouse brain-slice preparation, it was shown that while the removal of calcium from the superfusion medium completely shuts off the spontaneous firing of type I (large, spherical) neurons, bath-applied ASP and GLU have no effect on the spontaneous firing rate of these cells at physiologically relevant concentrations (see also Wenthold, Chapter 8, this volume).

V. THE MEDIAL SUPERIOR OLIVARY NUCLEUS

Despite the distinctive orientation of the laterally- and medially-directed bipolar dendrites of central-cell-band cells in the MSO (Schwartz, 1977), no distinctive labeling patterns were observed with any of the compounds tested (Schwartz, 1984a). After incubations with GABA and GLY, a consistently sparse distribution of labeled terminals was observed across the nucleus (Schwartz, 1982b, 1983a, c, 1984c). Close examination of these patterns revealed that labeled terminals were usually found perisomatically. Some

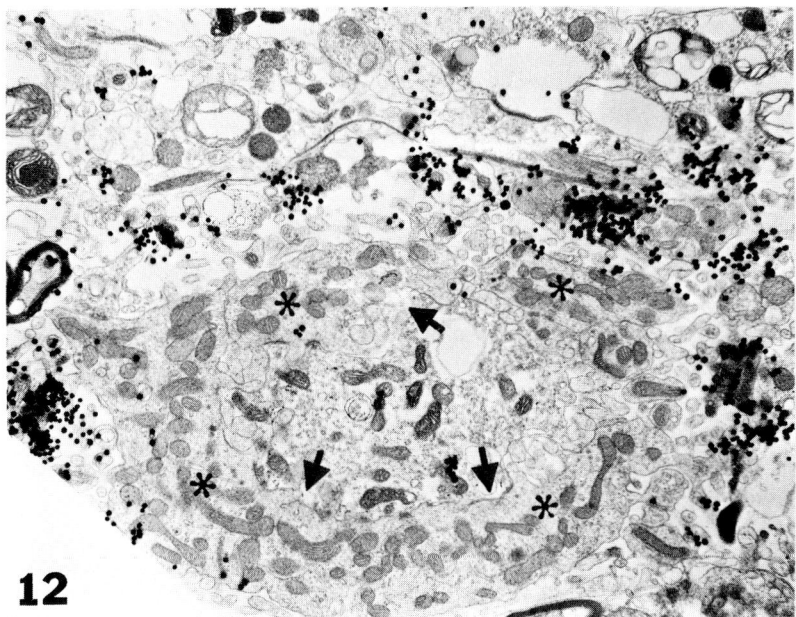

Figure 15-12. An electron-microscopic autoradiograph, illustrating a number of labeled terminals next to an unlabeled calyx (°) surrounding an AVCN spherical cell (arrows), following incubation with GABA. Cat IS320, 193-day exposure. × 9,700.

marginal cells, located at the edges of the MSO and oriented perpendicularly to its principal bipolar-cell dendrites, are contacted by more labeled endings than central-cellband cells (Fig. 13). The distribution of endings labeled with GABA and GLY is consistent with the distribution of the smaller, vesicle-containing type 2 and 3 endings previously demonstrated in the MSO with normal morphological material (Schwartz, 1981).

Although the LM autoradiographic patterns are similar, GABA-labeled terminals can be distinguished from GLY-labeled terminals on the basis of size, ultrastructural features, and competitive uptake studies (Schwartz, 1982b, 1983a, 1984b, c). The majority of GLY-labeled terminals are larger than the GABA-labeled terminals. When slices are incubated with ^{3}H–GLY in the presense of a tenfold excess of unlabeled GABA, the labeling pattern is unchanged. Similarly, the presence of an excess of unlabeled GLY has little effect on the GABA labeling pattern. Thus, even the small GLY-labeled terminals appear to be distinct from those labeled with GABA. On the basis of their morphology, the GLY-labeled endings probably belong to several classes. At least two groups can be distinguished on the basis of size. Figure 14 compares a typical large and small GLY-labeled ending. A constellation of features characterizes the group of larger endings (Schwartz, 1983c). The

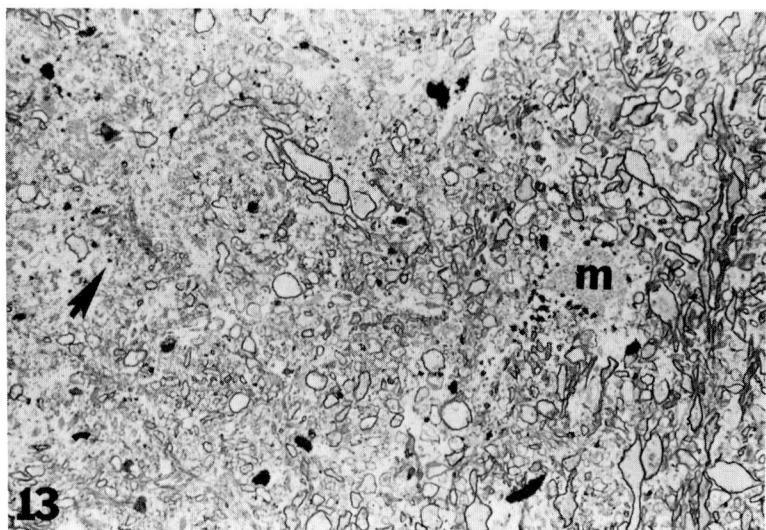

Figure 15-13. Sparse distribution of labeled endings are seen in the central cell band of the MSO following incubation with GABA. Several small clusters of silver grains are present in the central cell band. One is contacting a neuronal somata (arrow). A medial marginal cell (m) is contacted by a large number of labeled terminals. Cat IS254, 122-day exposure, × 110.

larger terminals usually have a spherical profile, are up to 7 μm in diameter, are usually located perisomatically, and are frequently found in apposition to other synaptic terminals. They have relatively symmetrical junctional appositions, and contain coated vesicles and neurofilaments (Fig. 15). The accumulation of label during GLY incubations is sodium dependent (Fig. 16) and virtually disappears when choline chloride is substituted for sodium salts in the incubation medium (Schwartz, 1983a, 1984c). Washing loaded slices with salt solutions containing elevated potassium concentrations (49 mM) in the presence of calcium causes a pronounced depletion of the accumulated label (Fig. 17) (Schwartz, 1983a, 1984c).

In some animals, a number of endings dispersed through the MSO were labeled after GLU uptake. Only a few endings were labeled after ASP or TAU incubations. All of the labeling patterns were conspicuously unlike the known distribution of CN efferents. No endings were labeled after ALA incubations.

VI. THE LATERAL SUPERIOR OLIVARY NUCLEUS

Of the amino acids which have been tested in the lateral superior olivary nucleus (LSO), GABA and GLY are the only ones which produced significant

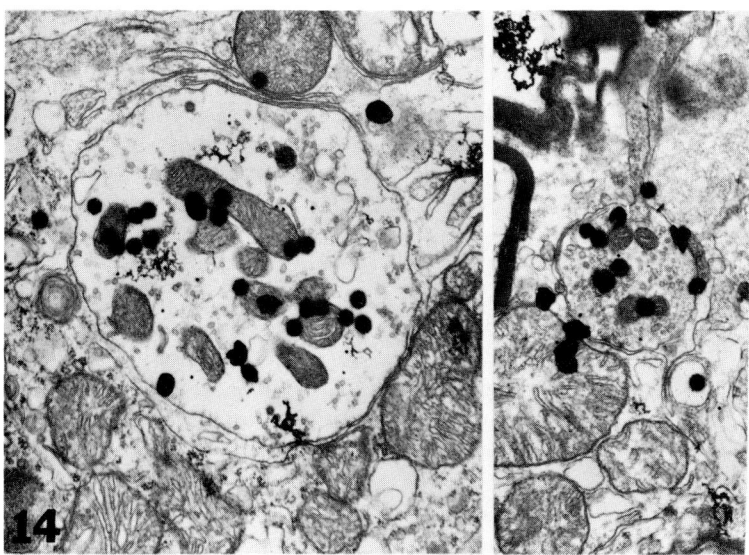

Figure 15-14. Relative size and typical ultrastructural characteristics of large and small terminals in the MSO of cat IS508 labeled after GLY incubations. Both terminals contact neuronal somata. A small unmyelinated axon emerges from the small terminal. 49-day exposure, × 13,500. (From Schwartz, 1984a).

terminal labeling. This result is similar to that found for the MSO. The size and structure of these labeled terminals is also similar to those observed in the MSO (Fig. 18) (Schwartz, 1982b, 1983a, c, 1984b, c). Incubations with ASP, GLU, PRO, and TAU resulted primarily in glial labeling (Schwartz, 1984a). In contrast to the MSO, GLY-labeled terminals occur in much greater numbers on LSO principal cells (Fig. 19). While the large GLY-labeled terminals occur primarily perisomatically, smaller GLY-labeled terminals occur in the neuropil, suggesting the presence of more than one class of GLY-labeled endings. A few, small, myelinated axons are seen after GLY incubations. Their location suggests that they give rise to the smaller GLY-labeled terminals in the neuropil.

GABA-labeled terminals are generally comparable in size to the small GLY-labeled terminals. As in the MSO, these are a separate population because they continue to label in the presence of excess unlabeled GLY, and vice versa. The sodium dependence of GLY uptake and the ability of high potassium concentrations to deplete GLY label are the same as seen in the MSO.

At the light-microscopic level, small cells labeled after L-ASP, D-ASP, and GLU incubations are hard to classify (Fig. 20), although at the electron-microscopic level, protoplasmic astrocytes are the most heavily-labeled glial

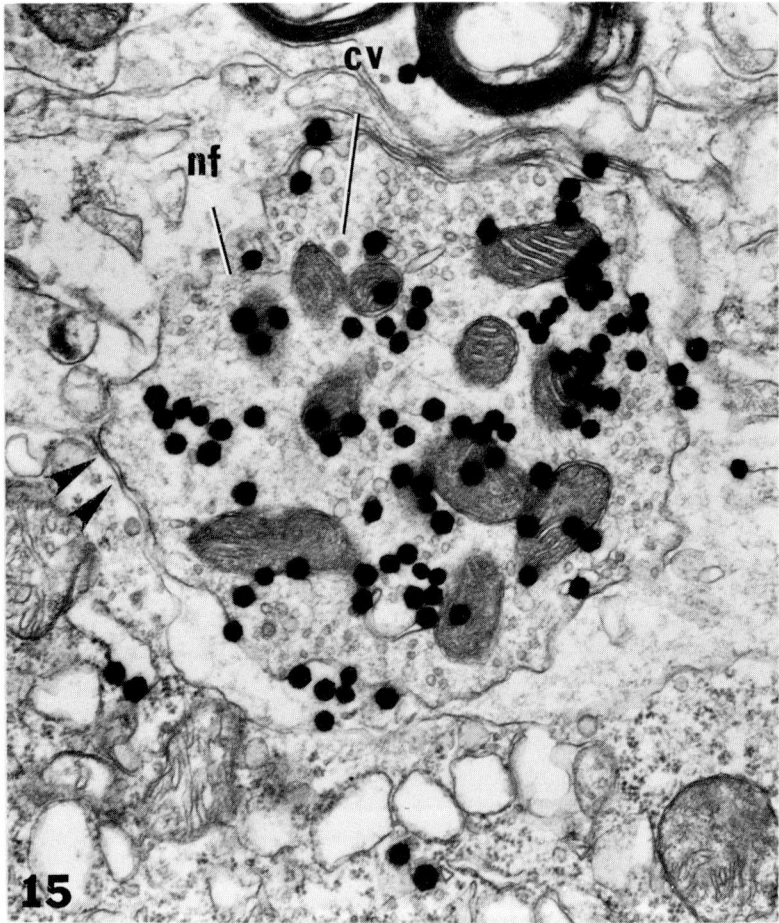

Figure 15-15. A large terminal from cat IS515, labeled after incubation in GLY in the presence of a tenfold excess of unlabeled GABA. Neurofilaments (nf), coated vesicles (cv), and symmetrical junctional appositions (arrows) are characteristic. 49-day exposure, × 27,300.

cells and no labeled neurons have yet been observed (Schwartz, 1984a).

The single most important criterion for the identification of a transmitter substance is the demonstration that it mimics the action of the naturally occurring transmitter (Werman, 1966). In the LSO, Moore et al. (1981) and Caspary et al. (Chapter 12, this volume) have shown that iontophoretic application of GLY to binaurally-driven units mimics the inhibitory effect of simultaneous sound stimulation to the contralateral ear, and strychnine, a GLY antagonist, blocks this effect. Information from the contralateral ear (and AVCN) reaches the LSO through a major projection to the principal cells of the contralateral medial nucleus of the trapezoid body (MNTB),

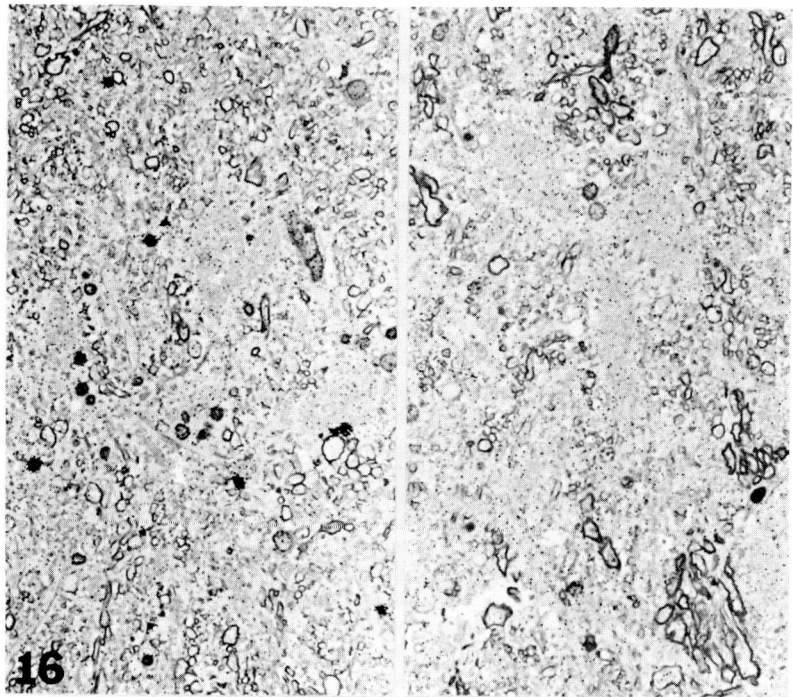

Figure 15-16. This illustration compares the degree of labeling with GLY in the presence (left) and absence (right) of sodium. Cat IS607, 219-day exposure, × 260.

whose axons cross the MSO on the way to their terminations in the LSO, while ipsilateral information goes directly from the AVCN to the ipsilateral LSO. Given the location and prominence of the GLY-labeled endings in the LSO, GLY's sodium-dependent uptake and potassium-stimulated depletion, and the physiological demonstration that GLY functions as a transmitter at some LSO terminals, it seems probable that the GLY-labeled terminals in the LSO also use GLY as a transmitter. Given the similarity of the large GLY-labeled terminals in the LSO and MSO, and the fact that our Golgi studies (Schwartz and Wittebort, 1976; Schwartz, 1984b) have shown that in neonatal animals a few of the largest axons crossing the MSO give off a small branch near the edges of the MSO, it is possible that the large GLY-labeled endings in both nuclei arise from MNTB principal-cell axons. Alternatively, both might arise from axons of AVCN neurons other than the spherical cells.

VII. SUMMARY AND CONCLUSIONS

Incubation of fresh cat-brain slices with low concentrations of high-specific-radioactivity compounds has been used to investigate the anatomical and

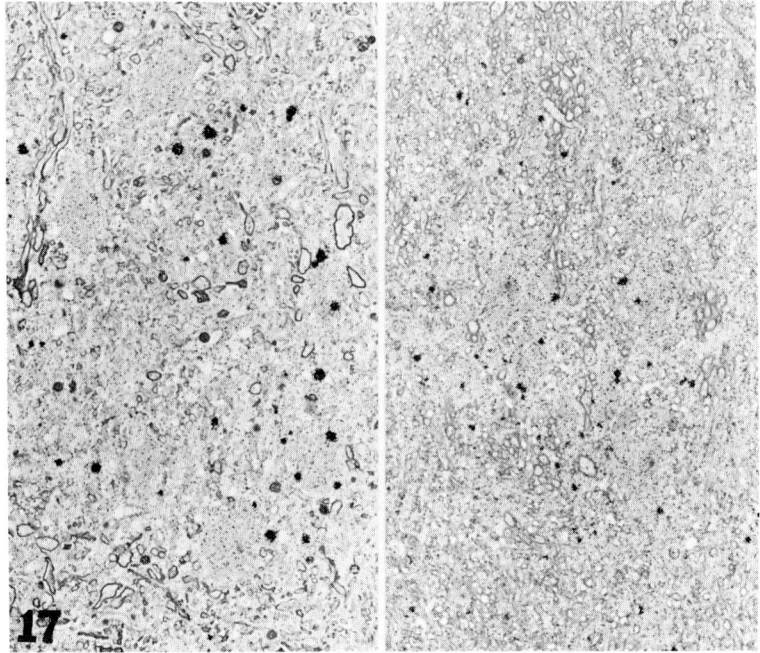

Figure 15-17. Depletion of label caused when a slice preloaded with GLY is rinsed in a solution containing no calcium and high (49 mM) potassium (left), or high potassium with normal calcium (right). at IS607, 219-day exposure, × 260.

biochemical properties of neural elements in the cochlear nucleus and superior olivary complex. The major observations of these studies are: (1) GABA- and GLY-labeled synaptic terminals are found in all regions examined, (2) GABA and GLY incubations label different populations of synaptic terminals, each of which has a characteristic distribution and morphology, (3) GLU distinctively and consistently labels a population of terminals in the DCN inner-molecular layer, but only occasionally labels terminals in other areas, (4) GLU, ASP, GLN, PRO, ALA, and TAU label glial cells, but no labeled neurons have been observed, (5) the morphology of the terminals labeled by GABA, GLY, and GLU is consistent with their being inhibitory or interneurons, and (6) the sodium dependence of uptake of GLY and the calcium-dependent depletion of label by high potassium concentrations are consistent with the expected behavior for a transmitter.

Thus, our findings suggest, but do not by themselves prove, that GABA, GLY, and GLU play a transmitter role in the terminals they label. Our findings with regard to GLU and ASP sound a cautionary note. While there is clearly a population of endings in the inner-molecular layer in the DCN, and possibly also in the AVCN, whose labeling with GLU is consistent with the

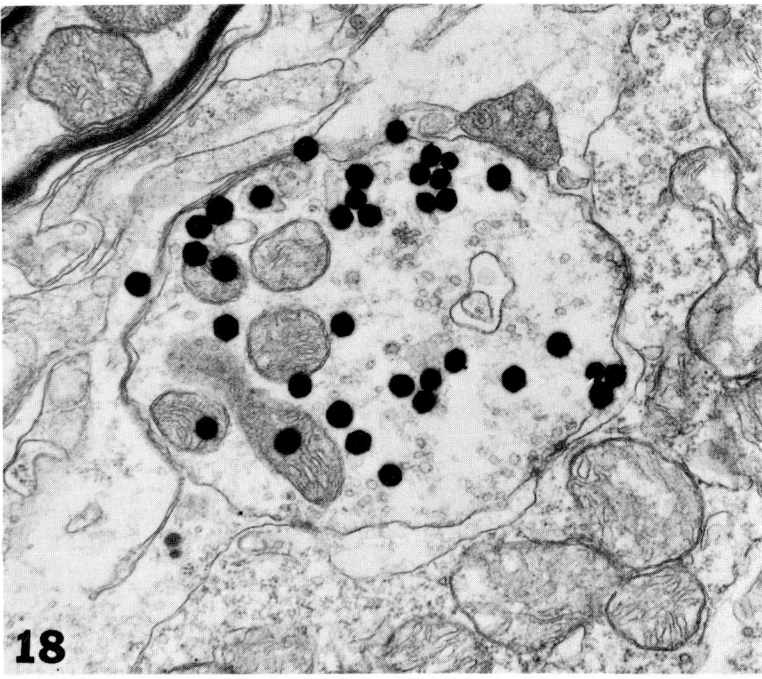

Figure 15-18. Large, GLY-labeled terminal from the LSO of cat IS515, with ultrastructural characteristics similar to the large, GLY-labeled endings seen in the MSO. 21-day exposure, × 23,000.

evidence about GLU being a transmitter in parallel fibers, the primary cochlear afferents are not labeled by GLU or ASP incubations. Our data are consistent with the possibility that if ASP or GLU were a transmitter, it might be taken up by glia for recycling in some other form. However, GLN, normally associated with recycling GLU from glia to neurons (Shank and Aprison, 1979), was also preferentially taken up into glia, but not into neurons.

ACKNOWLEDGMENTS

The technical assistance of M. Rita Watson and Gary Fink is gratefully acknowledged. This work was supported by NIH grants NS 09823 and NS 14503.

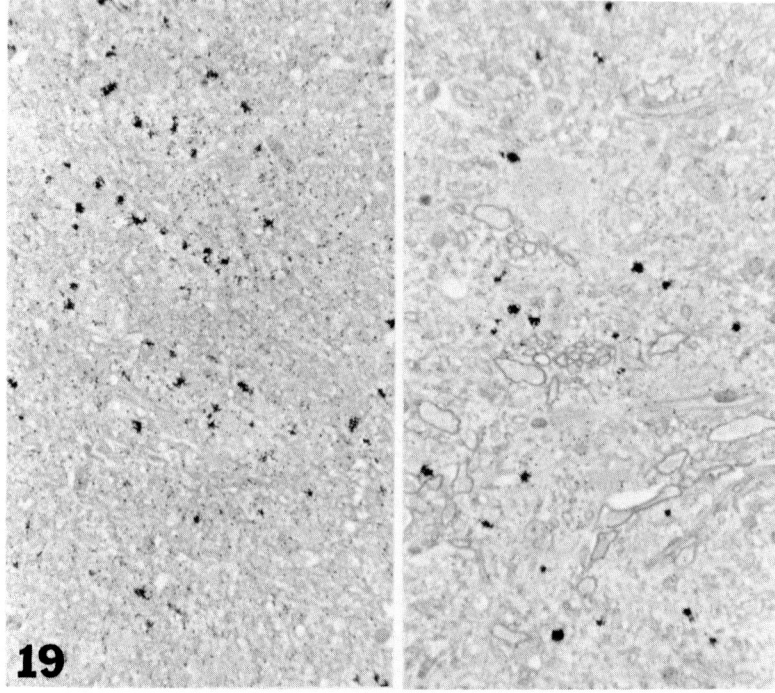

Figure 15-19. Greater density of GLY-labeled terminals observed in light-microscopic autoradiograms of the LSO (left), compared to the MSO (right). Both illustrations are from a single section of cat IS551, 42-day exposure, × 120.

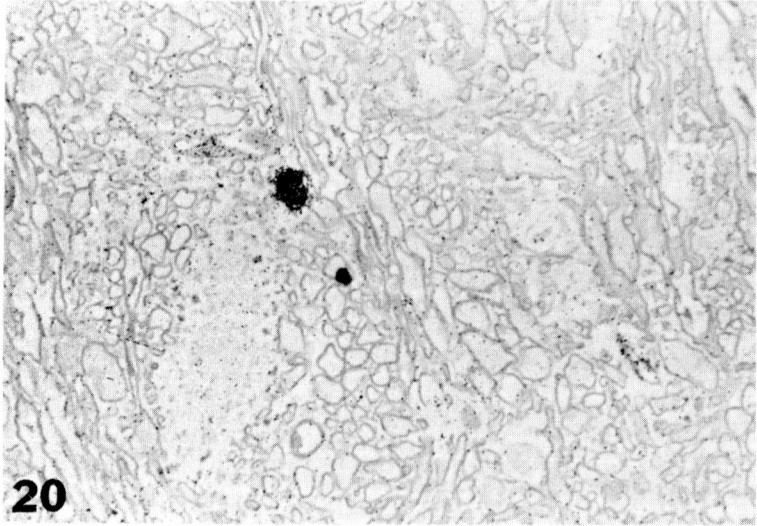

Figure 15-20. A small cell in the MSO of cat IS244, labeled after incubation with L-ASP. At the electron-microscopic level, no labeled neurons have been observed. 77-day exposure, × 110.

REFERENCES

Adams, J. C.: Multipolar cells in the ventral cochlear nucleus project to the dorsal cochlear nucleus and the inferior colliculus. *Neurosci. Lett. 37:* 205-298, 1983.

Adams, J. C., and Warr, W. B.: Origins of axons in the cat's acoustic striae determined by injection of horseradish peroxidase into severed tracts. *J. Comp. Neurol. 170:* 107-122, 1976.

Cant, N. B.: The fine structure of two types of stellate cells in the anterior division of the anteroventral cochlear nucleus of the cat. *Neuroscience 6:* 2643-2655, 1981.

Cant, N. B., and Morest, D. K.: The bushy cells in the anteroventral cochlear nucleus of the cat. A study with the electron microscope. *Neuroscience 4:* 1925-1945, 1979a.

Cant, N. B., and Morest, D. K.: Organization of the neurons in the anterior division of the anteroventral cochlear nucleus of the cat. Light microscopic observations. *Neuroscience 4:* 1909-1923, 1979b.

Cohen, E. S.; Brawer, J. R., and Morest, D. K.: Projections of the cochlea to the dorsal cochlear nucleus in the cat. *Exp. Neurol. 35:* 470-497, 1972.

Conlee, J. W., and Kane, E. S.: Descending projections from the inferior colliculus to the dorsal cochlear nucleus in the cat: an autoradiographic study. *Neuroscience 7:* 161-178, 1982.

Elverland, H. H.: Descending connections between the superior olivary and cochlear nuclear complexes in the cat, studied by autoradiographic and horseradish peroxidase methods. *Exp. Brain Res. 27:* 397-412, 1977.

Farley, G. R., and Warr, W. B.: Some recurrent projections of the superior olive to anteroventral and dorsal cochlear nucleus in the cat. *Soc. Neurosci. Abstr. 7:* 56, 1981.

Godfrey, D. A.; Carter, J. A., Berger, S. J., Lowry, O. H., and Matschinsky, F. M.: Quantitative histochemical mapping of candidate transmitter amino acids in cat cochlear nucleus. *J. Histochem. Cytochem. 25:* 417-431, 1977.

Guth, P. S., and Melamed, B.: Neurotransmission in the auditory system: a primer for physiologists. *Annu. Rev. Pharmacol. Toxicol. 22:* 383-412, 1982.

Guth, P. S.; Sewell, W. F., and Tachibana, M.: The pharmacology of the cochlear afferents and cochlear nucleus. In Brown, R. D., and Daigneault, E. A. (Eds.): *Pharmacology of Hearing: Experimental and Clinical Bases.* New York, John Wiley and Sons, Inc., 1981, pp. 137-151.

Hudson, D. B.; Valcana, T., Bean, G., and Timiras, P. S.: Glutamic acid: a strong candidate as the neurotransmitter of the cerebellar granule cells. *Neurochem. Res. 1:* 83-92, 1976.

Jones, D. R., and Casseday, J. H.: Projections to laminae in the dorsal cochlear nucleus in the tree shrew, Tupaia glis. *Brain Res. 160:* 131-133, 1979.

Kane, E. C.: Auditory neurons descending to caudal cochlear nucleus in cats: A horseradish peroxidase (HRP) study. *Am. J. Anat. 146:* 433-442, 1976a.

Kane, E. C.: Descending projections to specific regions of cat cochlear nucleus: a light microscopic study. *Exp. Neurol 52:* 372-388, 1976b.

Kane, E. S.: Descending input to the cat dorsal cochlear nucleus: an electron microscopic study. *J. Neurocytol. 6:* 583-605, 1977.

Kane, E. S., and Conlee, J. W.: Descending inputs to the caudal cochlear nucleus of the cat: degeneration and autoradiographic studies. *J. Comp. Neurol. 187:* 759-784, 1979.

Kane, E. S., and Finn, R. C.: Descending and intrinsic inputs to dorsal cochlear nucleus of cats: a horseradish peroxidase study. *Neuroscience 2:* 897-912, 1977.

Levi, G., and Gallo, V.: Glutamate as a putative transmitter in the cerebellum: stimulation by

GABA of glutamic acid release from specific pools. *J. Neurochem. 37:* 22–31, 1981.

Lorente de No, R.: Anatomy of eighth nerve. The central projection of the nerve endings of the internal ear. *Laryngoscope 43:* 1–38, 1933.

Lorente de No, R.: Central representation of the eighth nerve. In Goodhill, V. (Ed.): *Ear Diseases, Deafness and Dizziness.* New York, Harper and Row, 1979, pp. 64–86.

Lorente de No, R.: *The Primary Acoustic Nuclei.* New York, Raven Press, 1981.

Martin, M. R.: The effects of iontophoretically applied antagonists on auditory nerve and amino acid evoked excitation of anteroventral cochlear nucleus neurons. *Neuropharmacol. 19:* 519–528, 1980.

Moore, J. K.: Localization of GAD immunoreactivity in brainstem auditory nuclei. *Assoc. Res. Otolaryngolog. Abstr. 7:* 28, 1984.

Moore, M; Caspary, D. M., and Havey, D. C.: Iontophoretic application of putative inhibitory neurotransmitters into binaural units in the superior olive. *Soc. Neurosci. Abstr. 7:* 389, 1981.

Mugnaini, E.; Warr, W. B., and Osen, K. K.: Distribution and light microscopic features of granule cells in the cochlear nuclei of cat, rat and mouse. *J. Comp. Neurol. 191:* 581–606, 1980.

Oertel, D.: Cells in the anteroventral cochlear nucleus are insensitive to L-glutamate and L-aspartate; excitatory synaptic responses are not blocked by D-α-aminoadipate. *Brain Res. 302:* 213–220, 1984.

Sandoval, M. E., and Cotman, C. W.: Evaluation of glutamate as a neurotransmitter of cerebellar parallel fibers. *Neuroscience 3:* 199–206, 1978.

Schwartz, I. R.: A simple method for osmicating and flat-embedding large tissue sections for light and electron microscopy. *Stain Tech. 57:* 52–54, 1982a.

Schwartz, I. R.: Amino acid labeling patterns in fresh brain slices of the cat superior olivary complex, 1984a. In preparation.

Schwartz, I. R.: Autoradiographic evidence that glycine labeling of synaptic terminals in the superior olivary complex has transmitter-like properties. In Webster, W. W., and Aitken, L. M. (Eds.): *Mechanisms of Hearing.* Clayton, Australia, Monash University Press, 1983a, p. 147.

Schwartz, I. R.: Axonal organization in the cat medial superior olivary nucleus. In Neff, W. D. (Ed.): *Contributions to Sensory Physiology, Vol. 8..* New York, Academic Press, 1984b, pp. 99–129.

Schwartz, I. R.: Dendritic arrangements in the cat medial superior olive. *Neuroscience 2:* 81–101, 1977.

Schwartz, I. R.: Differential tritiated amino acid labeling of synaptic terminals in the cat medial superior olivary nucleus. *Assoc. Res. Otolaryngol. Abstr. 5:* 21, 1982b.

Schwartz, I. R.: Differential uptake of ^{3}H-amino acids in the cat cochlear nucleus. *Am. J. Otolaryngol. 4:* 300–304, 1983b.

Schwartz, I. R.: ^{3}H labeling following glycine incubations identifies a distinctive population of synaptic terminals in the cat medial superior olivary nucleus. *Assoc. Res. Otolaryngol. Abstr. 6:* 8–9, 1983c.

Schwartz, I. R.: The differential distribution of label following uptake of ^{3}H-amino acids in the dorsal cochlear nucleus of the cat: an autoradiographic study. *J. Exp. Neurol. 73:* 601–617, 1981.

Schwartz, I. R.: The differential distribution of synaptic terminal classes on marginal and central cells in the cat medial superior olive. *Am. J. Anat. 159:* 25–31, 1980.

Schwartz, I. R.: Uptake of glycine by endings in the cat medial superior olive, 1984c. In preparation.

Schwartz, I. R., and Bok, D.: Electron microscope localization of ^{125}I-α-bungarotoxin binding sites in the outer plexiform layer of the goldfish retina. *J. Neurocytol. 8:* 53-66, 1979.

Schwartz, I. R., and Wittebort, A. Z.: Axon terminals in the cat medial superior olivary nucleus. *Anat. Rec. 184:* 515, 1976.

Shank, R. P., and Aprison, M. H.: Biochemical aspects of the neurotransmitter function of glutamate. In Filer, L. J., Jr., Kare, M. R., Garattini, S., Reynolds, W. A., and Wurtman, R. J. (Eds.): *Glutamic Acid: Advances in Biochemistry and Physiology.* New York, Raven Press, 1979, pp. 139-150.

Wenthold, R. J.: Glutamate and aspartate as neurotransmitters for the auditory nerve. *Adv. Biochem. Psychopharmacol. 17:* 69-78, 1981.

Wenthold, R. J.: Neurochemistry of the auditory system. *Ann. Otol. Rhinol. Laryngol. Suppl. 74:* 121-131, 1980.

Werman, R.: Criteria for identification of a central nervous system transmitter. *Comp. Biochem. Physiol. 19:* 745-766, 1966.

PART III
BIOCHEMISTRY OF AUDITORY STRUCTURAL ELEMENTS

Chapter 16

THE ORGANIZATION OF ACTIN FILAMENTS IN THE STEREOCILIA OF THE HAIR CELLS OF THE COCHLEA

LEWIS G. TILNEY AND DAVID J. DEROSIER

I. Introduction
II. The Actin Bundle
III. The Role of Angular Disorder in Bundle Formation
IV. Determination of Cell Polarity
V. Sound Exposure Affects Crosslinking
VI. What Happens When a Stereocilium is Displaced, Such as Might Occur During Sound Stimulation
VII. Correlation of Stereociliary Length With Characteristic Frequency
References

I. INTRODUCTION

Located on the surface of the vertebrate cochlea are sensory cells or hair cells that transduce mechanical energy into an electrical response. Although the size of the cochlea and the number of hair cells per cochlea vary depending on the species—e.g., 150 hair cells in an alligator lizard (Mulroy, 1974), 10,000 in a bird (Tilney et al., 1984), and 13,000 in a cat (Lim, 1980)—in all cases transduction is thought to be carried out by the movement or displacement of projections that extend from the hair cells, the so-called stereocilia. What is remarkable is that the stereocilia are exceedingly stiff, and apparently capable, at threshold levels of stimulation, to detect displacement of only a fraction of an angstrom (Hudspeth, 1983). Whether or not this measurement is off by an order of magnitude or more, the fact remains that the stereocilia must be stiff enough to "detect" minute mechanical displacements. Our purpose in this brief review is to present evidence that, first, the organization of the actin filaments in the stereocilia gives rise to the mechanical rigidity; second, the resistance to mechanical displacement of the stereocilia, such as occurs during sound stimulation, is proportional to the number of crossbridging macromolecules between actin filaments; third, by a systematic variation in the number, length, and width of their stereocilia, hair cells located in different parts of the cochlea are fine tuned to different frequencies.

II. THE ACTIN BUNDLE

Within each stereocilium is an organized bundle of actin filaments that lie parallel to each other (Fig. 1). Careful examination of thin sections through these bundles reveals a periodic transverse striping (see Fig. 1). It is through a detailed analysis of this striping that we can begin to learn how this actin filament bundle can provide the stiffness to the stereocilium. Although striping patterns can be found in all vertebrate stereocilia so far examined (see Tilney, et al., 1980, 1983), it turns out that these patterns are most easily understood in the bird stereocilia, because in birds, as contrasted to lizards or mammals, the actin filaments are more regularly arranged, being hexagonally packed in transverse sections (Fig. 2). Yet to understand the striping pattern in bird stereocilia requires a close examination of the structure of an actin filament. Thus, in the next few paragraphs we will first consider the structure of an actin filament by using models; we will then integrate this information to "build" a compact bundle of actin filaments that corresponds to what is present in a bird stereocilium.

In Fig. 3 we have diagramatically depicted three actin filaments. Notice that each actin filament is schematically represented by a series of globular subunits which are arranged with helical symmetry. There are three obvious helical lines of subunits. First, there is a two start, right-handed helix indicated by the stippled and non-stippled rows of subunits that twist over each other; this helix has a half period of 375 Å and contains approximately 13 subunits per half turn. Second, there is a left-handed, single-start helix whose period is 59 Å, the so-called "genetic helix", and finally, there is a righthanded, one-start helix whose period is 51 Å. In this drawing (Fig. 3) we oriented each of the three actin filaments such that the narrowest profiles or "crossover points" of each filament are in transverse register. To understand this, consider the two-start helix; we can think of each actin filament as a somewhat flattened ribbon which first lies flat in front of us, and, as this "ribbon" twists in space, one of the component rows, e.g., the stippled one, will cross over the other, the non-stippled row. In stereocilia, as in this drawing, it turns out that the crossover points of adjacent filaments are in transverse register; the importance of this fact will be discussed subsequently. Since the subunits making up each actin filament are helically arranged, the position and hence the number of bridges between actin filaments is fixed by the geometry of the actin filaments and that of their packing in the bundle. More specifically, assume that a bridge connects two adjacent filaments together only when the subunits on two adjacent filaments point towards

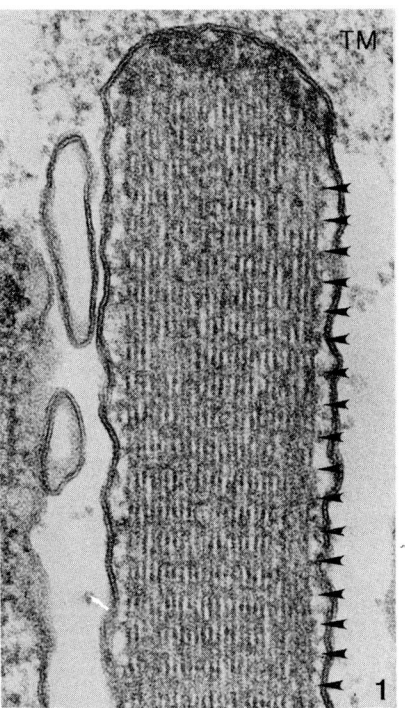

Figure 16-1. Longitudinal section through the tip of a stereocilium. The filament bundle terminates near the tip in some electron-dense material that is attached to the membrane. Also attached to the tip of this stereocilium is a portion of the tectorial membrane (TM). Close examination of the filament bundle reveals that there are periodic bands extending across the bundle (see arrows). (From Tilney et al., 1983.) × 152,000.

each other. Thus, in the balsa wood model in Fig. 4a, filaments A and B could be connected together by a bridge at position 2, but as filaments A and B rotate in space, these subunits will no longer point towards each other and thus could no longer be bridged together by a linear cross-connective. For example, filaments A and B could not be connected at subunit 7, however, we could connect filament A to filament C at subunit 7 by a bridge, because at this position, subunits 7 of A and C point towards each other. Likewise, filament C can be connected to filament B at subunit 11. In short, because of the helical symmetry of actin, the position of the crossbridges, assuming only one kind of bridge, is dictated by the geometry of the actin helix.

Now let us take this analysis one step further and examine a model which simulates a longitudinal section of a hexagonally-packed bundle of actin filaments such as might occur in a bird stereocilium. When all the actin filaments in the bundle have their crossover points in register, or to put it another way, when subunits 2 on each filament point in the same direction,

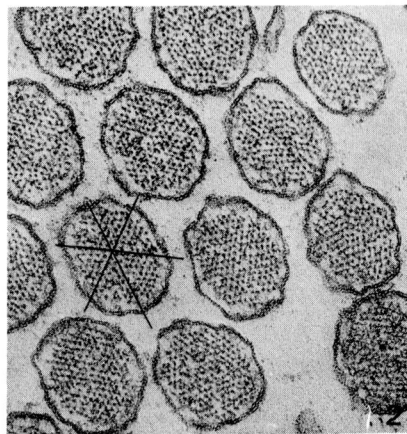

Figure 16-2. Transverse section through a number of stereocilia. The filaments within each stereocilium are arranged on a hexagonal lattice. This can most easily be appreciated by looking at the lines that mark the hexagonal lattice on which the filaments lie. (From Tilney et al., 1983.) × 81,000.

the bridges that form will line up to create striations. Thus, neighboring filaments would be crossbridged once per half turn of the two-start helix, or every 375 Å, and since there are three sets of neighbors in a hexagonally-packed bundle, there will be three striations per half turn at approximately 125 Å intervals (Fig. 4). We programmed the computer to draw this bundle as it would appear when examined in the electron microscope lying on a grid. There are two views of the same bundle which give rise to two different striation patterns. To understand how these patterns are generated, consider Fig. 5. What the reader should do is to take the bundle depicted in Fig. 5a and mentally flip it on to its back (Fig. 5c) so that filament A lies directly over filament D. In this view, known as the 1,1 view, we see that there are three crossbridges per half turn of the helix, spaced at 125-Å intervals. What is superficially confusing is that if we rotate the bundle by 30° (Fig. 5d), and then flip the bundle on its back (the 1,0 view), filament A will lie directly over filament C, now hiding from view the crossbridges between A and C. These hidden crossbridges are indicated on filaments A, B, and D in Fig. 5f by the filled circles. Thus, in the 1,0 view we would expect to see the filaments not only further apart than in the 1,1 view, but also we see two crossbridges separated by 125 Å, then a gap of 250 Å, and then two crossbridges, etc.

An examination of longitudinal thin sections of bird stereocilia at high resolution (Fig. 6), at a 30° angle of tilt, reveals the 1,1 and 1,0 views. In the case of the former we see 125-Å, periodic striations and in the latter we see a pair of striations separated by 125 Å followed by a gap of 250 Å, then a pair

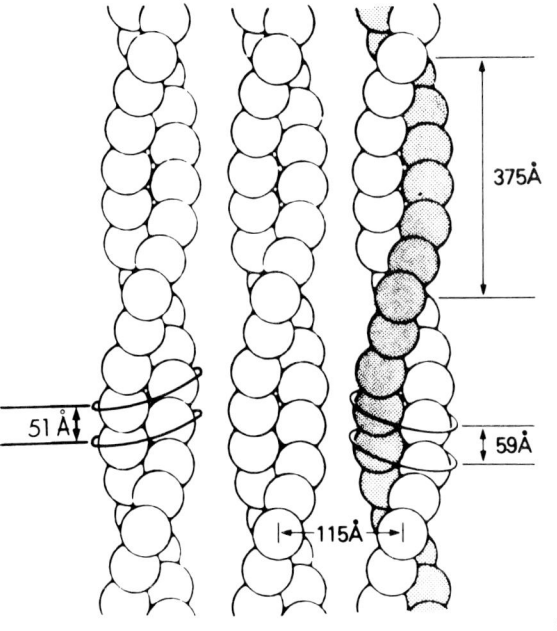

Figure 16-3. Schematic drawing depicting 3 actin filaments. Each actin filament is made up of globular subunits arranged on a helical pattern. The most obvious helical pattern is illustrated in the right-handed filament in which the filament can be considered to be composed of two strands, one stippled, the other unstippled, that wind around each other in the right-handed direction. Each half turn of the helix measures approximately 375 Å. Of interest is that in stereocilia, the point that the stippled strand crosses over the non-stippled strand of one filament is in transverse register to the filaments on all its sides. There are two other helical families, one a left-handed helix with a period of 59 Å, the other a right-handed helix with a period of 51 Å. (From Tilney et al., 1980.)

of striations, etc. In short, what we see in our micrographs corresponds exactly the to the geometry of the actin helix and the fact that the crossover points are in transverse register. In fact, we can prove that the crossover points of adjacent actin filaments are in transverse register by diffraction-pattern analysis of our thin sections (DeRosier and Tilney, 1984a, b; Tilney et al., 1980, 1983). However, this is beyond the scope of our brief review, and to belabor the point is not useful. Intuitively, one can see that this statement must be true because if the crossover points were not in transverse register and instead the actin filaments were translated relative to each other by some random amount, then subunits on neighboring filaments would not point towards each other and thus *no crossbridging whatsoever could occur.* There is one additional point that should be made in this regard. From these considerations we know not only approximately where the crossbridges

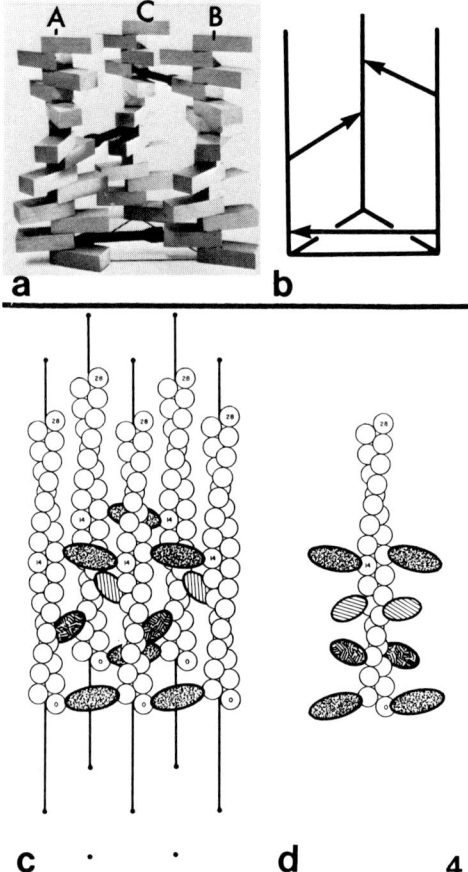

Figure 16-4. (a) Balsa wood model depicting the arrangement of crossbridges in a hexagonally-packed bundle of filaments. Three model filaments having the helical symmetry of actin are sitting on three vertices of a hexagonal lattice. The filaments are set so that the lowest subunit or balsa block on each is pointing in the same direction; thus the crossover points of adjacent actin filaments are in register. Crossbridges (black arrows) link neighboring filaments. Such a crossbridge appears between filaments A and B at subunit 2, the next to lowest subunit. Because of the helical symmetry of the models, a crossbridge also links filaments A and C at subunit 7 and filaments B and C at subunit 11. The crossbridges that link these three filaments will link together all the filaments in a hexagonal array.
(b) Schematic drawing of the crossbridge scheme depicted in (a).
(c) Five actin filaments linked together by bridges (stippled units). As in (a), the filaments lie on a hexagonal lattice. Because the crossover points are in register, the bridges lie on a series of planes running perpendicularly to the long axis of the filament. The crossbridges on different planes are stippled differently. If the whole bundle is looked at from afar, striations should appear. This can be seen in (d).
(d) One of the actin filaments depicted in (c), with its attached bridges. Note that there are three planes of crossbridges in every crossover. (From DeRosier and Tilney, 1980.)

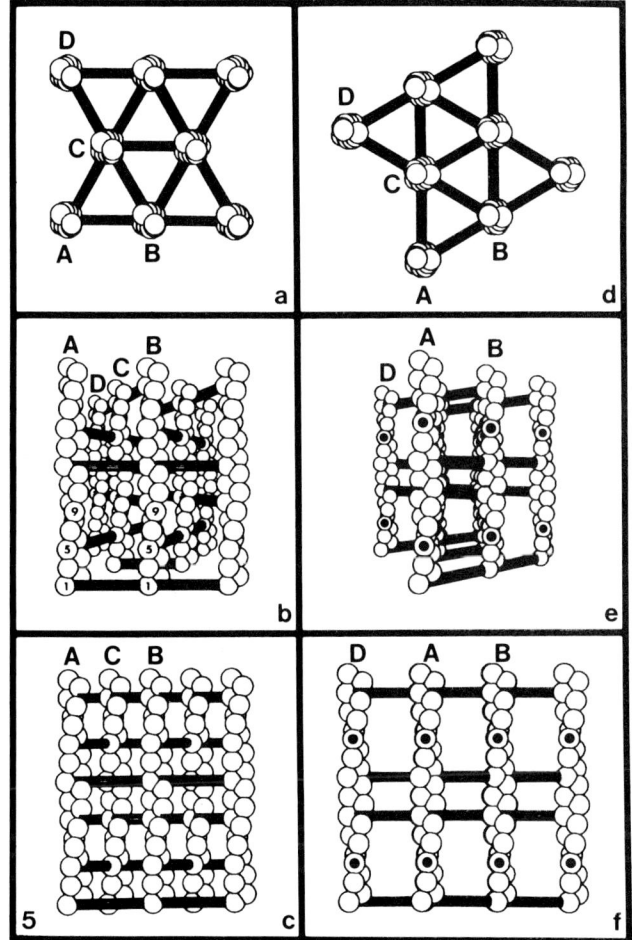

Figure 16-5. Computer-generated model of a crossbridged bundle of actin filaments. (a) Top view. (b) 1,1, view with perspective. If one flips the bundle seen in (a) on its back, one sees this view. The crossbridges are represented by the shaded lines. (c) 1,1, view without perspective. This image simulates what is seen in the electron microscope. A series of more or less evenly spaced transverse bands cross the filaments. (d) Top view. The same as in (a) but rotated 30°. (e) 1,0 view with perspective. If the bundle in (d) is now flipped, one generates this view instead of the 1,1 view. (f) 1,0 view without perspective. This image shows the characteristic two bands and a gap, rather than three evenly-spaced bands. It is easy to see that one band is missing because the crossbridges are viewed on end and appear hidden behind filaments. The dots indicate the positions of hidden crossbridges. (From Tilney et al., 1983.)

between adjacent actin filaments are located, but also how frequent they are or, more specifically, that in a hexagonally-packed bundle there will maximally be one crossbridge for every 4.5 actin subunits.

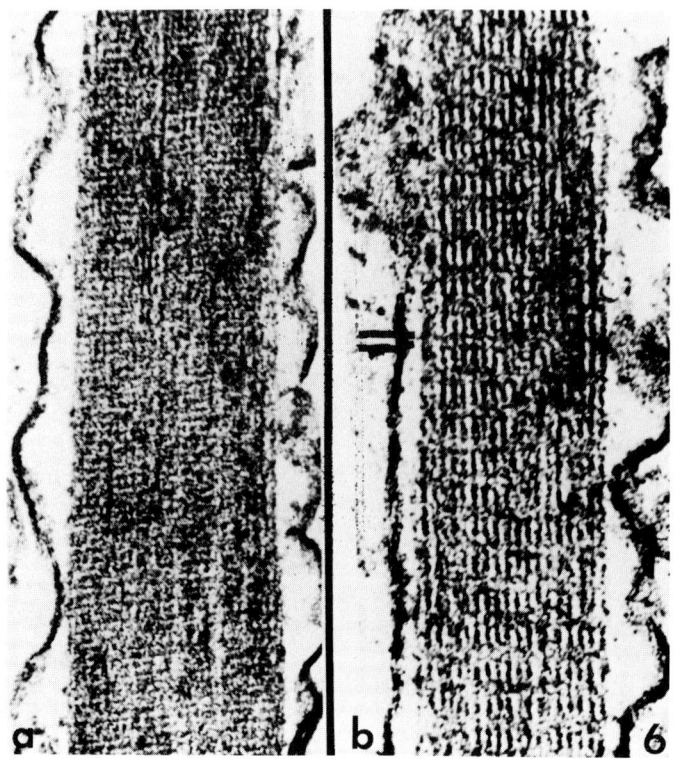

Figure 16-6. (a) Longitudinal section of a detergent-extracted stereocilium, cut to illustrate a 1,1 view similar to what was depicted in Fig. 5c. Prominent here are the horizontal striations, separated at 125-Å intervals, which are the crossbridging macromolecules. The longitudinal striations, due to the filaments, are not visible in this micrograph because the filaments are so close together. × 140,000. (b) Longitudinal section of a detergent-extracted stereocilium, cut to illustrate the 1,0 view, a view similar to what is depicted in Fig. 5f. The two dark lines depict horizontal striation, the crossbridging macromolecules. These are separated by 125 and 250 Å. The longitudinal striations are the actin filaments. (From Tilney et al., 1983.) × 194,000.

Although the above considerations, i.e., the bonding rules that we have developed, apply to lizard and mammal stereocilia as well as to those of the bird, the situation in lizard and mammal stereocilia is more complex and thus more difficult to quantitate. The essential difference between bird stereocilia and mammal and lizard stereocilia is that although all contain actin filaments whose crossover points are in transverse register, in lizard and mammal stereocilia the actin filaments *are not hexagonally packed* as seen when transverse sections are examined (Fig. 7). Instead, adjacent filaments in transverse section show no obvious relation with one another, even though they are closely spaced and are, in fact, crossbridged together, as can be shown by detergent-extracting the stereocilia (see Flock *et al.*, 1977; Tilney *et al.*, 1980). In the words of a crystallographer, they show liquid-like order in transverse section, but in longitudinal section they are paracrystalline, with their crossover points in transverse register. To understand the liquid-like order of the actin filaments in lizard and mammal stereocilia, examine Fig. 8, in which we have placed a balsa wood model of an actin filament at the center of a circle and a second model filament (B) on the circumference of the circle with its crossover points in transverse register with the first. Notice that in Fig. 8b we can connect these model filaments with a crossbridge at subunit 7. If we move filament B relative to filament A by a distance of about 10° or 10 Å, we can still connect filaments A and B, but now filaments A and B are connected together at subunit 6, etc. Similarly, if we move filament B another 10 Å around the circle, we can still connect filaments A and B, but now they are connected at subunit 5, and so on. Thus, provided the crossover points are in transverse register and adjacent filaments are separated by a distance equal to the length of the crossbridge, subunits on adjacent filaments can be linked together, but the exact subunits specified to be linked together depend on the position of the two filaments relative to each other. The important point for our consideration here is that hexagonal packing, such as occurs in bird stereocilia, maximizes the number of bridges between adjacent filaments and thus maximizes the rigidity of the actin bundle (see below) (DeRosier *et al.*, 1980). It also allows us to specify the ratio of the crossbridging molecules to actin monomers. With liquid-like order, such as occurs in lizard and mammal stereocilia, this ratio will be somewhat lower. In the lizard, the average number of nearest neighbors is 5. This corresponds to 2.5 crossbridges per repeat, or one crossbridge per 5.5 actin monomers, on the average.

III. THE ROLE OF ANGULAR DISORDER IN BUNDLE FORMATION

There is one further aspect of the actin filament that should be discussed. We have already mentioned that even though a crosslinked, hexagonally-packed bundle of actin exists, each actin filament does not display hexagonal symmetry, but rather, actin is made of helices of non-crystallographic symmetry. That is to say, the helical symmetry of actin is incompatible with crystallographically-regular, hexagonal symmetry. Thus, the question becomes, how can one produce a bundle of hexagonally-packed filaments with units that do not possess helical symmetry. To put this another way, to produce hexagonal packing, a distortion must be introduced into the bundle. Because of the nonhexagonal nature of the actin filament, in

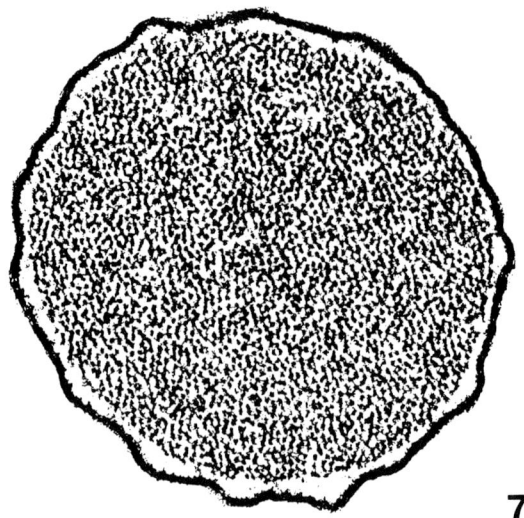

Figure 16-7. Transverse section through a lizard stereocilium. Of interest is that the actin filaments are not hexagonally packed and, in fact, show no obvious symmetric relation to one another. They are considered to display liquid-like order (From DeRosier et al., 1980.) × 112,000.

order to allow crosslinking between all adjacent filaments, either the crossbridge must be bendable or the actin subunit must tolerate an angular distortion by being able to locally rotate subunits azimuthally by about 7°. Recent studies by Egelman et al. (1982, 1983) indicate that there is about 10° of cumulative angular distortion in actin subunits in an isolated helix. This is sufficient to allow the hexagonal, tetragonal, or liquid-like bundles to form and to be stable (see DeRosier and Tilney, 1984b). This remarkable feature of the actin filament is depicted in Fig. 9. At the same time, we also know, and this fact will be discussed in detail further on, that the filaments are not easily stretched or compressed, even though they display angular disorder. It is as if the actin subunits are tightly threaded onto a stiff rod. They can be rotated using the rod as an axle, but cannot be moved up or down due to the tight packing of subunits (Fig. 9). The rotational flexibility, then, will allow bundles to form with hexagonal, tetragonal, or even liquid-like order. However, because of the angular disorder, depending upon the symmetry selected, the crossover distances within a filament will vary.

Obviously, for transduction of mechanical energy into electrical energy, the actin filament bundle within each stereocilium must be tightly coupled to the limiting plasma membrane. Little is known about this important topic, but what we do know is that in fast-frozen, deeply-etched, and rotary-shadowed preparations of stereocilia we find periodic connections that extend between the filament bundle and the plasma membrane (Hirokawa and

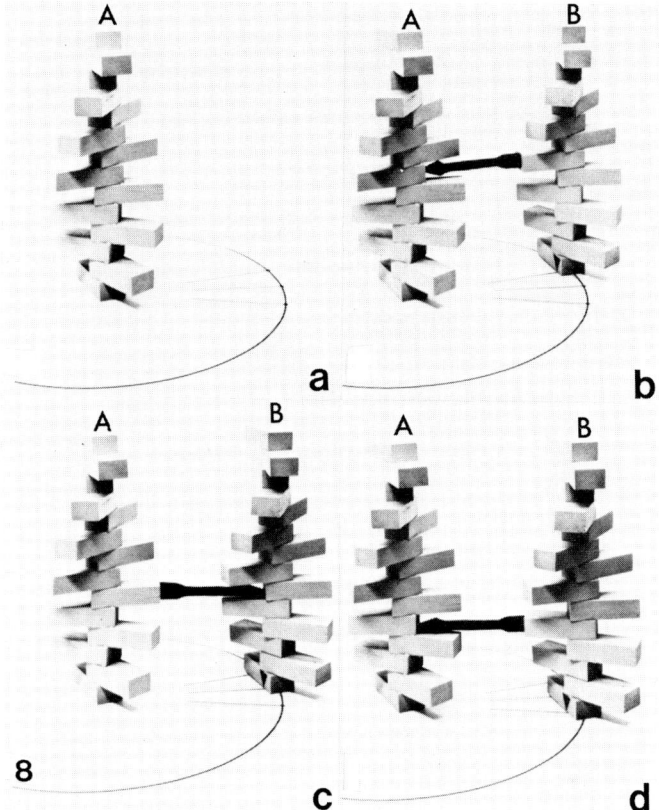

Figure 16-8. Models showing the potential of helices to be crossbridged, provided the crossover points are in register. In (a) a model filament (A) is placed at the center of a circle, indicating three positions separated by 10 Å. In (b) a second filament (B) is placed on the first position and it is crossbridged to the center filament (black arrow) at subunit 7. In (c) the crossbridge formed when the outer filament (B) is moved to the second position has now moved to subunit 6. Note that the outer filament is moved so that the two filaments keep their crossover points in register. In (d) the crossbridge expected when the outer filament is moved to the third position is at subunit 5. Continuing this process, the outer filament can be crossbridged to the inner one at any position along the circle if a modest amount of crossbridge bending (up to 13°) is allowed. The subunit crossbridged, however, will depend on the position. (From DeRosier et al., 1980.)

Tilney, 1982). Unfortunately, these connections are easily destroyed by inadequate fixation.

IV. DETERMINATION OF CELL POLARITY

The direction in which the stereocilia are displaced determines the sign and magnitude of the voltage change (see Hudspeth, 1983). The mechanism

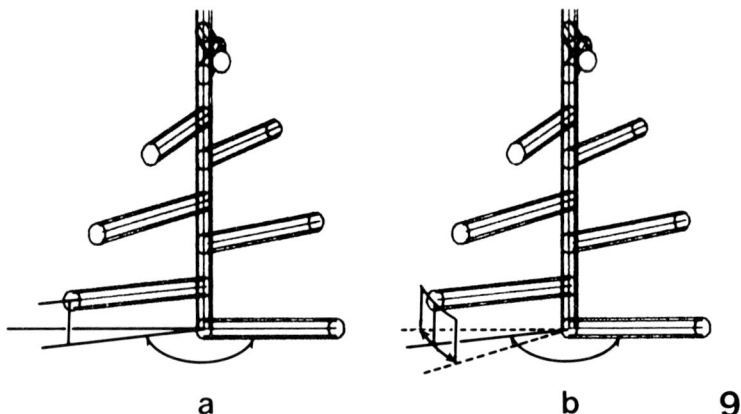

Figure 16-9. Drawing depicting the angular disorder in the actin subunits. It is as if the actin subunits could be rotated using the rod as an axle [see (b)] by an amount that is 10° or less. (From DeRosier and Tilney, 1984b.)

by which the movement is transduced is unknown, as is the source and directionality, or polarity. One can ask if the arrangement of actin filaments in a bundle might fix the polarity of the stereocilium. The answer, unfortunately, is no, because in the chicken, the bundle is hexagonally packed, indicating that there should be six equivalent directions of displacement. This is not found. Moreover, in the lizard where there is liquid packing, all directions are equivalent. The conclusion is that the packing of actin into bundles provides no basis for bundle polarity.

V. SOUND EXPOSURE AFFECTS CROSSLINKING

So far, we have described how the actin filaments are connected together to form compact bundles which, in detergent-extracted preparations, maintain their integrity and rigidity (see Flock et al., 1977; Tilney et al., 1980, 1983). The question at this point is: Are the actin filaments and their crossbridges essential for rigidity? Experiments in which organisms are subjected to intense noise stimulation bear directly on this question. Hunter-Duvar (1977) was the first to notice that in organisms in which there was a temporary threshold shift, the stereocilia were no longer rigid, but appeared limp and floppy, often draping themselves over the surface of the hair cell. To determine what caused this loss in stereociliary rigidity, we subjected lizards to intense noise stimulation which physiologically caused a temporary threshold shift (Fig. 10), then fixed the cochleae and examined the stereocilia in thin sections (Tilney et al., 1982). Although a number of changes were visible, such as blebbing, mitochondrial swelling, etc., there were two striking changes in the organization of the actin filaments in the stereocilia. The most common was a depolymerization of the actin filaments where the bases of the longest stereocilia made contact with the apical surface of the hair cell (Fig. 11). Because of this local depolymerization, the actin filaments in the stereocilia proper were no longer continuous with the rootlet filaments that extend into the cuticular plate region and thus anchor the stereocilium to the apical surface of the cell. This results in loss of stereociliary rigidity. It is as if one were to cut a tree at its base, thus severing the connection of the roots to the trunk. The slightest breeze would topple the tree. The second change was even more interesting, for it involved a loss in the crossbridges that connect neighboring filaments together (Fig. 12). Thus, the separation between adjacent actin filaments was variable, many of the filaments showing an undulatory profile. Interestingly, the loss of crossbridges is confined to only the tallest stereocilia in a bundle (Fig. 12, inset). Thus, if either the integrity of the actin filaments is compromised, or there is a reduction in crossbridging, the stereocilia lose their rigidity.

VI. WHAT HAPPENS WHEN A STEREOCILIUM IS DISPLACED, SUCH AS MIGHT OCCUR DURING SOUND STIMULATION

We next considered what happens to the actin filaments and their crossbridges when stereocilia are displaced, such as occurs during ordinary sound stimulation. What is exciting is that the transverse striations caused by the periodic nature of the crossbridges in bird stereocilia make it possible to interpret what happens during mechanical displacement. We will begin by considering two possible modes of bending a set of filaments (Fig. 13). In the first case (Fig. 13b), when the bundle is bent, filaments on the inside of the bend are compressed and those on the outside are stretched. The other extreme is one in which filaments are neither stretched nor compressed; instead there is slippage between the filaments (Fig. 13c). Selecting between these two possibilities is a relatively simple matter that can be resolved by carefully examining electron micrographs of bent bundles. In order to discern what morphological features of the image would arise from the two cases, examine Fig. 13a, which shows a drawing of three actin-like filaments with their crossover points in register. Lines between corresponding points on neighboring filaments are horizontal and, therefore, perpendicular to the bundle axis, which is vertical. If we now bend the bundle (Fig. 13b), we can maintain in the top part of the bundle the exact spatial relations present in the bottom half of the bundle by stretching and compressing filaments. That is to say, the local spatial relations between filaments in the portion of the

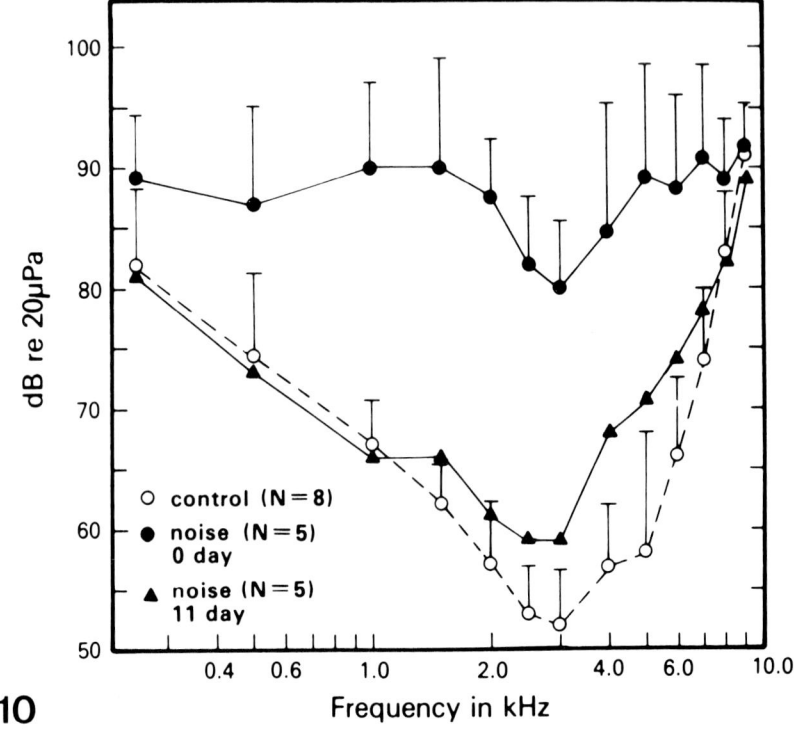

Figure 16-10. Lizards were subjected to intense noise stimulation which resulted in a temporary threshold shift (black dots), as recorded from the cochlear nerve. If the lizards were allowed to recover for 11 days, their hearing sensitivity returned to normal (triangular dots). (From Tilney et al., 1982.)

bundle before the bend and the portion of the bundle after the bend are exactly the same. We can see that the crossover points on adjacent filaments line up in rows perpendicular to the bundle axis. Thus, in the bottom half of the figure, the rows of crossover points are perpendicular to the lower bundle axis and lie on a horizontal plane, and in the top part of the bundle, beyond the bend, the rows of crossover points are perpendicular to the upper bundle axis but, of course, are no longer horizontal; rather, they are tilted at an angle determined by the angle of the bend. Now consider the second extreme in Fig. 13c in which there is neither stretching nor compression of filaments, but in which the bundle length is maintained by slippage between adjacent filaments. In such a case, the relation between crossover points in the upper part of the bundle is no longer maintained in the same way as that in the lower part of the bundle. To be specific, the rows are no longer perpendicular to the bundle axis; but remain horizontal. Thus, in such a bundle we expect to see a tilting of the rows of crossover points off-perpendicular to the

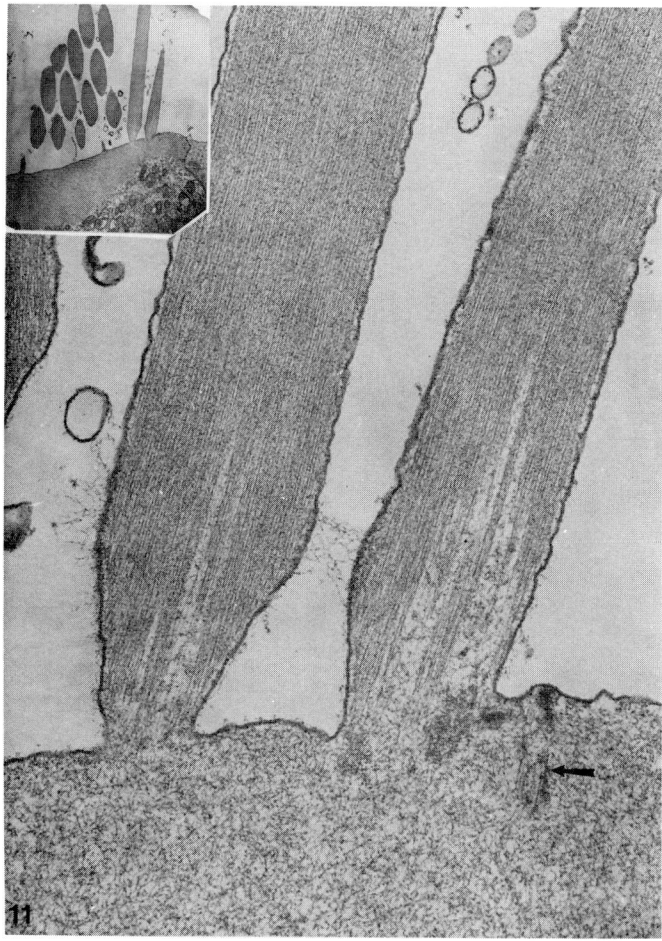

Figure 16-11. The cochleae of lizards subjected to intense noise stimulation, which physiologically caused a temporary threshold shift that was documented in Fig. 10, were fixed immediately after stimulation. Thin sections were cut and the longest stereocilia were examined. The most common lesion was local depolymerization of the actin filaments at the bases of the stereocilia. Thus, the actin filaments in the stereocilia proper are no longer continuous with the rootlet portion of the filaments in the cuticular plate region and are, thus, poorly anchored to the cuticular plate region. This results in loss of stereociliary rigidity. The arrow indicates a portion of the rootlet of the stereocilium. This rootlet was formerly attached to the stereocilium to the left of it. (From Tilney et al., 1982.) × 52,000. Inset: × 3,900.

bundle axis, and, more importantly, the rows following the bend should maintain themselves parallel to the rows preceding the bend. It is a relatively easy matter to examine electron micrographs and ask whether the situation is more like one extreme than the other and even to quantitate to what extent it is like one of the two extremes. In the case of the stereocilia of the

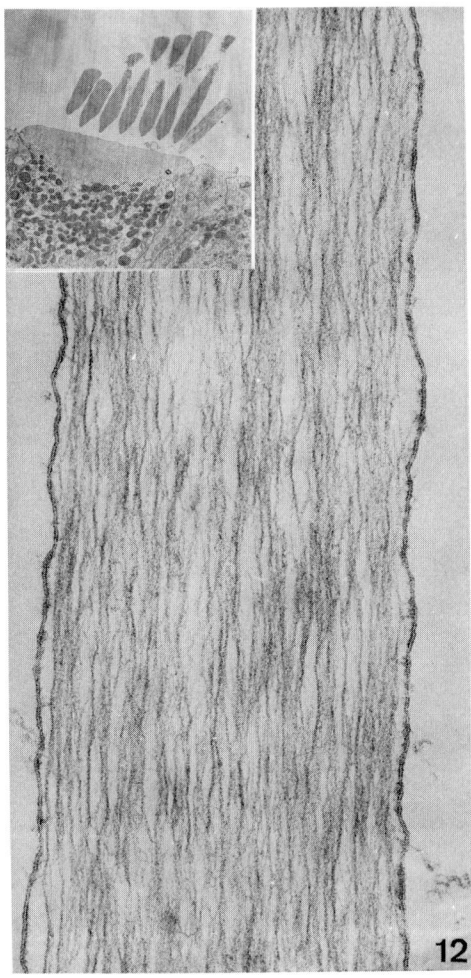

Figure 16-12. The cochleae of lizards subjected to intense noise stimulation, which physiologically caused a temporary threshold shift as documented in Fig. 10, were fixed immediately after stimulation. Thin sections were cut through the hair cells. Of considerable interest is that the actin filaments in the longest stereocilia show an undulatory profile, indicating that they are no longer crossbridged to each other. (From Tilney et al., 1982.) × 100,000. Inset: × 4,600.

bird, the situation in Fig. 13c is seen, namely, that there is predominantly, if not exclusively, filament slippage; there is no detectable stretching or compression of filaments.

Now we can ask what happens when a stereocilium is displaced so that it bends at its base rather than its middle. Presumably this is what would happen *in vivo* during sound stimulation. Looking at the micrographs of stereocilia bent at their bases, such as illustrated in Fig. 15 (compare Figs. 14

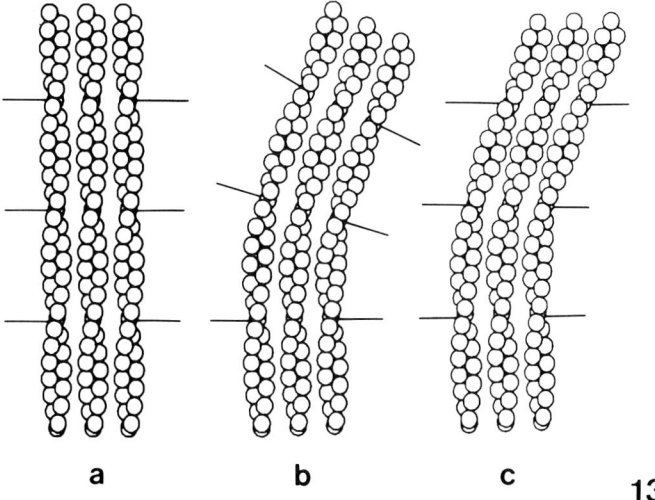

Figure 16-13. Two extreme possibilities for a bend. (a) A "bundle" of three filaments. The filaments are aligned with their crossover points in register, as indicated by the horizontal lines. (b) One possible bending mechanism. In this situation, the filament on the inside of the bend is compressed, while that on the outside is stretched. The consequence of this can be seen in the way the crossovers align. Note that the upper lines passing through the crossover points tilt relative to the lower ones; thus the crossover points remain perpendicular to the bundle axis. (c) A second possible mechanism. In this case, the filaments neither bend nor stretch, but slip relative to each other. In contrast to the case in (b), the lines connecting the crossovers do not tilt, but remain horizontal. (From DeRosier and Tilney, 1984).

and 15), we clearly see that the crossbridges remain parallel to the apical surface. Thus, when bending of a stereocilium occurs, filaments tend to slide past one another rather than being stretched or compressed. What is interesting and important is that even though bending of the stereocilium at its base involves only the 20 or so filaments at the point of bending, the result is somehow transmitted to *all* the filaments in the bundle so that *all the crossbridges in the bundle are affected* whether or not they are located at the point of bending.

Before we consider how this behavior affects the response of stereocilia during sound stimulation, let us ask whether this effect is observed in other vertebrates. As already mentioned, lizard stereocilia are more difficult to analyze because they lack the 125-Å periodicity caused by the crossbridges in a bundle in which the actin filaments are hexagonally packed, such as in the bird. Nevertheless, from the diffraction patterns of the stereocilia of the lizard cochlea, we have determined that the filaments slide past one another rather than being stretched or compressed (see the Discussion of Tilney *et al.*, 1980, for details). This is an extremely important conclusion because it

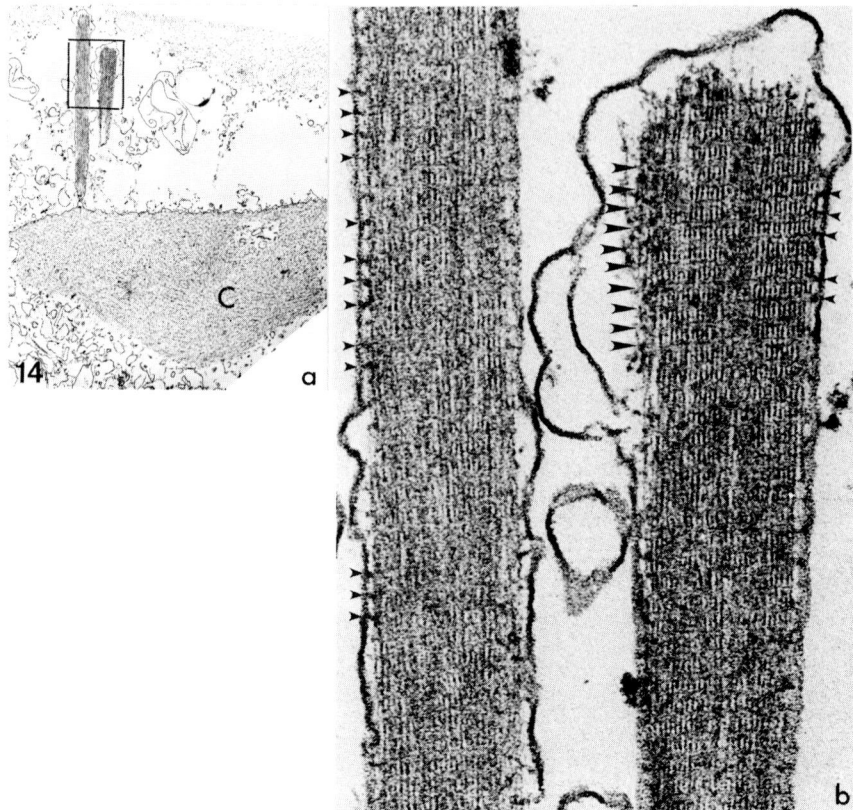

Figure 16-14. (a) Thin section through the apical end of a detergentextracted hair cell. Prominent is the cuticular plate (C) and two stereocilia that extend upwards from the plate. They lie perpendicular to the apical surface of the cuticular plate. × 8,300.
(b) Both stereocilia are illustrated at higher magnification. The characteristic banding pattern due to the crossbridges (see large arrows) is perpendicular to the long axis of this stereocilium, or parallel to the apical surface of the cuticular plate. The small arrows indicate points of crossbridging to the cell membrane. (From Tilney et al., 1983.) × 90,000.

means that the resistance to displacement of a stereocilium under a fixed torque will depend *on the total number of crossbridges being bent*, a variable that is systematically modified in cells at different parts of the cochlea, as will be discussed in detail later. Although we get the same results in numerous preparations, one might argue that our observations on the behavior of the bridges when a stereocilium is displaced is an artifact of the preparative procedure for electron microscopy. For example, perhaps in the native state the actin filaments are stretched and compressed and the crossbridges unbent, whereas it is only after fixation that the opposite is seen. Evidence from a different cell type bears directly on this uncertainty. Because the conclusion

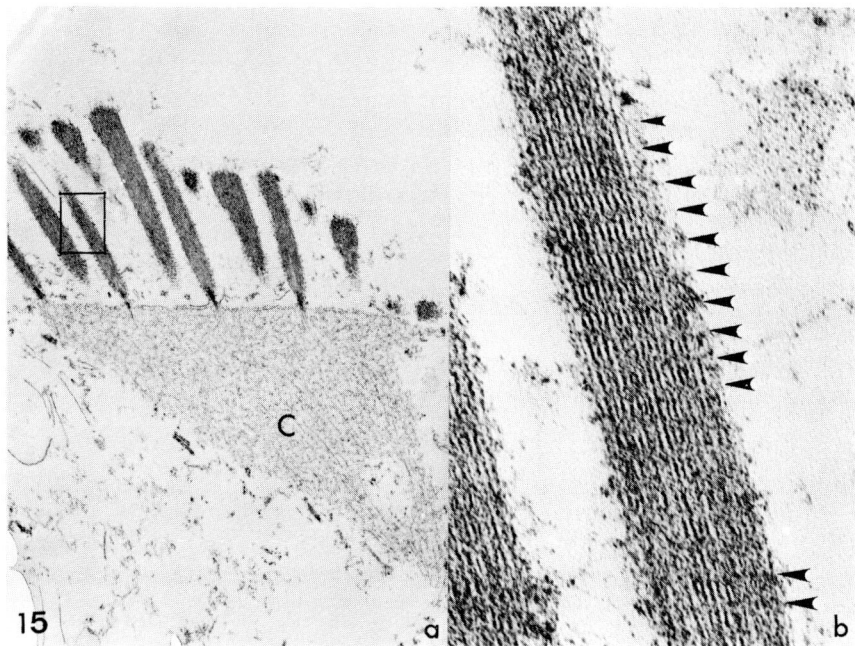

Figure 16-15. (a) Thin section through the apical end of a detergent extracted hair cell. Prominent are the cuticular plate (C) and the stereocilia. In this micrograph the stereocilia lie at an oblique angle to the apical surface of the cuticular plate. The square indicates a portion of a stereocilium, which is shown at higher magnification in (b). × 15,300.
(b) Of interest in this micrograph is that the banding pattern due to the crossbridges remains parallel to the apical surface of the cuticular plate. It is no longer perpendicular to the long axis of the stereocilium. Thus, when stereocilia are bent or displaced as with noise stimulation, the filaments do not stretch or compress, but rather tend to slide past one another, such as illustrated in model c of Fig. 13. (From Tilney *et al.*, 1983.) × 140,000.

is relevant to frequency tuning in the cochlea, we will digress briefly to mention this evidence, as it shows that what we have stated above is unlikely to be a preparative artifact.

The system we wish to mention is the sperm of the horseshoe crab, *Limulus*. In that system, as in the stereocilia, there is a crosslinked bundle of actin filaments, but unlike the actin bundle in the stereocilia, bends are built into the *Limulus* bundle as stable features of the bundle (DeRosier and Tilney, 1984a). To be more specific, in *Limulus* sperm, a 60-μm-long, hexagonally-packed bundle of actin filaments lies coiled up inside the sperm (Fig. 16a). This coil is not smoothly bent, but appears polygonal, consisting of sharp bends (elbows) separated by straight regions (arms) (Fig. 16b). Interestingly, this coil can be freed from the sperm and from all membranes by detergent extraction and still maintain its polygonal form. Using a variety of techniques to examine both the surface of the coil and its interior, we (DeRosier and Tilney, 1984a) found that the polygonal nature of the coil is not a property of some secondary scaffold, but rather is due to the rearrangement of interfilament crossbridges in the arms. The rearrangement, which makes use of specific bonding properties of the crossbridges, just as in the stereocilia, is made possible by the helical symmetry of actin, by the hexagonal packing of the filaments, and by the angular disorder of the actin subunits. Furthermore, the angle of each bend (154°) is determined by the interfilament separation and the axial spacing of actin subunits within a filament. Thus, what we find in the *Limulus* bundle is that in passing through an elbow, the actin filaments are not stretched or compressed, but just as in the stereocilia, they slip relative to each other. This slippage, and hence the elbow, is locked in by an axial shift of the crossbridges between adjacent rows of filaments (Fig. 17). Also, just as in the stereocilia, the angular disorder of the actin filaments seems important in accommodating the strain of this crossbridge rearrangement. In conclusion, then, in both stereocilia and *Limulus* sperm, which constitute two rather different situations not susceptible to the same preparative artifacts, bends in crosslinked bundles of actin occur without compression or stretching of the filaments. In the sperm, which is really the corollary to what happens when a stereocilium is bent, bends are actually built in as regular features of actin bundles and maintained as a stable feature of this bundle.

Let us now return to our conclusion that the resistance to displacement of a stereocilium under a fixed torque will depend on the total number of crossbridges being bent. One wonders if the cochlea somehow uses this information for the purpose of discrimination of frequency differences, such that cells located at one end of the cochlea are tuned to low frequencies, while those at the other end are tuned to higher frequencies. Although it is impossible at this point to make quantitative statements, there is considerable evidence to suggest that in fact hair cells located at different parts of the cochlea not only have different total numbers of crossbridges per stereocilium, but the *stereocilia themselves* are tuned, at least in lower vertebrates, to the characteristic frequency of the cell. Recall that in a hexagonally packed bundle of actin filaments, such as occurs in bird stereocilia, it is easy to calculate the number of crossbridges per stereocilium. More specifically, for every 375 Å of filament length there are three crossbridges. Thus, the total number of crossbridges per stereocilium depends on the number and length of filaments in that stereocilium. In the bird cochlea, a tall hair cell located at the distal or low frequency end of the cochlea has stereocilia 5.5 µm in length (Fig. 18) and 0.12 µm in width (Fig. 19). Accordingly, each stereocilium would contain 37,000 crossbridges. In contrast, at the proximal or high-frequency end, each stereocilium, by being only 1.5 µm in length (Fig. 18)

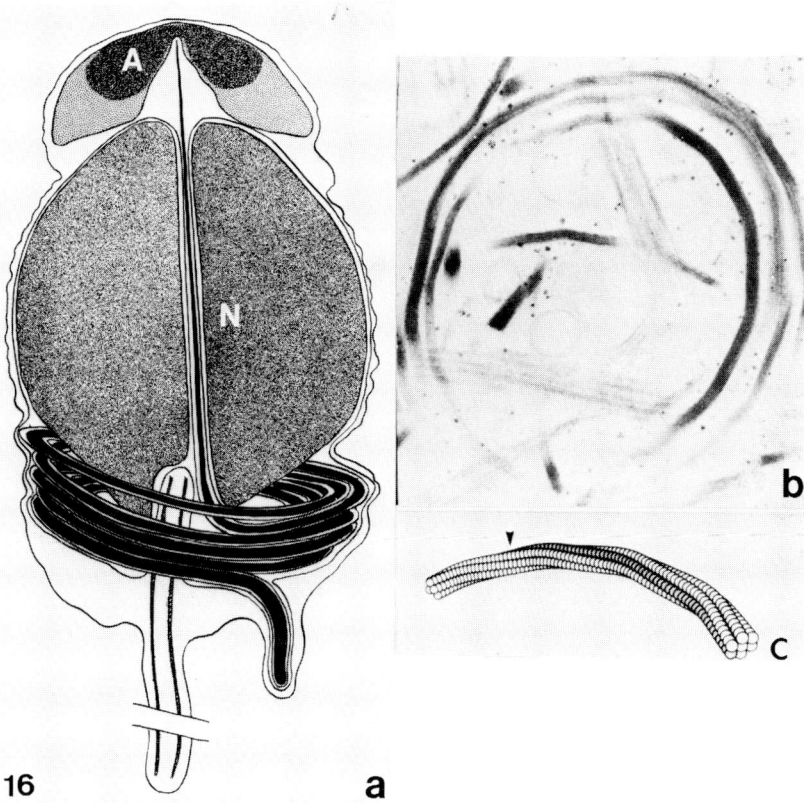

Figure 16-16. (a) Drawing of the sperm of the horseshoe crab, *Limulus*. Within this sperm is a 60-μm-long bundle of hexagonally-packed actin filaments which are cross-linked together by two crosslinking proteins. The bundle extends from the acrosomal vacuole (A) through a canal in the center of the nucleus (N), where it coils around the base of the sperm.
(b) Thin section through a portion of the coil of actin filaments at the base of an unreacted *Limulus* sperm. Of interest is that the bundle is not smoothly bent, but appears polygonal, consisting of fourteen straight regions, each 0.7 μm in length, separated by fourteen 154° bends or elbows.
(c) Model of a portion of the coil. Notice that the shaded filament indicated by the arrow begins on the inside edge of the first elbow, but it gradually moves to an outer edge at the fourth elbow. As it does so, the subunits on each filament slide relative to each other, as will be indicated in Fig. 17.

and 0.19 μm in diameter (Fig. 19), would contain only 10,000 crossbridges. Thus, the stereocilia from different ends of the cochlea should respond differently to an applied force. More impressive is that a careful examination of bird cochleae by scanning microscopy reveals that the length, width, and number of stereocilia, in fact, vary systematically from the distal to proximal ends. These variations in turn must affect the mechanical properties of the

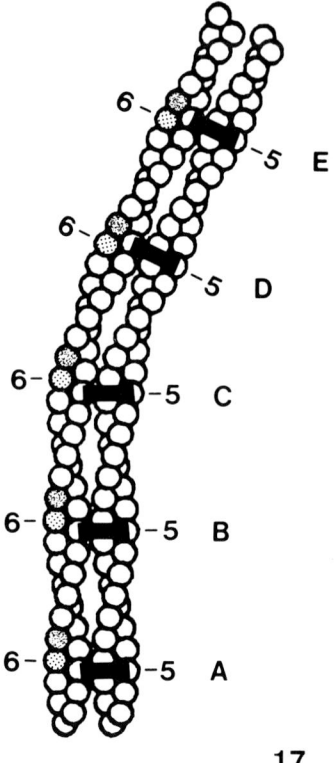

Figure 16-17. Model showing how a bend is built into a bundle of actin filaments. At the base of the diagram, the two filaments are in register and are crossbridged at subunit 5 in regions A and B. The crossbridges are indicated by black bars. The left-side filament is on a longer, outer track in the bend relative to the right-side filament. As a result, the two filaments slip out of register. If the bend is 154°, the amount of slippage is just enough to allow rebonding, in this case, subunit 7 on the left can be crossbridged to subunit 5 on the right (in regions D and E). Subunits 6 and 8 have been stippled in order to highlight the shift in crossbridge position. Note that in regions A, B, and C, the crossbridge is below the pair of stippled units, but in C and D it lies between them. Thus, in order to build a bend into a bundle, the filaments are not compressed or stretched. Instead they slide past one another and recrossbridge at the next subunit up. (From DeRosier and Tilney, 1984.)

stereocilia and hence the tuning of the cochlea. However, we were not prepared for the precision with which the cochlea "controls" these variables. As illustrated in Figs. 18-21, we see that not only within a single cochlea is the length, width, and number of stereocilia on hair cells at specific positions on the cochlea controlled to an accuracy of about 10%, but from cochlea to cochlea they are invariant. To emphasize this point, if we as investigators are given the length of the longest stereocilia on a hair cell and the width of the stereocilia, or total number of stereocilia per cell, we can pinpoint the position of the hair cell in two axes. Since each chicken cochlea is approximately 3.1 mm long and shows tonotopic organization as determined by lesions caused by intense noise stimulation (see Cotanche and Sulik, 1984), variables in stereocilia length, width, and number can be correlated with the numbers of crossbridges and with the frequency to which a particular hair cell is tuned.

VII. CORRELATION OF STEREOCILIARY LENGTH WITH CHARACTERISTIC FREQUENCY

Recently, it has been directly demonstrated that stereocilia from different hair cells are themselves mechanically tuned to the frequencies to which these hair cells respond (Frishkoff and DeRosier, 1984; Holton and Hudspeth, 1983). These studies were carried out on cochleae of the alligator lizard. (For most sound frequencies the basilar membrane in this organism does not show spatial variation associated with fine tuning, even though recordings from individual hair cells that sit on this membrane do show fine tuning.) What was done was to place the lizard cochlea in a small, fluid-filled chamber so that the stereocilia could be examined by a light microscope under stroboscopic illumination. The chamber was connected to a mechanical oscillator that was synchronized with the stroboscopic illuminator and the motion of the cochlea and the stereocilia that extend from different hair cells compared. What is fascinating and exciting is that at the characteristic frequency of the cell, the stereocilia will change the phase of their motion relative to the cochlea proper. For example, if an individual hair cell is tuned physiologically, for example to 1000 Hz, then with a sound of 500 Hz, its stereocilia will move in phase with respect to the cochlea and the hair cell body. Thus, relative to the hair cell, the stereocilia are stationary. At 1000 Hz, the phase of the motion changes and now the stereocilia bend back and forth relative to the cell, producing a change in its membrane potential. Thus, the stereocilia have resonating frequencies. Can we relate these fascinating results on the tuning of stereocilia to the length of the stereocilia and the number of crossbridges that connect adjacent actin filaments? Just as in the bird cochlea, the lengths of the tallest stereocilia in lizard cochleae vary in a systematic and predictable way depending upon their location in the cochlea. Thus, a cell tuned to 1000 Hz has stereocilia whose maximum lengths are 30 μm, while one tuned to 4000 Hz has stereocilia whose maximum lengths are 7 μm. What is even more interesting is that occasionally one encounters a cell whose stereocilia are longer or shorter than they should be, based upon their position in the cochlea. Significant is the fact that the stereocilia of these "strange" cells are mechanically tuned to a

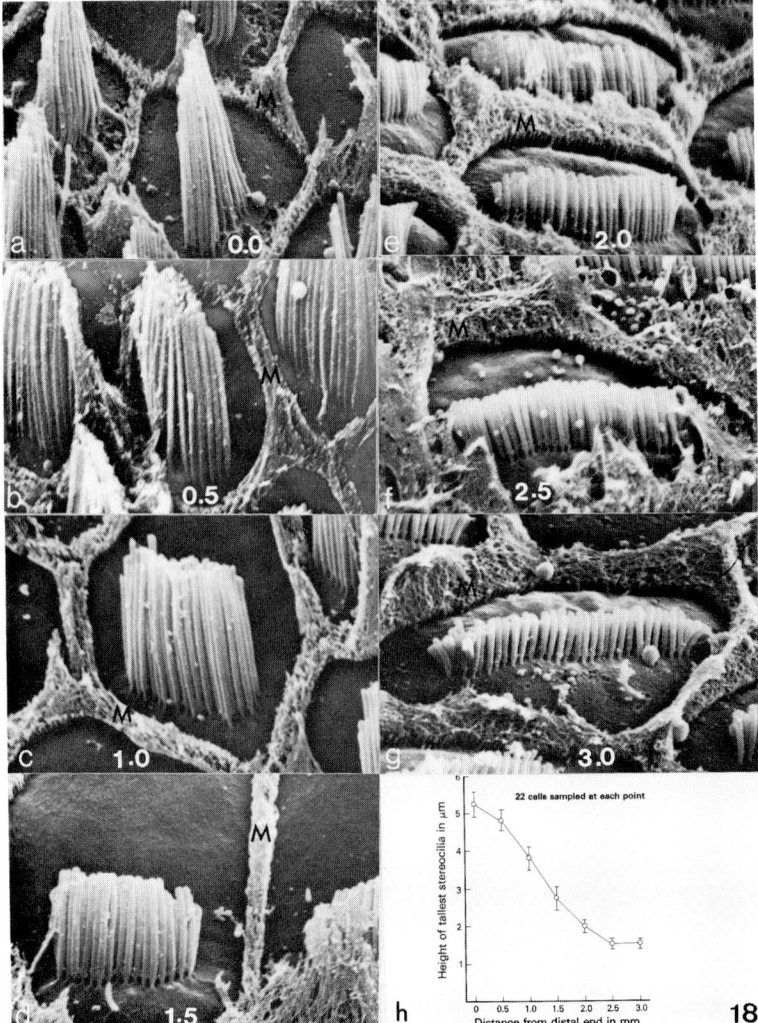

Figure 16-18. (a)–(g) Scanning electron micrographs of representative bundles of stereocilia that extend from the apical surface of hair cells located at discrete positions on the cochlea. The view illustrated here is the view of the stereocilia we would see when looking from the inferior margin of the cochlea towards its superior edge. These positions were determined by measuring the distance from the distal end of the cochlea to the hair cell in question. We have indicated this distance in millimeters on the micrographs (white letters). Cochleae are approximately 3.1 mm in length in 7-day neonatal chicks, and the micrographs included here are taken at 0.5-mm intervals. All of the micrographs are shown at the same magnification. The body of the hair cell is easy to distinguish because the apical surface is smooth. Adjacent hair cells are separated by a ridge of microvilli (M) that extend from the surface of supporting cells that separate adjacent hair cells. × 4,900.

(h) Graph of the height of the tallest stereocilia on a hair cell, plotted as a function of the distance from the distal end of the cochlea. This graph was derived from a single cochlea. Each point on the graph represents the data from at least 22 adjacent hair cells. Most of the micrographs illustrated in (a)–(g) were from this cochlea. (From Tilney and Saunders, 1983.)

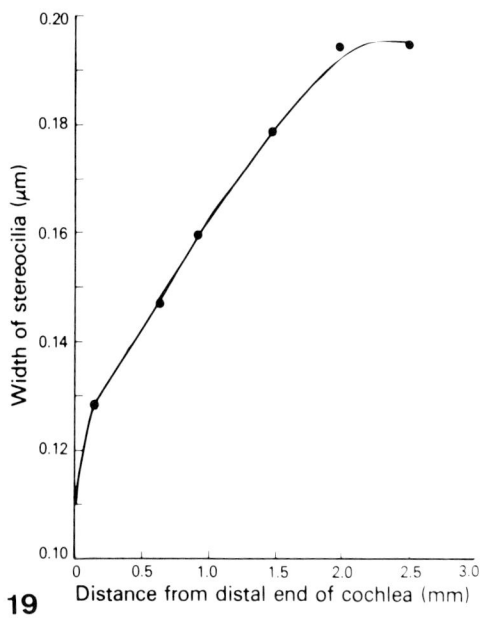

Figure 16-19. Graph depicting the width of the stereocilia on short hair cells of the bird cochlea as a function of their position from the distal end of the cochlea. Of interest is that near the distal end the stereocilia are thin, and they get progressively thicker as one approaches the proximal end.

frequency *that is related to the length of the stereocilia, not* to the location of that cell in the cochlea. Frischkopf and DeRosier (1984) and Holton and Hudspeth (1983) have amassed strong circumstantial evidence that the length of the stereocilia is directly related to their mechanical tuning.

Weiss and Leong (personal communications) have attempted to calculate tuning profiles for model stereocilia driven by periodic fluid motion, as must occur in the ear. They find a correlation between the phase shift and the length of the stereocilia. Their calculations of expected frequency and stereociliary length show good agreement with the measured results of Frischkopf and DeRosier (1984). Thus, for example, they would find a characteristic frequency of 1000 Hz for a 30 μm-long stereocilium. Weiss and Leong (personal communications) assumed the restoring force for a displaced stereocilium was independent of length. This may, or may not, be a good approximation, depending upon the stiffness of the rootlet relative to that of several thousand crossbridges. Clearly, model building of this type, and additional measurements of motion, will shed light on the relevant parameters of the stereocilia with regard to frequency tuning.

The extent to which the mechanical tuning of the stereocilia and of the

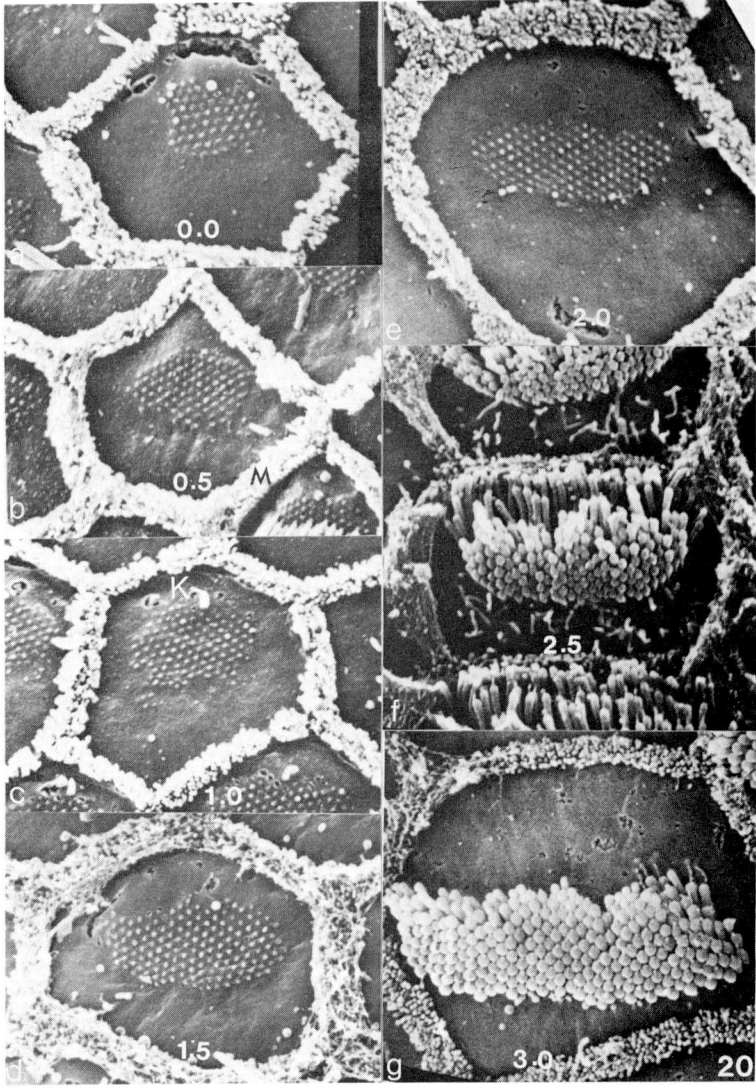

Figure 16-20. Scanning electron micrograph of the apical surfaces of hair cells located at 0.5-mm intervals from the distal end of the cochlea. These distances are indicated on the micrographs by the white numbers. In (a)-(e) the stereocilia have been "shaved" off, leaving scars. This allows us to count accurately the number of stereocilia per cell. In (f)-(g) the stereocilia have not been removed because they are very short and, therefore, extremely difficult to "shave" off. In all seven micrographs, the hair cell depicted is oriented in the same way. If the kinocilium (K) is present, or if the stub is present, it will be nearest the top of the page. Surrounding each hair cell is a tuft of microvilli (M) that extends up from the supporting cells that lie immediately beneath. By inspection of these micrographs, it is clear that the number of stereocilia per cell is lowest at the distal end and that there is a systematic increase in number as one travels towards the proximal end. (From Tilney and Saunders, 1983.) × 7,400.

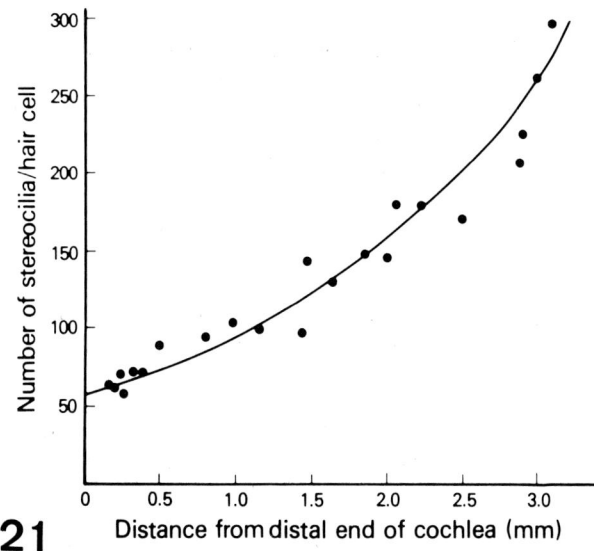

Figure 16-21. Graph depicting the number of stereocilia per hair cell as a function of the distance from the distal end of the cochlea. The information plotted here comes from seven cochleae. (From Tilney and Saunders, 1983.)

basilar and tectorial membranes accounts for the tuning of the hair cell remains unclear. On the one hand, Fettiplace and Crawford (1978) and Crawford and Fettiplace (1981), in a superb series of experiments, showed that hair cells in the turtle are electrically resonant at frequencies corresponding to the characteristic frequency of the cell. These resonant properties would appear to account for the fine-frequency selectivity of the hair cell. On the other hand, Frischkopf and DeRosier (1984), as well as Khanna (1983) and his collaborators, have shown that the stereocilia are mechanically tuned and, with an elegant set of measurements, have shown in the cat that the mechanical tuning curves in cochleae with reduced trauma approach the tuning curves of auditory nerve fibers.

To reconcile these two sets of observations, Weiss (1982) has proposed a system embodying both electromechanical and mechanoelectric resonant properties. Thus, mechanical resonance and electrical resonance are both manifestations of a common, underlying mechanism.

Whatever the resolution of this dilemma, the fact remains that the structural and mechanical integrity and properties of the stereocilium are important and that actin bundles play a role in maintaining and defining stereociliary shape.

REFERENCES

Cotanche, D. A., and Sulik, K. K.: Parameters of growth in the embryonic and neonatal chick basilar papilla. *Scann. Electr. Micros.*, 1984. In press.

Crawford, A. C., and Fettiplace, R.: An electrical tuning mechanism in turtle cochlear hair cells. *J. Physiol.* 312: 377-412, 1981.

DeRosier, D. J., and Tilney, L. G.: How to build a bend into an actin bundle. *J. Mol. Biol. 175:* 57-73, 1984a.

DeRosier, D. J., and Tilney, L. G.: The form and function of actin: a product of its unique design. In Shay, J. W. (Ed.): *Cell and Muscle Motility, Vol. 5.* New York, Plenum, 1984b, pp. 139-169.

DeRosier, D. J.; Tilney, L. G., and Egelman, E.: Actin in the inner ear: the remarkable structure of the stereocilium. *Nature 287:* 291-296, 1980.

Egelman, E. H.; Francis, N., and DeRosier, D. J.: F-actin is a helix with a random variable twist. *Nature 298:* 131-135, 1982.

Egelman, E. H.; Francis, N., and DeRosier, D. J.: Helical disorder and the filament structure of F-actin are elucidated by the angle-layered aggregate. *J. Mol. Biol. 166:* 605-629, 1983.

Fettiplace, R., and Crawford, A. C.: The coding of sound pressure and frequency in cochlear hair cells of the terrapin. *Proc. Roy. Soc. B 203:* 209-218, 1978.

Flock, A.; Flock, B., and Murray, E.: Studies on the sensory hairs of receptor cells in the inner ear. *Acta Otolaryng. 83:* 85-91, 1977.

Frischkopf, L. S., and DeRosier, D. J.: Mechanical tuning of freestanding stereociliary bundles and frequency analysis in the alligator lizard cochlea. *Hearing Res.* 1984. In press.

Hirokawa, N., and Tilney, L. G.: Interactions between actin filaments and between actin filaments and membranes in quick-frozen and deeply etched hair cells of the chick ear. *J. Cell Biol. 95:* 249-261, 1982.

Holton, T., and Hudspeth, A. J.: A micromechanical contribution to cochlear tuning and tonotopic organization. *Science 222:* 508-510, 1983.

Hudspeth, A. J.: Mechanoelectrical transduction by hair cells in the acousticolateralis sensory system. *Annu. Rev. Neurosci. 6:* 187-215, 1983.

Hunter-Duvar, I. M.: Morphology of the normal and acoustically damaged cochlea. *Scann. Electr. Micros. 2:* 421-428, 1977.

Khanna, S. M.: Interpretation of the sharply tuned basilar membrane response observed in the cochlea. In Fay, R. R., and Gourevitch, G. (Eds.): *Hearing and Other Senses: Presentations in Honor of E. G. Wever.* Groton, Connecticut, The Amphora Press, 1983, pp. 65-86.

Lim, O.: Cochlear anatomy related to cochlear micromechanics. A review. *J. Acoust. Soc. Amer. 67:* 1686-1695, 1980.

Mulroy, M. J.: Cochlear anatomy of the alligator lizard. *Brain Behav. Evol. 10:* 69-87, 1974.

Tilney, L. G.; DeRosier, D. J., and Mulroy, M. J.: The organization of actin filaments in the stereocilia of cochlear hair cells. *J. Cell Biol. 86:* 244-259, 1980.

Tilney, L. G.; DeRosier, D. J., Saunders, J. C., and Tilney, M. S.: Actin filaments, stereocilia, and hair cells of the bird cochlea III. The development and differentiation of hair cells and stereocilia in embryos, 1984. In preparation.

Tilney, L. G.; Egelman, L. S., DeRosier, D. J., and Saunders, J. C.: Actin filaments, stereocilia and hair cells of the bird cochlea II. Packing of actin filaments in the stereocilia and in the cuticular plate and what happens to the organization when stereocilia are bent. *J. Cell Biol. 96:* 822-834, 1983.

Tilney, L. G., and Saunders, J. C.: Actin filaments, stereocilia, and hair cells of the bird cochlea I. Length, number, width, and distribution of stereocilia of each hair cell are related to the position of the hair cell on the cochlea. *J. Cell Biol.* 96: 807–821, 1983.

Tilney, L. G.; Saunders, J. C., Egelman, E., and DeRosier, D. J.: Changes in the organization of actin filaments in the stereocilia of noise-damaged cochleae. *Hearing Res.* 7: 181–197, 1982.

Weiss, T. F.: Bidirectional transduction in vertebrate hair cells: a mechanism for coupling mechanical and electrical processes. *Hearing Res.* 7: 353–360, 1982.

Chapter 17

CONTRACTILE AND STRUCTURAL PROTEINS IN THE AUDITORY ORGAN

ÅKE FLOCK

I. Introduction
II. Vestibular System
III. Organ of Corti
   References

I. INTRODUCTION

For a number of years we have studied the mechanical properties of the sensory hairs of the receptor cells in the ear and have also examined the macromolecular structure and identity of the proteins in the mechanoreceptive region of these cells. This work began in the vestibular system, but has recently been taken to the organ of Corti in the guinea pig.

II. VESTIBULAR SYSTEM

In the vestibular system, we found that the mechanical-response properties of the sensory hair bundles is determined by the architecture of the stereocilia within the bundle (Flock and Orman, 1983). The sensory hairs are quite stiff. An inside cytoskeleton of filaments of the protein, actin (Flock et al., 1981), is responsible for the stiffness (Fig. 1). Actin is best known from muscle cells where it, together with other proteins, generates contraction. The motion of the sensory hairs was studied by directing a fluid jet against them. It was found that their motion became restricted under conditions which would promote contraction in a muscle fiber (Orman and Flock, 1983) (Fig. 2). It therefore appears possible, at least in the vestibular system, that the receptor cells are equipped with a mechanism for control of their mechanical-input properties, and thus for the sensitivity of the sense organ.

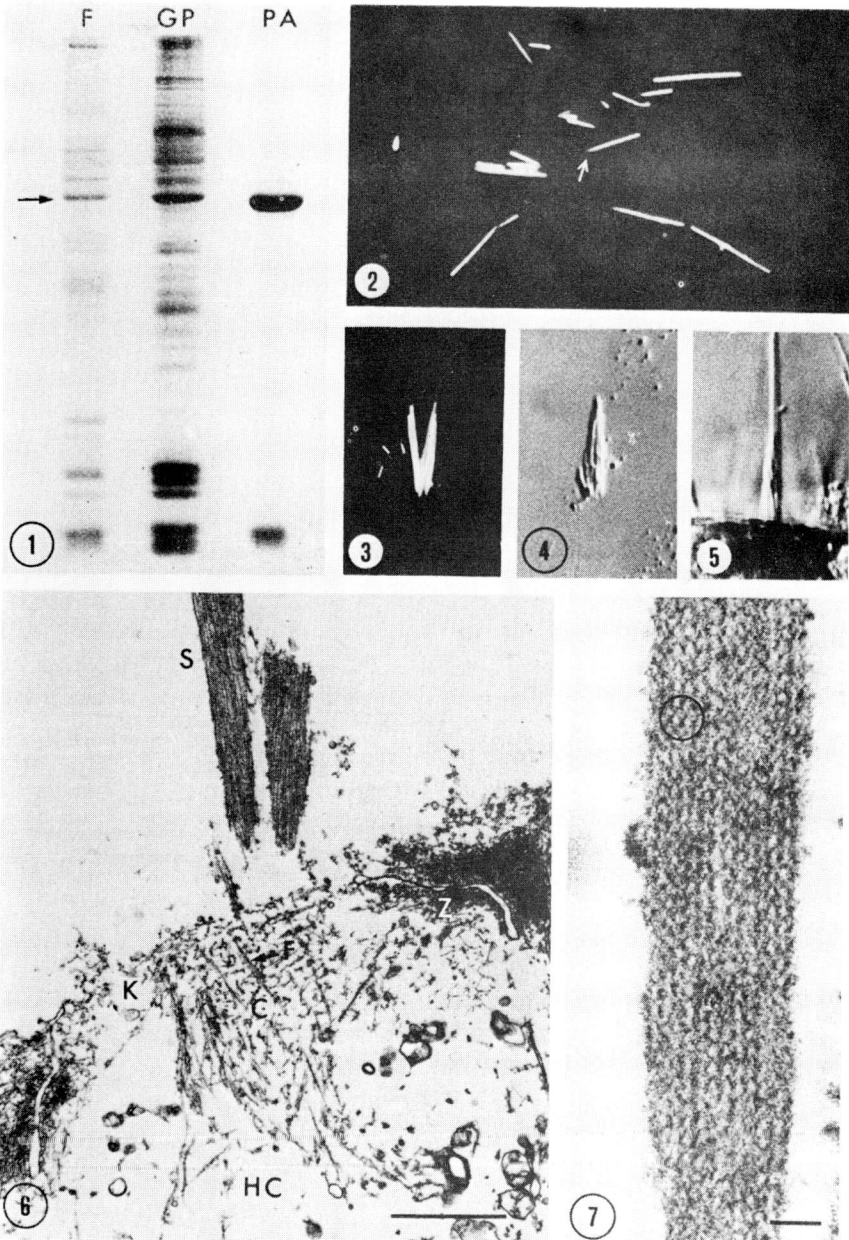

Figure 17-1. This display illustrates (1) the presence of actin in the crista ampullaris of frog (F) and guinea pig (GP) by gel electrophoresis (PA, pure actin), its location in stereocilia by immunofluorescence (2–5), and by electron microscopy via labeling with myosin S1 fragments (6,7). (From Flock et al., 1981.)

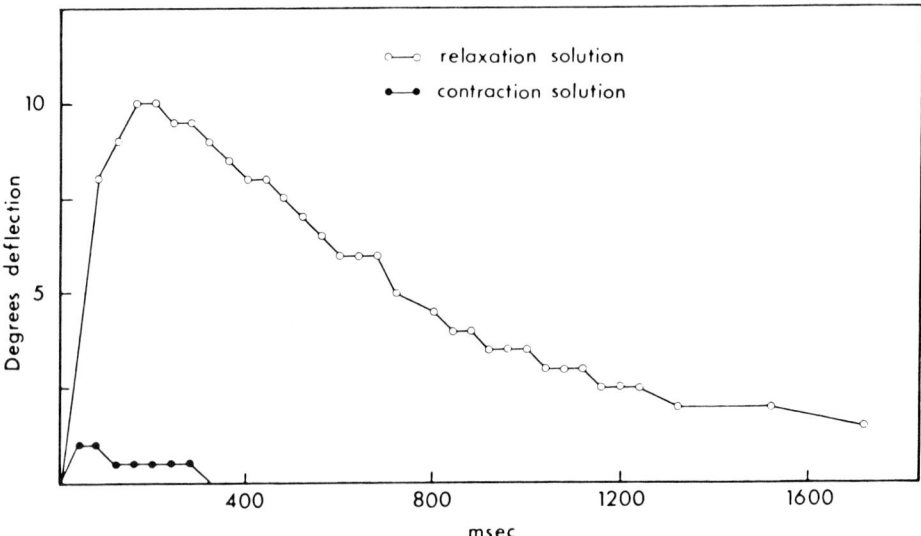

Figure 17-2. The motion pattern of a sensory-hair bundle in response to a fluid pulse becomes restricted in a medium that would give rise to contraction in a muscle fiber (filled circles), as compared to a relaxation solution (open circles). (From Orman and Flock, 1983.)

III. ORGAN OF CORTI

We have recently studied the organ of Corti in the guinea pig with respect to the distribution of various types of proteins in the sensory and supporting cells (Flock et al., 1982) (Figs. 3, 4) and the ultrastructural arrangement of these proteins. Myosin and tropomyosin have also been demonstrated (Flock, 1983). As in the vestibular system, the stereocilia and the cuticular plate contain actin. Another protein, named fimbrin, is also present in this region. It probably cross-binds actin filaments in a stable fashion. In the cuticular plate, myosin is present; in muscle cells this protein interacts with actin to cause contraction. This interaction is controlled by tropomyosin, a protein which we have now found to be present at the point in the cuticular plate where the stereociliary rootlets are inserted. Other combinations of proteins are seen in the supporting cells: actin together with tubulin forms rigid arches inside the Deiter's cells and pillar cells that encase the outer hair cells. The inner hair cells are conspicuously excluded from this framework, as if the two sets constitute two mechanically different entities. The inner hair cells are supplied with 95% of the afferent innervation, whereas the outer hair cells receive a heavy efferent innervation. It has recently been shown that excitation of the efferent fibers affects the mechanical characteristics of the cochlear partition (Siegel and Kim, 1971). Furthermore, stimulated (Kemp, 1978) and spontaneous (Zurek, 1981) acoustic emissions have been shown to originate in some inner-ear structure. Therefore, it is tempting to predict the existence in the organ of Corti of a "motor" system (outer hair cells) as well as a "sensory" system (inner hair cells).

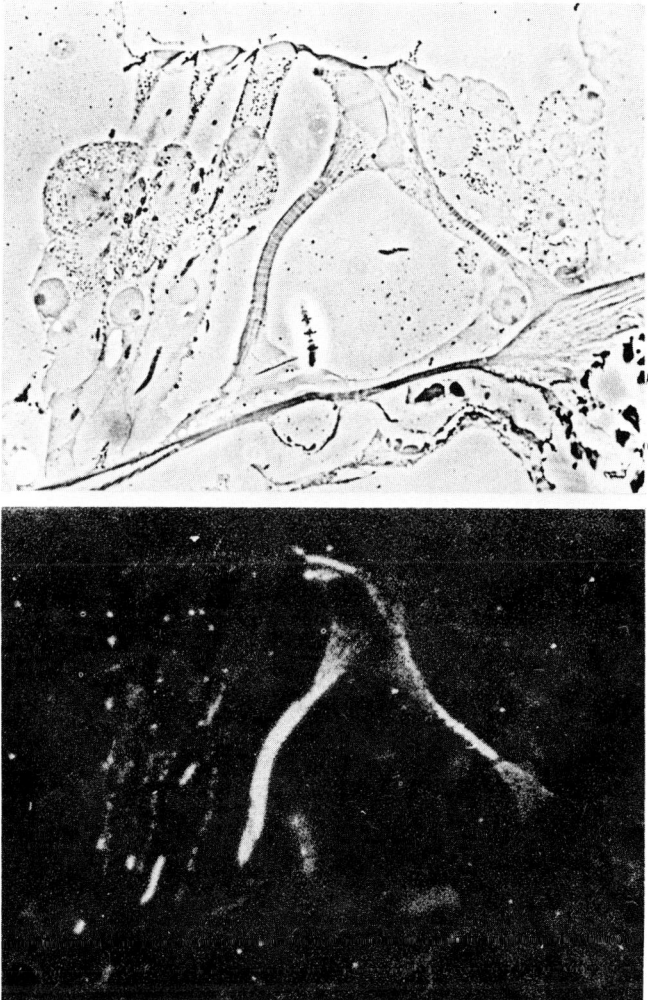

Figure 17-3. The distribution of tubulin in the organ of Corti is demonstrated here on 0.3–1.0-μm cryo-ultramicrotomy sections. (From Flock *et al.*, 1982.)

An expression of such a motor system is possibly seen from experiments performed on isolated turns of the guinea pig cochlea, where the compliance of the sensory hairs can be quantitatively measured (Strelioff and Flock,

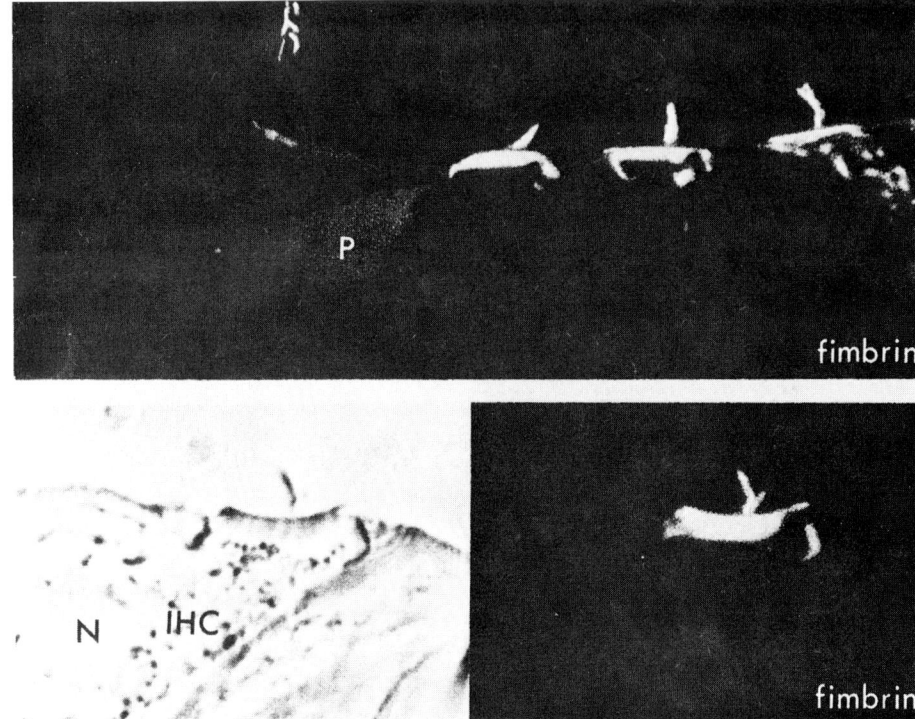

Figure 17-4. The distribution of fimbrin in the organ of Corti is demonstrated, in a manner similar to Fig. 3. (From Flock *et al.*, 1982.)

1984) (Fig. 5). An interesting finding is that the stiffness of the sensory hairs is about twice as large for displacement in the excitatory direction (stereocilia towards the basal body) as in the inhibitory direction. This asymmetry of stiffness may relate to the presence of contractile proteins described above, and could underlie at least some of the nonlinear phenomena observed in mechanical and electrophysiological responses of the cochlea. It indicates that the vulnerable, nonlinear "second filter" may be related to some property of the sensory hairs.

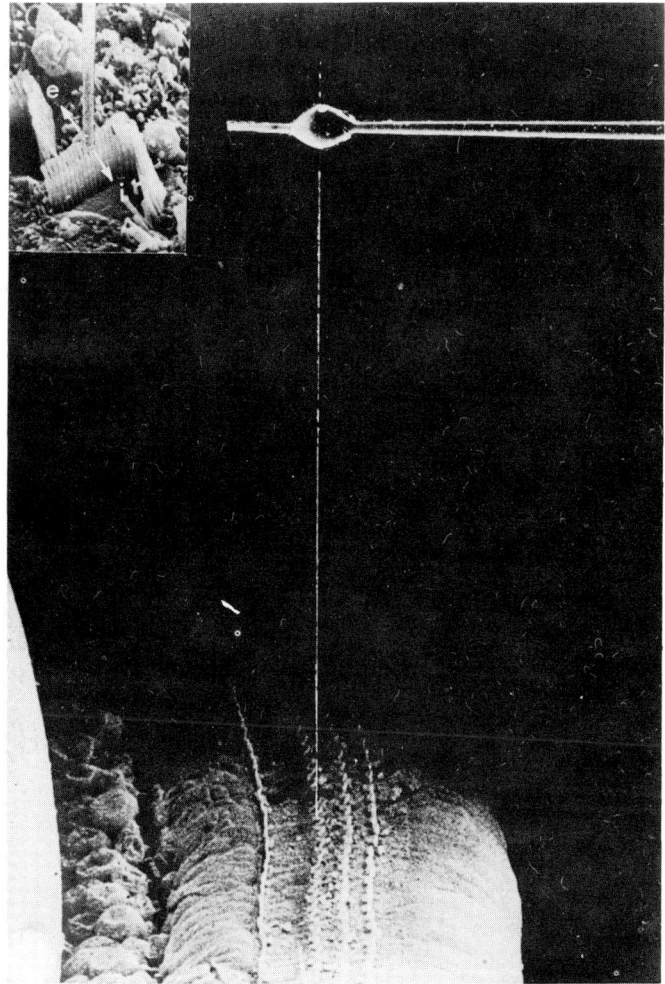

Figure 17-5. The stiffness properties of sensory hairs in the organ of Corti have been measured directly in isolated cochlear turns maintained in tissue culture medium, by applying force with a quartz glass fiber which is sufficient to displace the tip of the hairs 1 μm. The amount of bending provides a measure of torque stiffness. The stiffness was nonlinear in that twice as much force was needed in the excitatory direction (e) as in the inhibitory direction (i) (Inset). (From Strelioff and Flock, 1984).

REFERENCES

Flock, A.: Hair cells, receptor with a motor capacity? In Klinke, R., and Hartman, R. (Eds.): Hearing—Physiological Bases and Psychophysics. Berlin, Springer Verlag, 1983, pp. 2-9.
Flock, A.; Bretscher, A., and Weber, K.: Immunohistochemical localization of several cytoskeletal

proteins in inner ear sensory and supporting cells. *Hearing Res. 6:* 75-89, 1982.

Flock, A.; Cheung, H. C., Flock, B., and Utter, G.: Three sets of actin filaments in sensory cells of the inner ear. Identification and functional orientation determined by gel electrophoresis, immunofluorescence and electron microscopy. *J. Neurophysiol. 10:* 133-147, 1981.

Flock, A., and Orman, S.: Micromechanical properties of sensory hairs in receptor cells of the inner ear. *Hearing Res. 11:* 249-261, 1983.

Kemp, D. T.: Stimulated acoustic emissions from within the human auditory system. *J. Acoust. Soc. Am. 64:* 1386-1391, 1978.

Orman, S., and Flock, A.: Active control of sensory hair mechanics implied by susceptibility to media that induce contraction in muscle. *Hearing Res. 11:* 261-266, 1983.

Siegel, J. H., and Kim, D. O.: Efferent neural control of cochlear mechanics? Olivocochlear bundle stimulation affects cochlear biomechanical nonlinearity. *Hearing Res. 6:* 171-182, 1971.

Strelioff, D., and Flock, A.: Stiffness of sensory cell hair bundles in the isolated guinea pig cochlea. *Hearing Res.,* 1984. In press.

Zurek, P. M.: Spontaneous narrowband acoustic signals emitted by human ears. *J. Acoust. Soc. Am. 68:* 514-523, 1981.

Chapter 18

ACTIN, MYOSIN, AND ASSOCIATED PROTEINS IN THE VERTEBRATE AUDITORY AND VESTIBULAR ORGANS: IMMUNOCYTOCHEMICAL AND BIOCHEMICAL STUDIES

Detlev Drenckhahn, Thomas Schäfer, and Michael Prinz

I. Actin, Myosin, and Associated Proteins in Muscle and Nonmuscle Cells
II. Actin Filaments in Auditory and Vestibular Hair Cells
III. Immunocytochemical Location of Actin and Associated Proteins in the Vertebrate Inner-Ear Sensory Organs
   A. Actin
   B. Myosin
   C. Alpha-Actinin
   D. Tropomyosin
   E. Vinculin
   F. Fimbrin
   G. Villin
   H. Erythrocyte Membrane Proteins
IV. Biochemical Studies
V. Structural and Functional Implications
   References

I. ACTIN, MYOSIN, AND ASSOCIATED PROTEINS IN MUSCLE AND NONMUSCLE CELLS

In the past decade, actin, myosin, and several of the actin-associated polypeptides, which together constitute the contractile apparatus in muscle, have been demonstrated in a variety of nonmuscle cells. Actin and myosin are the major protein components of the contractile system in muscle. Their interaction is the basis for muscle contraction, which is the best-understood kind of cellular motility. Under appropriate ionic conditions, and using energy derived from ATP, thin actin filaments (5–7 nm) slide along thicker myosin filaments (10–15 nm) by ratchet-like, cyclic associations of myosin crossbridges (heads) with the actin filaments. In striated muscle, interaction between actin and myosin is regulated by calcium, which triggers contraction by an increase in concentration from less than 1 μM to about 10 μM. Calcium sensitivity is conferred by polypeptides associated with the actin filaments, i.e., tropomyosin (heterodimer of 33,000–37,000 dalton subunits) and the troponin complex (three subunits). However, in smooth muscle and nonmuscle cells, tropomyosin is probably not involved in regulating the calcium-dependent interaction between actin and myosin. Its role in these cells is thought to be to stabilize actin filaments (F-actin) and to modify the interaction of F-actin with other actin-binding polypeptides. In smooth and nonmuscle cells, calcium sensitivity of the actin-myosin interaction is regulated via a calcium-calmodulin dependent phosphorylation (via myosin light-chain kinase) of a 20,000-dalton polypeptide, which is the regulatory myosin light chain associated with myosin crossbridges. A further major protein component of the muscle actin-myosin filament system is the Z-line protein, α-actinin (dimer of M_r 200,000; M_r refers to the molecular weight of a protein determined electrophoretically, as a negatively-charged rod, after denaturation with the detergent sodium dodecyl sulfate), which interconnects actin filaments of adjacent contractile units and is also found at the plasmalemmal attachment sites of actin filaments in muscle and nonmuscle cells. At these attachment areas, another polypeptide is concentrated, named vinculin (M_r 130,000). Regarding the functional significance of these contractile proteins in nonmuscle cells, experimental data indicate that they may be involved in both structural support and various forms of cell motility, such as locomotion, cell division, endo- and exocytosis, and lateral mobility of surface receptors. For recent reviews on these subjects, see Goldman *et al.* (1979), Pollard (1982), Franzini-Armstrong and Peachy (1982), Gröschel-Stewart, and Drenckhahn (1982).

Recently, various components of the membrane cytoskeleton of erythrocytes have also been identified in nonerythroid cells. In erythrocytes, actin is attached to the plasma membrane through a meshwork of polypeptides comprised largely of the elongated protein spectrin, a heterotetramer of 220,000–240,000 dalton subunits. This cross-linked network of actin and spectrin is anchored by a further polypeptide, ankyrin (M_r 210,000), to the cytoplasmic domain of the anion-channel band 3, which is the major integral-membrane protein of the red-cell membrane. The demonstration in nonerythroid cells of some of the polypeptides indicates some general principles in the anchorage of the cytoskeleton in erythrocytes and nonerythroid cells (Branton, 1981; Kardami *et al.*, 1981; Burridge *et al.*, 1982; Cohen *et al.*, 1982; Drenckhahn *et al.*, 1984b).

II. ACTIN FILAMENTS IN AUDITORY AND VESTIBULAR HAIR CELLS

Stereocilia, which extend from the apical surface of hair cells, are thought to have a key role in the process of transduction of sound energy in electrical impulses. Each stereocilium is supported by a paracrystalline, axial bundle of actin filaments which have identical polarities and are extensively cross-linked (Flock and Cheung, 1977; De Rosier et al., 1980; Tilney et al., 1980; Chapter 16, this volume). The central filaments of this bundle extend downwards into the apical cytoplasm where they are anchored within a dense meshwork of richly-interconnected actin filaments, which together constitute the cuticular plate. Many of the actin filaments in the cuticular plate are connected to the apical plasma membrane by tiny, branching connecting units (Hirokawa and Tilney, 1982). Only a few studies have been performed to analyze the protein components associated with this distinct, actin-filament system (Flock et al., 1981, 1982; Drenckhahn et al., 1982; Sobin and Flock, 1983), and no data have been published so far on the occurrence and organization of contractile proteins in the remaining cell types of the inner-ear sensory organs.

III. IMMUNOCYTOCHEMICAL LOCATION OF ACTIN AND ASSOCIATED PROTEINS IN THE VERTEBRATE INNER-EAR SENSORY ORGANS

Immunocytochemical findings, to be described in this chapter, were obtained by immunostaining of 0.5–1.0 μm-thick tissue sections of quick-frozen, freeze-dried, and plastic-embedded pieces of the cochlea, basilar papilla, saccular macula, and lagena of the guinea pig, rat, mouse, golden hamster, chick and American caiman, as well as of the saccular macula of the goldfish. After partial removal of the resin, sections were incubated with antibodies, using the indirect immunofluorescence technique with corresponding controls as described (Drenckhahn et al., 1982; Drenckhahn and Mannherz, 1983). Fluorescent phalloidin, a specific probe for only the polymerized form of actin (F-actin) (Faulstich et al., 1983), was applied to frozen sections of the decalcified and perfusion-fixed labyrinth (see Drenckhahn et al., 1984a) of the rat. Controls were performed by preincubations of the sections with unlabeled phalloidin.

A. ACTIN

Cochlea. Antibodies to chicken-gizzard actin (isolated by a DNAase I column) and chicken microvillus actin (isolated by a partial purification followed by gel electrophoresis), as well as fluorescent phalloidin, produced a strong fluorescent staining of the stereocilia and cuticular plates of inner and outer cochlear hair cells in the guinea pig, rat, mouse, hamster, chicken, and American caiman (Figs. 1–6). Staining, frequently seen along the base of the mammalian outer cochlear hair cells (Figs. 1–4), may be due to binding of the antibodies and phalloidin to synapses (Drenckhahn and Kaiser, 1983; Drenckhahn et al., 1984a) or to the cup-shaped portion of Deiters' cells. In the inner and outer pillars, immunostaining was associated with the axial filament bundle, composed of microtubules and microfilaments extending from the strongly-stained basal cones (supporting hillock) to the head portions (Figs. 1, 4). Actin-specific fluorescence was also observed along the rather thin, microtubule-microfilament bundle in Deiters' cells. In the phalangeal processes, filamentous actin (visualized by phalloidin staining) was concentrated underneath the plasma membrane, whereas the central core bundle formed by microtubules was unstained (Fig. 3). Staining of the reticular lamina was inconsistent (strong to absent), probably depending on the planes of the sections. A rather strong immunostaining was always observed in association with the richly-interdigitated lateral and basal-cell borders of Boettcher's cells (Fig. 1). The cells of Hensen and Claudius and the inner sulcus cells displayed a faint fluorescence along the apical cell borders and, less conspicuously, along the lateral cell borders. Outside the organ of Corti, strong staining with both phalloidin and with antibodies to actin was seen in the interdental cells lining the limbus and in the basal-cell layer of the stria vascularis. As seen in Figs. 1 and 7, the bottle-shaped interdental cells were stained along the whole periphery. Fluorescence was strongest along the flattened apical portions, which serve as attachment sites for the tectorial membrane. The basal-cell layer, and probably also the intermediate cells of the stria vascularis, showed a bright fluorescence extending apically into the branching ascending processes, which appear to contact and surround intraepithelial capillaries (Fig. 9). The superficial layer (marginal cells) was more or less unstained, except for a faint fluorescence extending along the junctional complex (belt desmosomes). This junctional actin filament system was also seen in the epithelium lining the spiral prominence and Reissner's membrane (Figs. 8, 9).

Saccular macula, lagena. Vestibular hair cells showed the same bright staining of stereocilia and cuticular plates as described for the auditory hair cells (Fig. 10). A particularly high affinity for phalloidin and antibodies to actin was noted in the dense microfilament web located in the apex of supporting cells in the saccular macula. Similar observations were made with antibodies to actin in the utricular macula of the guinea pig (Sobin and Flock, 1983).

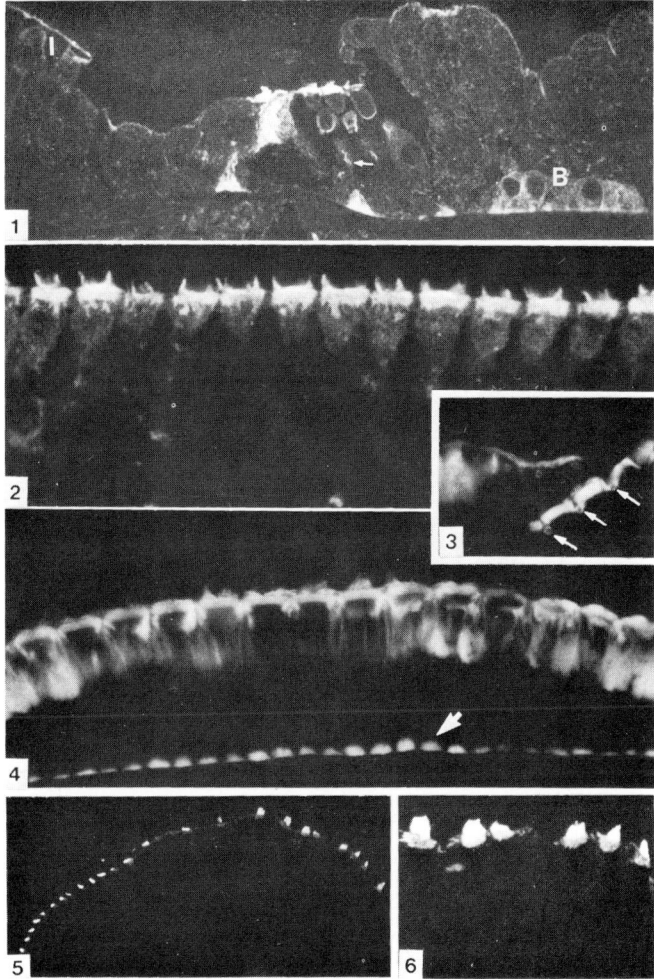

Figures 18-1–18-6. Distribution of actin in the mouse (Figs. 1, 2), rat (Figs. 3–4), and chicken auditory organ, as visualized by antibodies to actin (Figs. 1, 2, 5, 6) and fluorescent phalloidin (Figs. 3, 4). Actin is concentrated in the cuticular plate and stereocilia of inner hair cells (Figs. 1, 4, 5) (in Fig. 4 only the central two cells are not superimposed by the fluorescence of the inner pillars, the footplates of which are indicated by an arrow) and outer hair cells (Figs. 1, 2, 3, 5, 6) of the mammalian species and the chick. Arrow in Fig. 1 indicates the axial microfilament-microtubule bundle in the cells of Deiter, cells of Boettcher (B), interdental cells of the limbus (I). Arrows in Fig. 3 point to cross-sectioned phalangeal processes joining the reticular plate; the cores of the processes are unstained.

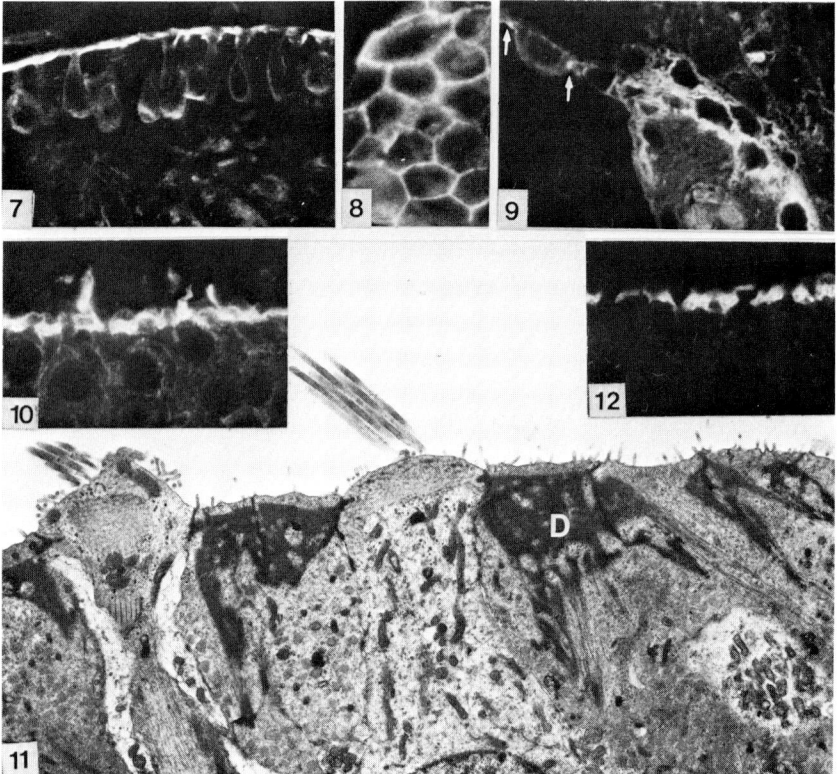

Figures 18-7–18-12. Typical pattern of actin in interdental cells (Fig. 7, rat, phalloidin), the terminal bar in spiral prominence epithelium (Fig. 8, rat, tangential section, phalloidin), the basal cell layer of the stria vascularis and the junctional complex of Reissner's membrane epithelium (arrows in Fig. 9, hamster, anti-actin), and saccular macula (Fig. 10, hamster, anti-actin). As seen in the corresponding electron micrograph (Fig. 11), the staining pattern of actin correlates with stereocilia, cuticular plates, and the dense apical web in supporting cells (D). Anti-α-actinin (Fig. 12) reacts only with the supporting cell web and the cuticuluar plates, but not with stereocilia.

B. MYOSIN

As described previously for the guinea-pig organ of Corti (Drenckhahn et al., 1982), antibodies to nonmuscle myosin from calf thymus labeled the cuticular plates of the inner and outer auditory hair cells as well as the cuticular plates of vestibular hair cells in the guinea pig, mouse, rat, and hamster (Figs. 13, 14, 16). Similar results were obtained with antibodies to human platelet myosin (Fig. 17). Myosin-like immunostaining was absent from all auditory and vestibular stereocilia, which also did not react with antibodies to smooth-muscle-type myosin from chicken gizzard and human uterus (Drenckhahn et al., 1982). The smooth-muscle-myosin-like fluorescence observed by MacCartney et al. (1980) in association with auditory stereocilia in whole mounts of the guinea pig organ of Corti may be due to an optical artifact (for discussion, see Egelman, 1981; Drenckhahn et al., 1982). The remaining epithelial cells of the cochlea did not react with any of the myosin antibodies tested, except for the basal and probably also the intermediate cell layer of the stria vascularis, which reacted strongly with both antibodies to smooth-muscle-type and nonmuscle-type myosin (Fig. 18). In the chicken basilar papilla, antibodies to mammalian thymus and platelet myosin gave only a faint, unconvincing immunostaining of the cuticular plates. The only indication of the presence of myosin in chick auditory hair cells was obtained with an antibody directed against chicken-gizzard myosin light chains (reacting with the 16,000 and 20,000 dalton components). These antibodies reacted rather weakly, but clearly above background level, with the cuticular plates of the chick inner and outer hair cells (Fig. 15).

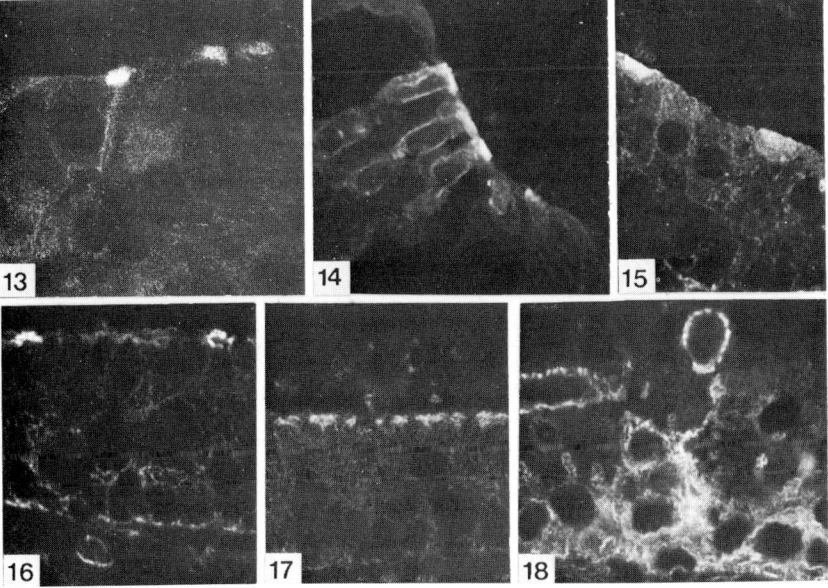

Figures 18-13–18-18. Antibodies to nonmuscle myosin from calf thymus (Figs. 13, 14, 16) and human platelets (Fig. 17), as well as antibodies to myosin light chains (Fig. 15), display a selective affinity for the cuticular plates of auditory hair cells in the mouse (Fig. 13), guinea pig (Fig. 14), chick (Fig. 15), as well as for vestibular epithelium (Fig. 17) of the mouse saccule. Strong myosin-specific staining of basal cells and perivascular elements is seen in the stria vascularis (Fig. 18, mouse).

C. ALPHA-ACTININ

The staining pattern observed in the auditory and vestibular sensory organs for antibodies to chicken gizzard α-actinin was virtually identical to the immunocytochemical distribution of actin (Figs. 12, 19–24). As described recently for the cochlear hair cells in the guinea pig (Drenckhahn et al., 1982), the distribution of α-actinin differed from that of actin only with respect to the auditory and vestibular stereocilia in all mammalian species investigated; mammalian stereocilia do not react with anti-α-actinin. Quite in contrast, stereocilia of all nonmammalian species investigated displayed a bright α-actinin-like immunofluorescence (Figs. 23, 24), which could be abolished by preabsorption of the antibody with purified α-actinin. This finding was unexpected and needs further confirmation because of the absence of α-actinin-like immunostaining in mammalian stereocilia and the absence of α-actinin and α-actinin-like immunoreactivity in intestinal microvilli of both mammalian and nonmammalian (Fig. 24, inset) species. (For a recent review regarding intestinal microvilli, see Mooseker, 1982, and Gröschel-Stewart and Drenckhahn, 1982).

D. TROPOMYOSIN

Antibodies to native chicken-gizzard tropomyosin displayed a strong affinity for the cuticular plates of both auditory and vestibular hair cells in the chick and caiman (Figs. 25–27). Immunostaining extended throughout the whole cuticular plates in a pattern similar to that described for antibodies to actin, myosin, and α-actinin. However, the antibodies to chicken smooth-muscle tropomyosin did not react with either hair cells or any other cell types in the mammalian sensory organs. This finding differs from a recent report indicating delicate tropomyosin-like immunoreactivity confined to the rootlets of the guinea pig auditory stereocilia (Flock, personal communication).

E. VINCULIN

The junctional complex of most of the cochlear-duct epithelia and epithelia of the saccular macula and lagena in the mammalian species and in the chick were faintly stained with anti-chicken-gizzard vinculin (Fig. 29). No staining was observed in the reticular plate of the mammalian organ of Corti, while the junctional complex of epithelia that line the saccular macula of the mouse and the basilar papilla of the chick were clearly labeled with anti-vinculin (Fig. 28). A strong immunostaining was seen in the basal-cell layer of the stria vascularis, where the vinculin immunoreactivity was concentrated along the plasma membrane of both the cell bodies and ascending processes (Fig. 30).

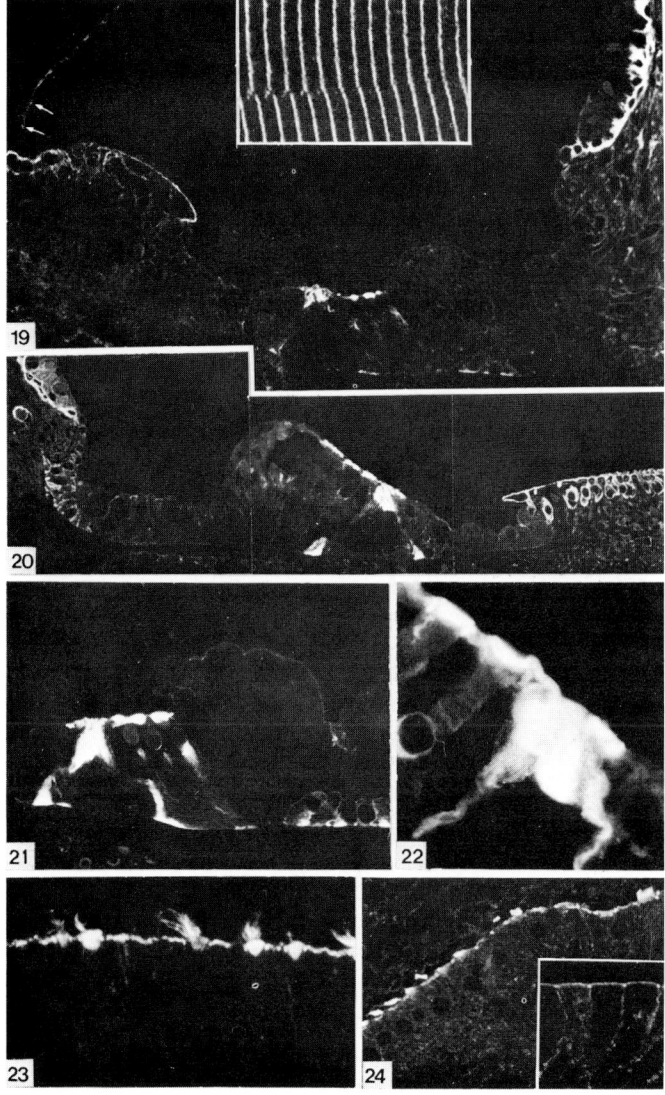

Figures 18-19–18-24. Distribution of α-actinin-specific immunofluorescence in the hamster (Fig. 19) and guinea-pig (Fig. 20) cochleas, and in the mouse (Fig. 21) and guinea-pig (Fig. 22) organ of Corti. Immunostaining is identical to the staining pattern obtained with anti-actin and phalloidin, except for the absence of staining of stereocilia. Arrows in Fig. 19 point to junctional immunoreactivity in the epithelium lining Reissner's membrane. Note staining of interdental cells of the limbus (Figs. 19, 20), the cells of Boettcher (Fig. 19, 21), the pillars (Fig. 19–22), and the reticular plate (Fig. 22). Immunoreactivity in the saccular macula (hamster) is shown in Fig. 12. In the nonmammalian sensory organs, such as the goldfish saccular macula (Fig. 23) and the chick basilar papilla (Fig. 24), anti-gizzard α-actinin reacts with both cuticular plates and stereocilia. Specificity of immunostaining in striated muscle (Z-line pattern) is shown in the inset of Fig. 19; in chicken intestinal epithelium (inset of Fig. 24), immunoreactivity is seen in the terminal web and is absent from the microvilli.

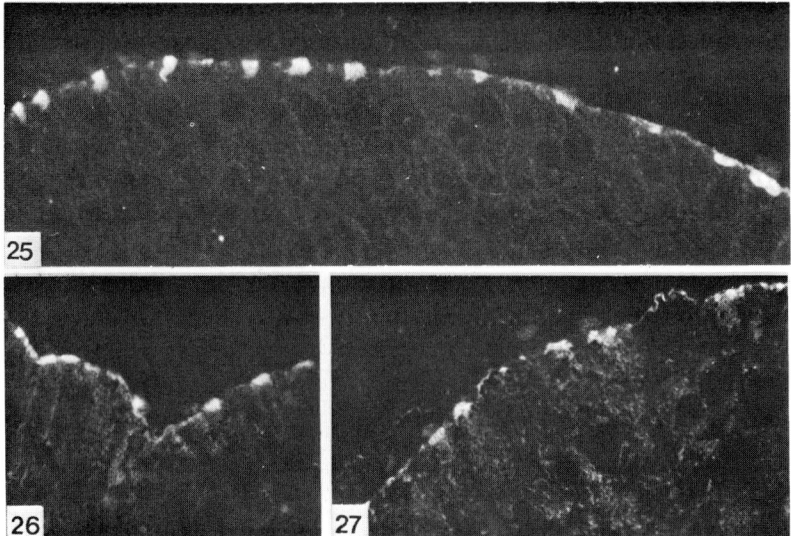

Figures 18-25–18-27. Immunostaining specific for chicken gizzard tropomyosin is confined to the cuticular plates of the auditory hair cells in the chicken (Fig. 25) and caiman (Fig. 27) and vestibular hair cells in the chicken lagena (Fig. 26).

F. FIMBRIN

Antibodies to chicken fimbrin, an actin-bundling polypeptide of intestinal microvilli (M_r 68,000), showed a selective affinity for the stereocilia of auditory and vestibular hair cells in all mammalian and nonmammalian species investigated (Figs. 31–36). In contrast to a recent report (Sobin and Flock, 1983), describing the presence of fimbrin in both stereocilia and cuticular plates ("with some failure") of the guinea-pig auditory hair cells, we did not observe any affinity of the fimbrin antibodies for the cuticular plates of either auditory or vestibular hair cells. This finding correlates well with ultrastructural studies, which did not reveal significant amounts of bundled actin filaments in the cuticular plate (Hirokawa and Tilney, 1983). With respect to the intensity of antibody fluorescence, there was a clear decrease in the staining intensity of stereocilia from the inner hair cells towards the outer hair cells. This decrease was not only noticed in the four mammalian species investigated, but was also seen in the chick basilar papilla, where the stereocilia of inner hair cells were strongly stained, whereas the outermost hair cells displayed only a very faint (hardly detectable) immunostaining with anti-fimbrin (Figs. 31, 33).

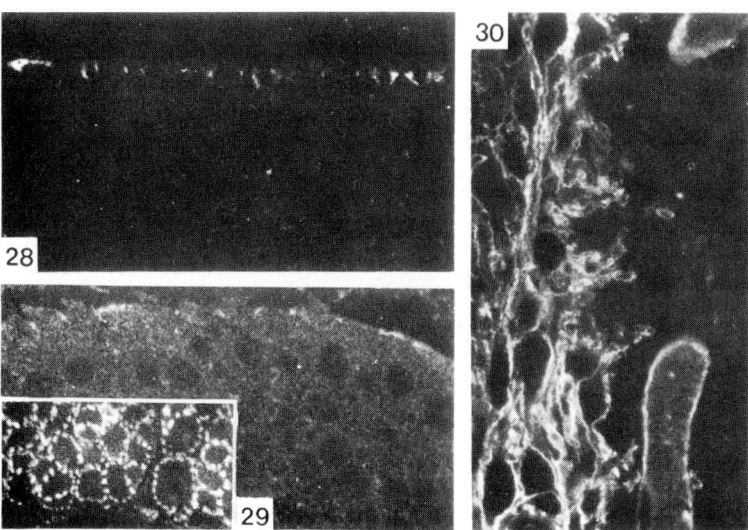

Figures 18-28–18-30. Anti-gizzard vinculin reacts with the apical-cell junctions in the mouse saccular macula (Fig. 28) and chicken basilar papilla (Fig. 29). A strong, membrane-associated staining is typical for the basal (intermediate) cells and their ascending processes in the stria vascularis (Fig. 30, mouse). The inset in Fig. 29 shows specificity of immunostaining in cross-sectioned intestinal smooth muscle (interrupted plasmalemmal staining).

G. VILLIN

As already described for the auditory hair cells in the guinea pig (Flock *et al.*, 1982; Drenckhahn *et al.*, 1982), antibodies to chicken intestinal microvillus villin (a polypeptide which bundles or disrupts actin filaments in a calcium-dependent manner) did not react with the stereocilia or with any other structure in the inner-ear organs of any of the vertebrates examined. However, in nitrocellulose replicas of electrophoretically-separated proteins of isolated chicken cochlear epithelium (see Section IV), a faint polypeptide band of 95,000 daltons was detected that cross-reacted with antibodies to native and SDS-denatured chicken intestinal villin. This indicates the presence of villin in cochlear epithelium. The local concentration of villin may be too low for detection by immunostaining.

H. ERYTHROCYTE MEMBRANE PROTEINS

Previous studies have demonstrated that many of the components of the erythrocyte membrane cytoskeleton also occur in nonerythroid cells (see Section I). The question as to whether the fine filaments connecting the actin

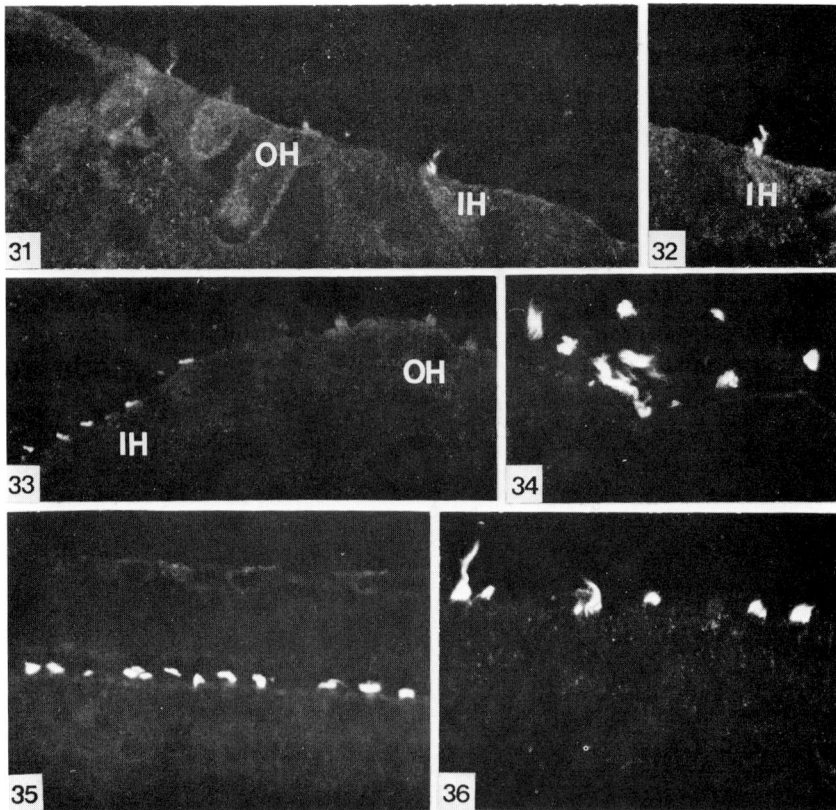

Figures 18-31–18-36. The actin-filament-bundling protein, fimbrin, is concentrated in stereocilia of the guinea pig (Fig. 31), mouse (Fig. 32), chicken (Fig. 33), and caiman (Fig. 34) auditory hair cells and in the vestibular hair cells in mouse saccule (Fig. 35) and chicken lagena (Fig. 36). Note the weak fluorescence in association with outer-hair-cell (OH) stereocilia and the clearly stronger fluorescence within inner-hair-cell (IH) stereocilia. Note the extreme length of the caiman auditory stereocilia (up to 100–150 μm).

filaments of the cuticular plate to the apical membrane (Section II) might be related to spectrin filaments was studied in the rat organ of Corti, using antibodies specific to rat erythrocyte spectrin (purified by SDS-polyacrylamide-gel electrophoresis). The spectrin antibodies (cross-reacting with the α and β subunits) displayed a well-defined affinity for the entire plasmalemmal zone of auditory hair cells and reacted also, albeit rather weakly, with the cuticular plates (Fig. 37). Staining was abolished by absorption with native spectrin extracted from rat erythrocyte ghosts by hypotonic lysis. Thus, it is reasonable to assume that the thin filaments connecting cuticular-plate actin filaments to the apical plasma membrane (Hirokawa and Tilney, 1982) are composed of spectrin-like polypeptides. The presence of an immunoreactive form of erythrocyte band 3 in nonerythroid cells (Kay *et al.*, 1983; Drenckhahn *et al.*, 1984b) and in isolated chicken cochlear epithelium (see Section IV) might indicate linkage of the subplasmalemmal web of actin and spectrin to a band-3-like membrane protein.

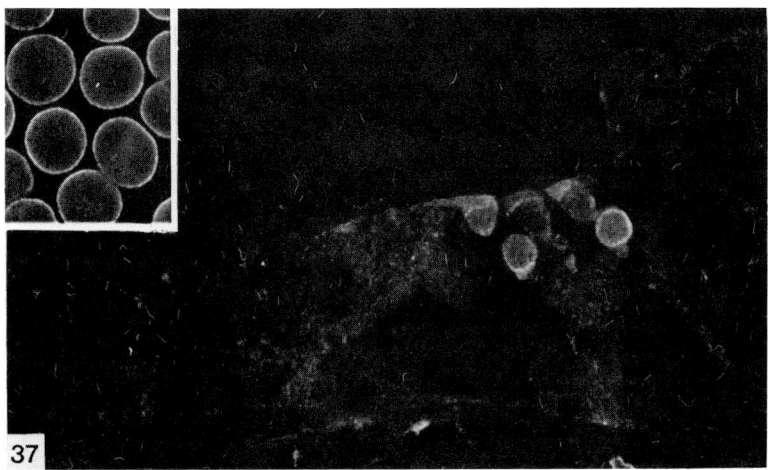

Figure 18-37. Anti-rat erythrocyte α,β spectrin binds to the periphery and the cuticular plate of rat cochlear hair cells. Inset shows immunostaining of rat erythrocytes.

IV. BIOCHEMICAL STUDIES

Several examples have been described showing that polyclonal and even monoclonal antibodies may cross-react with polypeptides sharing common antigenic sites, but being otherwise unrelated (e.g., cross-reaction of a monoclonal antibody to vimentin with tropomyosin; Blose *et al.*, 1981). To establish further the specificity of the immunocytochemical results described in Section III, immunoblotting studies were performed on electrophoretically-separated proteins of a crude preparation of the guinea pig organ of Corti (Fig. 38) and on isolated chicken cochlear epithelia enriched in apical fragments of hair cells (Fig. 39). Briefly, 50 cochleas were removed from 2-day-old male chickens, cut into two halves, and suspended for 30 min in buffer A (90 mM NaCl, 8 mM KH_2PO_4, 5.6 mM Na_2HPO_4, 1.5 mM KCl and 5 mM EDTA, pH 7). After agitation, the detached cells were filtered through a 200-μm-mesh nylon cloth and centrifuged at 400 g for 10 min. The cells were resuspended in buffer B (70 mM KCl, 5 mM $MgCl_2$, 2 mM EGTA, 10 mM imidazole, pH 7.2), disrupted by 5–10 passages through a 2-ml syringe (without needle), and then centrifuged at 1,000 g for 10 min. The fraction consisted of nuclei and cell fragments enriched in apical portions of hair cells, isolated cuticular plates, and stereocilia. Proteins were separated by SDS polyacrylamide-gel electrophoresis, transferred to nitrocellulose paper, and stained with antibodies as described (Drenckhahn et al., 1983). Antibodies to chicken-gizzard actin, tropomyosin, α-actinin, and to chicken intestinal brush border fimbrin, as well as to calf thymus myosin, displayed a selective affinity for the corresponding protein bands. No other cross-reactive bands were seen. Although no immunohistochemical staining was observed in the inner-ear epithelia with antibodies to villin, there was a distinct, albeit weak, cross-reactive protein band in the chicken cochelar epithelium at the molecular weight of villin (95,000 daltons) (see also Section III). Antibodies to the cytoplasmic domain of human erythrocyte band 3 (Drenckhahn et al., 1984b) were observed to bind to a 60,000–65,000-dalton polypeptide band showing electrophoretic mobility similar to a naturally-occurring fragment of band 3 in the red blood-cell membrane. Recently, a 60,000–64,000 dalton polypeptide has been identified as the major immunoreactive form of band 3 in nonerythroid cells (Kay et al., 1983; Drenckhahn et al., 1984b).

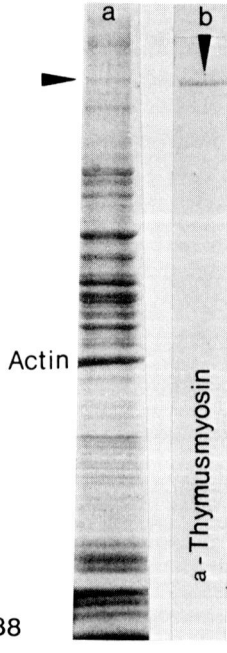

Figure 18-38. Anti-thymus myosin specifically binds to a polypeptide of Mr 200,000 (molecular weight of the myosin heavy chain) in a crude preparation of the guinea pig organ of Corti. Coomassie-blue stained (a) and corresponding immunostained (b) nitrocellulose sheets of electrophoretically-transferred SDS-polyacrylamide-gel proteins (5–15% gradient gel).

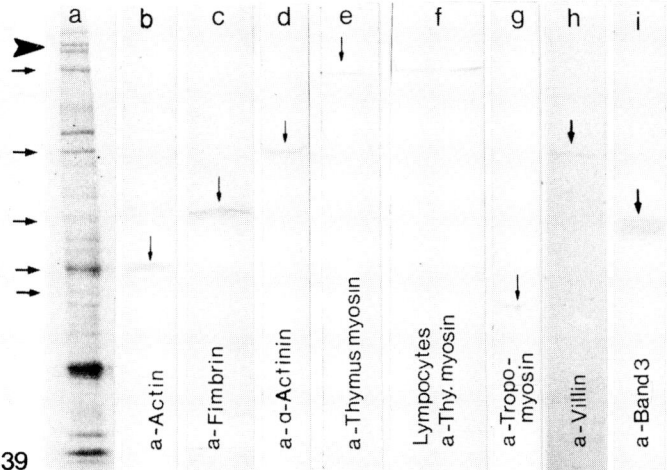

Figure 18-39. Immunoblotting of hair-cell-enriched epithelia of the chicken cochlea: Coomassie-blue stained track (a) and corresponding tracks stained with the antibodies as indicated (b–e, g–i). The myosin-immunoreactive band in (e) comigrates with lymphocyte myosin heavy chain, identified by immunoblotting of human lymphocytes (f). Arrows indicate immunoreactive polypeptide bands, most of which are seen as defined bands in the Coomassie-blue stained track (a). Arrowhead in (a) points to M_r 240,000–260,000 polypeptides, which may represent spectrin-like components.

V. STRUCTURAL AND FUNCTIONAL IMPLICATIONS

Recent physiological data indicate that the cochlea itself can produce vibration and sounds ("ringing in the ear") which might be important for modification of input-output characteristics (Kemp, 1978). Furthermore, there is experimental evidence that stimulation of efferent nerve fibers alters the mechanical response of outer hair cells (Mountain, 1980; Siegel and Kim, 1982). The molecular mechanism underlying these phenomena is still unknown. It has been suggested, for example, that changes in mechanical response might be produced by some kind of contractile mechanism, which, on the basis of the present studies, could be generated in the cuticular plate, the only site of myosin in the organ of Corti. Ultrastructural studies have demonstrated that the cuticular plate is composed of three different sets of actin filaments (Flock et al., 1981; Hirokawa and Tilney, 1982; Slepecky and Chamberlain, 1982), i.e., (1) a peripheral belt of filaments running parallel to the zonula adhaerens (belt desmosome), (2) a central, cuticular-plate web of less-well-ordered actin filaments, and (3) the vertically-oriented rootlet filaments of the stereocilia. This system of interconnected actin filaments has been shown in this and a previous study (Drenckhahn et al.,

1982) to be supplied with (1) myosin molecules, which may create tension in the peripheral belt and in the network of cuticular-plate filaments, (2) the muscle Z-line protein α-actinin, which most probably serves to cross-link the actin filaments, and (3) tropomyosin (so far, only documented in the chick hair cells), which may stabilize the actin filament and may regulate, in equilibrium with α-actinin, the degree of cross-linking of the cuticular-plate actin filaments. Calcium influx during excitatory hair deflection might trigger interaction between myosin and actin which, by virtue of the ultrastructurally demonstrated cross-links with rootlet filaments, might restrict or enhance motion of the stereocilia and thus modulate acoustic input-output characteristics (see hypothetical model presented in Fig. 40).

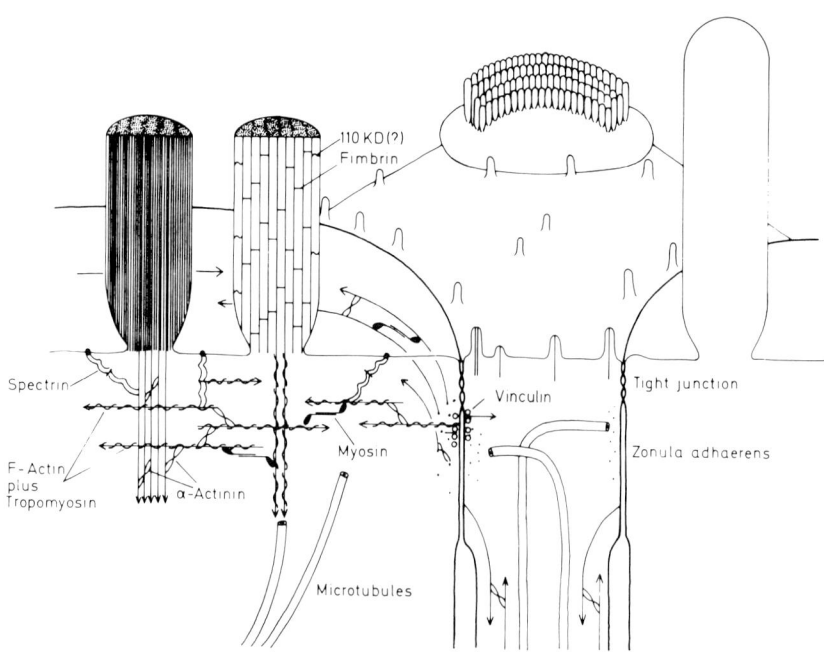

Figure 18-40. Hypothetical model for the structural and functional organization of the cytoskeleton in auditory hair cells (refer to Section V, this chapter).

Mechanical properties of the stereocilia may be further modified by different degrees of cross-linking in the stereociliary core bundle. In this respect, it is of interest that the intensity of immunostaining obtained with antibodies to the actin-bundling polypeptide, fimbrin (Matsudaira et al., 1983), displays considerable variation between inner and outer hair cells in both mammalian species and chick. These observations indicate a possible molecular basis for the well-studied differences in the stiffness of the stereocilia of inner and outer hair cells and of outer hair cells in different cochlear turns (see Flock and Strelioff, 1984).

While we detected strong actin and α-actinin-like immunostaining that was associated with the supporting cells (pillars and phalangeal cells of the organ of Corti and the interdental cells of the limbus), we were not able to detect myosin in these cells. It is reasonable, therefore, to assume that actin filaments, cross-linked by α-actinin, may have a structual role in these cells (together with the axial bundle of microtubules), and thus may serve to keep the distance between the basilar membrane and the tectorial membrane (which is proximally attached to the interdental cells) at constant values. No functional implications can be made at this stage as to the structural basis and functional significance of the strong, smooth-muscle-like immunofluorescence observed in the basal-cell layer of the stria vascularis.

Further studies are clearly needed to confirm our preliminary observations regarding spectrin and band 3-like polypeptides in the inner ear and to analyze by biochemical methods possible differences in the molecular composition of inner and outer stereocila as well as of stereocilia of mammalian and nonmammalian species (e.g., amount of fimbrin, presence of α-actinin). Finally, it still remains to be shown experimentally whether myosin and α-actinin serve merely to cross-link and tighten the web of actin filaments within the cuticular plate, or whether these proteins constitute a contractile filament system that might be able actively to modify mechanical properties of cochlear and vestibular hair cells.

ACKNOWLEDGMENTS

We thank Claudia Gerhardt and Heidi Schneider for skillful technical assistance in immunohistochemistry and photographic work. Excellent assistance in the biochemical studies was provided by Manfred Brandner. We are grateful to Dr. R. Klinke and Dr. J. Smolders (Frankfurt) for having provided two cochleas of an American caiman. Supported by grants from the Deutsche Forschungsgemeinschaft (Dr 91/3 2, 91/4-1).

REFERENCES

Blose, S. H.; Matsumura, F., and Lin, J. J.-C.: Structure of vimentin 10-nm filaments probed with a monoclonal antibody that recognizes a common antigenic determinant on vimentin and tropomyosin. *Cold Spring Harbor Symp. Quant. Biol.* 46: 455–463, 1981.

Branton, D.: Membrane cytoskeletal interactions in the human erythrocyte. *Cold Spring Harbor Symp. Quant. Biol.* 46: 1–5, 1981.

Burridge, K.; Kelly, P., and Mangeat, P.: Nonerythrocyte spectrins: Actin-membrane attachment proteins occurring in many cell types. *J. Cell Biol. 95:* 478–486, 1982.

Cohen, C. M.; Foley, S. F., and Korsgren, C.: A protein immunologically related to erythrocyte band 4.1 is found on stress fibres of nonerythroid cells. *Nature 299:* 648–650, 1982.

DeRosier, D. J.; Tilney, L. G., and Egelman, E.: Actin in the inner ear: the remarkable structe of the stereocilium. *Nature 287:* 291–296, 1980.

Drenckhahn, D.; Frotscher, M., and Kaiser, H. W.: Concentration of F-actin in synaptic formations of the hippocampus as visualized by staining with fluorescent phalloidin. *Brain Res. 300:* 381–387, 1984a.

Drenckhahn, D., and Kaiser, H. W.: Evidence for the concentration of F-actin and myosin in synapses and in the plasmalemmal zone of axons. *Eur. J. Cell Biol. 31:* 235–240, 1983.

Drenckhahn, D.; Kellner, J., Mannherz, H. G., Gröschel-Stewart, U., Kendrick-Jones, and Scholey, J.: Absence of myosin-like immunoreactivity in stereocilia of cochlear hair cells. *Nature 300:* 531–532, 1982.

Drenckhahn, D., and Mannherz, H. G.: Distribution of actin and the actin-associated proteins myosin, tropomyosin, alpha-actinin, vinculin, and villin in rat and bovine exocrine glands. *Eur. J. Cell Biol. 30:* 167–176, 1983.

Drenckhahn, D.; Zinke, K., Schauer, U., Appell, K. C., and Low, P. S.: Identification of immunoreactive forms of human erythrocyte band 3 in nonerythroid cells. *Eur. J. Cell Biol. 34:* 144–150, 1984b.

Egelman, E.: Problem of light piping in immunofluorescence studies. *Nature 294:* 674, 1981.

Faulstich, H.; Trischmann, H., and Mayer, D.: Preparation of tetramethylrhodaminyl-phalloidin and uptake of the toxin into short-term cultured hepatocytes by endocytosis. *Exp. Cell Res. 144:* 73–82, 1983.

Flock, A.; Bretscher, A., and Weber, K.: Immunohistochemical localization of several cytoskeletal proteins in inner ear sensory and supporting cells. *Hearing Res. 6:* 75–89, 1982.

Flock, A., and Cheung, H. C.: Actin filaments in sensory hairs of inner ear receptor cells. *J. Cell Biol. 75:* 339–343, 1977.

Flock, A., and Strelioff, D.: Graded and nonlinear mechanical properties of sensory hairs in the mammalian hearing organ. *Nature 310:* 597–599, 1984.

Flock, A.; Cheung, H. C., Flock, B., and Utter, G.: Three sets of actin filaments in sensory cells of the inner ear. Identification and functional orientation determined by gel electrophoresis, immunofluorescence and electron microscopy. *J. Neurocytol. 10:* 133–147, 1981.

Franzini-Armstrong, C., and Peachy, L. D.: Striated muscle-contractile and control mechanisms. *J. Cell Biol. 91:* 166s–186s, 1981.

Goldman, R. D.; Milsted, A., Schlosser, J. A., Starger, J., and Yerna, M. J.: Cytoplasmic fibers in mammalian cells. *Annu. Rev. Physiol. 41:* 703–722, 1979.

Gröschel-Stewart, U., and Drenckhahn, D.: Muscular and cytoplasmic contractile proteins; biochemistry, immunology, structural organization. *Collagen Rel. Res. 2:* 381–463, 1982.

Hirokawa, N., and Tilney, L. G.: Interactions between actin filaments and between actin filaments and membranes in quick-frozen and deeply etched hair cells of the chick ear. *J. Cell Biol. 95:* 249–261, 1982.

Kardami, E.; Alexis, M., Paz, P. de la, and Gratzen, W.: Phosphorylation and the binding of calcium and magnesium to skeletal myosin. *Eur. J. Biochem. 110:* 153–160, 1981.

Kay, M. M. B.; Tracey, C. M., Goodman, J. R., Cone, J. C., and Bassel, P. S.: Polypeptides immunologically related to band 3 are present in nucleated somatic cells. *Proc. Natl. Acad. Sci. U.S.A. 80:* 6882–6886, 1983.

Kemp, D. T.: Stimulated acoustic emissions from within the human auditory system. *J. Acoust. Soc. Am. 64:* 1386–1391, 1978.

MacCartney, J. C.; Comis, S. D., and Pickles, J. O.: Is myosin in the cochlea a basis for active motility? *Nature 288:* 491-492, 1980.

Matsudaira, P.; Mandelkow, E., Renner, W., Hesterberg, L. K., and Weber, K.: Role of fimbrin and villin in determining the interfilament distances of actin bundles. *Nature 301:* 209-214, 1983.

Mooseker, M. S.: Actin binding proteins of the brush border. *Cell 35:* 11-13, 1983.

Mountain, D. C.: Changes of endolymphatic potential and crossed olivocochlear bundle stimulation alter cochlear mechanics. *Science 210:* 71-72, 1980.

Pollard, T. D.: Cytoplasmic contractile proteins. *J. Cell Biol. 91:* 156s-165s, 1981.

Siegel, J., and Kim, D.: Efferent neural control of cochlear mechanics? Olivocochlear bundle stimulation affects cochlear biomechanical nonlinearity. *Hearing Res. 6:* 171-182, 1982.

Slepecky, N, and Chamberlain, S. C.: Distribution and polarity of actin in the sensory hair cells of the chinchilla cochlea. *Cell Tissue Res. 224:* 15-24, 1982.

Sobin, A., and Flock, A.: Immunohistochemical identification and localization of actin and fimbrin in vestibular hair cells in the normal guinea pig and in a strain of the waltzing guinea pig. *Acta Otolaryng. 96:* 407-412, 1983.

Tilney, L. G.; DeRosier, D. J., and Mulroy, M. J.: The organization of actin filaments in the stereocilia of cochlear hair cells. *J. Cell Biol. 86:* 244-259, 1980.

Chapter 19

AUDITORY NERVE PROTEINS

Robert J. Wenthold

I. Introduction
II. Axonal Transport in the Auditory Nerve
III. Rapidly Degraded Proteins of the Auditory Nerve
IV. Auditory Nerve Proteins After Hair Cell Loss
V. Discussion
   References

I. INTRODUCTION

Axons and axon terminals lack the components necessary for protein synthesis and, therefore, all proteins must be supplied to these structures from other sources. There may be a small amount of protein synthesis by synaptic mitochondria and transfer of proteins to axons from glial cells, but most axonal proteins are supplied from the cell body and transported to the axon and axon terminal by a process known as axonal transport. Axonal transport was first identified by Weiss and Hiscoe (1948), when they ligated an axon and observed an accumulation of material on the proximal side of the ligature. Since that time, many aspects of axonal transport have been studied, and it has been found that anterograde axonal transport occurs at several distinct rates, ranging from less than 1 mm/day up to 400 mm/day. Each rate group contains different proteins and can be linked to the transport of a different cellular structure (Lasek and Hoffman, 1976; Grafstein, 1977; Schwartz, 1979; Wilson and Stone, 1979; Grafstein and Forman, 1980). For example, the slower groups contain the cytoskeletal proteins, while the fast-transport group is composed of membrane proteins and glycoproteins. It has also been shown that some of these proteins are returned to the cell body by way of retrograde axonal transport (Kristensson, 1978; Bisby, 1982). The components and function of retrograde transport are poorly defined because of the current technical limitations involved in studying this process. Anterograde axonal transport, with the use of confined radioactive precursor injections and analysis of selective regions corresponding to a specific pathway, can be used to characterize axonal proteins synthesized by a single group of neurons. Furthermore, because different types of proteins are transported at different rates, the population of proteins within a specific neuronal cell type can be selected for study. Such an approach has an advantage over other methods of studying neuronal proteins, such as subfractionation, in that single neuronal populations can be investigated. Axonal transport, therefore, can be used to study presynaptic and axonal components of specific neuronal populations under both normal conditions and, for example, states of development, regeneration, and degeneration. Proteins which may play critical roles in such processes can then be identified for further characterization.

II. AXONAL TRANSPORT IN THE AUDITORY NERVE

Most axonal transport studies have been carried out on neurons with long axons. Such a system allows injection of radioactive precursor near the cell bodies without diffusion to the site of analysis, and also provides transport over a distance sufficient to give adequate separation of the different rate groups. A popular system for studying axonal transport is the retinal ganglion cell and optic nerve and tract, which *in toto,* provides a convenient injection site with physical barriers to prevent diffusion of radioactive precursor from the retina. While the spiral ganglion cells of the inner ear have comparatively short axons, we found that radioactive precursors are sufficiently contained within the cochlea after injection to allow the analysis of proteins which are transported to the cochlear nucleus (Tytell *et al.,* 1980). Injections can conveniently be done by replacing the perilymph with artificial perilymph containing the radioactive precursor. To insure uniform distribution of label, a volume of artificial perilymph somewhat larger than that of the perilymph is injected. It should be noted that a basic assumption made in any study utilizing a radioactive amino acid, which is incorporated into proteins, is that the incorporated radioactivity accurately represents the protein. For example, if a particular amino acid is very abundant in a protein, use of the amino acid as a radioactive precursor may seem to indicate that this protein is present in much higher concentrations than it actually is. This should not affect the basic interpretation of axonal-transport data if the same amino acid is used in all experiments and if it is not used as an absolute measure of abundance. Most experiments on the auditory nerve were done using ^{35}S-methionine as precursor. While methionine is only moderately abundant in most proteins, ^{35}S-methionine offers the advantages of availability at high specific activity and at a higher energy than tritium-labeled compounds, making it more suitable for use in experiments that involve autoradiographic detection in polyacrylamide gels.

Transported proteins can be analyzed in the axon as they are being transported, and in the innervated tissue when they reach the synapse or proximal part of an axonal ligature or cut, where the proteins accumulate. Because proteins of fast transport pass through the axon rapidly, this group is best studied in the innervated tissue or with an axonal cut or ligature. More slowly-transported groups often are rapidly degraded when they reach the terminal and are, therefore, studied in the axon. Because most of our studies on the auditory nerve were done on fast transport, proteins were usually analyzed after they reached the terminals of the auditory nerve in the cochlear nucleus.

After injection of ^{35}S-methionine into the cochlea, a pattern of radioactive proteins in the cochlear nucleus is obtained, as shown in Fig. 1. It is readily apparent that this pattern changes with time after injection; the proteins labeled early are lost and additional labeled proteins appear, due to arrival of the more slowly-transported groups. At 1 and 3 h, the major components are a 25 kilodalton (kd) and a 140 kd protein; the radioactivity in the 140 kd component is lost rapidly, so that it is only a minor labeled band at 12 h, while the 25 kd protein is still present at 10 days after injection. The earliest labeling seen in whole cochlear nucleus is at 1 h after injection. While only two bands are seen at 1 h in Fig. 1, other labeled components can be visualized by longer exposure or by the use of more isotope.

Our research has concentrated on the most rapidly-transported group of proteins in the auditory nerve. In other neurons, it has been shown that this group is transported at rates from 200–400 mm/day and is composed mostly of membrane-associated proteins and glycoproteins (Grafstein, 1977; Grafstein and Forman, 1980). We have studied the early events involving these proteins in the anteroventral cochlear nucleus (A), the interstitial nucleus (I), the auditory nerve (N), and the spiral ganglion (G). It can be seen that radioactive proteins are present in G, N and I before they reach the synaptic terminals in A (Fig. 2). Thirty min after injection, a complex pattern of labeled proteins is found in G, with major proteins of the fast-transport group being distinguishable. Such a pattern is expected since synthesis of proteins in all transport groups, as well as those not transported, occurs at the same time. However, the change in transport rate can be seen in the labeling patterns in N, I and A. At 30 min after injection, no labeling is seen in A, very slight labeling is present in I, and somewhat heavier labeling is in N. More slowly-transported proteins are not seen by 6 h in A, but are seen at this time in I, as evidenced, for example, by the appearance of a band of about 30 kd; this band is already present in N at 1 h after injection. From these studies, we can estimate the rate of fast transport in auditory nerve. The length of N, from leaving the cochlea to entering A, is about 2.5 mm. We have not determined the length of time required for protein synthesis, but estimate it to be about 30 min, based on values in the literature and the time required for appearance of labeled proteins in N. Therefore, a maximum of

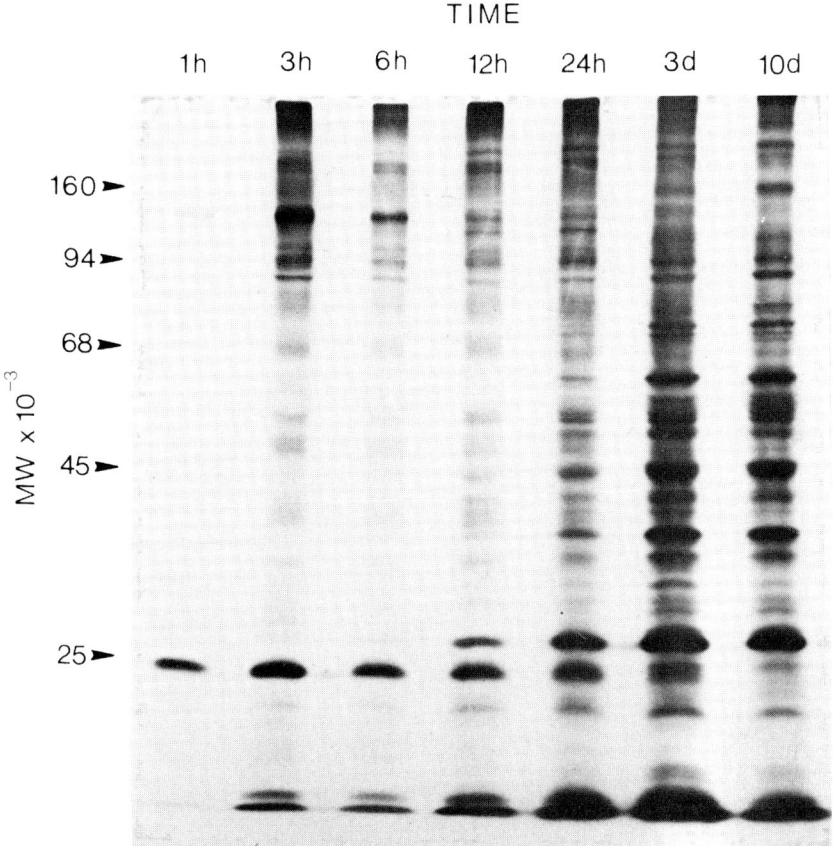

Figure 19-1. Fluorograph of labeled proteins in the cochlear nucleus after injection of ^{35}S-methionine into the cochlea. Guinea pigs were injected with 125 μCi of isotope and killed at times indicated. The whole cochlear nucleus was removed and analyzed on 10% acrylamide gels (Laemmli, 1970.)

30 min is required for transport of 2.5 mm, which is a minimum rate of 120 mm/day. This is somewhat low compared to the rate determined for other neurons, but is certainly an underestimate, because we have not accounted for transport within the cochlea or for the delay required for sufficient accumulation of radioactive protein in A for detection in a polyacrylamide gel.

The sequential appearance of individual proteins in structures en route to the cochlear nucleus is strong evidence that these proteins are, in fact, synthesized by spiral ganglion cells and transported to the cochlear nucleus, rather than their being synthesized locally by cells within the cochlear nucleus. Other evidence also supports this claim. If the auditory nerve is cut, the transported proteins will accumulate proximal to the cut and the gel

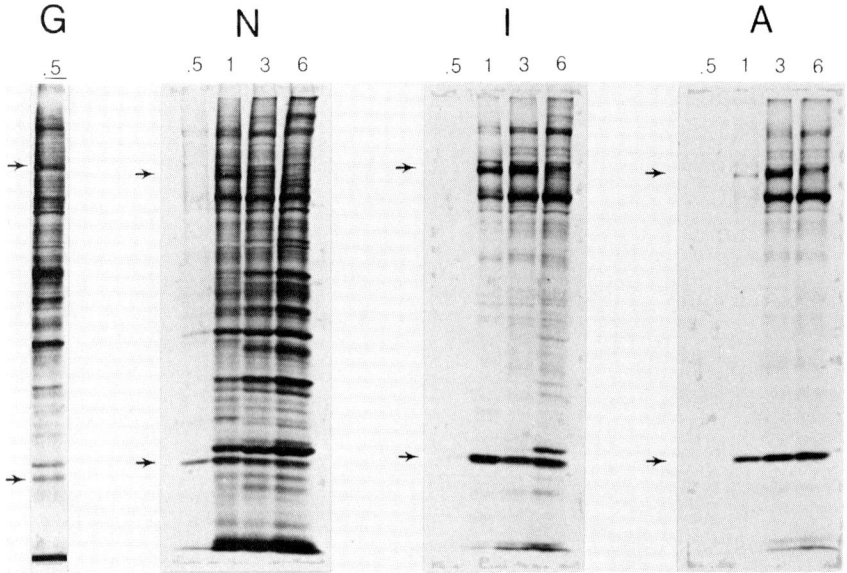

Figure 19-2. Fluorograph of labeled proteins in the spiral ganglion (G), auditory nerve (N), interstitial nucleus (I), and anteroventral cochlear nucleus (A), after injection of ^{35}S-methionine (250 µCi) into the cochlea. Animals were killed 0.5 to 6 h after injection and analyzed on 10% acrylamide gels. Arrows indicate the position of the major 25-kd and 140-kd rapidly-transported proteins.

pattern of these proteins will be the same as that obtained if transport were to continue to the cochlear nucleus (Tytell et al., 1980). The injection of isotope directly into the cochlear nucleus produces a pattern of labeled proteins totally unlike that seen after injection into the cochlea (Tytell et al., 1980). Finally, when we compare rapidly-transported proteins of the auditory nerve with those of longer axon systems, many proteins appear the same, as discussed below (Wenthold and McGarvey, 1982a).

III. RAPIDLY DEGRADED PROTEINS OF THE AUDITORY NERVE

As mentioned above, fast-transported proteins lose radioactivity at varying rates after they reach the cochlear nucleus. Because the radioactivity is associated with an amino acid which has been incorporated into the protein, the loss of radioactivity from an individual protein provides a measure of the biological half-life of that protein. The most rapidly-turning over, fast-transported protein in the auditory nerve is the major 140 kd protein, which displays a half-life of less than 3 h (Wenthold and McGarvey, 1982a). Two-dimensional gel analysis reveals that this protein exists as a single major component with an isoelectric point of 4.7 (Fig. 3). Studies with fucose labeling and lectin binding show the 140 kd protein is also the major rapidly-transported glycoprotein in the auditory nerve, containing about 25% of the total fucose incorporated into transported proteins. If the fast component of auditory nerve proteins is labeled with ^{35}S-methionine, solubilized, and passed through a column of immobilized concanavalin A (a lectin which selectively binds glycoproteins containing glucose and mannose residues), the major retained protein is the 140-kd protein (Fig. 4). Based on these properties, we have called this protein "Rapidly Turned-Over Glycoprotein" (RTGP) (Tytell et al., 1980). Comparison of one-dimensional gel patterns for the fast components of auditory and optic nerves shows that RTGP is *not* a major protein of fast transport in the optic nerve (Fig. 5). If RTGP were absent from the optic nerve, it may suggest a specialized role to the auditory nerve. To determine if RTGP is absent from the optic nerve or present in small amounts there, two-dimensional gel analysis was done on rapidly-transported proteins of both optic and auditory nerves (Fig. 6). These studies showed that RTGP is present in the optic nerve, but as a very minor protein; the identity of this minor, optic-nerve protein to RTGP has been verified by peptide-mapping studies (unpublished observations). We found, however, that a number of other proteins with molecular weights and isoelectric points similar to those of RTGP were present in the optic nerve (Fig. 6, proteins 1–7). Detailed analysis revealed that these proteins are also present in the auditory nerve, but as very minor components (Wenthold and McGarvey, 1982a). In both the auditory nerve and the optic nerve, these proteins all appear to be rapidly lost when they reach the presynaptic terminal. This suggests that RTGP may be only one of several proteins with similar molecular weights, isoelectric points, and turnover rates. Our recent data suggest that these proteins may also be related with respect to solubility characteristics, and, based on peptide mapping studies (unpublished observations), some may be modifications of a single protein.

The unique properties of this family of proteins raise interesting questions concerning their functions. The differences in amounts of the different forms, between the optic and auditory nerve, may suggest that they have similar or identical functions in the two nerves, but that the functions are expressed differently, for example, due to local environmental differences. In some cases, the different amounts may reflect various degrees of post-translational modification. A major question concerns their rapid turnover rates in relation to their function. A rapidly-turning-over protein has the capability of quickly changing levels in response to changes in synthesis or degradation rates. Such a protein would be well suited to serve a regulatory role in a neuron, for example, being involved in communication between the cell body and axon terminals. Another possibility is that RTGP and the

Figure 19-3. Fluorograph of labeled proteins in the anteroventral cochlear nucleus 3 h after injection of ^{35}S-methionine into the cochlea. Two-dimensional analysis, done with electrofocusing in the first dimension (horizontal) and electrophoresis in 10% acrylamide in the second (vertical). Arrow shows position of rapidly-transported glycoprotein (RTGP) and arrowhead indicates major 25-kd rapidly-transported proteins.

related proteins are involved only in guiding membrane components to the axon terminal; in such a role, they would have completed their task when they had reached the terminal and would then be rapidly degraded.

We have sought to determine the fate of rapidly-turning-over proteins when they reach the axon terminal, in order to gain insights into the function of RTGP and related proteins. Transported proteins can be degraded at the terminal, released into the extracellular space, or retrogradely transported back to the cell body. Since up to 70% of rapidly-transported proteins are reported to be returned to the cell body (Grafstein and Forman, 1980), we have initially studied retrograde transport as a possible route for loss of RTGP from the cochlear nucleus. In these studies, we label the auditory nerve proteins by cochlear injection, wait until the transported proteins reach the terminal, and then cut the auditory nerve and allow retrogradely-transported proteins to accumulate distal to the cut for several hours. Analysis of these proteins shows no accumulation of RTGP in the axon distal to the cut (Fig. 7). The most straightforward explanation of these results is that

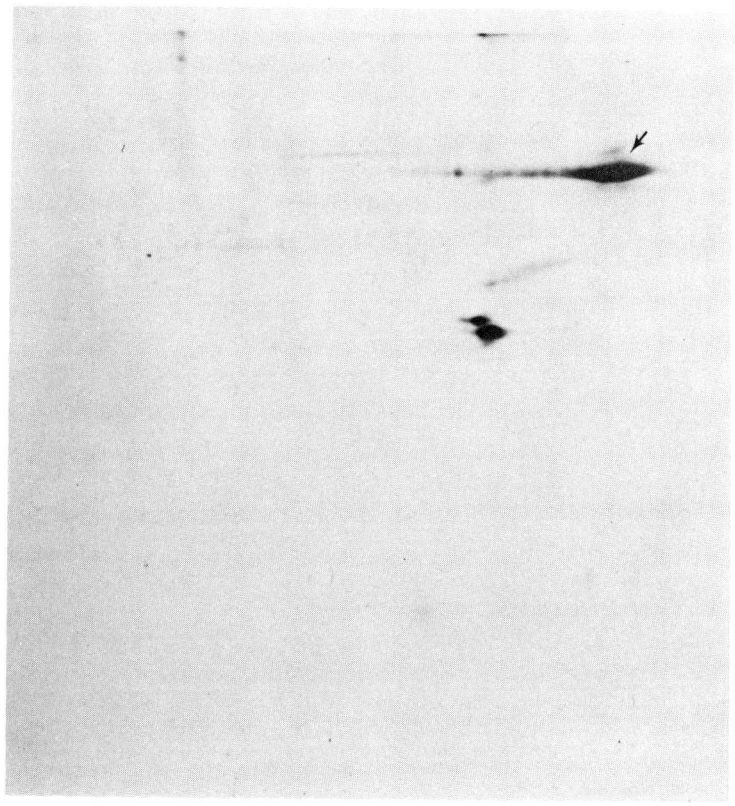

Figure 19-4. Fluorograph of concanavalin-A binding fraction of labeled proteins in the cochlear nucleus 3 h after injection of ^{35}S-methionine into the cochlea. The cochlear nucleus proteins were separated on a concanavalin A affinity column as previously described (Wenthold and McGarvey, 1982a). Arrow indicates RTGP.

RTGP is not retrogradely transported in the auditory nerve. However, it must be considered that the "label, lesion, distal-accumulation" paradigm is not satisfactory for studying retrograde transport. While it is known that retrograde transport continues for some time in the distal segment of a cut axon, it has not been shown that the transport is identical to retrograde transport in an intact axon. Also, RTGP may be retrogradely transported, but lost from the cut, distal end of the nerve, either through degradation or leakage. In analogous studies of anterograde transport, however, RTGP accumulates proximal to the axon cut. These results do not support retrograde transport as the major route of elimination of RTGP from the cochlear nucleus. We are presently studying the other routes of RTGP elimination, i.e., degradation at the synaptic terminal and release into the extracellular space.

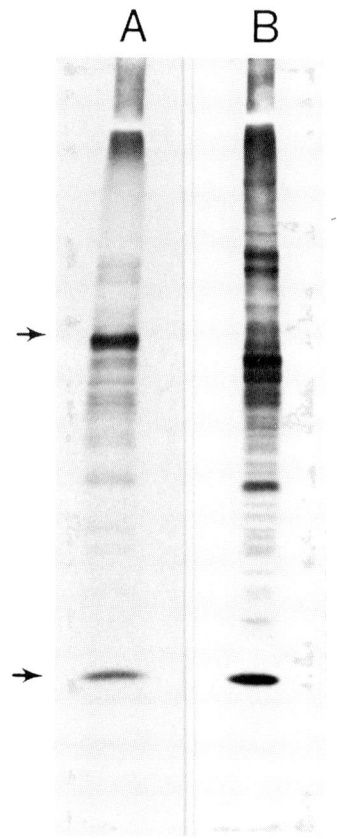

Figure 19-5. Fluorograph of rapidly-transported proteins in the auditory nerve (A) and optic nerve (B) after labeling with ^{35}S-methionine. Labeling and tissue processing were done as previously described (Wenthold and McGarvey, 1982a). Analysis was done on a 5–17% acrylamide gradient gel. Arrows show RTGP (upper) and major 25-kd protein (lower).

IV. AUDITORY NERVE PROTEINS AFTER HAIR CELL LOSS

It has been widely reported that spiral ganglion cells degenerate after hair cell loss, resulting, for example, from noise damage, genetic lesion or ototoxic drugs (Spoendlin, 1979; 1981). It has further been reported that, while hair cell loss can occur quickly in response to these insults, spiral-ganglion-cell loss occurs more slowly, and a population of spiral ganglion cells may persist indefinitely. Factors which determine the survival of spiral ganglion cells under these conditions are not known, but are obviously important, especially in light of recent advances in the development of cochlear prostheses. Morphological changes have been reported in the spiral-ganglion-cell bodies

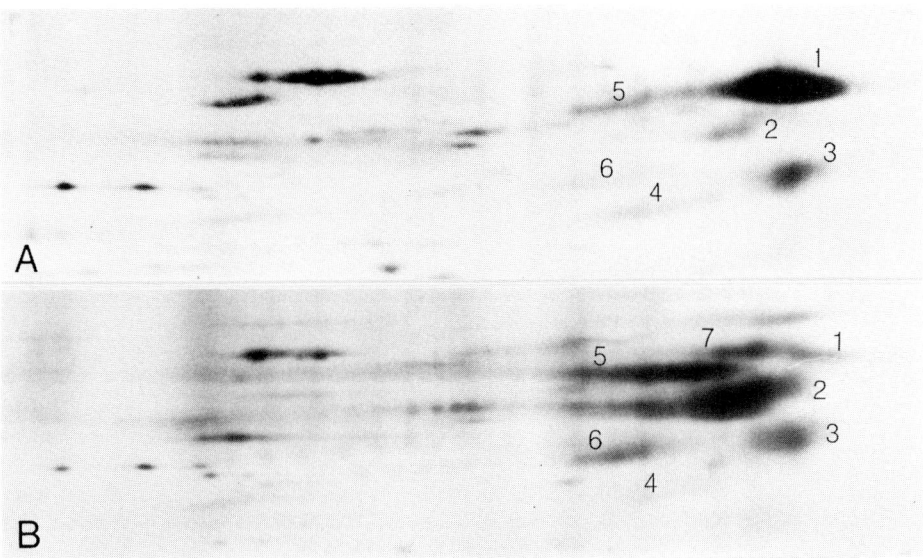

Figure 19-6. Fluorograph of rapidly-transported proteins of the auditory nerve (A) and optic nerve (B) after labeling with ^{35}S-methionine. Only the portion of the two-dimensional gel showing RTGP and related proteins is shown. Proteins with properties similar to those of RTGP are indicated by numbers, with co-migrating proteins in the two nerves having the same number. RTGP is number 1.

as well as in their presynaptic terminals in the cochlear nucleus after hair-cell loss (Gulley et al., 1978; Koitchev et al., 1982). We sought to determine whether or not damage to, or loss of, hair cells affects the rapidly-transported proteins of the auditory nerve. This was studied both in the waltzing guinea pig, which displays an age-dependent loss of hair cells (Ernston, 1971), and in the normal guinea pig after treatment with neomycin, an ototoxic antibiotic. In both cases, loss of hair cells is accompanied by a change in the two-dimensional gel pattern of rapidly-transported proteins of the auditory nerve (Fig. 8 compared to Fig. 3). There is a significant increase in two series of polypeptides of average molecular weights 27 kd and 36 kd. These proteins appear as a series of several labeled spots with slightly different isoelectric points and molecular weights. Such a pattern is characteristic of glycoproteins, with each spot representing a slightly modified form of the protein, often involving the sugar residues (Baumann and Doyle, 1979). Preliminary peptide-mapping studies of these proteins in the auditory nerve suggest that these multiple spots represent modifications of the same protein (unpublished observation). Under normal conditions, these proteins are only very lightly labeled, but under conditions of hair-cell loss, they become major, transported proteins. Of the different conditions studied,

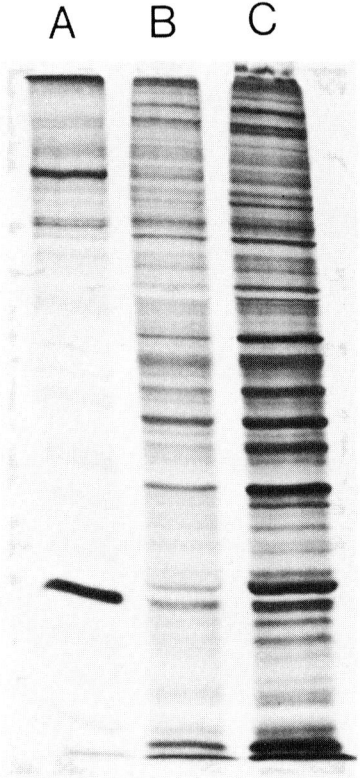

Figure 19-7. Fluorograph of labeled proteins in the cochlear nucleus 3 h after injection (A), in the distal portion of the auditory nerve 9 h after injection and 6 h after cutting the nerve (B), and in the intact auditory nerve 9 h after injection (C). Arrow indicates the position of RTGP. Note the lack of significant accumulation of RTGP in the distal segment of the auditory nerve (B), compared to the intact nerve (C).

maximum increase in these proteins was 6-fold compared to normal, a situation which was found in the 80-day-old waltzing guinea pig. Further characterization showed these proteins to be glycoproteins and at least partially associated with the synaptic membrane (Wenthold and McGarvey, 1982b). Because the amount of these two proteins that is transported in the auditory nerve appears dependent on the presence of hair cells, we have labeled these "hair cell-controlled" proteins (HC proteins).

The increased axonal transport of the HC proteins by spiral ganglion cells could arise in several ways, e.g., by the loss of a trophic factor associated with hair cells, changes in supporting cells of the organ of Corti, damage to afferent fibers in the organ of Corti, or changes in the activity of the auditory nerve. Because these proteins are elevated in the 10-day-old waltzing guinea

Figure 19-8. Fluorograph of labeled proteins in the cochlear nucleus of a neomycin-treated guinea pig 3 h after injection of ^{35}S-methionine into the cochlea. Animals were treated with neomycin for one week as previously described (Wenthold and McGarvey, 1982b) and transported proteins were studied two weeks later. Note the dramatic increase in labeling of two sets of proteins (arrows), referred to in the text as HC proteins. For comparison to the normal animal, see Fig. 3.

pig, an age before there is hair-cell loss, it appears that hair-cell loss is not required for their increased expression, but rather that the auditory nerve can be affected by apparent abnormalities in the hair cell. This may suggest that the genetic abnormality of the waltzing guinea pig is associated with afferent synapse. On the other hand, it may indicate that a normally-functioning hair cell is necessary for the normal expression of the HC proteins in spiral ganglion cells. The results do not differentiate between an increased transport of these proteins in the auditory nerve and an increased synthesis with increased transport. For example, the HC proteins may normally be destined for spiral-ganglion-cell dendrites, but after hair-cell damage, they may be routed to the axon. Our preliminary results, from two-dimensional analyses of spiral-ganglion-cell bodies, show that HC proteins are not normally synthesized at high levels, suggesting the changes after hair-cell damage are due to an increased synthesis. Our idea of the function of the HC proteins will be much clearer when we determine the events which cause their increase. The changes we see in transported proteins in the auditory nerve with hair cell-damage, namely, a rather large alteration of only a few proteins, are similar to those observed in other neurons undergoing development or regeneration. In these cases, only a very few, minor proteins change and these proteins have been termed "growth-associated proteins" (Skene and Willard, 1981a, b, c). It is possible that the HC proteins in the auditory nerve belong to this class of proteins. It is necessary to determine if the HC proteins are elevated under other conditions, such as during development. It must also be determined if all spiral ganglion cells express HC proteins. Perhaps the capability of expressing these proteins is necessary for the cell to survive hair-cell loss.

V. DISCUSSION

Our studies of auditory nerve proteins have employed axonal transport as a method of selectively identifying those proteins synthesized by spiral ganglion cells and transported down their axons to the cochlear nucleus. This approach provides a mechanism for monitoring the synthesis, transport, and degradation of a large number of proteins under normal conditions and under conditions where the function of the auditory nerve is disrupted. While such an approach can identify proteins in a specific population of neurons, this method is limited by the fact that proteins are identified only as bands or spots on acrylamide gels, with no indication of their function. Furthermore, analysis is dependent on the cumbersome "label, transport, analyze-by-gel" technique, which cannot be conveniently or widely applied to analysis of a large number of samples. Therefore, our objective is to use axonal transport as a first step to identify auditory nerve proteins which, based on their synthesis, transport, and degradation properties, warrant further study. With this approach, we have identified two populations of proteins for characterization: RTGP and related proteins, and the HC proteins. Both of these appear to be very minor protein populations, as is the case for most of the rapidly-transported proteins. However, their purification is feasible, especially by using the recently developed technique of production of monoclonal antibodies. While extremely little is currently known about the biochemistry of the auditory nerve, a great deal of information is available on its anatomy and physiology. This information, in addition to the relative ease of studying the auditory nerve, makes it an excellent system for biochemical analysis. Identified proteins can be analyzed anatomically by immunocytochemistry, and their function may be addressed physiologically and pharmacologically, a critical aspect in characterizing a neuronal protein. Furthermore, events, such as development and degeneration after hair-cell loss, can be charted for the auditory nerve, which allows correlations to be made with the biochemical processes.

The types of proteins chosen for characterization depend, of course, on the cellular structure or function to be studied. For example, we have chosen to study selected, rapidly-transported proteins because we are interested in the synaptic membrane and the known association of this group of proteins with that structure. In general, our knowledge of the biochemistry of the neuron is elementary. Of the several thousand proteins known to be expressed in mammalian neurons, those that have been biochemically characterized represent only a very small percentage. Those molecules involved in events such as development, regeneration and formation of synapses have not been identified, and, consequently, the molecular bases of these events are not understood. Through monoclonal and recombinant DNA studies, it has been shown that there are a large number of brain-specific proteins, and that specific groups of neurons contain specific antigens (Barnstable, 1980; Trisler et al., 1981; Zipser and McKay, 1981; Sutcliffe et al., 1983). Because the auditory nerve originates from two populations of neurons, the transported proteins we study could arise from either one or both of the two cell types. This raises the distinct possibility that these two cell types and their processes can be distinguished by analyzing their specific antigens. These are only a few of the questions that should be addressed concerning the biochemistry of the auditory nerve.

REFERENCES

Barnstable, C. J.: Monoclonal antibodies which recognize different cell types in the rat retina. Nature 286: 231–235, 1980.

Baumann, H., and Doyle, D.: Localization of membrane glycoproteins by in situ neuraminidase treatment of rat hepatoma tissue culture cells and two dimensional gel electrophoretic analysis of the modified proteins. J. Biol. Chem. 254: 2542–2550, 1979.

Bisby, M. A.: Functions of retrograde axonal transport. Fed. Proc. 41: 2307–2311, 1982.

Ernston, S.: Cochlear morphology in a strain of the waltzing guinea pig. Acta Otolaryng. 71: 469–482, 1971.

Grafstein, B.: Axonal transport: The intracellular traffic of the neuron. In Kandel, E. R. (Ed.): Handbook of Physiology: The Nervous System I. Cellular Biology of Neurons. Washington, D. C., American Physiological Society, 1977, pp. 691–717.

Grafstein, B., and Forman, D. S.: Intracellular transport in neurons. Physiol. Rev. 60: 1167–1283, 1980.

Gulley, R. L.; Wenthold, R. J., and Neises, G. R.: Changes in the synapses of spiral ganglion cells in the rostral anteroventral cochlear nucleus of the waltzing guinea pig following hair cell loss. Brain Res. 158: 279–294, 1978.

Koitchev, K.: Guilhaume, A., Cazals, Y., and Aran, J. M.: Spiral ganglion changes after massive aminoglycoside treatment in the guinea pig. Acta Otolaryngol. 94: 931–938, 1982.

Kristensson, K.: Retrograde transport of macromolecules in axons. *Annu. Rev. Pharmacol. Toxicol. 18:* 97–110, 1978.

Laemmli, U. K.: Cleavage of structual proteins during the assembly of the head of bacteriophage T4. *Nature. 227:* 680–685, 1970.

Lasek, R. J., and Hoffman, P. N.: The neuronal cytoskeleton, axonal transport and axonal growth. In Goldman, R.; Pollard, T., and Rosenbaum, J. (Eds.): *Cell Motility.* Vol. 3. Cold Spring Harbor, Cold Spring Laboratory, 1976, pp. 1021–1049.

Schwartz, J. H.: Axonal transport: Components, mechanisms and specificity. *Annu. Rev. Neurosci. 2:* 467–504, 1979.

Skene, J. H. P., and Willard, M.: Changes in axonally transported proteins during axon regeneration in toad retinal ganglion cells. *J. Cell Bio. 89:* 86–95, 1981a.

Skene, J. H. P., and Willard, M.: Axonally transported proteins associated with axon growth in rabbit central and peripheral nervous systems. *J. Cell Biol: 89:* 96–103, 1981b.

Skene, J. H. P., and Willard, M.: Characteristics of growth-associated polypeptides in regenerating toad retinal ganglion cell axons. *J. Neurosci. 1:* 419–426, 1981c.

Spoendlin, H.: Neural connections of the outer haircell system. *Acta Otolaryng. 87:* 381–387, 1979.

Spoendlin, H.: Differentiation of cochlear afferent neurons. *Acta Otolaryng. 91:* 451–456, 1981.

Sutcliffe, J. G.; Milner, R. J., Shinnick, T. M., and Bloom, F. E.: Identifying the protein products of brain-specific genes with antibodies to chemically synthesized peptides. *Cell 33:* 671–682, 1983.

Trisler, G. D.; Schneider, M. D., and Nirenberg, M.: A topographical gradient of molecules in retina can be used to identify neuron position. *Proc. Nat. Acad. Sci. 78:* 2145–2149, 1981.

Tytell, M.; Gulley, R. L., Wenthold, R. J., and Lasek, R. J.: Fast axonal transport in auditory neurons of the guinea pig: A rapidly turned-over glycoprotein. *Proc. Nat. Acad. Sci. 77:* 3042–3046, 1980.

Weiss, P. A., and Hiscoe, H. B.: Experiments on the mechanism of nerve growth. *J. Exp. Zool. 107:* 315–396, 1948.

Wenthold, R. J., and McGarvey, M. L.: Different polypeptides are rapidly transported in auditory and optic neurons. *J. Neurochem. 39:* 27–35, 1982a.

Wenthold, R. J., and McGarvey, M. L.: Changes in rapidly transported proteins in the auditory nerve after hair cell loss. *Brain Res. 253:* 263–269, 1982b.

Wilson, D. L., and Stone, G. C.: Axoplasmic transport of proteins. *Annu. Rev. Biophys. Bioeng. 8:* 27–45, 1979.

Zipser, B., and McKay, R.: Monoclonal antibodies distinguish identifiable neurons in the leech. *Nature. 289:* 549–554, 1981.

Chapter 20

COMPOSITION AND PROPERTIES OF THE MAMMALIAN TECTORIAL MEMBRANE

KAREN P. STEEL

I. Introduction
II. Ultrastructure of the Tectorial Membrane
III. Histochemistry of the Tectorial Membrane
IV. Biochemical Analysis of the Tectorial Membrane
   A. Proteins
   B. Carbohydrates
V. Properties of the Whole Tectorial Membrane
VI. Conclusions
   References

I. INTRODUCTION

Of all the structural elements in the cochlea, the tectorial membrane (TM) might be expected to be the most straightforward to describe biochemically. It is an extracellular matrix consisting of just two main types of fibril. Because it is acellular, its biochemical analysis is not complicated by an enormous number of interdependent components which are inevitably present in whole cells. Samples of TM can be isolated by simple dissection, and a relatively large amount of material is available for analysis (in the mouse, for example, there is approximately 4 μg dry weight TM per cochlea).

The TM might also be expected to be one of the simpler cochlear components to understand at the molecular level, because it is extracellular. It is, to a large extent, surrounded by fluid, which limits the possibilities for direct influence by cellular components. If the molecular structure of the TM were known, then its behavior in response to mechanical or chemical changes in its environment might be predictable. These features make the properties and function of the TM amenable, in principle, to a biochemical approach.

However, the simple appearance of the TM conceals a complex behavior, and there is still no consensus about the composition of the TM. The purpose of this chapter is to review what is known about the biochemistry of the TM and to relate this to some of its physical properties.

II. ULTRASTRUCTURE OF THE TECTORIAL MEMBRANE

The TM is conventionally considered to be composed of fibrils embedded in an "amorphous ground substance," but by using special, prolonged staining, the "amorphous" material can be resolved into a second class of fibril (Kronester-Frei, 1978b). These two classes of fibrils were named by Kronester-Frei as type A and type B, respectively. Type A fibrils are straight and unbranched, and tend to run in bundles throughout the TM. They are found particularly in the basal layer of the TM (Fig. 1A). Type B fibrils (the "amorphous" material) are more irregular in their alignment, and may be branched. They are found throughout the TM and occur in two different states: loosely packed (highly hydrated) and tightly packed (weakly hydrated). Weakly-hydrated type B fibrils form the cover net and Hensen's stripe (Fig. 1B).

III. HISTOCHEMISTRY OF THE TECTORIAL MEMBRANE

A consistent observation regarding the histochemical reactions of the TM is that the TM gives a strongly-positive reaction to the periodic acid-Schiff (PAS) staining procedure (e.g., Iurato, 1960; Igarashi and Alford, 1969; Schätzle, 1971; Ross, 1974). This staining technique demonstrates adjacent hydroxyl (CHOH–CHOH) groups or adjacent hydroxylamino groups (CHOH--CHNH$_2$). The main classes of bio-polymeric substances which contain these groups are glycogen and glycoproteins (Rambourg, 1971). Because the TM gives a positive PAS reaction after amylase treatment, which should remove glycogen from the treated TM sections (Plotz and Perlman, 1955; Wislocki and Ladman, 1955; Ross, 1974), the substance responsible for the positive reaction of the TM may be a glycoprotein.

Periodic acid-reactive groups also occur in other types of biopolymeric materials, such as mucopolysaccharides and proteins containing large amounts of hydroxylysine, or in proteins with serine or threonine as terminal amino acids. However, mucopolysaccharides and proteins yield a positive PAS reaction only if the periodic-acid treatment is unduly prolonged (Rambourg, 1971); therefore, these macromolecules are unlikely to account for the strong TM staining reaction. It would, nonetheless, be interesting to know whether the TM contains large amounts of the appropriate PAS-positive amino acids (hydroxylysine, serine, and threonine).

The PAS-positive substance is an insoluble component of the TM, because the TM still gives a positive reaction after thorough washing *in situ* with distilled water, saline, or artificial endolymph. Several authors have reported

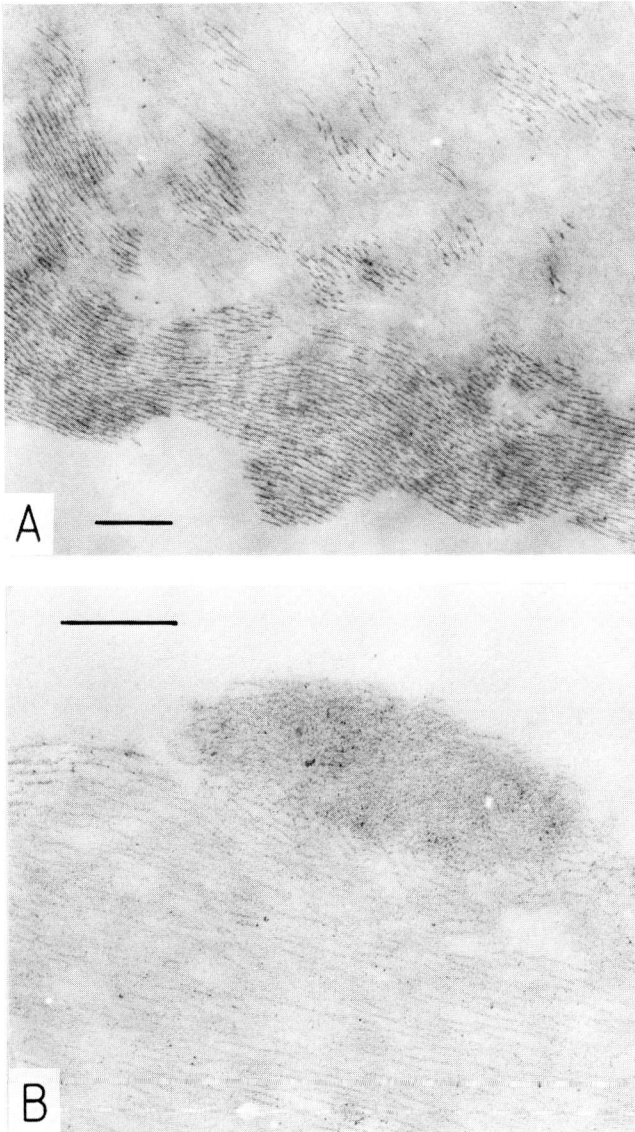

Figure 20-1. A. Transmission electron micrograph of the basal layer of the TM, showing the dense array of type A fibrils at the lower border, with more type A fibrils running in bundles amongst the type B fibrils in the upper half of the picture. Scale bar = 0.5 μm.
B. Upper surface of the TM, showing highly hydrated type B fibrils in the lower half of the picture and weakly hydrated type B fibrils forming the cover net. Scale bar = 0.5 μm.

that it is the fibrils of the TM (presumably type A fibrils) which appear to be periodate-reactive, while the "amorphous" interfibrillar material is poorly stained (e.g., Kuttner, 1977).

Other histochemical staining methods have been applied to the TM, with mixed results—so mixed that it is not possible to draw any reasonable conclusions. For example, while some authors report a metachromatic reaction with toluidine blue, others report no signs of metachromasia (see Steel, 1983b). Immunofluorescence techniques have also proved disappointing as a tool for studying the composition of the TM, because the TM apparently binds fluorescently-labeled anti-IgG antibody in a non-specific fashion (e. g., Fex et al., 1982).

Iurato (1967) studied the sensitivity of isolated TM's to enzymes and other agents, and found that TM's could be dissolved in pepsin at acid pH or trypsin at pH 7.4, suggesting that protein was an important component. On the other hand, neither hyaluronidase nor liposolvents had apparent effects (Iurato, 1967), minimizing, perhaps, the importance of carbohydrates or lipids in maintaining structural integrity of the TM.

IV. BIOCHEMICAL ANALYSIS OF THE TECTORIAL MEMBRANE

A. PROTEINS

Biochemical studies of the TM have shown that carbohydrates are only a minor component, while proteins form the major part of the TM. Iurato (1960, 1967) found the total nitrogen in the TM is 15%, indicating that the TM is largely protein in composition.

The only data available on amino-acid composition are those of Naftalin et al. (1964) and Iurato (1960), obtained over 20 years ago by the use of paper chromatography. Both reports concluded that the TM was not composed of collagen, but may consist of a protein of the keratin, epidermin, myosin, or fibrinogen group. The reason for this conclusion is illustrated in Table 1. Collagen is characterized by the presence of hydroxyproline, a large amount of glycine, and the absence of cysteine. The tectorial membrane, on the other hand, contains no hydroxyproline, some cysteine, and only a moderate amount of glycine.

Amino-acid analysis of the TM proteins should be repeated, using the more sensitive techniques which are available today (e.g., as described in Chapter 4). The type of protein present in the TM fibrils may have important functional consequences for the way in which the TM responds to basilar membrane displacement. For example, fibrils of collagen are rigid, strong, and not very extensible, and are often found in locations where force needs to be efficiently transmitted, such as in tendons (e.g., Dickerson and Geis, 1969). If the TM fibrils were collagenous in nature, then the whole TM could be rigidly pushed upwards by the stereocilia during sound stimulation. If the TM were relatively flexible, then the stereocilia might instead deform the

TABLE 20-I
COMPARISON OF AMINO ACID COMPOSITIONS

Amino acids	Tectorial membrane	Keratin	Collagen	Fibroin
GLY	13.7	8.1	33.0	44.6
ALA	11.7	5.0	10.7	29.4
VAL	7.5	5.1	2.3	2.2
LEU + ILE	13.9	9.7	3.3	1.2
ASP + ASN	11.2	6.0	4.5	1.3
GLU + GLN	12.5	12.1	7.1	1.0
THR	5.5	6.5	2.0	0.9
LYS	8.4	2.3	2.7	0.3
TYR	2.9	4.2	0.4	5.2
PHE	4.6	2.5	1.2	0.5
SER	7.0	10.2	4.3	12.2
CYS	1.0	11.2	0	0
HYPRO	0	0	9.4	0
PRO	—	7.5	12.2	0.3
ARG	—	7.2	5.0	0.5

Principal amino acids of the tectorial membrane and of three typical fibrous proteins, expressed as percent of total quantified amino-acid residues. Typical fibrous proteins were merino wool keratin, rat tail tendon collagen, and *Bombyx mori* silk fibroin (Dickerson and Geis, 1969). TM data were calculated from Iurato (1960). Naftalin *et al.* (1964) described a similar distribution of amino acids in their TM samples, except that cysteine was reported to be a major component.

lower layer of the TM, and (to speculate further) if there were a sufficiently strong, elastic restoring force in the TM fibrils, then the TM might spring back to its usual position, pushing the sterocilia back again. Keratin is a fairly elastic protein. A simplified view of the structure of the α-keratin of wool consists of an α-helix with weak hydrogen bonds between successive turns of the helix, making the structure relatively extensible and flexible. The α-helices are cross-linked between chains with strong disulfide bonds, which resist stretching and provide a restorative force when tension is released. The strength of this restoring force will depend on the number of disulfide bonds, which in turn depends on the number of cysteine residues in the protein chain.

The α-keratins fall into two categories, "soft" and "hard", depending on their sulfur (cysteine) content: low-sulfur keratins, such as those in skin, are more flexible and extensible than are high-sulfur keratins, as found in horn or hooves (Dickerson and Geis, 1969). Therefore, if the TM fibrils are a keratin-like protein, it would be important to determine how much cysteine this protein contains. Iurato (1960), in his analysis of the TM amino acids, reported only a small amount of cysteine, but Naftalin *et al.* (1964) estimated that cysteine was relatively abundant in TM samples.

Another approach to identifying the protein(s) present in the TM is to determine the molecular weights of protein sub-units by SDS (sodium dodecyl sulfate) polyacrylamide-gel electrophoresis. Isolated TM's are dissociated into protein subunits with the detergent SDS, and the sample is then subjected to electrophoresis using standard procedures (Steel, 1980). Subunits migrate through the gel at different rates, depending on their molecular weights, and form bands in the gel which may be subsequently stained with the protein stain, Coomassie blue. The molecular weight of the subunits in each band may be estimated by comparing their migration distances with distances traveled by known standard proteins in adjacent gels (Weber and Osborn, 1969). Figure 2 illustrates the banding pattern produced by TM samples, with estimated molecular weights marked on the left. The most intensely-stained bands correspond to molecular weights of 145,000, 155,000, and 165,000. Up to 22 individual bands can be resolved in some gels.

The TM subunit molecular weights may be compared with molecular weights of other fibrous proteins which have been described previously. If the molecular weight of a particular TM subunit corresponds to that of a known protein, the subunit band can be extracted and analysed further, to establish a possible identity between it and a known fibrous protein. However, none of the major TM bands corresponds exactly with the predicted band positions of several of the more common fibrous proteins found in mammalian tissue (marked on the right of Fig. 2). The TM proteins, therefore, do not seem to be identical with any of these well-described fibrous proteins, although analysis of subunit molecular weight does not rule out the possibility that there may be qualitative similarities with any of these common proteins.

B. CARBOHYDRATES

The observation of a strongly-positive PAS reaction suggests that carbohydrates might be a major component of the TM, as discussed previously in Section III. However, recent biochemical analysis indicates that there is in fact probably little carbohydrate present. There is probably not sufficient carbohydrate, in terms of percent dry weight, for either of the two main fibril types to be composed of polysaccharide. Saito and Daly (1970) reported that acid mucopolysaccharides accounted for only 0.1% of the dry weight of the TM. Nonetheless, carbohydrate may be functionally relevant as a minor component. For example, the carbohydrate side-chains of glycoproteins may modify the nature of the basic protein chain by altering its conformation, overall charge density, or degree of hydration.

The distribution of periodate-reactive moities within the TM suggests that the type A fibrils are the most likely site of any TM glycoproteins. Because the basal layer of the TM is largely composed of type A fibrils, and this is the surface which is closest to the stereocilia of the hair cells, it would be

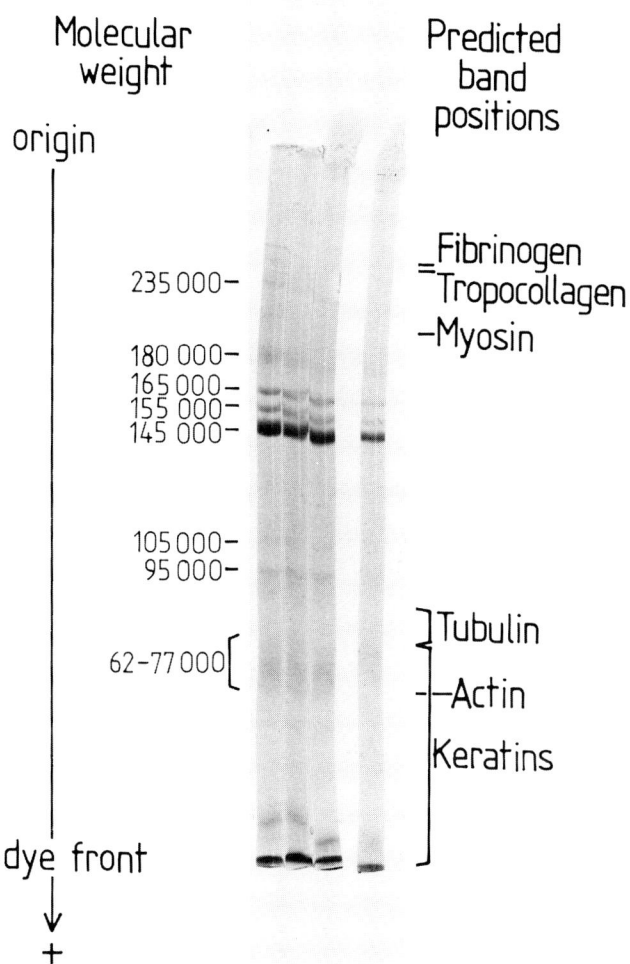

Figure 20-2. SDS polyacrylamide-gel electrophoresis of dissociated TM subunits. Bands were stained with Coomassie blue. The molecular weights of the main bands are marked on the left, and were calculated from standard proteins subjected to electrophoresis at the same time. On the right, the predicted positions of bands of several common proteins are marked.

particularly interesting to discover which (if any) of the 22 bands seen in polyacrylamide gels represents glycoprotein, and might, therefore, be a component of the type A fibrils. Two approaches have been used in attempts to demonstrate a glycoprotein band.

The first approach is to directly stain the polyacrylamide gels after electrophoresis of the TM subunits, using either the PAS stain (Fairbanks et al., 1971) or dansyl hydrazine, a fluorescent stain reported to be a sensitive indicator of periodic acid-reactive groups (Eckhardt et al., 1976). Neither stain succeeded in staining any of the bands, which were subsequently visualized with Coomassie blue (Steel, 1980). Known glycoproteins were used as quantitative controls for the stains, and it was thus calculated that 0.8 µg or less of suitable carbohydrate could be readily detected. (Approximately 5–10 µg of TM material was used in each test gel.)

The second approach was based on the observation by Segrest et al. (1971) that glycoproteins migrate at rates which are not proportional to their molecular weights during SDS electrophoresis and that the apparent molecular weight varies according to the concentration of acrylamide in the gel. Normal proteins, unassociated with carbohydrate, bind SDS in amounts directly proportional to the protein molecular weight, and since the charge on the resulting SDS-protein complex is dominated by the charge of the SDS, each protein subunit, irrespective of its molecular weight, will have a constant charge/mass ratio (Reynolds and Tanford, 1970). Therefore, during SDS polyacrylamide-gel electrophoresis, the different rates of migration of the different proteins will be due solely to the different sizes of the protein-SDS complexes, with large complexes being held back by the sieving effect of the gel to a greater extent than small complexes.

In the case of glycoproteins, SDS is presumed to bind only to the protein part of the molecule, leaving the carbohydrate moiety free (Pitt-Rivers and Ambesi Impiombato, 1968). This produces a charge/mass ratio which is lower than that of standard proteins, thus introducing an additional variable. The lower charge-to-mass ratio appears to retard the migration of the SDS-glycoprotein complex during electrophoresis, and this effect would be most pronounced at low concentrations of gel, where the sieving effect is least. Thus, when the mobility of a glycoprotein is compared with that of standard proteins used in calibration, the apparent molecular weight of the glycoprotein calculated from its mobility at low gel concentrations will be much higher than its apparent molecular weight calculated using higher gel concentrations (Segrest et al., 1971).

In order to test whether any of the TM bands behave in an anomalous way in gels of different concentrations, four pairs of gels were prepared, with acrylamide concentrations of 5%, 7%, 10%, and 12%. An SDS-denatured TM sample was applied to one gel of each pair, and the other gel was used for a mixture of standard proteins of known molecular weights. The resulting gels, stained with Coomassie blue, are shown in Fig. 3A. The mobilities of the standard proteins were used to produce calibration curves for each gel concentration, and these curves were used to estimate the apparent molecular weights of six of the main TM bands. These estimates are plotted against gel acrylamide concentration in Fig. 3B. The observed molecular weights do not appear to be larger at low gel concentration, indicating that these particular bands either represent proteins alone, or contain quantities of carbohydrates so small that there is no detectable change in charge/mass ratio in complexes with SDS.

Although neither of the two approaches described above has succeeded in identifying any of the TM bands as glycoprotein, some very recent biochemical analyses (in collaboration with Dr. Z. Ellis and Prof. F. Hemming of the University of Nottingham Medical School) have confirmed that carbohydrate is present, but only in small quantities. Samples of TM's were hydrolysed to reduce the polysaccharides present to monosaccharides, which were then trimethylsilylated and subjected to gas-liquid chromatography (GLC) using standard techniques (Bhatti et al., 1970). Preliminary results are presented in Table II. The total amount of carbohydrate detected was approximately 2.5% of the dry weight of the TM samples. Glucose, galactose, mannose, and N-acetylglucosamine were reliably detected. The presence of mannose is a strong indicator that glycoprotein is present in the TM, since it is rarely found in polysaccharides and is present only in very small quantities in proteoglycans.

V. PROPERTIES OF THE WHOLE TECTORIAL MEMBRANE

An improved understanding of the biochemistry of the TM may help to explain several puzzling aspects of the behavior of the whole TM. The TM has no conventional cell membrane, and, therefore, has traditionally been considered to be electrically and ionically transparent. It certainly has close contact with the fluids bathing it, because tracer substances introduced into scala media are later found within the TM matrix (e.g., Rudert, 1969; Ilberg, 1968). However, there is some evidence that the TM is able to act as a separate phase, maintaining its own particular ionic and electrical character. Two lines of evidence suggest this conclusion.

The first line of evidence comes from studies of the elemental profile of the TM *in situ*, using x-ray microprobe analysis. Both Anniko and Wroblewski (1980) and Ross (1974) reported that there was a gradient in the distribution of sodium and potassium within the TM, with relatively higher levels of potassium near the upper border and relatively higher levels of sodium near the lower boundary of the TM. Lim (1977) concluded that the TM did not appear to be saturated with endolymph, but had a lower potassium content than endolymph. Anniko (1981) reported that the developing TM showed an elemental content different from that of the surrounding developing endolymph.

The second indication that the TM behaves as a separate phase comes from the finding that the isolated TM *in vitro* maintains a potential difference relative to the bathing fluid (Steel, 1983a). When pieces of TM are allowed to equilibrate in various bathing fluids and are then impaled with glass microelectrodes, a stable, negative potential may be recorded, which is

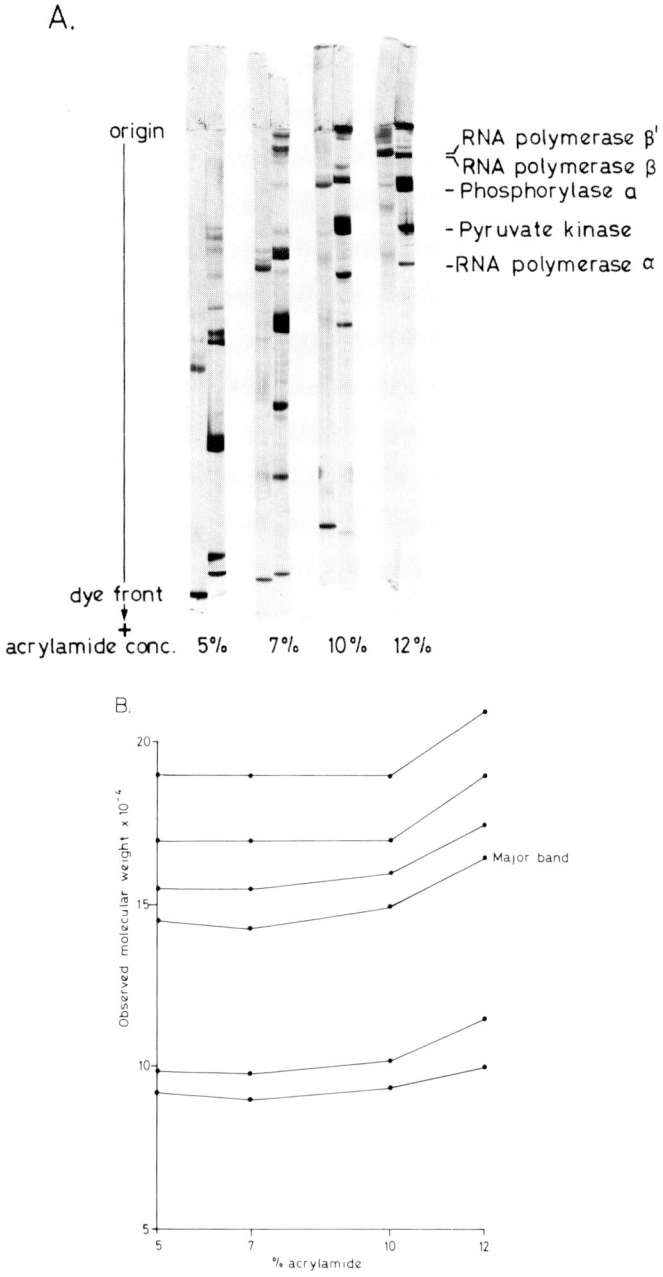

Figure 20-3. A. Four pairs of gels of differing acrylamide concentrations. On the left of each pair is the gel containing TM subunits, and on the right is the gel with the standard proteins used for calibration. Standard proteins used are marked on the right side of the figure. B. Observed molecular weights of six readily-identifiable TM bands, calculated by comparing mobilities with those of the standard proteins, plotted as a function of acrylamide concentration in the gel. The slight increases in observed molecular weights in the 12% gels are probably due to the difficulty in making accurate measurements of the small distances traveled by the bands, complicated by swelling of the tops of the running gels.

TABLE 20-II
CARBOHYDRATE CONTENT OF THE TECTORIAL MEMBRANE

Glucose	1%
Galactose	0.4%
Mannose	0.35%
N-Acetylglucosamine	0.7–1.1%

The above values are preliminary estimates obtained from gas-liquid chromatographic analysis. Total carbohydrate is approximately 2.5% dry weight.

dependent upon both the pH and the ionic strength of the bathing medium (Fig. 4). The most likely explanation for this observation is that a Donnan equilibrium is set up, as has been reported to occur in several other acellular biological matrices (see Steel, 1983a for references). The hypothesis is that when the macromolecules forming the TM are at a pH other than their isoelectric point, they will carry net charge and these charges will be fixed in the TM matrix, even without the presence of a conventional bounding membrane. The TM will thus form a distinct phase containing fixed charges within a bathing fluid containing mobile ions. Because the isoelectric point of the TM has been reported to be approximately 4.3 (Kronester-Frei, 1979), the TM matrix will be negatively charged at pH 7.43, which is the pH of endolymph (Bosher, 1979). The electrical gradient will tend to pull cations into the matrix, and the resulting concentration gradient will tend to pull them out, and a Donnan equilibrium will thus be established, with a net negative charge inside the TM. The reduction in the potential recorded in the TM with decreasing pH is exactly what would be expected to occur as the macromolecules approach their isoelectric points and the fixed charges are consequently reduced. The dependence of the potential on ionic strength is also what would be predicted if a Donnan equilibrium were established, since higher concentrations of mobile ions between the charged TM fibrils will tend to screen the effect of these fixed charges (Steel, 1983a).

One consequence of the Donnan equilibrium is that the concentration of ions within the TM will be higher than in its bathing fluid, and the Nernst equation can be used to calculate predicted internal ion concentrations (Steel, 1983a). Whether higher ion concentrations or a potential difference exist *in vivo* in the TM remains to be determined, but these *in vitro* experiments do indicate that the TM is capable of acting as a separate phase.

The fixed charges on the TM fibrils, indicated by the finding of a negative potential in the matrix at neutral pH, will be related to the composition of the TM. A high proportion of acidic amino acids, such as aspartate or glutamate, could explain the balance of the fixed charges, or alternatively, sulfate groups in the carbohydrate component might add to the fixed charge.

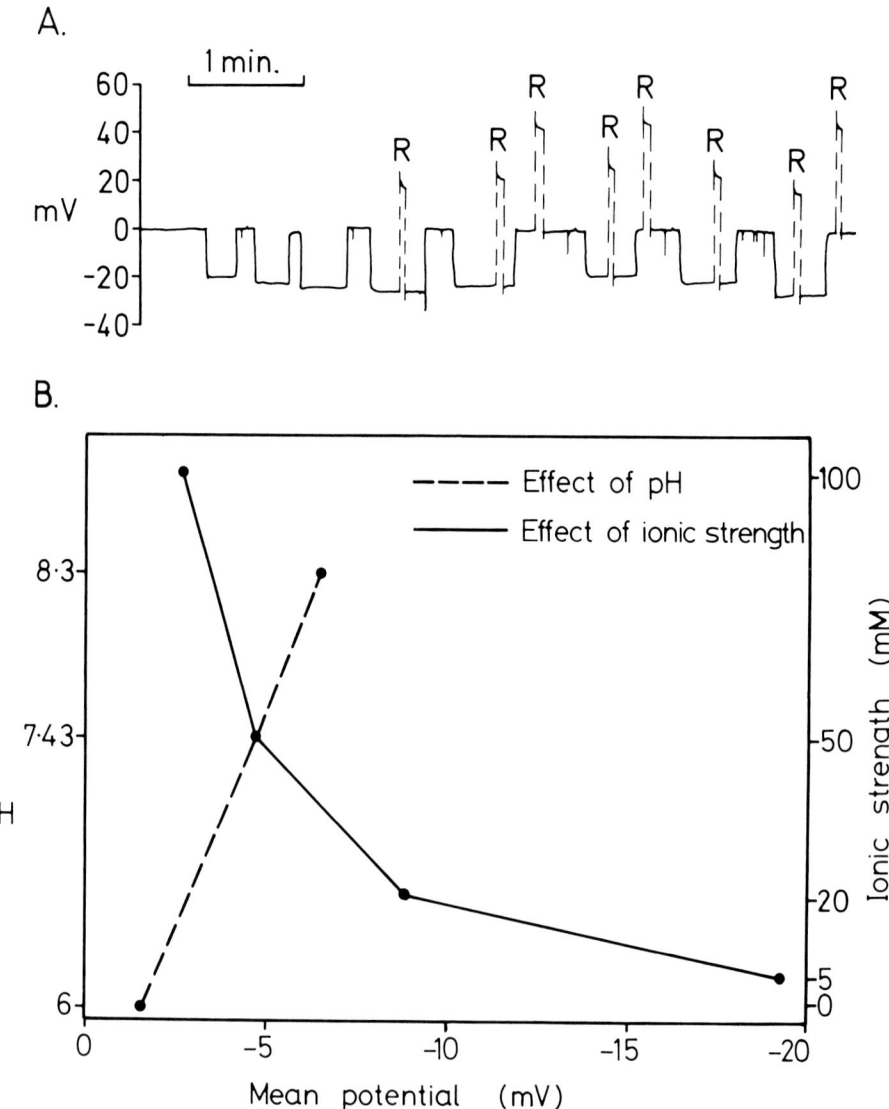

Figure 20-4. A. Trace of the potential recorded at eight different locations along a piece of TM equilibrated in 5 mM KCl pH 7.43. Note that the recording returns to zero after each impalement. R = electrode resistance test; 1 mV = 1 MΩ. B. Effect of pH and ionic strength of the bathing medium (phosphate-buffered KCl) on the mean potential recorded within the TM. For pH measurements, ionic strength was held constant at 50 mM, and for ionic strength recordings, the pH was 7.43.

The polyelectrolyte nature of the TM matrix may well be important in determining the physical characteristics of the matrix, particularly its resilience. If the TM fibrils are highly charged, they will be resistant to being pushed closer together, since like charges repel. The internal osmotic pressure, which would be expected within the TM as a result of the establishment of a Donnan equilibrium and a higher ionic concentration inside the matrix (Steel, 1983a), and which would presumably normally be counteracted by crosslinks within the TM, may also influence the overall elasticity of the structure.

VI. CONCLUSIONS

The features of the TM which have been described here may have a number of functional implications. The observations suggest lines of inquiry which might be usefully pursued, as well as provide evidence against certain models and assumptions about the TM. One model that can be ruled out is the model based on displacement potentials, which would require the presence of quantities of mucopolysaccharides in the TM (e.g., Jensen et al., 1954; Christiansen, 1961; Barrett, 1975). The assumption that the TM is electrically and ionically transparent *in vivo* should also be regarded with some suspicion.

One consequence of the reports that the TM may contain ionic concentrations different from those of endolymph is that the TM might be useful as a reservoir for important ions (such as potassium or calcium), or might help to buffer ion concentrations in the small amount of fluid in the sub-tectorial hair-cell region. The TM may, in this way, modify the microenvironment at the top of the hair cells, irrespective of whether this area is sealed. Modification of the hair-cell environment is a function which has also been ascribed to the lateral-line cupulae of the *Xenopus* frog (Russell and Sellick, 1976; McGlone et al., 1979).

Kronester-Frei (1978a, 1979) has reported that the TM shrinks and swells in different ways, depending on the particular ion to which it is exposed. For example, it shrinks irreversibly in the presence of sodium, and shrinks or swells reversibly when calcium is added or removed from the bathing fluid. This differential response to different ions, together with the finding that a Donnan equilibrium is set up by the TM matrix, suggest that the TM may be acting as an ion exchanger (Steel, 1983a). If this were true, the TM might be able to exert a more specific influence on the ionic content of the fluid bathing the hair-cell stereocilia than has previously been suspected.

Finally, if the TM does shrink and swell reversibly when calcium is added or removed from its environment, it would be interesting to ask whether the reverse might not occur: if sound stimulation results in some deformation of the lower surface of the TM, then is calcium released from binding sites on the TM matrix in response? The release of calcium into the sub-tectorial fluid may then be important for hair-cell function.

ACKNOWLEDGMENTS

I am grateful to Dr. Z. Ellis, Biochemistry Department, University of Nottingham Medical School, for her help in obtaining the GLC carbohydrate measurements, and to Prof. F. W. Hemming, Prof. M. P. Haggard, and Dr. G. R. Bock for their comments on the manuscript. Part of the work described was supported by the Deutsche Forschungsgemeinschaft, under SFB 50.

REFERENCES

Anniko, M.: Elemental analysis of the early postnatal tectorial membrane. *Hearing Res.* 4: 1-9, 1981.

Anniko, M., and Wroblewski, R.: Elemental composition of the mature inner ear. *Acta Otolaryng.* 90: 425-430, 1980.

Barrett, T. W.: Hyaluronic acid salt—A mechanoelectrical transducer. *Biochim. Biophys. Acta* 385: 157-161, 1975.

Bhatti, T.; Chambers, R. E., and Clamp, J. R.: The gas chromatographic properties of biologically important N-acetylglucosamine derivatives, monosaccharides, disaccharides, trisaccharides, tetrasaccharides and pentasaccharides. *Biochim. Biophys. Acta* 222: 339-347, 1970.

Bosher, S. K.: The nature of the negative endocochlear potentials produced by anoxia and ethacrynic acid in the rat and guinea pig. *J. Physiol. (Lond.)* 293: 329-345, 1979.

Christiansen, J. A.: An attempt to explain the microphonic effect of the inner ear by means of displacement potentials. *Acta Otolaryng. Suppl.* 163: 76-79, 1961.

Dickerson, R. E., and Geis, I.: *The Structure and Action of Proteins.* New York, Harper and Row, 1969.

Eckhardt, A. E.; Hayes, C. E., and Goldstein, I. J.: A sensitive fluorescent method for the detection of glycoproteins in polyacrylamide gels. *Anal. Biochem.* 73: 192-197, 1976.

Fairbanks, G.; Steck, T. L., and Wallach, D. F. H.: Electrophoretic analysis of the major polypeptides of the human erythrocyte membrane. *Biochemistry* 10: 2606-2617, 1971.

Fex, J.; Altschuler, R. A., Wenthold, R. J., and Parakkal, M. H.: Aspartate aminotransferase immunoreactivity in the cochlea of the guinea pig. *Hearing Res.* 7: 149-160, 1982.

Igarashi, M., and Alford, B. R.: Cupula, cupular zone of otolithic membrane, and tectorial membrane in the squirrel monkey. *Acta Otolaryng.* 68: 420-426, 1969.

Ilberg, C. von: Elektronenmikroskopische Untersuchungen über Diffusion und Resorption von Thoriumdioxyd an der Meerschweinchenschnecke. 3. Limbus spiralis. *Arch. Klin. Exp. Ohren-Nasen-Kehlkopfheilkd.* 192: 163-175, 1968.

Iurato, S.: *Submicroscopic Structure of the Inner Ear.* London, Pergamon Press, 1967.

Iurato, S.: Submicroscopic structure of the membranous labyrinth. I. The tectorial membrane. *Z. Zellforsch.* 52: 105-128, 1960.

Jensen, C. E.; Koefoed, J., and Vilstrup, T.: Flow potentials in hyaluronate solutions. *Nature (Lond.)* 74: 1101-1102, 1954.

Kronester-Frei, A.: Sodium dependent shrinking properties of the tectorial membrane. *Scann. Electr. Micros.* II: 943-948, 1978a.

Kronester-Frei, A.: The effect of changes in endolymphatic ion concentrations on the tectorial membrane. *Hearing Res.* 1: 81-94, 1979.

Kronester-Frei, A.: Ultrastructure of the different zones of the tectorial membrane. *Cell Tiss. Res.* 193: 11-23, 1978b.

Kuttner, K.: Ultrahistochemical studies to demonstrate mucopolysaccharides in the organ of Corti of the guinea pig. In Portmann, M., and Aran, J.-M. (Eds.): Inner Ear Biology. Paris, INSERM, 1977, pp. 201-208.

Lim, D. J.: Current review of SEM techniques for inner ear sensory organs. *Scann. Electr. Micros. II:* 401-408, 1977.

McGlone, F. P., Russell, I. J., and Sand, O.: Measurement of calcium ion concentrations in the lateral line cupulae of Xenopus laevis. *J. Exp. Biol. 83:* 123-130, 1979.

Naftalin, L.; Spencer Harrison, M., and Stephens, A.: The character of the tectorial membrane. *J. Laryngol. Otol. 78:* 1061-1078, 1964.

Pitt-Rivers, R., and Ambesi Impiombato, F. S.: The binding of sodium dodecyl sulphate to various proteins. *Biochem. J. 109:* 825-830, 1968.

Plotz, E., and Perlman, J. B.: A histochemical study of the cochlea. *Laryngoscope 65:* 291-312, 1955.

Rambourg, A.: Morphological and histochemical aspects of glycoproteins at the surface of animal cells. *Int. Rev. Cytol. 31:* 57-114, 1971.

Reynolds, J. A., and Tanford, C.: The gross conformation of protein-sodium dodecyl sulfate complexes. *J. Biol. Chem. 245:* 5161-5165, 1970.

Ross, M. D.: The tectorial membrane of the rat. *Am. J. Anat. 139:* 449-482, 1974.

Rudert, H.: Experimentelle Untersuchungen zur Resorption der Endolymphe im Innenohr des Meerschweinchens. 3. Elektronenmikroskopische Untersuchungen zur Feinstruktur des Saccus endolymphaticus sowie zur Speicherung von Ferritin in häutigen Labyrinth. *Arch. Klin. Exp. Ohren-Nasen-Kehlkopfheilkd. 193:* 201-235, 1969.

Russell, I. J., and Sellick, P. M.: Measurement of potassium and chloride ion concentrations in the cupulae of the lateral lines of Xenopus laevis. *J. Physiol. (Lond.) 257:* 245-255, 1976.

Saito, H., and Daly, J. F.: Quantitative analysis of acid mucopolysaccharides in the normal guinea pig cochlea. *Acta Otolaryng. 69:* 333-340, 1970.

Schätzle, W.: *Histochemie des Innenohres.* Munich, Urban and Schwarzenberg, 1971.

Segrest, J. P.; Jackson, R. L., Andrews, E. P., and Marchesi, V. T.: Human erythrocyte membrane glycoprotein: a re-evaluation of the molecular weight as determined by SDS polyacrylamide gel electrophoresis. *Biochem. Biophys. Res. Commun. 44:* 390-395, 1971.

Steel, K. P.: Donnan equilibrium in the tectorial membrane. *Hearing Res. 12:* 265-272, 1983a.

Steel, K. P.: The proteins of normal and abnormal tectorial membranes. *Acta Otolaryng. 89:* 27-32, 1980.

Steel, K. P.: The tectorial membrane of mammals (Review). *Hearing Res. 9:* 327-359, 1983b.

Weber, K., and Osborn, M.: The reliability of molecular weight determinations by dodecyl sulfate-polyacrylamide gel electrophoresis. *J. Biol. Chem. 244:* 4406-4412, 1969.

Wislocki, G. B., and Ladman, A. J.: Selective and histochemical staining of the otolithic membranes, cupulae and tectorial membrane of the inner ear. *J. Anat. 89:* 3-12, 1955.

Chapter 21

COLLAGENASE AND BONE RESORPTION IN CHRONIC OTITIS MEDIA

Cheng-Chun Huang, Hiroshi Moriyama, and Maxwell Abramson

I. Introduction
II. Purification of Collagenase
III. Preparation of the Anti-Collagenase Sera
IV. Tissue Sources for Immunocytochemical Studies
V. Immunocytochemical Stainings
   A. Human Middle-Ear Cholesteatoma
   B. Animal Models
      1. Chronic Otitis Media in Rats
      2. Keratin-Induced Chronic Inflammation
      3. Effect of Pressure on Chronic Inflammation
VI. Collagenase and Prostaglandin Production from Cell Cultures
VII. Conclusions
   References

I. INTRODUCTION

Otitis media is defined as an inflammation or infection of the middle-ear cavity. Bone destruction is a common feature of chronic otitis media, with or without cholesteatoma.[1] Hearing loss, associated with bone destruction, is a frequent complication of this ear disease. The mechanism of bone resorption in chronic otitis media is not well understood. Walsh et al. (1955) postulated that bone-dissolving enzymes were involved in bone resorption. Because collagen constitutes more than 90% of bone protein, collagenase appears to be a major factor in the degradation of organic matrix of bone. Abramson (1969) demonstrated high collagenase activity in cultures of human chronic otitis media, especially those with cholesteatoma (Table I).

Demineralization seems to be the first step in bone resorption, because hydroxyapatite, the calcium phosphate-hydroxide crystal of bone, normally protects collagen from denaturation and enzymatic degradation (Glimcher et al., 1957). Thomsen et al. (1977) proposed that bone resorption was carried

TABLE 21-I
COLLAGENASE ACTIVITY OF CHOLESTEATOMA EPITHELIUM MEASURED BY RELEASE
OF ^{14}C-PEPTIDES FROM ^{14}C-COLLAGEN

	Culture	Collagenase Activity (counts/min)
Experiment 49		
	a	719
	b	2,658
	c	866
	d	2,615
Experiment 37		
	a	1,750
	b	769
	c	2,826
	d	2,532

Each culture contained 20 mg of tissue. 0.1 mg of ^{14}C-collagen represents 20,000 counts/min; the counting efficiency was the same for all samples. (From Abramson, 1969.)

out by the activity of lysosomal enzymes, such as acid phosphatase, and through the local accumulation of metabolic acids, such as lactic and citric acids. Parathyroid hormone, osteoclast-activating factor from stimulated lymphocytes (Raisz, 1965), endotoxin from gram-negative microorganisms (Hausmann et al., 1972), and prostaglandins (Robinson and McGuire, 1975) have all been shown to increase bone decalcification in vitro.

Clinical evidence (Tos, 1979; Sade et al., 1982; Thomsen et al., 1982) and studies with animal models (Abramson et al., 1975) have shown that chronic otitis media with cholesteatoma causes greater bone resorption than chronic otitis media without cholesteatoma. This suggests that keratinized epithelium in the cholesteatoma sac, or products of the epithelium, play a role in enhancing bone destruction. The keratinized epithelium may enhance bone resorption through pressure due to accumulation of keratin debris in the middle-ear cholesteatoma sac (Fig. 1) and also through an inflammatory stimulus which induces the formation of destructive granulation tissue. The granulation tissue adjacent to bone seems to be responsible for bone resorption in chronic otitis media, either with or without cholesteatoma.

In this chapter, we will discuss the preparation of anti-collagenase serum and the immunocytochemical localization of collagenase in tissue from human chronic otitis media and from relevant animal models. We will also discuss the possible relation between bone resorption and inflammatory granulation tissue and pressure caused by epithelial debris. From these studies, we hope

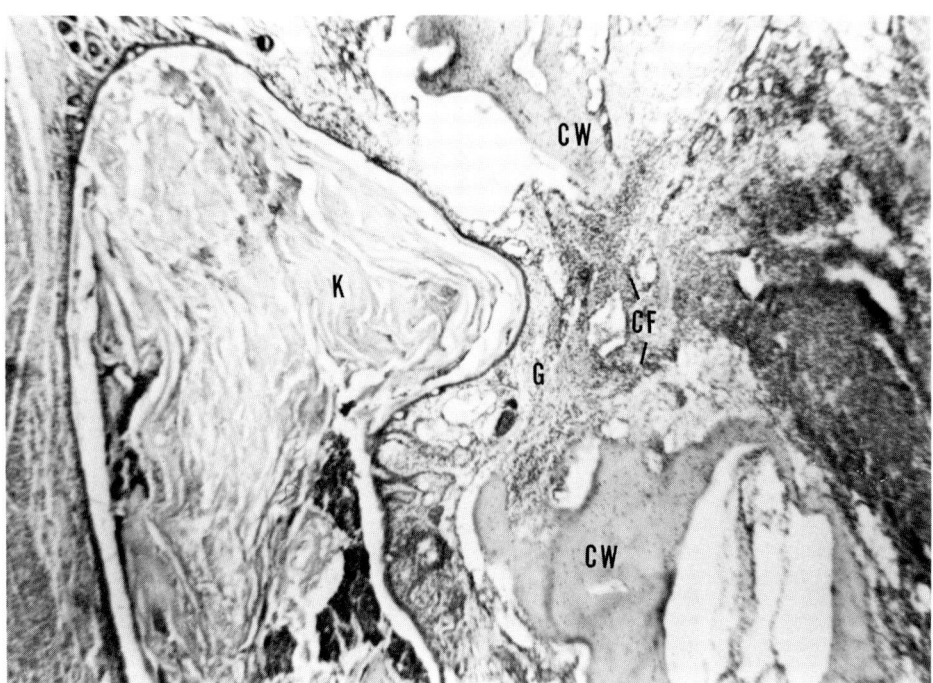

Figure 21-1. Experimental cholesteatoma in a guinea pig. A large epidermal cyst in the middle ear is shown, which contains keratin (K). Granulation tissue (G) is found adjacent to the cochlear wall (CW), and extensive bone resorption (cochlear fistula, CF) can be seen. Hematoxylin-eosin stain. × 45.

to add to our understanding of the mechanisms of bone resorption in chronic otitis media.

II. PURIFICATION OF COLLAGENASE

Collagenase produced from tissue cultures of human middle-ear cholesteatoma (Abramson, 1969) has properties which are similar to those of collagenases obtained from healing wounds (Grillo and Gross, 1967), rheumatoid synovia (Evanson et al., 1967), and gingiva of periodontitis (Geiger and Harper, 1980). Collagenases make a single cleavage through the tropocollagen molecule at a locus three-fourths of the distance from the amino terminal to the carboxyl terminal end (Fig. 2). In vivo, the degraded collagen products are further broken down into small peptides and amino acids. The collagenase activity is inhibited by serum, cysteine, and EDTA.

Normal skin and tumor tissues, either from humans or from animals, are sources providing a large quantity of crude collagenase for purification. In the present studies, tissues were minced and then cultured in serum-free,

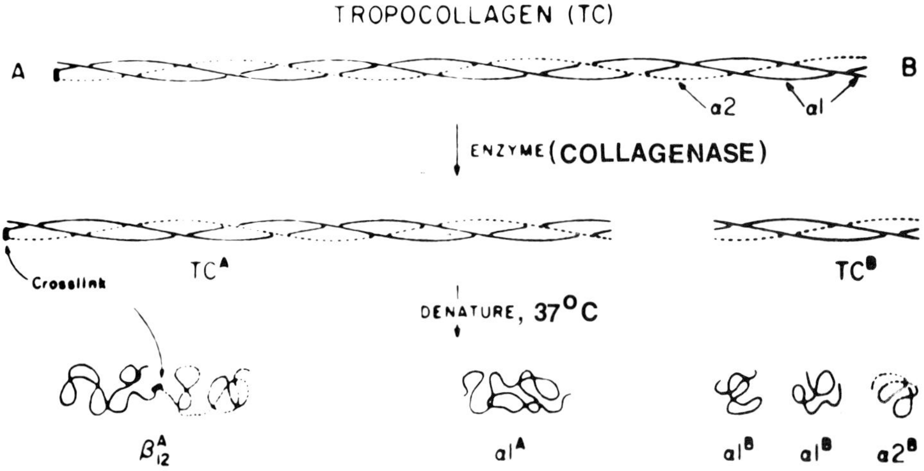

TROPOCOLLAGEN (TC)

Figure 21-2. Diagram of the triple-helical structure of the tropocollagen (TC) molecule and its cleavage into 75% TC^A and 25% TC^B fragments by collagenase. At 37°C, the triple-helical fragments are denatured to randomly-coiled fragments.

Dulbecco's modified Eagle medium (MEM) for eight days in an atmosphere of 95% oxygen and 5% carbon dioxide. The medium was changed daily and pooled. The crude collagenase was passed through a column of glass beads (CPG-250, Pierce Chemical Company, Rockford, IL). The collagenase activity in the eluant was assayed by a method which depends on the release of ^{14}C-glycine-containing peptides from reconstituted ^{14}C-labeled collagen (Huang and Abramson, 1975). Most of the collagenase activity was retained on the glass beads, while other proteases, such as gelatinase and leucine aminopeptidase, passed directly through the column (Huang et al., 1979). After washing the column with different concentrations of Tris buffers (Huang et al., 1979), collagenase was eluted with 2M lithuim bromide at pH 8.2. Collagenase was further purified by gel-filtration chromatography, first with Sephadex G-150 (data not shown), and then with superfine Sephadex G-100 (Fig. 3). Active fractions were concentrated by ultrafiltration, using an Amicon UM-10 filter. The homogeneity on polyacrylamide gels of purified rat-tumor collagenase is shown in Fig. 4. The molecular weights of human skin collagenase and rat tumor collagenase were found to be 45,000 and 71,000 daltons, respectively.

III. PREPARATION OF THE ANTI-COLLAGENASE SERA

Anti-collagenase sera were produced by injecting white rabbits, weighing 4–5 kg, with the purified human skin collagenase or rat tumor collageanse.

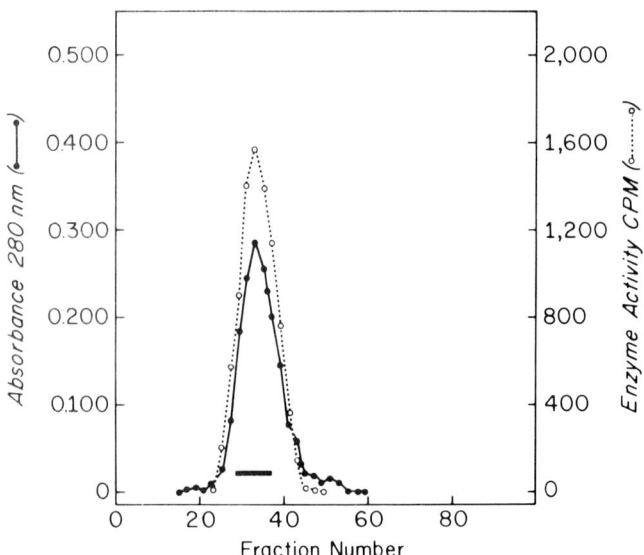

Figure 21-3. Gel-filtration chromatography of rat-tumor collagenase. The sample (10 mg protein) was applied to a superfine Sephadex G-100 column (1.5 × 100 cm) and the column was eluted with 0.05 M Tris-HCl, 0.2 M NaCl, and 0.005 M $CaCl_2$ (pH 7.6). The flow rate was 3.6 ml/h, and 1.4 ml fractions were collected. A 100-μl aliquot from each fraction was used for the collagenase activity assay.

The animals were initially immunized by means of a subcutaneous injection of an emulsion of 0.2 mg enzyme in 1 ml complete Freund's adjuvant. Two booster injections were made, each separated by an interval of two weeks. Sera were collected from the ear veins of the rabbits two weeks after the last booster injection. The serum samples were heated to 56°C for 30 min to destroy complements. The immune sera were then absorbed with red blood cells or with liver acetone powder to remove any nonspecific antibody. Monospecificity of anticollagenase sera was demonstrated by immunodiffusion (Figs. 5 and 6) and immunoelectrophoresis (Fig. 7). Immunodiffusion analysis on agar gel, carried out according to the method of Ouchterlony (1958), indicated that the rat-tumor collagenase cross-reacted with the rat-granuloma collagenase, but did not cross-react with the human-skin collagenase. This demonstrates species specificity of the collagenase.

IV. TISSUE SOURCES FOR IMMUNOCYTOCHEMICAL STUDIES

Specimens of middle-ear cholesteatoma tissues and granulation tissues were obtained from humans during middle-ear surgery. Frozen sections were cut at 4 μm. For animal models, otic capsules were obtained from rats with

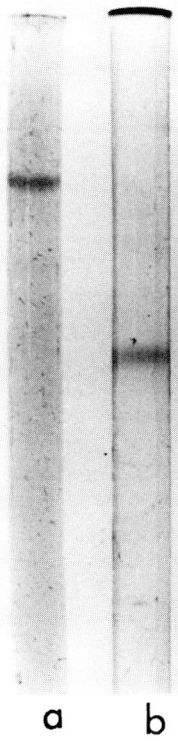

Figure 21-4. Polyacrylamide-gel electrophoresis of rat tumor collagenase: (a) with SDS and (b) without SDS. The gels were stained with Coomassie blue, showing a single band in each case.

naturally occuring otitis media and from rats whose tympanic cavities had been implanted with keratin debris or with laminaria (an expanding seaweed material). Using a ventral incision to the neck (Moriyama et al., 1984), keratin or laminaria was implanted through the bulla into the rat's right middle ear. Left ears served as controls.

Keratin was prepared from excised, dorsal skin of male Fisher rats by suspending the skin overnight in 2M sodium bromide at 4°C (Huang et al., 1975). The epidermis was separated from the dermis and treated with 0.5% trypsin to remove the epithelial cells. A layer of keratin was obtained, which was air dried and ground into a powder. Laminaria was obtained from Milex Products, Chicago, IL.

After sacrifice of the animals, the temporal bones were fixed with Perfix (Fisher Scientific Company, Springfield, NJ). Decalcification was carried out in 5% trichloroacetic acid for 4 days. Specimens were embedded in paraffin blocks and sections were cut at 8 μm.

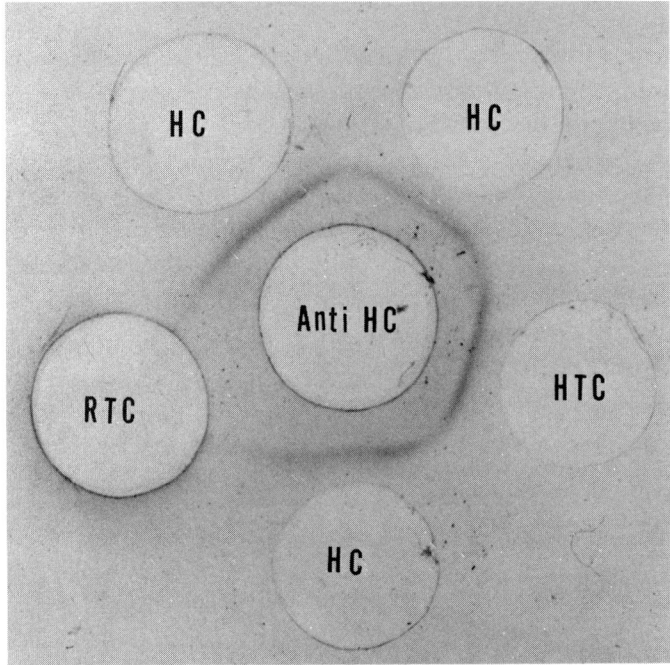

Figure 21-5. Immunodiffusion analysis of anti-human-skin collagenase (Anti HC) against purified human-skin collagenase (HC), human-tumor collagenase (HTC), and rat-tumor collagenase (RTC). No cross-reaction can be seen between anti-human collagenase and rat-tumor collagenase.

V. IMMUNOCYTOCHEMICAL STAININGS

Human tissue sections were stained by indirect immunofluorescence (Reddick *et al.*, 1974), as shown in Fig. 8. Rat temporal bone sections were stained by the unlabeled peroxidase-anti-peroxidase (PAP) method (Fig. 8), according to Sternberger (1970). These two methods included the following steps: 1. Tissue sections were first incubated with rabbit anti-collagenase serum. 2. The tissue sections were labeled with either fluorescein-isothiocyanate (FITC)-conjugated goat anti-rabbit IgG (for indirect immunofluorescence), or with peroxidase-antiperoxidase complex (for the PAP method). For the PAP method only, sections were further reacted with 3,3′-diaminobenzidine (0.2 mg/ml in Tris buffer, pH 7.6) containing 0.003% hydrogen peroxide, and were then counterstained with hematoxylin.

To test the specificity of the immune reaction, the following control experiments were performed: 1. Normal rabbit serum was substituted for rabbit anti-collagenase serum. 2. The anti-collagenase serum was reacted

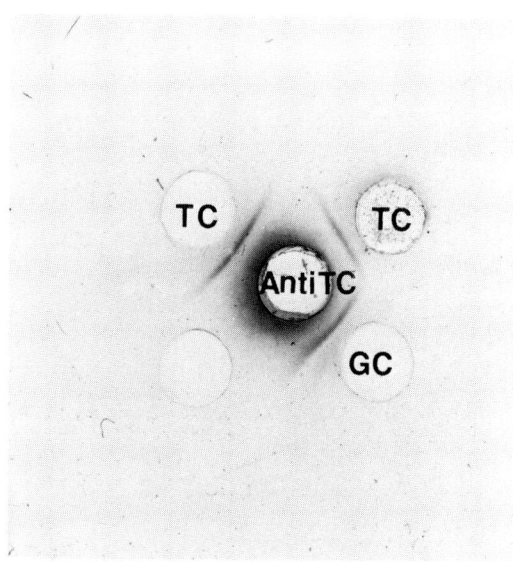

Figure 21-6. Immunodiffusion analysis of anti-rat-tumor collagenase (Anti TC) against purified rat-tumor collagenase (TC) and rat-granuloma collagenase (GC).

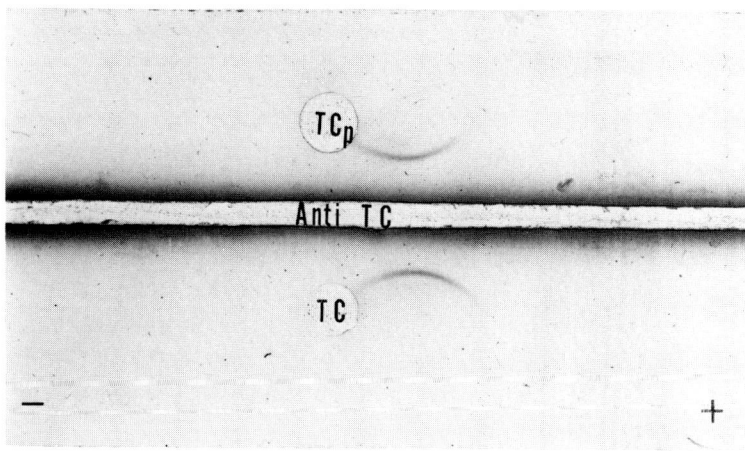

Figure 21-7. Immunoelectrophoresis of rat-tumor collagenase (TC). Anti TC, anti-rat-tumor collagenase serum; TC_p, partially-purified rat-tumor collagenase.

with excess collagenase. (This step binds all the receptor sites of the antibody and blocks its binding to tissue collagenase.) 3. The rabbit anti-collagenase serum was eliminated and the tissue section was labeled directly with FITC

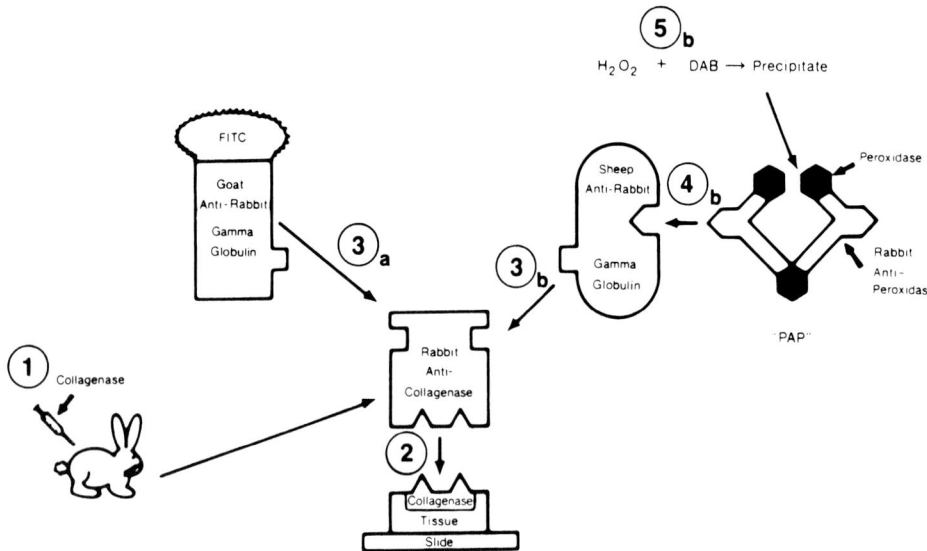

Figure 21-8. Immunocytochemical reaction of the tissue collagenase with anti-collagenase. Rabbit anti-collagenase serum is incubated with tissue collagenase. The collagenase-anti-collagenase complex is then labeled with either FITC or PAP for visualization (see text).

or PAP. All these controls showed an absence of staining, indicating that the immune reactions were specific.

A. HUMAN MIDDLE-EAR CHOLESTEATOMA

The immunofluorescent staining of human middle-ear cholesteatoma tissue is shown in Fig. 9. Collagenase immunofluorescence, appearing as a bright green under the microscope, can be seen in the granulation tissue, especially on collagen fibers and on the basement membrane. No staining can be seen in the epithelial layer and the keratinized layer. In a tympanic membrane with cholesteatoma, bright staining of fibroblasts in the connective tissue is visible (Fig. 10). Neither the epitheluim of the tympanic membrane nor the epithelium of the cholesteatoma showed staining for collagenase. In Fig. 11, a cluster of macrophages in a section of middle-ear cholesteatoma shows intracellular fluorescence.

B. ANIMAL MODELS

1. Chronic Otitis Media in Rats

A section of the temporal bone of a rat with naturally-occuring otitis media, stained with PAP, is shown in Fig. 12. Collagenase immunoreactivity,

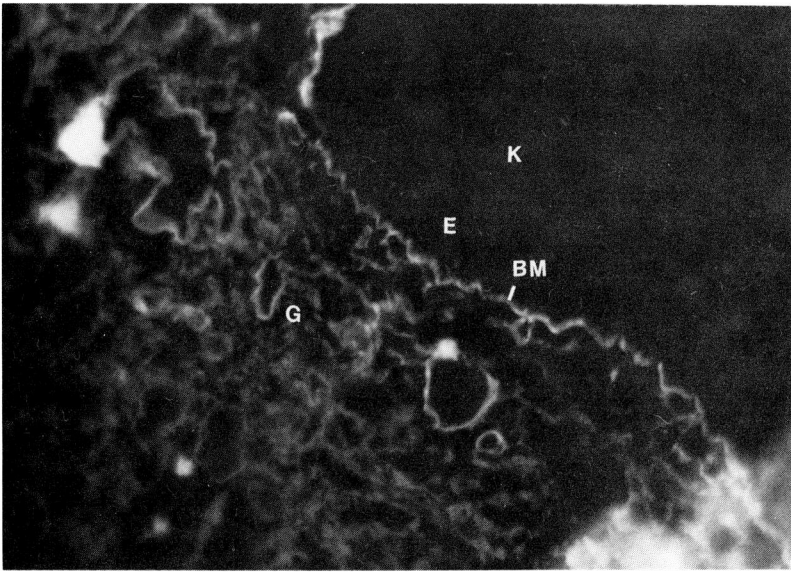

Figure 21-9. Collagenase staining of human middle-ear cholesteatoma by the indirect fluorescent method. Bright staining is seen in the granulation tissue (G) and at the basement membrane (BM). No staining can be seen in the epithelial layer (E) and in the keratinized layer (K). × 88.

originally appearing as dark brown, can be seen in the fibroblasts, mononuclear cells, and osteoclasts in the area of bone resorption. Fibroblasts at other sites were weakly stained. This suggests that direct cell contact between fibroblasts and macrophages may be important in collagenase production. The granulation tissue appears somewhat less stained, compared to the area of bone resorption. Osteocytes close to (but not in) the area of bone resorption did not show staining for collagenase. This is in contrast to the findings of previous study (Gantz et al., 1978), where osteocytes were found to be stained by anti-collagenase. It is possible that osteocytes can produce collagenase only in very rapid bone resorption. e.g., as in inflammation induced by oxalic acid (Gantz et al., 1978).

In tympanic membranes with chronic otitis media, collagenase appeared only in fibroblasts adjacent to epithelial basal cells and not in fibroblasts at other sites remote from epithelial basal cells (Fig. 13). This observation again suggests that direct cell contact between fibroblasts and epithelial cells is a possible mode of collagenase production.

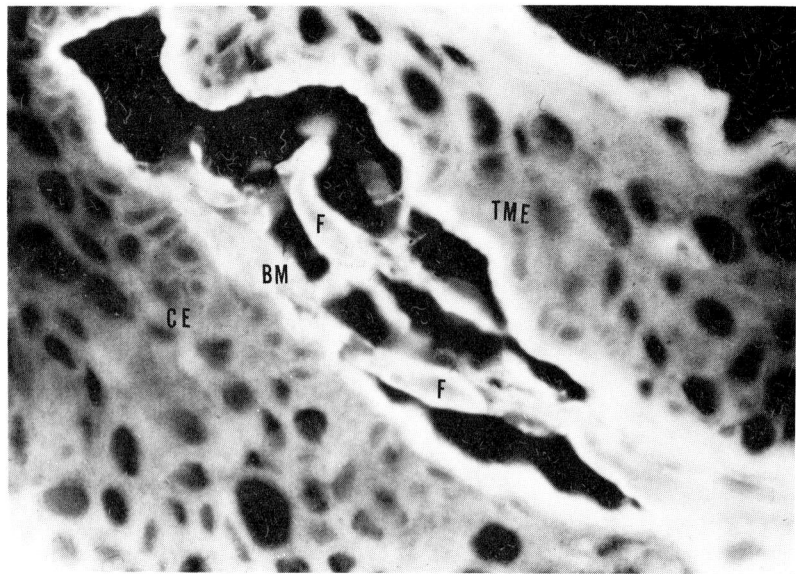

Figure 21-10. Collagenase staining of human tympanic membrane. Bright staining can be seen in the basement membrane (BM) and fibroblasts (F). No staining can be seen at the tympanic membrane epithelium (TME) and the cholesteatoma epithelium (CE). × 396.

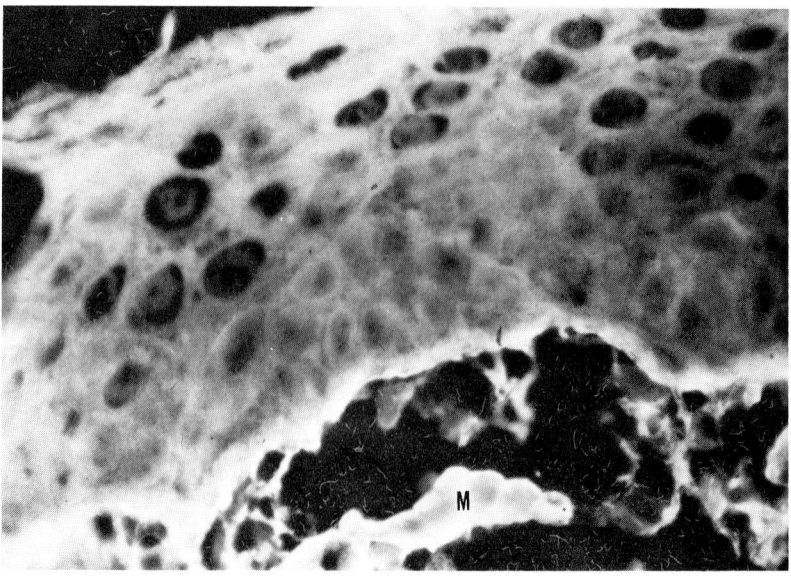

Figure 21-11. Collagenase staining of human middle-ear cholesteatoma. Bright staining can be seen in the region of the macrophages (M). × 399.

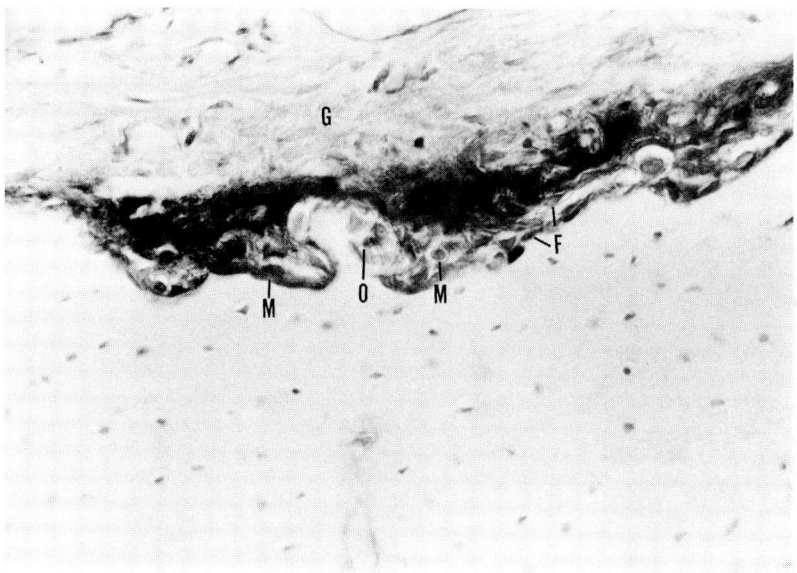

Figure 21-12. Area of bone resorption of naturally-occurring rat otitis media. The tissue section was stained by the peroxidase-antiperoxidase method. Fibroblasts (F), macrophages (M), and osteoclasts (O) show collagenase immunoreactivity, appearing as dark brown under a light microscope. G, granulation tissue. × 127.

2. Keratin-Induced Chronic Inflammation

Kaneko et al. (1980) and Yuasa et al. (1982) have noted that epithelial debris and keratin in human cholesteatoma can induce foreign-body granulomas in the middle ear. We have studied the effect of epithelial debris in the middle ears of rats. Keratin (used as "epithelial debris"), placed in the rat tympanic cavity, produced large granulomas containing numerous macrophages and foreign-body giant cells after two weeks of implantation (Fig. 14). Bone resorption was observed in the cochlear wall. Similarly to naturally-occurring otitis media, macrophages, fibroblasts, and osteoclasts were seen in the area of bone resorption, and these cells stained for collagenase (Fig. 15).

3. Effect of Pressure on Chronic Inflammation

Ruedi (1956), Baron (1967), Tos (1979), and Thomsen (1982) have suggested that pressure is able to induce bone resorption in cholesteatoma. Pressure can indeed stimulate the resorption of bone by bone cells in certain situations. Since bone is a crystal, deformation of bone has been reported to involve a transduction of mechanical energy into electrical energy through a

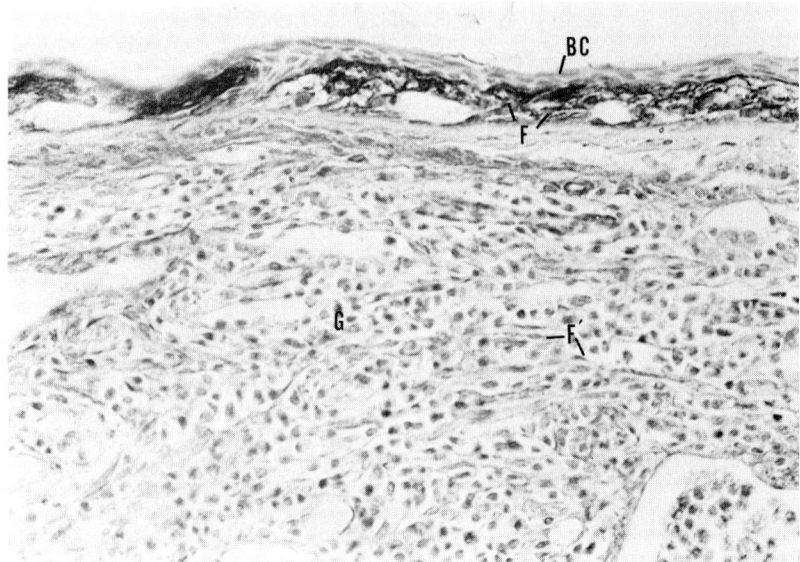

Figure 21-13. Collagenase staining, by the peroxidase-antiperoxidase method, of the tympanic membrane of a rat with chronic otitis media. Collagenase appears in fibroblasts (F) next to epithelial basal cells (BC), originally dark brown. Fibroblasts (F') at other sites do not show collagenase staining. G, granulation tissue. × 127.

piezo-electric effect (Bassett, 1968). In the latter study, relative electron positivity was shown experimentally to result in bone resorption. Yamasaki et al. (1982) induced bone resorption and formation of osteoclasts by insertion of a piece of orthodontic elastic band between two molars.

We have demonstrated the effect of pressure in the rat middle ear by inserting a piece of laminaria (4 mm in length and 2 mm in diameter, weighing 35 mg) into the bullas of rats. Although complete swelling of laminaria occured within 24 h, 2–3 weeks were required to cause extensive bone resorption of the cochlea and bulla (Figs. 16 and 17), with formation of a layer of granulation tissue. By the peroxidase-anti-peroxidase method, collagenase appeared to be localized in osteoclasts, fibroblasts, and macrophages. The ruffled border of osteoclasts in the area of bone resorption showed staining for collagenase (Fig. 18).

It seems that a layer of granulation tissue adjacent to the bone surface is necessary to cause bone resorption, and that pressure itself does not directly cause the resorption. This hypothesis is supported by our present studies (Fig. 19) in which indomethacin, an anti-inflammatory agent and an inhibitor of prostaglandin synthesis, was given to animals after insertion of laminaria. A previous study of ours (Huang, 1981) and studies of others (Wahl et al., 1977) have demonstrated that prostaglandin E_2 is able to stimulate collagen-

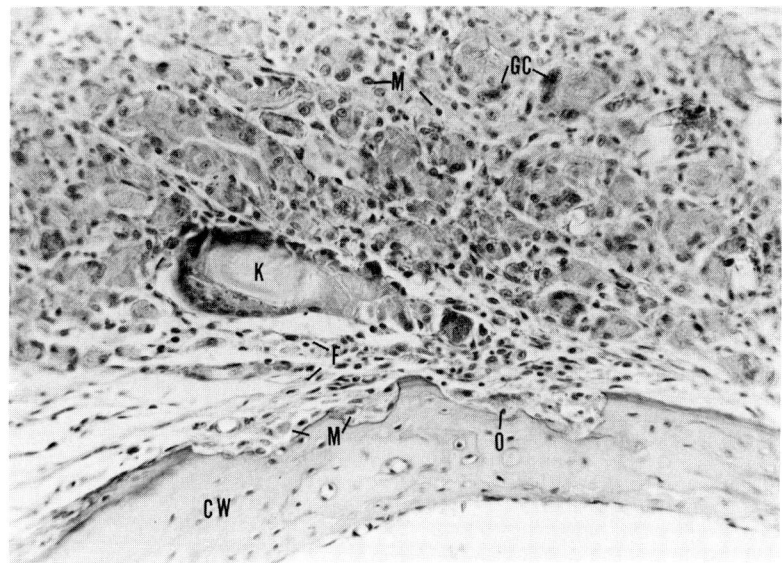

Figure 21-14. Rat temporal bone two weeks after placing keratin (K) in the bulla, showing bone resorption of the cochlear wall (CW). The granulation tissue contains numerous macrophages (M) and giant cells (GC). Osteoclasts (O), fibroblasts (F), and macrophages can be seen in the area of bone resorption. Hematoxylin-eosin stain. × 62.

ase production. When indomethacin was given to rats subcutaneously (5 mg/kg body weight) on a daily basis, after implantation of laminaria, formation of granulation tissue and bone resorption were inhibited (Figs. 19 and 20). This finding suggests that physical force (pressure) induces the formation of inflammatory granulation tissue, which then causes bone resorption, but that the pressure itself does not cause the bone resorption.

VI. COLLAGENASE AND PROSTAGLANDIN PRODUCTION FROM CELL CULTURES

Macrophages may play an important role in chronic inflammation because of their phagocytic activity, secretory function, and participation in immunological events (Davis and Allison, 1972). We prepared macrophages by inducing their formation in the rat peritoneal cavity with injections of mineral oil (Wahl et al., 1974). When macrophages were cultured in the presence of endotoxin (20 μg/ml), the prostaglandin E_2 level reached a maximum of 90 ng/10^7 cells during 12-24 h of culture (Moriyama et al., 1984). Prostaglandin E_2 concentration was determined by radioimmunoassay (Jaffe and Behrman, 1974). However, macrophages alone, even when stimu-

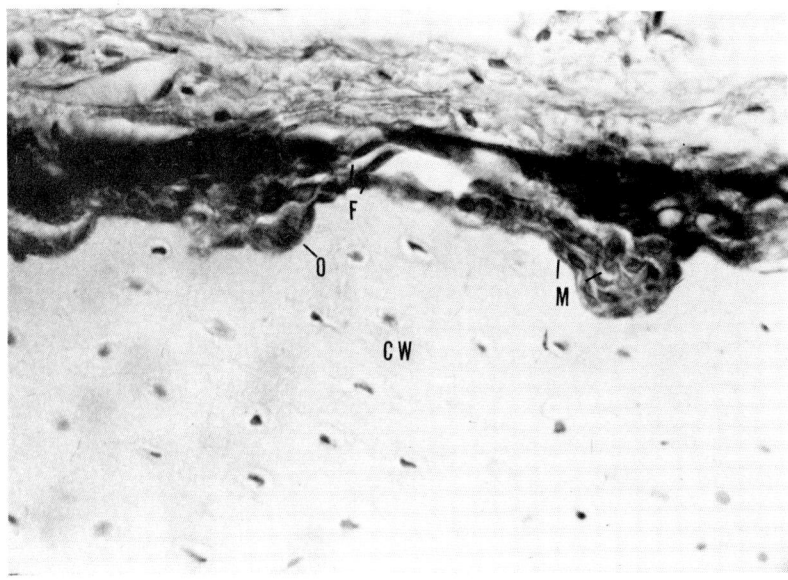

Figure 21-15. Collagenase staining of the rat cochlear wall (CW), two weeks after implanting keratin as described in the text. Fibroblasts (F), macrophages (M), and osteoclasts (O) show collagenase immunoreactivity, which originally appeared as dark brown in the area of bone resorption. × 127.

lated by endotoxin, produced a very low activity of collagenase (less than 10 units/10^5 cells; one unit is defined as the amount of enzyme degrading 1 µg of collagen in 20 h at 37°C). Macrophages may require other agents to stimulate collagenase production, such as a lymphokine, which is a soluble protein factor released by stimulated lymphocytes (Wahl et al., 1975).

On the other hand, macrophages may function as effector cells to stimulate collagenase production in fibroblasts. Our study showed that the "conditioned" medium obtained from endotoxin-activated macrophages considerably enhances fibroblast collagenase production (Table II). Since the infiltration of macrophages and fibroblasts is always seen to be associated with bone resorption in chronic inflammation of the middle ear, it is possible that cellular interaction between macrophages and fibroblasts is necessary to produce collagenase, which, in turn, appears to cause the resorption of bone.

VII. CONCLUSIONS

Studies of collagenase immunocytochemistry in chronic otitis media, in human and animal models, suggest the following:

1. Collagenase is localized in tissue sections of human middle-ear cholesteatoma and in tissues of naturally-occurring and experimental otitis media in rats.

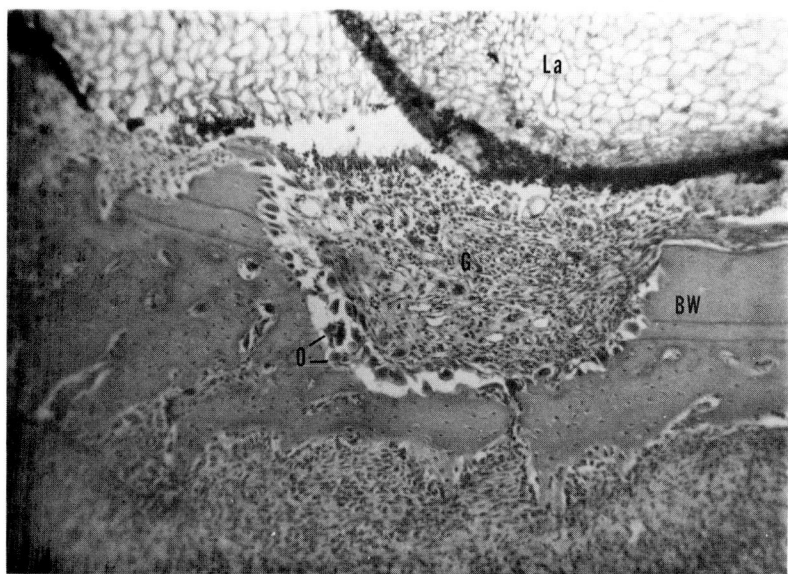

Figure 21-16. Rat temporal bone, three weeks after insertion of laminaria (La) (see text). The bullar wall (BW) is eaten away by granulation tissue (G). Numerous osteoclasts (O) are seen in the resorbing margin. Hematoxylin-eosin stain. × 31.

2. Collagenase is found in granulation tissue but not in epithelial debris.

3. Collagenase appears to originate within macrophages, fibroblasts, and osteoclasts.

4. Granulation tissue, as a layer adjacent to the bone surface, induces the process of bone resorption.

5. Keratin ("epithelial debris"), acting as a foreign body, is able to induce the formation of granulation tissue.

6. Pressure, produced by the epidermal cyst in cholesteatoma, may stimulate granulation tissue to cause synthesis of prostaglandins, collagenase, and other resorbing factors, which, in turn, may cause osteoclastic bone resorption.

7. The mechanism of cellular interaction between macrophages and fibroblasts and between epithelial basal cells and fibroblasts for the production of collagenase appears to be an important causative factor in bone resorption of chronic otitis media.

The mechanism of cellular interaction by which collagenase is produced for degradation of organic matrix of bone is the subject of our ongoing investigations.

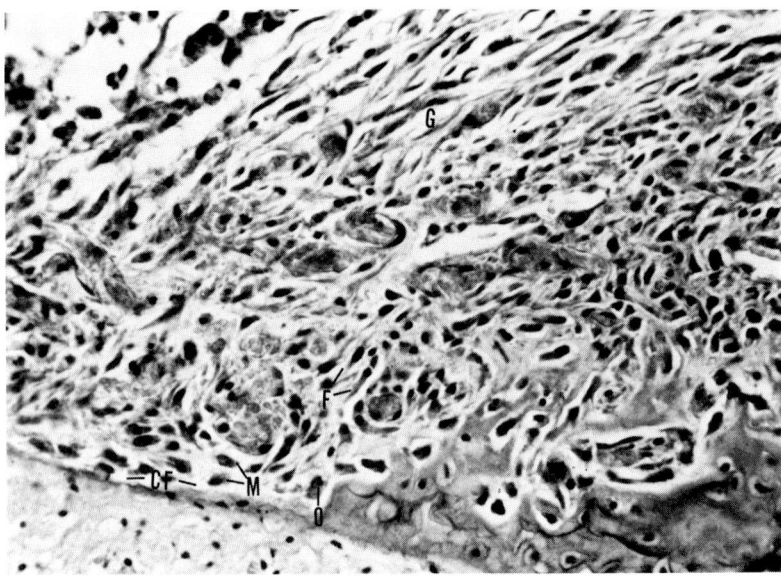

Figure 21-17. Rat temporal bone, two weeks after insertion of laminaria, showing complete bone resorption of the cochlear wall (cochlear fistula, CF). Osteoclasts (O), macrophages (M), and fibroblasts (F) are seen in the area of fistula. Hematoxylin-eosin stain. × 79.

ACKNOWLEDGMENTS

We wish to thank Miss Doris Look for excellent technical assistance and Dr. Diego Saporta for advice during the preparation of the manuscript. This investigation was supported by the Deafness Research Foundation and by NIH grant NS 13951.

NOTE

[1]Cholesteatoma refers to a cystlike lesion comprised of an irregular lining of proliferating, keratinizing, stratified squamous epithelium and an associated inflammatory reaction in a chronically-infected tympanic (middle-ear) cavity.

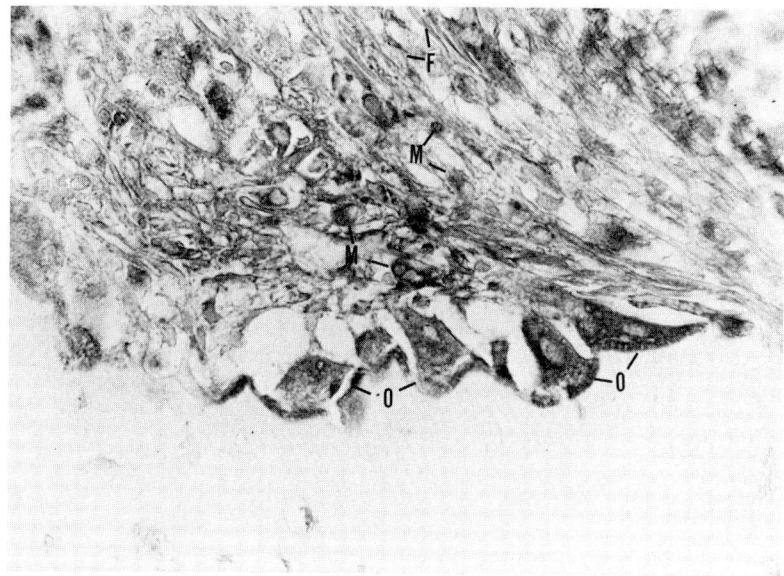

Figure 21-18. Collagenase staining, by the peroxidase-antiperoxidase method, of the area of bone resorption two weeks after insertion of laminaria. Osteoclasts (O), fibroblasts (F), and macrophages (M) show collagenase immunoreactivity. × 316.

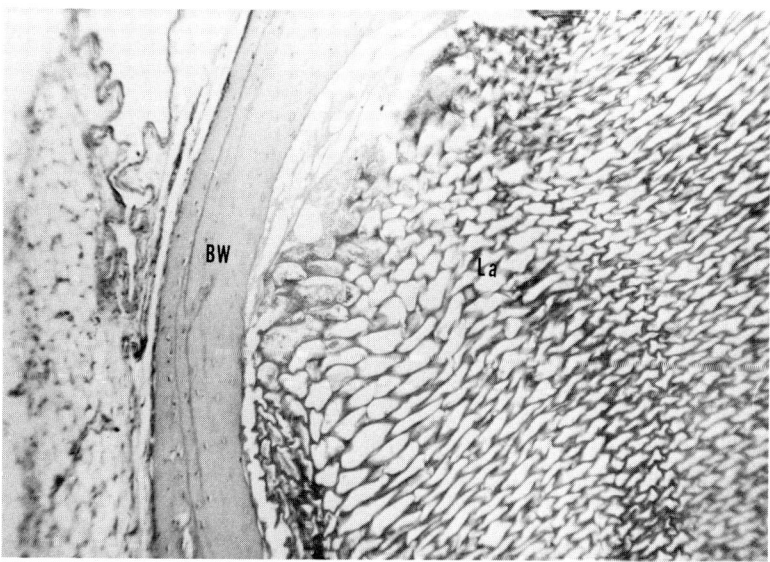

Figure 21-19. Rat temporal bone, two weeks after insertion of laminaria (La) accompanied by administration of indomethacin (see text), shows neither granulation-tissue formation nor bone resorption of the bullar wall (BW). × 48.

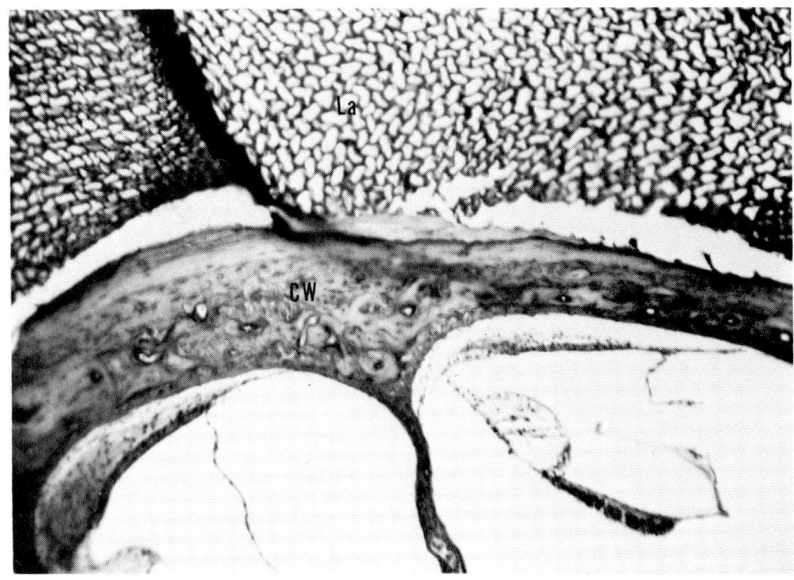

Figure 21-20. Rat temporal bone, two weeks after insertion of laminaria (La) accompanied by administration of indomethacin, again shows neither granulation-tissue formation nor bone resorption of the cochlear wall (CW). × 51.

TABLE 21-II
CUMULATIVE COLLAGENASE PRODUCTION IN CELL CULTURES

	Culture	Collagenase (units/10^5 cells)
Fibroblasts:		
	+EMCM	259
	+MCM	<10
	−MCM	<10
Macrophages:		
	+Endotoxin	<10
	−Endotoxin	<10

Values in right-hand column represent the average of triplicate experiments performed on a four-day culture. MCM = macrophage-conditioned medium; EMCM = endotoxin-activated, macrophage-conditioned medium. One unit is defined as the amount of enzyme degrading 1 mg of collagen in 20 h at 37° C.

REFERENCES

Abramson, M.: Collagenase activity in middle ear cholesteotoma. *Ann. Oto. Rhinol. Laryngol.* 78: 112-126, 1969.

Abramson, M.; Asarch, R. G., and Litton, W. B.: Experimental aural cholesteatoma causing bone resorption. *Ann. Otol. Rhinol. Laryngol. 84:* 425-433, 1975.

Baron, S. H.: Preservation of the cholesteatoma matrix in the modified radical mastoidectomy. *Laryngoscope 78:* 905-911, 1967.

Bassett, C. A. L.: Biologic significance of piezo-electricity. *Calcif. Tiss. Res. 1:* 252-273, 1968.

Davis, P., and Allison, A. C.: The secretion of macrophage enzymes in relation to the pathogenesis of chronic inflammation. In Nelson, D. S. (Ed.): Immunology of Macrophages. New York, Academic Press, 1972, pp. 427-461.

Evanson, J. M.; Jeffrey, J. J., and Krane, S. M.: Human collagenase: Identification and characterization of an enzyme from rheumatoid synovium in culture. *Science 158:* 499-502, 1967.

Gantz, B. J.; Abramson, M., and Huang, C. C.: Localization of collagenase in chronically inflamed guinea pig temporal bone. *Otolaryngol. 86:* 236-240, 1978.

Geiger, S., and Harper, E.: Human gingival collagenase in periodontal disease: The release of collagenase and the breakdown of endogenous collagen in gingival explants. *J. Dental Res. 59:* 11-16, 1980.

Glimcher, M. J.; Hodge, A. J., and Schmitt, F. D.: Macromolecular aggregation states in relation to mineralization: The collagen hydroxyapatite system as studied in vitro. *Proc. Natl. Acad. Sci. U.S.A. 43:* 860-867, 1957.

Grillo, H. C., and Gross, J.: Collagenolytic activity during mammalian wound repair. *Devel. Biol. 15:* 300-317, 1967.

Huang, C. C.: Prostaglandin E_2, collagenase and protease levels in culture of rat tumor. *Otolaryngol. Head Neck Surg. 89:* 564-568, 1981.

Huang, C. C., and Abramson, M.: Purification and characterization of collagenase from guinea pig skin. *Biochim. Biophys. Acta. 384:* 484-492, 1975.

Huang, C. C.; Wu, C. H., and Abramson, M.: Collagenase activity in cultures of rat prostate carcinoma. *Biochim. Biophys. Acta. 570:* 149-156, 1979.

Huang, L. Y.; Stern, I. B., Clagett, J. A., and Chi, E. Y.: Two polypeptide chain constituents of the major protein of the cornified layer of newborn rat epidermis. *Biochemistry 14:* 3573-3580, 1975.

Hausmann, E.; Weinfeld, N., and Miller, W. A.: Effect of lipopolysaccharides on bone resorption in tissue culture. *Calcif. Tissue Res. 9:* 272-282, 1972.

Jaffe, B. M., and Behrman, H. R.: Prostaglandins E, A, and F. In Jaffe, B. M., and Behrman, H. R. (Eds.): *Methods of Hormone Radioimmunoassay.* New York, Academic Press, 1974, pp. 19-34.

Kaneko, Y.; Yuasa, R., Jse, Y., Shikawa, H., Rokugo, M., Tanioka, S., and Shibahara, Y.: Bone destruction due to rupture of a cholesteatoma sac: A pathogenesis of bone destruction in aural cholesteatoma. *Laryngoscope 90:* 1865-1871, 1980.

Moriyama, H.; Huang, C. C., Shirahata, Y., and Abramson, M.: Effects of keratin on bone resorption in experimental otitis media. *Arch. Otorhinolaryngol. 230:* 61-81, 1981.

Ouchterlony, O.: Diffusion-in-gel methods for immunological analysis. *Prog. Allergy 5:* 1 18, 1958.

Raisz, L. G.: Bone resorption in tissue culture: Factors influencing the response to parathyroid harmone. *J. Clin. Invest. 44:* 103-116, 1965.

Reddick, M. E.; Bauer, E. A., and Eisen, A. Z.: Immunocytochemical localization of collagenase in human skin and fibroblasts in monolayer culture. *J. Invest. Dermatol. 62:* 361-366, 1974.

Robinson, D. R., and McGuire, M. B.: Prostaglandins in the rheumatoid diseases. *Ann N. Y. Acad. Sci. 256:* 318-329, 1975.

Ruedi, L.: Cholesteatosis of the attic. *J. Laryngol.* 72: 593-609, 1958.

Sade, J.; Berco, E., Buyanover, D., and Brown, M.: Ossicular damage in chronic middle ear inflammation. In Sade, J. (Ed.): *Proceedings of the Second International Conference on Cholesteatoma and Mastoid Surgery.* Amstelveen, The Netherlands, Kugler, 1982, pp. 347-358.

Sternberger, I. A.; Hardy, P. H., Cuculis, J. J., and Meyer, H. G.: The unlabeled antibody-enzyme method of immunochemistry. Preparation and properties of soluble antigen-antibody complex (horseradish peroxidase) and its use in identification of spirochetes. *J. Histochem. Cytochem.* 18: 315-333, 1970.

Thomsen, J.; Bretlau, P., Jorgensen, M. B., and Kristensen, H. K.: Bone resorption in chronic otitis media. In Sade, J.; McCabe, B., and Abramson, M. (Eds.): *Proceedings of the First International Conference on Cholesteatoma.* Birmingham, Alabama, Aesculapius Publishing, 1977, pp. 136-148.

Thomsen, J.; Tos, M., Nielsen, M., and Jorgensen, M. B.: Bone destruction in inflammatory diseases of the ear. In Sade, J. (Ed.): *Proceedings of the Second International Conference on Cholesteatoma and Mastoid Surgery.* Amstelveen, The Netherlands, Kugler Publications, 1982, pp. 397-411.

Tos, M.: Pathology of the ossicular chain in various chronic middle ear disease. *J. Otolaryngol.* 93: 769-780, 1979.

Wahl, L. M.; Wahl, S. M., Mergenhagen, S. E., and Martin, G. R.: Collagenase production by endotoxin-activated macrophages. *Proc. Natl. Acad. Sci. U.S.A.* 71: 3598-3601, 1974.

Wahl, L. M.; Olsen, C. E., Sandberg, A. L., and Mergenhagen, S. E.: Prostaglandin regulation of macrophage collagenase production. *Proc. Natl. Acad. Sci. U.S.A.* 74: 4955-4958, 1977.

Walsh, T. E.; Covel, W. P., and Ogura, J. H.: The effect of cholesteatosis on bone. *Ann. Otol. Rhinol. Laryngol.* 60: 1100-1113, 1951.

Yamasaki, K.; Miura, F., and Suda, T.: Prostaglandin as a mediator of bone resorption induced by experimental tooth movement in rats. *J. Dental Res.* 59: 1635-1642, 1982.

Yuasa, R.; Kaneko, Y., Takasaka, T., Jino, Y., Tanioka, S., and Hanjima, T.: The significance of keratinization in the mechanism of bone destruction in cholesteatoma. In Sade, J. (Ed.): *Proceedings of the Second International Conference on Cholesteatoma and Mastoid Surgery.* Amstelveen, The Netherlands, Kugler, 1982, pp. 419-427.

PART IV
BIOCHEMISTRY OF LABYRINTHINE SUPPORT SYSTEMS

Chapter 22

NOISE-INDUCED CHANGES OF COCHLEAR ENERGY METABOLISM

Jochen Schacht and Barbara Canlon

I. Introduction
II. Energy Metabolism
III. Deoxyglucose Trapping
IV. Cochlear Glucose Utilization
    A. Deoxyglucose Uptake in Cochlear Structures
    B. Effects of Sound Stimulation
V. Summary
VI. Outlook
    References

I. INTRODUCTION

The search for biochemical reactions triggered or modulated by sound dates back to the earliest investigations of cochlear metabolism and has since been a focal point of auditory biochemistry. One of the first molecular theories of transduction was that of Vinnikov and Titova (1964) concerning stimulus-induced diffusion of endolymphatic acetylcholine to receptor sites on stereocilia. Almost all conceivable cellular constituents or processes—e.g., nucleic acids, proteins and protein synthesis, enzymatic activities, intermediary metabolites, membrane structures, transport and permeability—were subsequently analyzed for a possible involvement in transduction mechanisms, without providing clues for a rational hypothesis.

Energy metabolism in various aspects proved to be a recurring theme in studies of perilymphatic glucose and lactate, intracellular glycogen and ATP, or activities of enzymes involved in glycolysis. Schnieder's findings (Schnieder and Janzer, 1969; Schnieder, 1974) of increased lactate after sound stimulation spurred discussions of "anaerobic glycolysis" of the organ of Corti and "lactate poisoning" as a mechanism of acoustic trauma. Subsequently, detailed analyses (Scheibe *et al.,* 1976, 1981), laid these speculations to rest by establishing that perilymphatic glucose, pyruvate and lactate remained uninfluenced by sound exposure. The most exacting and direct determinations of the energy state of cochlear tissues have been contributed over many years by Ruediger Thalmann and his group at Washington University, St. Louis. Various conditions of noise exposure, including those leading to asymptotic threshold shift, were found not to alter ATP levels or other biochemical parameters measured in the sensory cells of the organ of Corti (Thalmann, 1976). While the results of many earlier histochemical and semiquantitative biochemical studies may be criticized on technical grounds, the investigations by Scheibe and Thalmann are above such critique because of their exact methodology. Therefore, it has to be accepted as established that levels of cochlear energy metabolites are stable during stimulation by sound.

For expert reviews on cochlear energy metabolism, and a critical assessment of techniques, animal models and previous results, the reader is referred to Thalmann (1976) and Thalmann et al. (1980). It is the aim of this review to summarize recent developments in cochlear physiology and biochemistry concerning the application of the "deoxyglucose technique" to the inner ear. Before we discuss these studies, however, a review of the concept of "energy metabolism" is appropriate.

II. ENERGY METABOLISM

Energy metabolism, the synthesis and utilization of ATP, is a strictly-regulated cellular process. Its rate varies with physiological demands on a tissue, which may range from mechanical work, such as muscle contraction, to hormone- or neurotransmitter- stimulated secretory events, and, neural processing. Any functional load will lead to a higher rate of energy consumption, i.e., utilization of ATP. In turn, regulatory feedback mechanisms will match a higher rate of utilization with increased ATP production in order to maintain constant levels of this high-energy compound (except for the extremes, e.g., anoxia, where such mechanisms break down and ATP levels decrease). As a consequence, "energy metabolism" needs to be measured as the rate of turnover of ATP or rate of utilization of the ultimate sources of energy: carbohydrates, fatty acids, amino acids, or intracellular stores of glycogen. Any or all of the latter substrates may contribute to the generation of ATP (Fig. 1), depending on the physiological state of the organism (e.g., feeding vs. starvation) and the tissue itself. For example, brain relies almost exclusively on glucose (ketone bodies may serve as additional substrates, but not fatty acids or amino acids) while liver or heart draw more than half of their energy needs from fat. This diversity, and the technical difficulties of the measurements involved, make the assessment of energy metabolism of a tissue highly complex. This complexity has to be borne in mind in the following discussions.

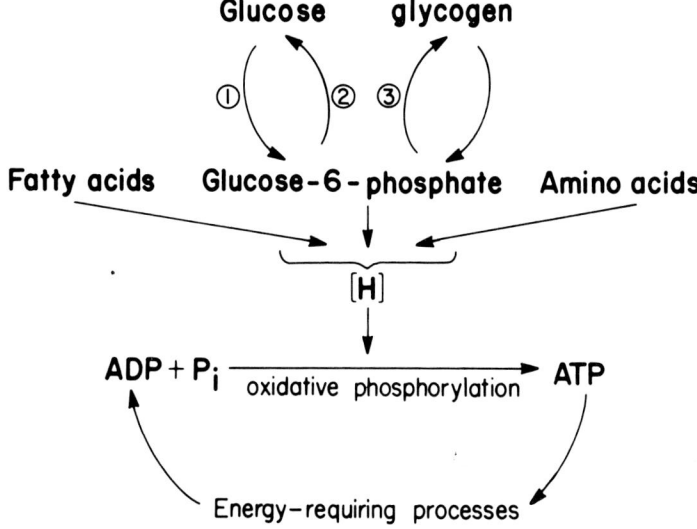

Figure 22-1. Major pathways of energy metabolism. Fatty acids, carbohydrates and amino acids can be catabolized to yield reduced electron-transporting coenzymes ("[H]") and, consequently, ATP. The relative contribution of these energy sources varies with the tissue. Energy-requiring processes include synthesis of metabolic intermediates, synthesis of structural components, synthesis of neurotransmitters, membrane transport, and maintenance of intracellular potentials. Specific pathways: 1. Uptake of serum glucose into the cell and phosphorylation by hexokinase. 2. Dephosphorylation by glucose-6-phosphatase to glucose. 3. Synthesis of intracellular glycogen stores.

III. DEOXYGLUCOSE TRAPPING

The development of the "deoxyglucose technique" considerably advanced the quantitation of glucose-derived energy metabolism (Sokoloff, 1977; Sokoloff et al., 1977). The glucose analogue, 2-deoxy-D-glucose, shares systems for uptake and phosphorylation with its natural counterpart, glucose (Pathway 1, Fig. 1), but is not subsequently metabolized in glycolysis. If radiolabeled deoxyglucose is administered to an organism, labeled deoxyglucose-6-phosphate will accumulate in the tissues because this charged molecule cannot leave the cell by diffusion. The rate of accumulation will be proportional to the utilization of glucose, and, if certain kinetic parameters are known, the absolute rate of glucose utilization can be calculated.

The validity of this technique for assessing energy metabolism rests on several assumptions that seem largely met—but not unchallenged—for brain. These assumptions need to be examined for the inner ear, as we have pointed out previously (Canlon and Schacht, 1981, 1983). These assumptions are:

1. Glucose must be the only (or, at least, the major) source of energy. While the stria vascularis appears to utilize glucose preferentially (Marcus et al., 1978; Kambayashi et al., 1982a, b), hair cells may have different nutritional requirements (Zenner, 1984).

2. Deoxyglucose-6-phosphate must not be dephosphorylated (Pathway 2, Fig. 1). However, glucose-6-phosphatase has considerable activity in inner-ear structures, as demonstrated histochemically in the guinea pig (Küttner, 1975) and biochemically in the mouse (Canlon and Schacht, 1983). Dephosphorylation of deoxyglucose-6-phosphate can lead to diffusion of deoxyglucose out of the cell, both during the experiment and during postmortem dissection, to gain access to the cochlea.

3. Deoxyglucose-6-phosphate must not participate in other reactions. However, Pujol et al. (1981) have shown that glycogen in some inner-ear structures of the cat can become labeled by deoxyglucose (Pathway 3, Fig. 1). This will trap a portion of the deoxyglucose by a route which is not related to energy utilization.

4. Arterio-venous differences of glucose and deoxyglucose have to be analyzed for the determination of kinetic constants. While arterial values can be measured, experimental access to venous blood from cochlear structures is not possible. Even if the vein of the cochlear aqueduct were to be considered accessible, it would only yield information about the auditory and vestibular system as a whole. Therefore, a "lumped constant," needed for the conversion of deoxyglucose uptake to glucose utilization (Sokoloff et al., 1977), cannot be calculated for individual inner-ear tissues.

From these considerations, it is clear that methods established for the brain cannot uncritically be applied to the inner ear. However, if the above cautions are incorporated into the experimental design and into the interpretation of results, the technique of deoxyglucose trapping can yield valuable information on the auditory system. Indeed, results to date have presented new and exciting insights into cochlear metabolism and physiology.

IV. COCHLEAR GLUCOSE UTILIZATION

A. DEOXYGLUCOSE UPTAKE IN COCHLEAR STRUCTURES

There is presently substantial agreement that a differential pattern of labeled deoxyglucose uptake exists in cochlear structures. Sans et al. (1980) and Pujol et al. (1981) demonstrated the highest concentration of label in stria vascularis of cat and guinea pig, while Ryan et al. (1982) reported equal uptake into spiral ligament, spiral prominence, and stria vascularis of the gerbil. Technical difficulties left the latter studies unsatisfactory, as both diffusion and loss of radioactivity occurred during preparation of tissues for radioautography. A refined technique (Ryan and Sharp, 1982; Ryan, 1984; Chapter 23, this volume) conclusively documented preferential accumulation in stria vascularis, while the organ of Corti, including the hair cells, lacked significant labeling. A high deoxyglucose uptake into stria vascularis is consistent with the preferential utilization of glucose and the high metabolism ascribed to this tissue. Quantitative comparisons of metabolic rates of individual cochlear structures should, however, not be inferred, for the reasons outlined previously.

B. EFFECTS OF SOUND STIMULATION

The initial confusion about the effects of sound on deoxyglucose trapping in the inner ear may reflect the problems that we have mentioned above. Meyer zum Gottesberge and Orsulakova (1980) reported that glucose utilization in the membranous lateral-wall tissues of the rat cochlea decreased from moderate to high intensities of noise. Lane (1980), in a study marred by technical difficulties, found small sound-induced increases in the dissected "whole cochlea" of the guinea pig and inconsistent effects in the chick inner ear. Ryan et al. (1982) observed sound-increased uptake in the eighth nerve and central pathways, but no significant effects in cochlear structures of the gerbil.[1]

Our laboratory critically evaluated the deoxyglucose technique for the ear and documented a low rate of deoxyglucose phosphorylation, short half-life of the tracer in the cochlea, and high cochlear phosphatase activity (Canlon and Schacht 1980, 1981, 1983), important parameters that were discussed before. Using techniques adapted to the specific situation of the cochlea, we were able to demonstrate a complex but consistent pattern of deoxyglucose uptake in the mouse inner ear: an increase with moderate stimulation and a decrease from the maximum at higher levels of sound exposure, while serum kinetics of deoxyglucose levels of phosphorylation, and glucose concentrations were unchanged (Canlon and Schacht, 1981, 1983; Schacht and Canlon, 1981). In these studies, a steep rise in deoxyglucose uptake occurred in the organ of Corti and the lateral-wall tissues, from silence to approximately 55 dBA, followed by almost a plateau up to 85 dBA (Fig. 2). Since the spectrum of stimulus was broad (wide-band noise from 0.5–20 kHz, and up to approximately 40 kHz at 20 dB attenuation), it is evident that little energy is required to excite metabolism in the auditory periphery maximally. This confirms for the cochlea that physiological activity is accompanied by increased energy utilization and is analogous to previous observations of highly-active central auditory structures in the rat (Sokoloff, 1977) and in the mouse (Nowaczyk and DesRosiers, 1981). Our own experiments revealed a similar uptake pattern for the eighth nerve and the inferior colliculus. The results also suggested a reconsideration of some earlier experiments on the auditory system: "control" animals at ambient sound levels are, in fact, already "stimulated". This may have bearing, not only on biochemical, but also on morphological or physiological studies, such as those on the effects of sound on cochlear microcirculation.

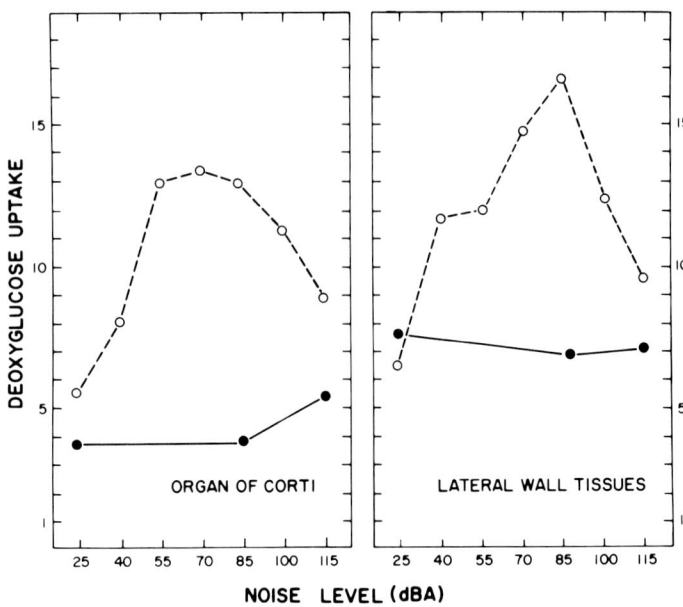

Figure 22-2. Response of deoxyglucose uptake to noise exposure. Mice received a pulse of 5 mCi [^{3}H]-deoxyglucose/kg body weight and were killed after 60 min of exposure to noise at the levels indicated. For details, see Canlon and Schacht (1983, 1984). ○—○ Young CBA mice ("normal"); •—• aged CBA mice (36 mo), with sensorineural hearing loss.

In apparent contrast to earlier reports (Ryan et al., 1982), we found that deoxyglucose uptake in the inner ear of the gerbil is increased by sound (Canlon et al., 1984). Recent radioautographic analyses in this species have confirmed and extended our findings (Ryan, 1984; Ryan et al., 1984): inner but not outer hair cells are highly responsive to sound stimulation. Inner hair cells are considered to be the primary processors of afferent information, while outer hair cells may modulate the electrical or mechanical properties of the basilar membrane (Dallos, 1981; Siegel and Kim, 1983). The differential labeling by deoxyglucose could then be of significance for a theory of inner- vs. outer-hair-cell function. We have to keep in mind, however, both the limitations of the deoxyglucose technique and the possibility that the observed difference may be due to different primary energy sources for the hair cells. Indeed, Zenner reported (1984) that outer hair cells in short-term culture require amino acids but not glucose as substrates. Only experiments with isolated inner or outer hair cells will eventually shed more light on this issue.

A unique and intriguing feature of cochlear metabolism is the response to sound of the lateral-wall tissues (Canlon and Schacht, 1983; Fig. 2), recently confirmed by radioautography in the gerbil and localized to the stria vascularis (Ryan, 1984). How can a non-sensory and non-neural tissue respond in parallel to the sensory structures? The stria vascularis is generally assumed to maintain the ionic composition of endolymph ($K^+ >> Na^+$) and the endolymphatic potential. During stimulation of the organ of Corti, the impedance of the basilar membrane is altered, presumably due to hair-cell activity, resulting in increased K^+-fluxes through the inner hair cells (Strelioff et al., 1971). A higher rate of strial metabolism is then required to maintain the K^+/Na^+ ratio and the endolymphatic potential, which would necessitate a greater supply of substrate through increased blood flow. The absence of feedback mechanisms from stria vascularis on cochlear microcirculation poses an apparent problem. However, cochlear blood flow is probably not controlled at the level of individual tissues, but by autonomic innervation up to the modiolar level. This would be before or near the branching of the spiral modiolar artery to the radiating arterioles, which supply the lateral wall, the vessels of the osseous spiral lamina, and the area of the basilar membrane (Tereyama et al., 1966; Spoendlin and Lichtensteiger, 1967; Ross, 1971). Thus, sympathetic control, triggered by excitation of hair cells, should provide global increased blood flow to all needy cochlear structures, perhaps via a mechanism such as that suggested in Fig. 3.

This model implies that acoustic stimulation may modulate cochlear microcirculation. In the central nervous system it is well established that stimulated metabolism is accompanied by elevated local blood flow (Reivich, 1974). For the auditory periphery, however, this has been a matter of controversy. Early studies reported that sound increased the directly-observed cell velocity in the lateral wall (Perlman and Kimura, 1962), but experiments with microspheres as markers of total cochlear blood flow could not demonstrate an effect of sound (Hultcrantz, 1979). Recently, however, Prazma (1984) found increased cochlear blood flow in the gerbil following one h of sound exposure. These divergent results may not be contradictory, but reflect different experimental conditions in terms of the duration of sound exposure (6 min for microsphere experiments), the area of the cochlea under investigation, and the handling of "control" animals, as pointed out above.

At higher sound intensities, deoxyglucose uptake in the cochlea of the mouse decreases from the maximum. Whether or not this is a consequence of decreased blood flow is currently unresolved. While vasoconstriction has

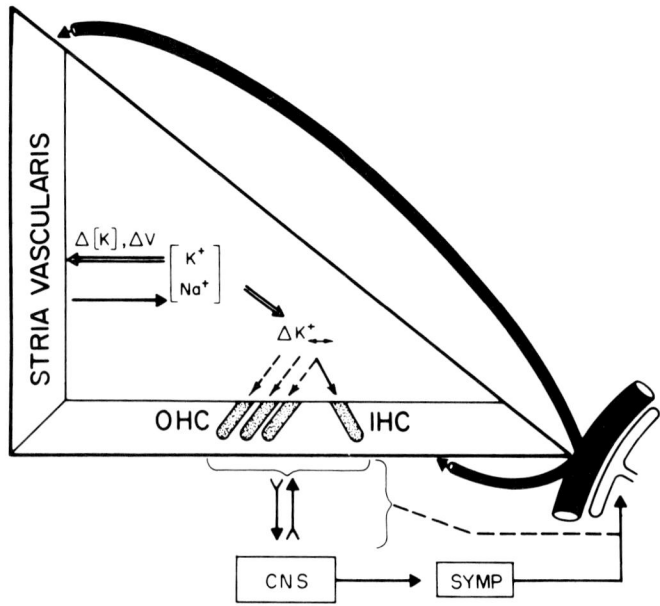

Figure 22-3. A mechanism for the metabolic coupling of the stria vascularis to stimulation of the organ of Corti. Excitation of hair cells alters K^+ flux ($\Delta K^+\Leftrightarrow$) across the basilar membrane. Blood flow is mainly regulated in response to afferent activity via the sympathetic nervous system (SYMP). The stria vascularis responds to changes of endolymphatic K^+ concentration ($\Delta[K]$) or of endolymphatic potential (ΔV) by increasing its metabolism to maintain equilibrium. CNS, central nervous system; IHC, inner hair cells; OHC, outer hair cells.

been suggested as an effect of intense noise (Hawkins, 1971), this suggestion remains controversial (Axelsson et al., 1981). Alternatively, mechanical damage to the cochlea may account for the decrease in metabolism. It is possible that species differences exist regarding this decrease, as exposure to these sound levels did not produce the biphasic response in the gerbil (Canlon et al., 1984).

In contrast to the initial controversy about the effects of sound on deoxyglucose uptake in organ of Corti and stria vascularis, the results for the cochlear neural structures were consistent. Spiral ganglion or eighth nerve were stimulated by pure tones or wide-band noise in the gerbil (Ryan et al., 1982) and the mouse (Canlon and Schacht, 1983). The autoradiographic studies in the gerbil also indicated that the anatomical location of the increase depended on the frequency of exposure.

V. SUMMARY

The major studies of glucose utilization in the cochlea of the mouse and the gerbil are now in agreement: sound stimulation increases the metabolic rate in the organ of Corti (inner hair cells) and in the stria vascularis. The report of Lane (1980) of a minor stimulation in the guinea pig and an inconsistent effect in the chick, and the study by Meyer zum Gottesberge and Orsulakova (1980) in the rat, remain difficult to evaluate, because sufficient detail is lacking in these brief reports and because follow-up studies on the respective species do not exist. However, the finding of Meyer zum Gottesberge and Orsulakova (1980) that functional overload decreases glucose utilization from its maximum at moderate intensities appears to be in good agreement with the observations in the mouse.

VI. OUTLOOK

Studies of glucose utilization have posed a number of interesting questions of cochlear physiology. What is the biochemical difference between inner and outer hair cells? Why does metabolism decrease with intense sound? What is the mechanism of the metabolic coupling of the stria vascularis? But another major and tantalizing area of research remains to be explored: A recent development of the deoxyglucose technique applied to the central nervous system is positron-emission-tomographic (PET) scanning. This technique is capable of detecting positron emissions from appropriate deoxyglucose derivatives (^{11}C-deoxyglucose or ^{18}F-fluorodeoxyglucose) as indicators of local metabolism in the conscious human subject. Sensory stimulation has been shown to cause increased labeling in the corresponding central-nervous-system structures, and pathologies are evident by the lack of appropriate responses (Alavi *et al.*, 1981; Rottenberg and Cooper, 1981). The resolution of the detector system is presently barely adequate to resolve labyrinthine structures, but investigations of peripheral pathologies remain exciting possibilities. Questions, such as whether sudden hearing loss is due to a lack of blood flow and metabolism or whether a "metabolic" presbycusis exists, may eventually be answered. We have recently investigated deoxyglucose uptake in animals with different forms of hearing loss in order to determine whether cochlear pathologies manifest themselves in altered deoxyglucose uptake (Canlon *et al.*, 1984). Both genetically-determined hearing loss in C57/BJ mice, as well as sensorineural hearing loss in aged CBA mice, show distinctly abnormal response patterns: the curves reveal a lack of stimulated metabolism (Fig. 1). Thus, an altered metabolic response is indicative of cochlear pathologies, providing a basis for future studies of the human ear.

ACKNOWLEDGMENTS

The authors wish to acknowledge Dr. A. L. Nuttall for helpful discussions. The authors' research cited here was supported by NIH Program Project Grant NS 05785.

NOTE

[1] Most studies cited here analyze deoxyglucose uptake by radioautography. Our laboratory quantitates radioactivity by scintillation counting after dissection of tissues. For further details, the reader should consult the original articles.

REFERENCES

Alavi, A.; Reivich, M., Greenberg, J., Hand, P., Rosenquist, A., Rintelmann, W., Christman, D., Fowler, J., Goldman, A., MacGregor, R., and Wolf, A.: Mapping of functional activity in brain with ^{18}F-Fluoro-deoxyglucose. *Sem. Nuclear Med. 11:* 24–31, 1981.

Axelsson, A.; Vertes, D., and Miller, J.: Immediate noise effects on cochlear vasculature in the guinea pig. *Acta Otololaryngol. 91:* 237–246, 1981.

Canlon, B., and Schacht, J.: *Assessment of Cochlear Energy Metabolism with the Deoxyglucose Method.* 17th Workshop on Inner Ear Biology, Stockhom, Sweden, 1980.

Canlon, B., and Schacht, J.: The effect of noise on deoxyglucose uptake into inner ear tissues of the mouse. *Arch. Otorhinolarygol. 230:* 171–176, 1981.

Canlon, B., and Schacht, J.: Acoustic stimulation alters deoxyglucose uptake in the mouse cochlea and inferior colliculus. *Hearing Res. 10:* 217–226, 1983.

Canlon, B.; Takada, A., and Schacht, J.: Glucose utilization in the auditory system: Cochlear dysfunctions and species differences. *Comp. Biochem. Physiol.* In press, 1984.

Dallos, P.: Cochlear physiology. *Ann. Rev. Psychol..32:* 153–190, 1981.

Hawkins, J. E., Jr.: The role of vasoconstriction in noise-induced hearing loss. *Ann. Otol. 80:* 903–913, 1971.

Hultcrantz, E.: The effect of noise on cochlear blood flow in the conscious rabbit. *Acta Physiol. Scand. 106:* 29–37, 1979.

Kambayashi, J.; Kobayashi, T., DeMott, J. E., Marcus, N. Y., Thalmann, I., and Thalmann, R.: Effect of substrate-free vascular perfusion upon cochlear potentials and glycogen of the stria vascularis. *Hearing Res. 6:* 223–240, 1982a.

Kambayashi, J.; Kobayashi, T., Marcus, N. Y., DeMott, J. E., Thalmann, I., and Thalmann, R.: Minimal concentrations of metabolic substrates capable of supporting cochlear potentials. *Hearing Res. 7:* 105–114, 1982b.

Küttner, K.: Ultrahistochemical demonstration of glucose-6-phosphatase in hair cells of the guinea pig organ of Corti. *Arch. Otorhinolaryngol. 209:* 169–177, 1975.

Lane, R. J.: A new technique for studying cochlear metabolism using 2-deoxy-D-[^{14}C] glucose: Localization of glucose metabolism with sound stimulation. *Trans. Penn. Acad. Ophthalmol. Otolaryngol. 34:* 80–85, 1981.

Marcus, D. C.; Thalmann, R., and Marcus, N. Y.: Respiratory quotient of stria vascularis of guinea pig in vitro. *Arch. Otorhinolaryngol. 221:* 97–103, 1978.

Meyer zum Gottesberge, A., and Orsulakova, A. *Uptake and Utilization of 14-C-deoxyglucose in the Cochlea.* 17th Workshop on Inner Ear Biology, Stockholm, Sweden, 1980.

Nowaczyk, T., and DesRosiers, M. H.: Application of the 2-deoxy-D-[^{14}C]-glucose method to the mouse for measuring local cerebral glucose utilization. *European Neurol. 20:* 169-172, 1981.

Perlman, H. B., and Kimura, R.: Cochlear blood flow in acoustic trauma. *Acta Otolaryngol. 54:* 99-110, 1962.

Prazma, J.: The effect of one hour of noise exposure on cochlear blood flow. *Assoc. Res. Otolaryngol. Abstr. 7:* 39, 1984.

Pujol, R.; Sans, A., and Calas, A.: High resolution radioautographic study of the inner ear following in vivo tritiated deoxyglucose administration. *Eur. Neurol. 20:* 157-161, 1981.

Reivich, M: Blood flow metabolism couple in brain. *Res. Publ. Ass. Res. Nerv. Ment Dis. 53:* 125-140, 1974.

Ross, M. D.: Fluorescence and electron microscopic observations of the general visceral, efferent innervation of the inner ear. *Acta Otolaryngol. Supp. 286:* 1-18, 1971.

Rottenberg, D. A., and Cooper, A. J. L.: Positron emission tomography of the central nervous system. *Trends Biochem. Sci. 6:* 120, 1981.

Ryan, A. F.: Anatomical measures of physiological parameters in the cochlea. In Berlin, C. (Ed.): *Recent Advances in Hearing Science.* San Diego, College Hill Press, in press, 1984.

Ryan, A. F.; Goodwin, P., Woolf, N. K., and Sharp, F.: Auditory stimulation alters the pattern of 2-deoxyglucose uptake in the inner ear. *Brain Res. 234:* 213-225, 1982.

Ryan, A. F., and Sharp, F. R.: Localization of [^{3}H]2-deoxyglucose at the cellular level using freeze-dried tissue and dry-looped emulsion. *Brain Res. 252:* 177-180, 1982.

Ryan, A. F.; Woolf, N. K., and Sharp, F. R.: Deoxyglucose uptake patterns in the auditory pathway: metabolic response to sound stimulation in the adult and neonate. *Assoc. Res. Otolaryngol. Abstr. 7:* 143, 1984.

Sans, A.; Pujol, R., Carlier, E., and Calas, A.: Cellular detection of the in vivo incorporation of tritiated 2-deoxyglucose. Radioautographic study in the inner ear. *CR Acad. Sci. (Paris) 290:* 1225-1227, 1980.

Schacht, J., and Canlon, B.: The effect of noise exposure on deoxyglucose uptake in the inner ear of the mouse. *Neurosci. Abstr. 7:* 535, 1981.

Scheibe, F.; Haupt, H., and Hache, U.: Vergleichende Untersuchungen der Laktatkonzentration von Perilymphe, Blut und Liquor cerebrospinalis normaler und schallbelasteter Meerschweinchen. *Arch. Otorhinolaryng. 214:* 19-25, 1976.

Scheibe, F.; Haupt, H., Rothe, E., and Hache, U.: Zur Glukose-, Pyruvat- und Laktatkonzentration von Perilymphe, Blut and Liquor cerebrospinalis unbelasteter und schallbelasteter Meerschweinchem in Athylurethan-narkose. *Arch. Otorhinolaryngol. 233:* 89-97, 1981.

Schnieder, E. A.: A contribution to the physiology of the perilymph. Part III: On the origin of noise-induced hearing loss. *Ann. Otol. Rhinol. Laryngol. 83:* 406-412, 1974.

Schnieder, E. A., and Janzer, A.: Vergleichende elektrophysiologische, histologische und biochemlsche Untersuchungen beim Meerschweinchen nach Schallbelastung. *Arch. Klin Exp. Ohren-Nasen-Kehlkopfheilkl. 194:* 579-583, 1969.

Siegel, J. H., and Kim, D. O.: Efferent neural control of cochlear mechanics? Olivocochlear bundle stimulation affects cochlear biomechanical nonlinearity. *Hearing Res. 6:* 171-182, 1982.

Sokoloff, L.: Relation between physiological function and energy metabolism in the central nervous system. *J. Neurochem. 29:* 13-26, 1977.

Sokoloff, L.; Reivich, R. M., Kennedy, C., DesRosiers, M. H., Patlack, C. S., Pettigrew, K. D., Sakurada, O., and Shinohara, M.: The ^{14}C-deoxyglucose method for measurement of local cerebral glucose utilization. *J. Neurochem. 28:* 897-916, 1977.

Spoendlin, H., and Lichtensteiger, W.: The sympathetic nerve supply to the inner ear. *Arch.*

Klin. Exp. Ohren-Nasen-Kehlkopfheilk. 189: 346–359, 1967.

Strelioff, D.; Haas, G., and Honrubia, V.: Sound-induced electrical impedance changes in the guinea pig cochlea. *J. Acoust. Soc. Amer. 51:* 617–620, 1971.

Tereyama, Y.; Holz, E., and Beck, C.: Adrenergic innervation of the cochlea. *Ann. Otol. 75:* 69–86, 1966.

Thalmann, R. R.: Quantitative biochemical techniques for studying normal and noise-damaged ears. In Henderson, D.; Hamernik, R. P., Dosanjh, D. S., and Mills, J. H. (Eds.): *Effects of Noise on Hearing.* New York, Raven Press, 1976, pp. 129–154.

Thalmann, R.; Marcus, D. C., and Thalmann, I.: Biochemistry of the inner ear. In Gorlin, R. J. (Ed.): *Morphogenesis and Malformation of the Ear.* New York, Alan R. Liss, Inc., 1980, pp. 83–105.

Vinnikov, Ya. A., and Titova, L. K.: *The Organ of Corti: Its Histopathology and Histochemistry.* New York, Consultants Bureau, 1964.

Zenner, H. P. Short-time culture of isolated outer hair cells from guinea pig cochlea. *Assoc. Res. Otolaryngol. Abstr. 7:* 12, 1984.

Chapter 23

DEOXYGLUCOSE UPTAKE PATTERNS IN THE AUDITORY SYSTEM: METABOLIC RESPONSE TO SOUND STIMULATION IN THE ADULT AND NEONATE

ALLEN F. RYAN, NIGEL K. WOOLF, ANDREW CATANZARO,
STUART BRAVERMAN, AND FRANK R. SHARP

I. Introduction
II. Deoxyglucose Uptake in the Auditory System
  A. Cochlea
    1. Adult
    2. Neonate
  B. Central Auditory Pathway
    1. Stimulus Intensity
    2. Stimulus Frequency
    3. Behavior
    4. Neonate
III. Conclusions
    References

I. INTRODUCTION

The application of the 2-deoxy-D-glucose (2-DG) autoradiographic technique to the investigation of neural function, which was initiated in the 1970's (Kennedy et al., 1975), helped to bridge the gap between the disciplines of neurophysiology and neuroanatomy. Within the constraints of the method, this technique allows direct visualization of the distribution of presumed metabolic activity, and thus indirectly of neural activity, in brain tissue. For sensory systems, this technique has demonstrated functional activity in a number of central sensory pathways, including central auditory pathways, but application to the sensory end organs has not occurred as rapidly. In the cochlea, this has, in part, been due to the technical problems of obtaining access to inner-ear tissues through the bony cochlear capsule.

The biochemical basis of the 2-DG technique has been described in detail by Schacht in Chapter 22. Briefly, the method is based upon the incorporation of the glucose analogue, 2-DG, into tissue via the same uptake mechanisms by which glucose itself is transported into cells. Most of the 2-DG is converted into 2-deoxy-D-glucose-6-phosphate. At this point, the tracer cannot proceed along the glucose metabolic pathways and is trapped in the tissue with a half-life of about 24 h. Some of the 2-DG remains unconverted in the tissue, and a small fraction is also incorporated into glycogen.

2-Deoxy-D-glucose, injected as an intravascular pulse, is rapidly removed from the circulation, so that within 45–60 min only a small percentage of the tracer remains in circulation. Tissue uptake of 2-DG depends upon the serum concentration of the tracer, the serum concentration of glucose with which 2-DG competes for transport, and the level of metabolic activity of the tissue in question. The net effect is that tissues with high metabolic activity incorporate more 2-DG than tissues with low metabolic activity. Metabolism in the brain is almost entirely aerobic. Activation of neural pathways usually results in dramatic elevations in 2-DG uptake in the associated neural structures. This can be demonstrated visually if the 2-DG is radiolabeled and autoradiographs are prepared from tissue sections. The relation between 2-DG uptake and neural activity has not been fully illucidated. Under many conditions, uptake of the tracer appears to parallel the discharge rate of neurons, presumably driven by ion transport, transmitter synthesis, and other correlates of impulse and synaptic transmission (Sokoloff, 1977). However, it has been reported that neural inhibition may either decrease (Silverman, 1979; Serviere and Webster, 1983; Sharp and Ryan, 1984) or increase (Ackerman, 1984) 2-DG uptake.

II. DEOXYGLUOCOSE UPTAKE IN THE AUDITORY SYSTEM

A. COCHLEA

1. Adult

Application of the 2-DG technique to the cochlea presents several difficulties not present in brain. The primary problems result from the highly water-soluble nature of 2-DG. Once cells containing the tracer are perturbed or disrupted, the 2-DG diffuses freely from its *in vivo* location. In brain, this problem is solved by freezing, frozen-sectioning, and rapid drying of the sections. Such a technique is not applicable to the cochlea because of the presence of a bony capsule and large fluid spaces adjacent to all of the relevant tissues. Frozen sections would be difficult to obtain, and upon drying, the large extracellular fluid pools would lead to tracer diffusion.

The following protocol was developed to avoid the above difficulties. After injection of ^{14}C-2-DG (16.7 μCi/100 g of body wt) followed by a 1-h incubation period either in silence or with acoustic stimulation, gerbils were sacrificed and their inner ears were rapidly dissected from the temporal bones without breaking the round or oval windows. The intact cochleae were then frozen in a Freon-12 slush, cooled to $-159°C$ in liquid nitrogen, lyophilized at $-40°C$ and 0.01 Torr for 72 h, vaporfixed over 4% osmium tetroxide followed by acrolein, and embedded in Spurr resin using only organic solvents. The embedded cochleas were cut in half along the modiolus, using a bone-cutting lathe, and exposed on LKB Ultrafilm. All embedding fluids were counted in a liquid-scintillation counter, to ensure that no loss of radioactive tracer occurred at any step of the procedure.

This technique was used to assess the effects of acoustic stimulation upon deoxyglucose uptake in inner-ear tissues (Ryan *et al.*, 1982a). The results of this investigation are presented in Fig. 1. The figure shows typical autoradiographs prepared from cochleas which were kept in silence or exposed at 85 dB SPL to one of four stimulus conditions: wide-band noise, or one of three different pure tones.

The major feature of the 2-DG uptake pattern observed for the condition of silence is the very high level of uptake observed in the lateral-wall structures, both spiral ligament and stria vascularis, when compared to the remaining cochlear tissues. No marked difference in uptake between the spiral ligament and the stria vascularis can be observed. During exposure to wide-band noise at an intensity of 85 dB SPL, 2-DG uptake in the spiral ganglion and the eighth nerve increased dramatically, when compared to lateral-wall uptake. A slight increase in the uptake of 2-DG in the stria vascularis, as compared to the spiral ligament, can be seen in some animals. Relatively little change is observed in the organ of Corti. Localized regions of relatively high 2-DG uptake are observed in tonotopically-appropriate areas of the cochlea during exposure to pure tones. With a 0.75 kHz tone, uptake is greatest in the spiral ganglion of the apical turn, and in a band of the eighth nerve along the lateral edge. During exposure to a 3.0-kHz tone, uptake is greatest in the lower-middle-turn spiral ganglion and in a band of the eighth nerve along its medial edge. With a 12.0-kHz tone, uptake is greatest in the lower-basal-turn ganglion, and in a small region of the eighth nerve close to the cochlear base.

The effects of 85-dB-SPL, wide-band noise on 2-DG uptake were examined quantitatively by obtaining serial serum samples throughout the 1-h, 2-DG incubation period. Specific activity curves for 2-DG were determined for three noise-exposed and three control subjects, using techniques similar to those of Sokoloff *et al.* (1977). This allowed a quantitative comparison of tissue concentrations of 2-DG, obtained by microdensitometry from cochlear autoradiographs, across all subjects. The results of this analysis are illus-

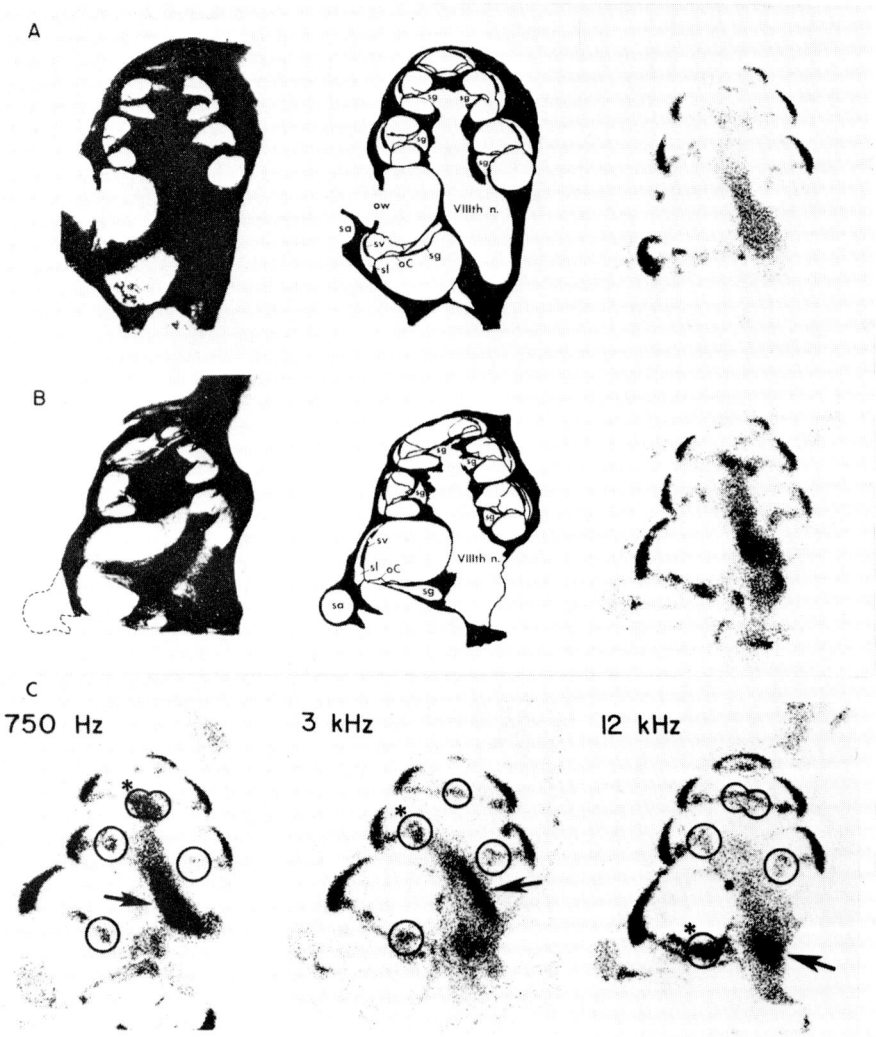

Figure 23-1. Typical autoradiographs obtained from cochleae maintained in silence (A), exposed to wide-band noise (B), or to one of three pure tones, as indicated (C), for 1 h following injection of ^{14}C-2-DG. The light micrographs in A and B show the corresponding plastic sestions, which were approximately 200 μm in thickness. The camera lucida drawings show the structures at the surface of each section which were exposed to the film; sv, stria vascularis; sl, spiral ligament; oC, organ of Corti; sg, spiral ganglion; sa, stapedial artery; ow, oval window. (From Ryan et al., 1982a.)

trated in Fig. 2, which compares 2-DG uptake in various tissues of the first cochlear turn, for silence versus 85-dB-SPL, wide-band noise. It is apparent from the figure that acoustic activation has little effect upon 2-DG uptake in the spiral ligament. Some increase in uptake is observed in the stria vascularis during noise, although the increase is not statistically significant. A modest increase in uptake occurs in the organ of Corti. In the spiral ganglion and eighth nerve, wide-band noise induces dramatic increases in 2-DG uptake.

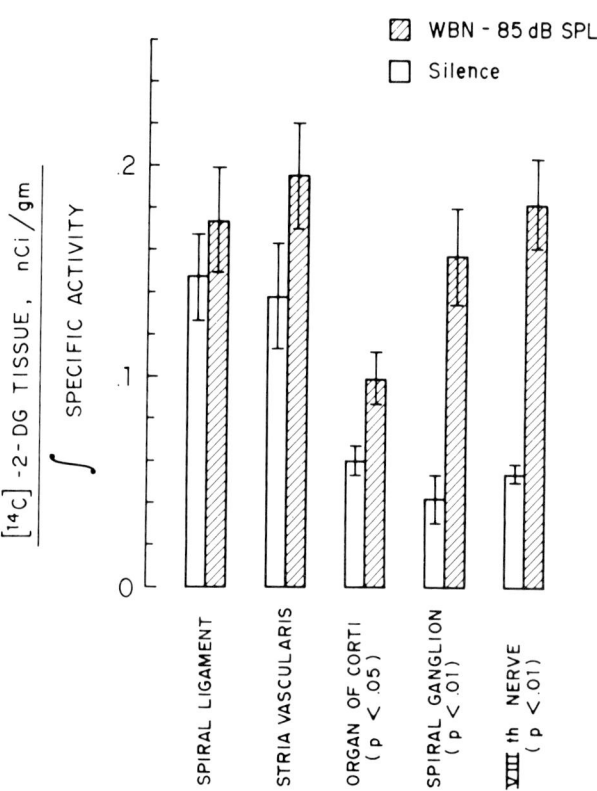

Figure 23-2. Mean 2-DG levels in cochlear tissues of the lower basal turn from inner ears in silence and in 85-dB-SPL, wide-band noise. Tissue levels have been normalized for each subject by the integral of the plasma specific activity curve. Each mean represents six cochleae; vertical bars show one standard deviation about each mean. (From Ryan et al., 1982a.)

When the pattern of 2-DG uptake in the cochlea was examined across several intensities of wide-band noise, it was found that a relative increase in uptake in the ganglion and nerve occurred gradually between 0 and 65 dB SPL, with a very sharp increase at 85 dB SPL. Little additional increase in relative uptake occurred between 85 and 105 dB SPL. In the organ of Corti, a gradual increase in uptake occurred between 0 and 105 dB SPL.

One question raised by these observations was whether or not the small increases in 2-DG uptake observed in the stria vascularis and organ of Corti represent a more substantial increase of the metabolism of a small portion of these structures. In order to define the pattern of cochlear 2-DG uptake with greater precision, a method for the use of ^{3}H-labeled 2-DG was developed (Ryan and Sharp, 1982). Essentially, the same procedures as employed for the ^{14}C-labeled tracer were used. However, a dose of 1.67 mCi/100 g body weight was employed to compensate for the weaker beta emissions produced by tritium decay. After bisection of the plastic-embedded cochlea, individual turns were separated, the bony capsule was removed, and 3–5 µm sections were cut with glass knives. The sections were flattened on glass slides with xylene or anhydrous glycerol, covered with nearly-dry emulsion by the dry-loop technique (Caro and Tubergen, 1962; Ryan and Sharp, 1982), and exposed for 125 days. Sections from the lower basal turn of noise-exposed cochleae (85-dB-SPL, wide-band noise) were compared with those from cochleae of animals kept in silence.

Resolution of tracer was achieved at the cellular and subcellular levels, as illustrated in Fig. 3. For cochleae of animals kept in silence, the greatest number of grains were observed over the stria vascularis and spiral ligament, with far fewer grains over other tissues. These results are in agreement with the previous ^{14}C-2-DG results (Ryan *et al.*, 1982a). In cochleae exposed to noise, the number of grains over spiral ganglion cells and eighth nerve fibers increased, as had been observed with ^{14}C-2-DG. However, the increased resolution of the ^{3}H-2-DG technique permitted new observations. In silence, the uptake of 2-DG in the stria vascularis was approximately equivalent to that seen in the spiral ligament. Uptake in all three strial cell types was equal. During noise exposure, 2-DG uptake increased significantly and equally in all three cell types of the stria vascularis, but not in the spiral ligament (Fig. 3a). In the organ of Corti, increased uptake was observed in inner hair cells, but not in supporting cells or outer hair cells. Increased uptake was apparent in the region of the nerve endings underneath the outer hair cells (Fig. 3d). Uptake of 2-DG increased in spiral ganglion cells and in eighth nerve fibers in the modiolus, as had been observed with ^{14}C-2-DG. However, with ^{3}H-2-DG, it was apparent that wide variations in uptake occurred from one individual spiral ganglion cell to the next (Fig. 3b). Increased uptake in individual afferent dendrites within the osseous spiral lamina was also observed (Fig. 3c).

In summary, acoustic stimulation increases the incorporation of 2-DG into certain cochlear tissues, namely, the stria vascularis, inner hair cells, and cochlear neurons. This is consistent with the reports of Canlon and Schacht (1981, 1983, Chapter 22, this volume), who observed increases in 2-DG uptake in microdissected samples of lateral wall, organ of Corti, and the modiolous during noise exposure. The increase in 2-DG uptake in the stria vascularis is consistent with Davis' (1965) battery model of cochlear trans-

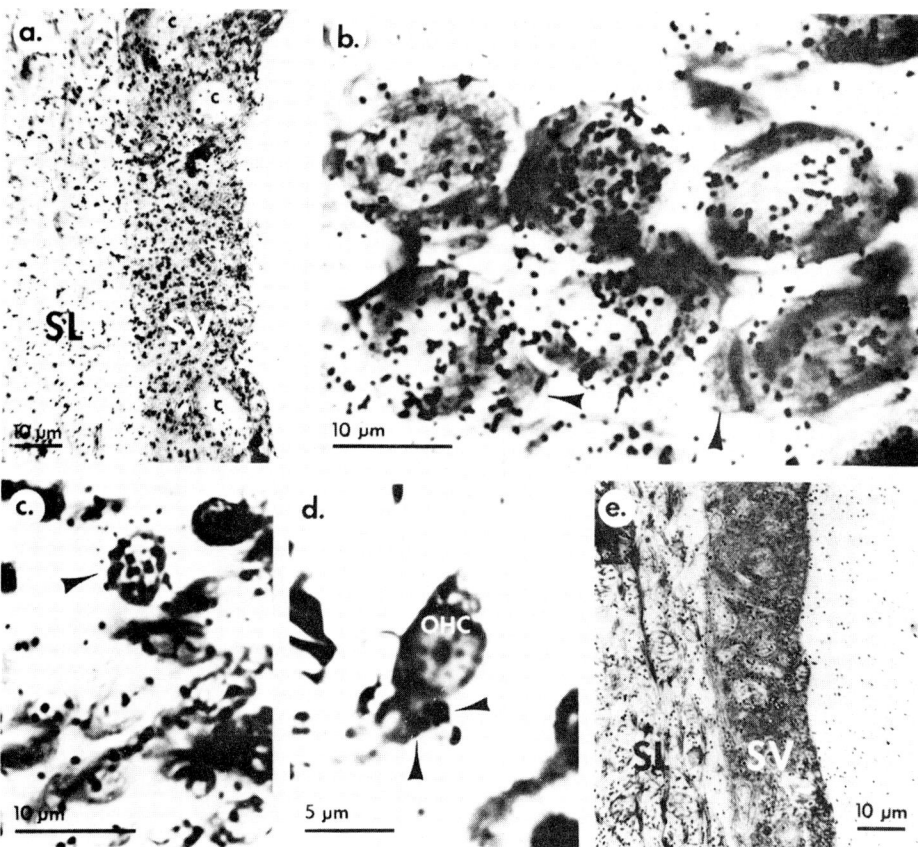

Figure 23-3. Dry-loop autoradiographs of cochlear tissues following injection with ^{3}H-2-DG and exposure to 85-dB-SPL, wide-band noise. a. Stria vascularis (SV) and spiral ligament (SL). b. Spiral ganglion neurons and satellite cells (arrows). c. Nerve fibers in the osseous spiral ganglion. Note the fiber with particularly high uptake (arrow). d. Dense accumulations of grains (arrows) over the synaptic region under the outer hair cells. e. Autoradiograph of a section which was dipped in emulsion rather than dry-looped, showing the loss of localized label due to diffusion of the 2-DG. (From Ryan and Sharp, 1982.)

duction, in which the stria provides the energy for the generation of the cochlear potentials via the endocochlear potential. In the organ of Corti, the inner hair cell is the only type which shows a metabolic response to stimulation. This is consistent with the fact that 90–95% of eighth-nerve afferent fibers synapse on this hair-cell population. The lack of increase in 2-DG uptake during sound stimulation in the outer hair cells suggests that this cell type may derive the energy for its role in transduction indirectly, perhaps via the endocochlear potential. Alternatively, the outer hair cells may utilize an energy source other than glucose. For example, it has been reported that amino acids are more important than glucose in maintaining hair cells in culture (Zenner, 1984), and that preferental incorporation of the amino acid glutamine by hair cells has been observed *in vivo* (Ryan and Schwartz, 1984). Other possibilities for biochemically supporting the outer hair cells include anaerobic and lipid metabolism. The high level of 2-DG uptake observed in the spiral ligament in the present studies was unexpected, and suggests that this tissue may play a more important role in cochlear function than has previously been appreciated. Since 2-DG uptake in the ligament is not affected by acoustic stimulation, this role may be one of homeostasis, perhaps of the cochlear perilymph.

2. Neonate

Cochlear 2-DG uptake has been investigated with ^{14}C autoradiography in the neonate, from the initial appearance of auditory function to the age at which adult thresholds are achieved. Cochlear microphonic responses can first be elicited in the gerbil by acoustic stimulation at 12 days after birth (DAB), with thresholds of about 100 dB SPL. Thresholds rapidly decrease and approach adult sensitivity by 18 DAB (Woolf and Ryan, 1984). While the cochlear endolymph is essentially adult-like in ionic composition by 12 DAB, the endocochlear potential undergoes a period of rapid maturation between 12 and 18 DAB (Ryan and Woolf, 1983). The uptake of 2-DG in the cochlea during this period is illustrated in the autoradiographs of Fig. 4. At 12 DAB, 2-DG uptake in the lateral wall tissues is uniformly low, in marked constrast to that of the adult. Uptake of 2-DG in the bone of the cochlear capsule, which is virtually nil in the adult, is slightly higher than that seen in soft tissues. By 13 DAB, uptake in the stria vascularis is higher than that seen in adjacent tissues. Between 14 and 18 DAB, uptake of 2-DG in the lateral-wall structures gradually increases, while that in the bony capsule decreases.

The appearance of auditory function is correlated with an increase in 2-DG uptake, and thus presumably of metabolism, in the lateral wall. This increase appears to match closely the maturation of the endocochlear potential. This strongly suggests that the majority of the metabolism of the lateral wall is devoted to the production of this potential. Interestingly, the generation of the cochlear endolymph does not seem to require a high level of metabolism in the lateral wall, because this fluid is adultlike at 12 DAB when 2-DG uptake is very low.

B. CENTRAL AUDITORY SYSTEM

Several investigators have shown that the 2-DG technique can be used to identify functional activation of the central auditory pathway (Silverman *et*

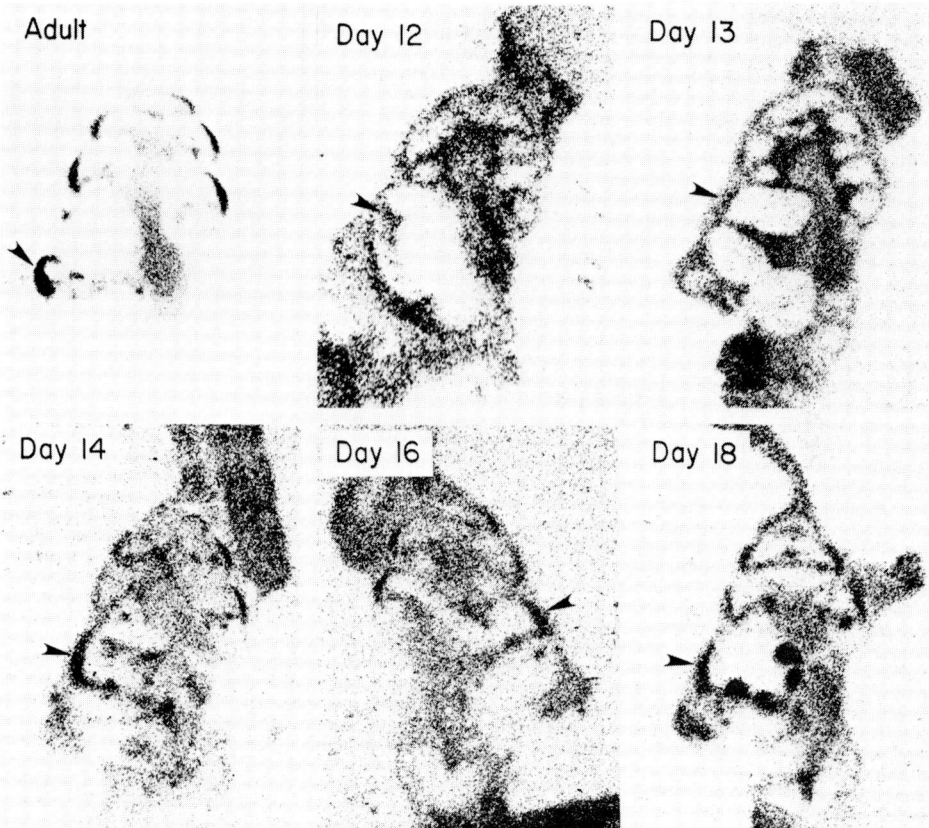

Figure 23-4. Autoradiographs illustrating the uptake of 2-DG in the cochlea of neonatal gerbils at 12, 13, 14, 16, and 18 days after birth. Note the sequential development of a high level of 2-DG uptake in the membranous lateral wall (arrows) over this period.

al., 1977; Webster et al., 1978; Jones and Disterhoft, 1979; Scheich et al., 1979; Hungerbuhler et al., 1981; Scheich et al., 1983). We have explored the effects of acoustic stimulation upon 2-DG uptake in the gerbil central auditory pathway in some detail (Sharp et al., 1981; Ryan et al., 1982b, c), using the standard techniques which have been developed for brain tissue. Briefly, awake animals were injected with 16.7 μCi of ^{14}C-labeled 2-DG. They were kept for 1 h inside a double-walled, sound-attenuated room, in the stimulus condition of choice. Following this period, the brain was removed, frozen and sectioned. The sections were rapidly dried and autoradiographs were generated by exposing the sections to x-ray film for 7 days. The density of the autoradiographic image of a brain structure provided a measure of 2-DG uptake during the 1-h period following injection of the tracer.

1. Stimulus Intensity

Almost all central auditory structures show increases in 2-DG uptake during acoustic stimulation. Figure 5 shows the effects of wide-band noise stimulation on 2-DG uptake in several auditory structures, and illustrates the basic types of responses observed. For most auditory nuclei, 2-DG uptake increases regularly with increasing stimulus intensity up to about 85 dB SPL. Relatively little additional increase is observed between 85 and 105 dB SPL. However, in the inferior colliculus, 2-DG uptake increases to very high levels between 0 and 25 dB SPL, and then increases more slowly at higher intensities. By far the greatest degree of stimulus-associated 2-DG uptake is observed in the inferior colliculus. The smallest increase in 2-DG uptake observed in a central auditory structure during stimulation occurs in the medial geniculate nucleus and auditory cortex. No change in 2-DG uptake occurred in response to acoustic stimulation in the medial nucleus of the trapezoid body. The reasons for the high degree of apparent metabolic response in the inferior colliculus, and the absence of response in the medial nucleus of the trapezoid body, are not clear. The relatively modest response observed in the medial geniculate and auditory cortex, on the other hand, may relate to the greater sensitivity of these structures to such variables as attention and physiologic state, as discussed below.

2. Stimulus Frequency

When pure tones are used as stimuli, 2-DG uptake increases in different regions of central auditory structures, depending upon the frequency of the stimulus employed. This is illustrated in the autoradiographs of Fig. 6A, which show the patterns of 2-DG uptake observed in the posterior cochlear nuclear complex following stimulation with wide-band noise or one of three pure tones. In each of the three divisions of the complex, a pure tone produces a simple, focal region of elevated 2-DG uptake in the tonotopically appropriate area, as shown for the dorsal and posterior ventral cochlear nuclei. In Fig. 6, the patterns of 2-DG uptake produced by each of the pure tones have been semiquantitated by measuring optical densities along the tonotopic axis of the autoradiographic image of the structure on a microdensitometer.

The 2-DG response to pure tones was not as simple in the inferior colliculus. As illustrated in Fig. 6B, 2-DG uptake in the central nucleus of the colliculus did not occur uniformly even with broad-band stimulation. Rather, 2-DG uptake occurred in a series of 3–4 bands oriented dorsomedially to ventrolaterally. The response to pure tones was also in accordance with this banding pattern. As shown by the densitometry functions, each pure tone

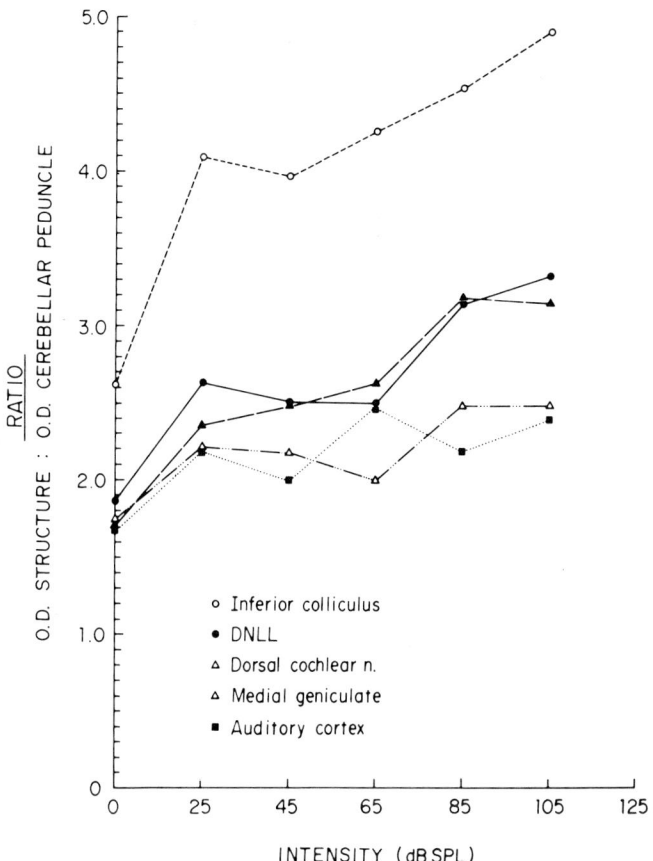

Figure 23-5. Increase in relative ^{14}C-2-DG uptake in various central auditory structures, as a function of wide-band noise intensity. Note that the inferior colliculus shows a much greater response to stimulation than the typical auditory structure, illustrated by the response of the dorsal cochlear nucleus and the dorsal nucleus of the lateral lemniscus. In contrast, the 2-DG response of the medial geniculate nucleus and auditory cortex are markedly less pronounced than those of the brainstem and midbrain auditory nuclei. (From Sharp et al., 1981.)

activated a tonotopically appropriate portion of the inferior colliculus, but only up to the "ceiling" represented by the wide-band-noise, uptake response. There is thus a metabolic pattern of response in the inferior colliculus which is rigidly defined and within which other responses tend to occur. The basis for this pattern is not clear. Neither the cytoarchitecture nor the functional characteristics of neurons in the inferior colliculus have been shown to display any pattern comparable to the banding in the 2-DG response. The pattern is not of binaural origin, because it occurs unchanged in the inferior colliculus of monauralized animals.

The response to pure-tone stimulation at various intensities may shed

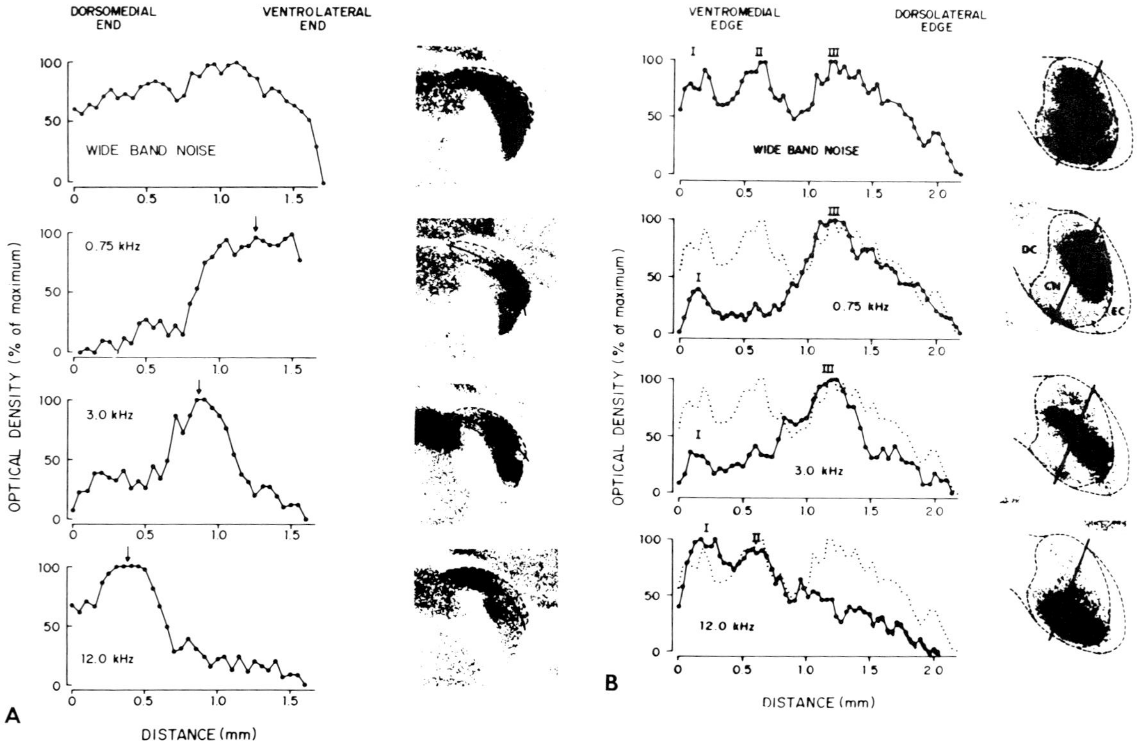

Figure 23-6. A. Variation in 2-DG uptake, expressed as optical density of the autoradiographic image, along the tonotopic axis of the dorsal cochlear nucleus, during stimulation with wide-band noise or one of three pure tones, as indicated. The arrow shows the location of the center of each band of elevated 2-DG uptake in the pure-tone functions. The solid lines in the autoradiographs show the location of optical density measurements. B. Variation in 2-DG incorporation along the tonotopic axis of the inferior colliculus, during stimulation with wide-band noise or a pure tone. The dotted line in each pure-tone function represents the response to wide-band noise. Note the agreement between the maximum optical density in each pure-tone condition and the tonotopically-appropriate portion of the response to noise.

some light on this question. Gerbils were exposed to a 3.0-kHz signal at intensities of 25, 45, 65, 85, or 105 dB SPL. The patterns of 2-DG uptake produced by these stimuli in the posterior cochlear nuclear complex and in the inferior colliculus are shown in Fig. 7. In the cochlear nuclear complex, 3.0-kHz stimulation produces a narrow region of high 2-DG uptake at 25 dB SPL (best illustrated in the dorsal cochlear nucleus, DCN). This region broadens only slightly at 45 and 65 dB SPL. The band widens significantly at 85 dB, expanding primarily in the dorso-medial direction, that is, toward the area of higher-frequency representation in the DCN. At 105 dB SPL, the area of high 2-DG uptake spreads to fill the entire dorso-medial DCN. Relatively little spread toward the ventral-lateral (low frequency) DCN is observed. The area of relatively high 2-DG uptake produced by pure-tone stimulation in the DCN is homogeneous. A more complex response to pure tones is observed in the inferior colliculus. A band of relatively high 2-DG uptake is observed in the inferior colliculus in response to 25-dB–SPL, 3.0-kHz stimulation. As in the DCN, this band increases only slightly in width at 45 and 65 dB SPL, and more so at 85 dB SPL. However, on either side of the band of elevated 2-DG uptake are bands of 2-DG uptake which are of lower density than those of the rest of the inferior colliculus. These bands of lower 2-DG uptake increase in width with increasing stimulus intensity, until at 85 dB SPL, they fill almost the entire inferior colliculus on either side of the band of elevated 2-DG uptake. At 105 dB SPL, relatively high 2-DG uptake spreads to fill the entire ventromedial (high-frequency) end of the colliculus, as in the DCN. However, the pattern of higher uptake is broken into bands, similar to those seen with wide-band noise stimulation in Fig. 6B.

Silverman (1979) and Serviere and Webster (1983) have also reported bands of relatively low 2-DG uptake associated with bands of higher uptake in the inferior colliculus following pure-tone stimulation. Such "suppressive" bands appear to be related to the inhibition of single-unit discharges, both above and below the characteristic frequencies, as reported in electrophysiological studies in this nucleus (e.g., Ryan and Miller, 1978). The fact that bands of relatively low 2-DG uptake are not observed in the cochlear nucleus or other brainstem auditory nuclei suggests that inhibition may play a more important role in the inferior colliculus than in lower auditory nuclei. The banding observed at 105 dB SPL may also reflect the complex interplay of excitation and inhibition.

3. Behavior

As mentioned previously, the very low level of 2-DG uptake which is observed in the medial geniculate nucleus and auditory cortex during expo-

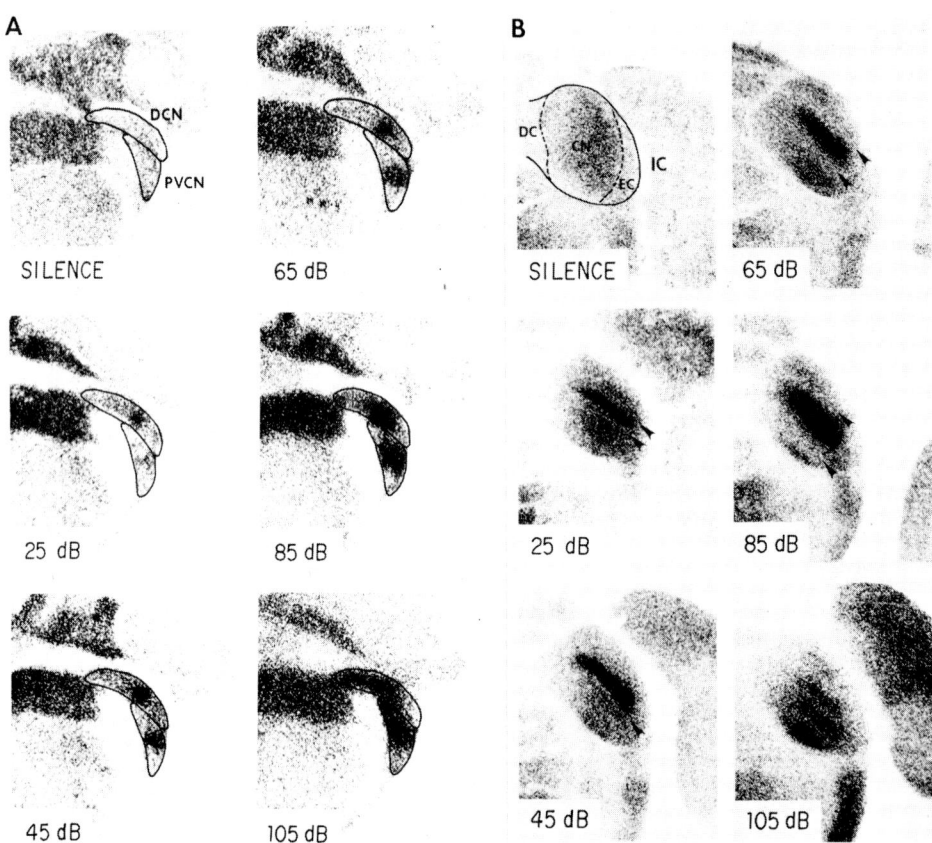

Figure 23-7. A. Patterns of 2-DG uptake in the dorsal cochlear nucleus during silence or during exposure to a 3.0-kHz tone at various intensities. Note that the band of relatively high 2-DG uptake produced by the tone increases only slightly in width from 25–65 dB SPL, then broadens toward the area of high-frequency representation at 85 and 105 dB SPL. DCN, dorsal cochlear nucleus; PVCN, posterior ventral cochlear nucleus. B. Similar data from the inferior colliculus. Note that bands of relatively low 2-DG uptake occur on either side of the band of relatively high 2-DG uptake at stimulus intensities of 25–85 dB SPL (arrows). These bands broaden with increasing stimulus intensity. At 105 dB SPL, the area of relatively high 2-DG uptake is broken into a series of bands.

sure to acoustic stimuli (Fig. 5) may be related to attentional variables. Miller et al. (1972) and Ryan et al. (1984) have reported that neurons in the higher auditory centers are more responsive to repetitive acoustic stimuli when animals are trained to attend to the signals. Skinner and Yingling (1977) have suggested that attention may "gate" sensory signals at the level of the thalamus. The repetitive signals used in our 2-DG experiments might be filtered out by such a "thalamic gate". To investigate this possibility, we trained animals in an intensity discrimination task to detect small variations in the intensity of a wide-band noise signal set at 85 dB SPL. 2-DG was then injected, just before a 1-h session in which the subjects performed in this behavioral task, and thus, were actively attending to the acoustic signal. 2-DG uptake during behavioral performance was compared to uptake during identical noise exposure in animals which had not been trained. As a control for general arousal effects associated with behavioral performance, animals were trained to detect small changes in the illumination level of the experimental chamber, during the presence of the wide-band noise signal. The uptake of 2-DG was then assessed during performance in the visual discrimination task. The relative uptake of 2-DG in the three groups was similar in all auditory nuclei, up to and including the inferior colliculus (Fig. 8). Relative uptake in the medial geniculate nucleus and auditory cortex was significantly greater for the animals performing in the auditory intensity-discrimination task, but not for those performing in the visual intensity-discrimination task. These data suggest that the lower levels of 2-DG uptake, observed in these higher auditory nuclei of naive animals, may be related to a lack of attention to the acoustic signals, and that the attentional effects are greater in the medial geniculate nucleus and auditory cortex.

4. Neonate

There is extensive evidence that the processing of acoustic signals during auditory ontogeny is different from that seen in adults (Rubel, 1984). This may reflect the varying rates at which different components of the central auditory system develop. Morest (1969) has shown that the anatomical development of the central auditory pathway proceeds more rapidly in the peripheral auditory nuclei than in the higher auditory centers. Other investigators (Purpura et al., 1965, 1968) have suggested that inhibitory phenomena may play a more important role early in the developmental process. To determine whether the spatial representation of acoustic stimuli in the neonate is different from that seen in adults, we assessed the uptake of 2-DG in the central auditory pathway in gerbils during the period of rapid functional development of the auditory system. Neonates at ages of 12, 14, 16, and 18 DAB were injected with 2-DG and exposed to 105-dB–SPL, wide-band noise. This intensity was chosen in order to compensate for the very high thresholds of animals at younger ages. A significant increase in 2-DG uptake was observed only in the ventral cochlear nucleus at 12 DAB. At 14 DAB, as illustrated in Fig. 9, noise stimulation produced increased 2-DG uptake in all auditory nuclei up to the level of the ventral nucleus of the lateral lemniscus. Higher auditory nuclei showed no stimulus-evoked increase. At 16 DAB, stimulus-evoked increases were seen through the inferior colliculus,

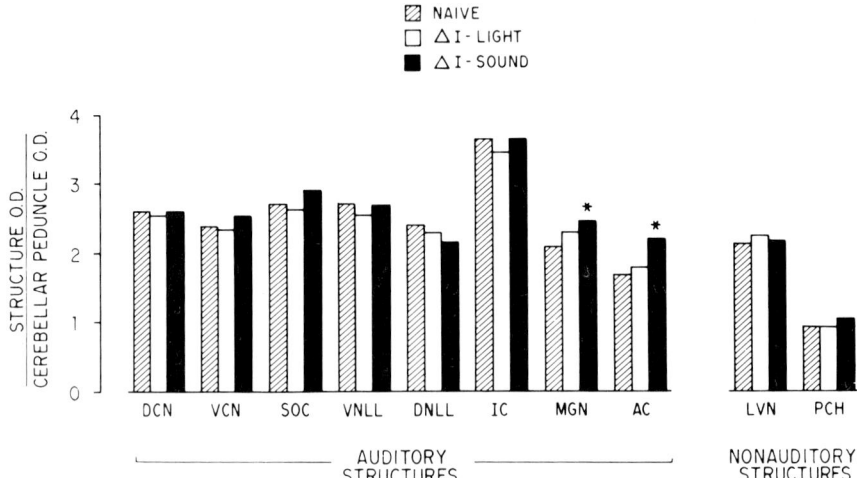

Figure 23-8. Relative 2-DG uptake, expressed as optical densities from autoradiographic images, in various structures of the central auditory pathway during exposure to 85-dB-SPL, wide-band noise. Uptake in naive animals is compared to that seen in animals performing in either an auditory (ΔI–SOUND) or a visual (ΔI–LIGHT) intensity-discrimination task. 2-DG uptake in the medial geniculate nucleus and in the auditory cortex is significantly higher (*) during performance in the auditory discrimination task than in either of the other two groups. No effect of performance is seen in lower auditory nuclei, in the lateral vestibular nucleus, or in the posterior cerebellar hemisphere.

at 18 DAB through the auditory cortex. Auditory function, as reflected by stimulus-evoked, presumed metabolic activity, thus appears sequentially along the neuraxis. This may reflect sequential maturation of the central auditory pathway, as observed anatomically by Morest (1969). Alternatively, if inhibitory responses mature before excitatory processes, initial responses at higher centers may be weighted more toward inhibition than excitation, because inhibition is more prevalent in higher auditory centers than in more peripheral nuclei.

Rubel and Ryals (1983) and Lippe and Rubel (1983) have presented evidence that frequency representation in the chick is different from that in the adult chicken. To determine whether this is also the case in mammals, we assessed the response of neonatal gerbils to pure tones at 14 DAB. Pure tones at frequencies of 0.75, 3.0, or 12.0 kHz were used, at an intensity of 105 dB SPL. Responses to pure tones in the neonates occurred at different locations than in adults (Fig. 10). Areas of higher 2-DG uptake in the 14 DAB gerbils were displaced to regions which, in the adult, represent frequencies 1–2 octaves higher. The changes in frequency representation observed in these experiments indicate that the frequency re-

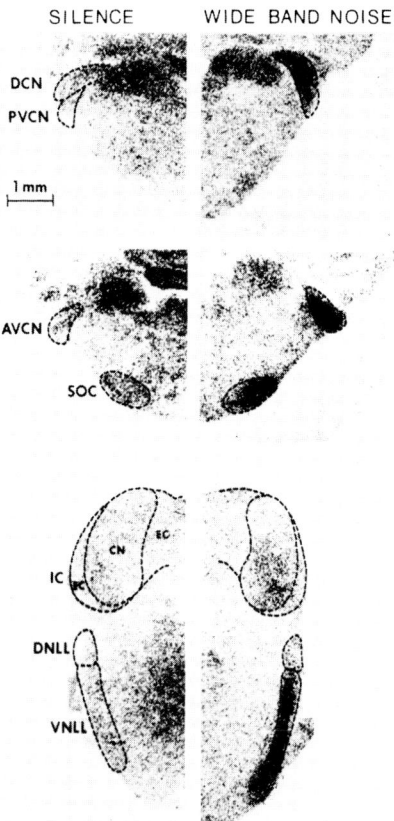

Figure 23-9. Patterns of 2-DG uptake in central auditory structures in silence and during exposure to 105-dB-SPL, wide-band noise in a neonatal gerbil at 14 DAB. Stimulus-evoked increases in 2-DG uptake are visually apparent in all auditory nuclei, up to and including the ventral nucleus of the lateral lemniscus (VNLL), but not in the dorsal nucleus of the lateral lemniscus (DNLL), or higher nuclei. (From Ryan et al., 1982c.)

sponse of the basilar membrane may be significantly different in neonates than in adults.

III. CONCLUSIONS

The 2-DG autoradiographic technique has proven to be a powerful tool for the investigation of apparent metabolic activity and auditory function in both the cochlea and the central auditory pathway. It has provided new insights into both peripheral and central auditory physiology.

In the cochlea, the high level of 2-DG uptake in the spiral ligament suggests that this structure plays a more important role in cochlear function

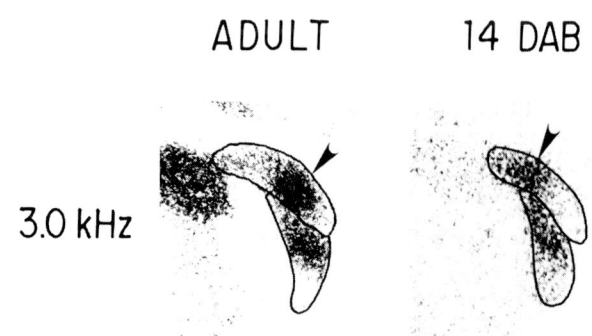

Figure 23-10. Patterns of 2-DG uptake produced in the posterior cochlear nucleus complex by a 3.0-kHz tone in an adult gerbil, and at 14 DAB. An 85-dB–SPL tone was used for the adult, and a 105-dB-SPL tone for the neonate. Note the difference in position of the bands of higher 2-DG uptake (arrows) between the two ages.

than has previously been recognized. Since sound stimulation does not increase 2-DG uptake in the spiral ligament, the function of this tissue may be related more to cochlear homeostatis than to transduction. The lack of stimulus-evoked increases in 2-DG uptake in outer hair cells is an important difference between this class of sensory cell and the inner hair cell, and indicates that the two cell types may operate in fundamentally different ways in the auditory transduction process. Increases in the aerobic metabolism of the stria vascularis during auditory development are closely correlated with the rise of the endocochlear potential, rather than with the concentration of endolymphatic potassium. This suggests that quite different processes generate these two aspects of the endolymphatic compartment.

2-DG autoradiography is ideal for investigating the spatial distribution of activity, including both tonotopic organization and the comparison of 2-DG uptake in different auditory structures. This technique has revealed patterns of neural activity in the central auditory pathway which have not been observed with other methodologies. The banding which is commonly observed in the inferior colliculus in response to sound stimulation is not closely related to any of the known morphological characteristics of the inferior colliculus. It may be a result of the complex interplay of inhibition and excitation which is known to occur in this nucleus.

As the 2-DG technique is developed further, it promises to provide more insights into function in auditory structures. For example, the application of high-resolution, ^{3}H-2-DG autoradiography to the study of the cochlea may help clarify the functional status of Type II spiral ganglion cells. In the central auditory pathway, the technique may also help identify the cells and synapses involved in inhibitory processing, such as may occur in the inferior colliculus.

ACKNOWLEDGMENTS

This research was supported by NIH grants NS 14945 and NS 00176, by the Research Service of the Veterans Administration, and by the Duaei Hearing Research Fund. The technical expertise of Delores DiPietro and Charles Graham contributed significantly to the completion of these experiments, and is gratefully acknowledged.

REFERENCES

Ackerman, S. K.: Increased glucose metabolism during long-duration recurrent inhibition of hippocampal pyramidal cells. *J. Neurosci.* 4: 251–264, 1984.

Canlon, B., and Schacht, J.: The effect of noise on deoxyglucose uptake into inner ear tissues of the mouse. *Arch. Otorhinolaryngol.* 230: 171–176, 1981.

Canlon, B., and Schacht, J.: Acoustic stimulation alters dexyglucose uptake in the mouse cochlea and inferior colliculus. *Hearing Res.* 10: 217–226, 1983.

Caro, L. G., and Tubergen, R. P. Van: High resolution autoradiography. Section I. Methods. *J. Cell. Biol.* 15: 173–188, 1962.

Davis, H.: A model for transducer action in the cochlea. *Cold Spr. Harb. Symp. Quant. Biol.* 30: 181–190, 1965.

Hungerbuhler, J. P.; Saunders, J. C., Greenberg, J., and Reivich, M.: Functional neuroanatomy of the auditory cortex studied with ^{14}C-2-deoxyglucose. *Exper. Neurol.* 71: 104–121, 1981.

Jones, L. S., and Disterhoff, J. F.: Visualizing the rabbit auditory pathway with ^{14}C 2-deoxy-D-glucose. *Soc. Neurosci. Abstr.* 5: 23, 1979.

Kennedy, C.; Des Rosiers, M. H., Jehle, J. W., Reivich, M., Sharp, F. R., and Sokoloff, L.: Mapping of functional neural pathways by autoradiographic survey of local metabolic rate with ^{14}C- 2-deoxyglucose. *Science* 287: 850–853, 1975.

Lippe, W., and Rubel, E. W.: Development of the place principle: tonotopic organization. *Science* 219: 514–516, 1983.

Miller, J.; Sutton, D., Pfingst, B., Ryan, A., Beaton, R., and Gourevitch, G.: Single cell activity in the auditory cortex of rhesus monkeys: behavioral dependency. *Science* 177: 449–451, 1972.

Morest, D. K.: The growth of dendrites in the mammalian brain. *Z. Anat. Entwikl.-Gesch.* 128: 190–316, 1969.

Purpura, D.; Prelevic, S., and Santini, M.: Postsynaptic potential and spike variations in the feline hippocampus during postnatal ontogenes. *Exp. Neurol.* 22: 408–422, 1968.

Purpura, D.; Shofer, R., and Scarff, T.: Properties of synaptic activities and spike potentials of neurons of immature neocortex. *J. Neurophysiol.* 28: 925–942, 1965.

Rubel, E. W.: Ontogeny of auditory system function. *Ann. Rev. Physiol.*, 1984. In press.

Rubel, E. W., and Ryals, B. M.: Development of the place principle: acoustic trauma. *Science* 219: 512–514, 1983.

Ryan, A. F.; Goodwin, P., Woolf, N. K., and Sharp, F. R.: Auditory stimulation alters the pattern of 2-deoxyglucose uptake in the inner ear. *Brain Res.* 234: 213–225, 1982a.

Ryan, A. F., and Miller, J. M.: Single unit responses in the inferior colliculus of the awake and performing rhesus monkey. *Exp. Brain Res.* 32: 389–408, 1978.

Ryan, A. F.; Miller, J., Pfingst, B. E., and Martin, G. K.: Effects of reaction time performance upon single unit discharge in the central auditory pathway of the rhesus monkey. *J. Neurosci. 4:* 298–308, 1984.

Ryan, A. F., and Schwartz, I. R.: Preferential glutamine incorporation by cochlear hair cells; implications for the afferent cochlear transmitter. *Brain Res. 290:* 376–379, 1984.

Ryan, A. F., and Sharp, F. R.: Localization of (^{3}H) 2-deoxyglucose at the cellular level using freeze-dried tissue and dry-looped emulsion. *Brain Res. 252,* 177–180, 1982.

Ryan, A. F., and Woolf, N. K.: Energy dispersive x-ray analysis of inner inner ear fluids and tissue during the ontogeny of cochlear function. *Scann. Electr. Micros. 1:* 201–207, 1983.

Ryan, A. F.; Woolf, N. K., and Sharp, F. R.: Functional ontogeny in the central auditory pathway of the mongolian gerbil: a deoxyglucose study. *Exper. Brain Res. 47:* 428–436, 1982b.

Ryan, A. F.; Woolf, N. K., and Sharp, F. R.: Tonotopic organization in the central auditory pathway of the monoglian gerbil: a 2-deoxyglucose study. *J. Comp. Neurol. 207,* 369–380, 1982c.

Scheich, H.; Bock, W., Bonke, D., Langner, G., and Maier, V.: Acoustic communication in the guinea fowl (Numida meleagris). In Ewert, J.-P; Capranica, R. C., and Jingle, D. J. (Eds.): *Advances in Vertebrate Neuroethology.* New York, Plenum, 1983, pp. 731–782.

Scheich, H.; Bonke, B. A., Bonke, H., and Langner, G.: Functional organization of some auditory nuclei in the guinea fowl demonstrated by the 2-deoxyglucose technique. *Cell Tiss. Res. 204:* 17–27, 1979.

Serviere, J., and Webster, W. R.: Excitatory and inhibitory tonotopic contours in the inferior colliculus of the cat: a ^{14}C-2-deoxyglucose study. In Webster, W. R., and Aitkin, L. M. (Eds.): *Mechanisms of Hearing.* Clayton, Victoria, Australia, Monash University Press, 1983, pp. 77–82.

Sharp, F. R., and Ryan, A. F.: Regional ^{14}C-2-deoxyglucose uptake during forelimb movements evoked by rat motor cortex stimulation: Pons, cerebellum, medulla, spiral cord, muscle. *J. Comp. Neurol. 224:* 286–306, 1984.

Sharp, F. R.; Ryan, A. F., Goodwin, P., and Woolf, N. K.: Increasing intensities of wide-band noise increase ^{14}C-2-deoxyglucose uptake in gerbil central auditory structures. *Brain Res. 230:* 87–96, 1981.

Skinner, J. E., and Yingling, C. D.: Central gating mechanisms that regulate event-related potentials and behavior. A neural model for attention. *Prog. Clin. Neurophysiol. 1:* 30–69, 1977.

Silverman, M. S.: Deoxyglucose demonstration of a new auditory projection field within the inferior colliculus. *Soc. Neurosci. Abstr. 5:* 30, 1979.

Silverman, M. S.; Hendrickson, A. E., and Clopton, B. M.: Mapping of the tonotopic organization of the auditory system by uptake of radioactive metabolites. *Soc. Neurosci. Abstr. 3:* 11, 1977.

Sokoloff, L.: Relation between physiological function and energy metabolism in the central nervous system. *J. Neurochem. 29:* 13–26, 1977.

Sokoloff, L.; Reivich, M., Kennedy, C., DesRosiers, M. H., Patlak, C. S., Pettigrew, K. D., Kakaurda, O., and Shinohara, M.: The deoxyglucose method for the measurement of local cerebral glucose utilization: theory, procedure, and normal values in the conscious and anesthetized albino rat. *J. Neurochem. 28:* 13–36, 1977.

Webster, W. R.; Serviere, J., Batini, C., and Laplante, S.: Autoradiographic demonstration with 2-(^{14}C)-2-deoxyglucose of frequency selectivity in the auditory system of cats under conditions of functional activity. *Neurosci. Lett. 10:* 43–48, 1978.

Woolf, N. K., and Ryan, A. F.: The development of auditory function in the cochlea of the mongolian gerbil. *Hearing Res.*, 1984. In press.

Zenner, H. P.: Short-time culture of isolated outer hair cells from guinea pig cochlea. *Assoc. Res. Otolaryngol. Abstr.* 7: 12, 1984.

Chapter 24

PERSPECTIVES IN THE PHYSIOLOGICAL CHEMISTRY OF THE COCHLEAR DUCT

RUEDIGER THALMANN AND DANIEL C. MARCUS

I. Introduction
II. Electrochemical Properties of the Endolymph
III. Origin and Access Pathways of Potassium and Glucose Utilized by the Stria Vascularis
IV. Possible Types and Locations of Ion-Transporting Mechanisms
V. Ionic Conductances of Labyrinthine Membranes: Relation to Cation Transport
References

I. INTRODUCTION

In this chapter, we will review some of the generally-accepted concepts concerning the formation of endolymph and the endolymphatic resting potentials, and will present some recent findings about these processes. It should be stated at the outset that no attempt is being made to present a comprehensive treatment of the subject, because the endolymphatic system is a highly complex entity. An enormous amount of experimental work has been carried out to date, and yet, there still exists no cohesive theory. Some of the recent work done in this and other laboratories has raised a number of intriguing questions.

Because of the high degree of complexity that has been revealed by these studies, it has become necessary to resort to special techniques in order to examine the system effectively, such as perfusion of the perilymphatic, endolymphatic, and vascular spaces, either alone or in combination.

It has also been found expedient to resort to the study of simpler, related systems such as the vestibular portion of the inner ear. Experimental approaches analogous to those used in the characterization of secretory epithelia in other systems are far more readily applied to the vestibular labyrinth than to the cochlea. In turn, with due caution, the situation in the vestibular labyrinth can be extrapolated to the cochlea, or at least can serve as a guide to the design of more targeted studies in the much more complex, auditory part of the inner ear.

An example of this approach will be outlined in some detail later in this chapter; the studies using barium (Ba^{++}, a blocker of K^+ channels) will serve to illustrate the variety of experimental schemes that must be implemented in order to arrive at a more complete understanding of corresponding processes in different portions of the inner ear.

II. ELECTROCHEMICAL PROPERTIES OF THE ENDOLYMPH

The modern era of endolymph physiology was ushered in by Békésy (1952), who, some thirty years ago, demonstrated that the endolymphatic space of the cochlea exhibits an electrical potential (termed the endolymphatic potential, EP) which is positive by some 80-90 mV with respect to perilymph and blood. Shortly thereafter, Smith et al. (1954) demonstrated that endolymph has a surprisingly high concentration of K^+ (approximately 150 mEq/L); however, not until much later was it firmly established that the Na^+ concentration of cochlear endolymph is unusually low (1 mEq/L or less; Bosher and Warren, 1968). Simple calculations demonstrated that both of these ions are displaced from electrochemical equilibrium, which in turn suggested that K^+ was transported into, and Na^+ out of, endolymph by active processes powered by metabolic energy.

The luminal resting potentials of the vestibular endolymphatic spaces are much lower than that of the cochlea (about +5 mV in the utricle and the ampullae), but the K^+ concentration is about as high as that of the cochlea (Sellick and Johnstone, 1975). The Na^+ concentration in vestibular endolymph, however, is substantially higher than in the cochlea, ranging between 10-15 meq/L. Nevertheless, because of the low resting potentials, the existing electrochemical gradients indicate that, as in the cochlea, K^+ must be actively transported into and Na^+ out of the endolymph.

Several lines of evidence suggested an association between K^+ secretion and the generation of the lumen-positive electrical potentials. The stria vascularis was identified by Tasaki and Spyropoulous (1959) as the tissue responsible for generation of the EP. These investigators opened the cochlear duct near the apex, drained most of the endolymph, and lightly touched the surface of different tissues with the tip of a microelectrode. The surface of the stria was the only tissue exhibiting a positive potential with respect to a remote electrode. Although this relatively crude experiment was first carried out in 1959, it still remains the most direct piece of evidence indicating that the stria is the tissue responsible for the generation of EP.

By analogy, it is generally agreed that in the vestibular labyrinth, secretion and absorption of K^+ and Na^+ and generation of the endolymphatic potentials are carried out by the dark-cell regions surrounding the sensory areas of the utricle and the ampullae (Kimura, 1969). The saccule contains no dark cells and its luminal potential has been shown to be a remote, passive manifestation of the cochlear EP (Sellick and Johnstone, 1975).

Since *in vitro* techniques offer numerous advantages in the characterization of secretory tissues, several investigators have attempted to apply this type of approach to the elucidation of strial function. Indeed, it was found possible to isolate the stria vascularis and to maintain it in a viable state *in vitro* for several hours. In this way, several fundamental metabolic, ionic, and pharmacological properties of this tissue have been characterized (Chou and Rodgers, 1962; Marcus *et al.*, 1978). However, all conclusions derived to date from *in vitro* experiments are tempered by the fact that the normally existing exposure of the luminal and contraluminal surfaces of the stria vascularis to fluids of fundamentally different ionic composition (and electrical polarization) could not be maintained *in vitro*. All attempts at mounting the stria vascularis between two half chambers, in analogy to an Ussing chamber, have failed so far. Such a preparation would allow characterization of the transport properties of the stria vascularis in a controlled fashion and would allow rather definitive conclusions to be drawn. This represents a major target area for future exploration.

In view of the lack of sufficiently powerful, *in vitro* techniques, we have no choice at this time but to rely on more indirect approaches. Sellick and Bock (1974), for instance, attempted to determine the interrelation between the rate of K^+ secretion and the magnitude of the sensory component of the EP, using endolymphatic perfusion techniques with artificial media as their primary investigative tool. In these experiments, the authors replaced endolymph with a solution of low K^+ concentration and determined the behavior of the K^+ activity and the EP as a function of time after the substitution. On the basis of several assumptions, they were able to infer from these data the rate of K^+ secretion. In this way, a strong positive correlation between the rate of K^+ secretion and the magnitude of the secretory component of the EP was demonstrated. The assumptions used in these calculations, however, leave the door open to more rigorous proof of the relation between the two phenomena. For instance, the authors have not established definitively whether the EP can be maintained at normal levels during endolymphatic perfusion under control conditions. It has been the experience of auditory researchers that it is extremely difficult to maintain normal levels of cochlear potentials during perfusion of the endolymphatic space with artificial media, even when the medium closely resembles the composition of natural endolymph. In addition, in the reported experiments (Sellick and Bock, 1974), endolymphatic perfusion was stopped as soon as the perfusate had reached the location of the recording electrode. This raises the possibility of the occurrence of artifacts due to longitudinal diffusion from nonperfused regions.

Experimental work in several laboratories confirms that the observed ion-transport phenomena depend upon metabolic energy. Following an interruption of the blood supply, cochlear oxygen tension drops rapidly, in close correlation with a precipitous decline of strial ATP and of the EP (Konishi et al., 1961; Thalmann et al., 1973). In addition, a pronounced decline in K^+ secretion during anoxia has been demonstrated (Konishi et al., 1978).

III. ORIGIN AND ACCESS PATHWAYS OF POTASSIUM AND GLUCOSE UTILIZED BY THE STRIA VASCULARIS

All of these experiments, combined with a consideration of the histologic properties of the different cells bordering the endolymphatic spaces, suggest that the strial marginal cells in the cochlea and the dark cells in the vestibular labyrinth are responsible for the described transport processes (Kimura, 1969). Further, because the stria vascularis, as the name implies, is highly vascularized, and because its capillaries are isolated from the perilymphatic compartment by the basal-cell layer (Rodriguez-Echandia and Burgos, 1965; Kimura, 1969) it appeared, *a priori*, most plausible that the vascular route would provide the most direct access for different chemical agents to the generator of the EP. However, several experimental observations cast doubt upon this concept. In order to test this and related questions in a definitive way, we developed a technique for arterial perfusion of the surviving inner ear with artificial media (Wada et al., 1979a; Kobayashi et al., 1984).

During the course of validating this technique we found that several pharmacological substances, such as oubain, loop diuretics, etc., do indeed exert their action upon the EP at much faster rates when applied via the vascular route than via the perilymphatic route (Kusakari et al., 1978; Wada et al., 1979b). In fact, application of furosemide by arterial perfusion produces an almost immediate decline of the EP (Thalmann et al., 1982), which is similar to the dramatic effect upon the tubular potential of the loop of Henle of the kidney, when furosemide is applied directly to the luminal membrane (Greger et al., 1983). However, quite unexpectedly, if K^+, the presumed ionic substrate of the generator of the EP, is omitted from the vascular perfusate, no immediate effect upon the potential is seen (Wada et al., 1979a). Only after some 20–40 min of K^+-free vascular perfusion does the EP start to decline. This finding, considered in isolation would, of course, tend to discredit the notion that electrogenic transport is the mechanism of the generation of the EP. However, when perilymph is replaced by a K^+-free medium, the EP *does* decline rapidly (Marcus et al., 1981; Konishi, 1973), which, while re-affirming the role of K^+ in the generation of the EP, would further suggest that the K^+ is derived from perilymph, rather than blood. Although such a process would appear to be somewhat circuitous, other data from our laboratory, dealing with the ionic and ATP content of the stria vascularis during K^+-free perilymphatic perfusion (Marcus et al., 1981) also support this concept. More direct supporting evidence was provided by radiotracer studies by Sterkers et al. (1982) (see also Sterkers, Chapter 26, this volume) and Konishi et al. (1978).

Although these results have identified the transport pool for K^+ rather definitively, they give no indication whatsoever about the exact *route* of K^+ transport within the stria vascularis; this has generated a new set of questions. Is perilymphatic K^+ taken up by the basal cells and then relayed in some fashion to the marginal cells to be ultimately secreted into endolymph? Or, alternatively, is K^+ capable of flowing through the junctional complexes between the basal cells? The question concerning the route of K^+ transport is very difficult to test and may remain a mystery for quite some time.

It is of interest in this context that glucose is likewise capable of crossing the basal cell layer at a rate sufficient to satisfy the metabolic requirements of the generator of the EP (Kambayashi et al., 1982). During glucose-free vascular perfusion, the EP does not begin to decline until the perilymphatic glucose pool has been exhausted. In contrast to the situation with K^+, however, the elimination of perilymphatic glucose, in the presence of adequate blood glucose, does not lead to significant impairment of the EP; this indicates that under normal, *in vivo* conditions, the *metabolic* requirements of the generator of the EP are supplied via the vasculature (and that the perilymphatic source contributes only under exceptional circumstances).

IV. POSSIBLE TYPES AND LOCATIONS OF ION–TRANSPORTING MECHANISMS

Next, the possible membrane processes which may contribute to K^+ secretion and Na^+ absorption deserve consideration. In essence, most of the experimental evidence on this subject has been derived from experiments in which the ionic composition of the inner-ear fluids was systematically varied or in which pharmacological agents were used to alter ion transport properties of the cells in certain, relatively well-defined ways. The EP served as the dependent variable in the majority of these experiments.

Essentially two basic models of K^+-transport by the strial marginal cells have been proposed on the basis of this type of experimentation in the last ten to fifteen years (Marcus et al., 1983). As diagrammed in Fig. 1, both models exhibit the ubiquitous Na^+/K^+ exchange pump in the basolateral membrane. In the model shown in the lower panel of Fig. 1 ("single-pump model"), the basolateral pump is the sole driving force for K^+-secretion, with passive flow of the ion through the apical membrane. The upper panel shows the "dual-pump" model. This model features an apical, rheogenic (i.e., generating ionic current) K^+-pump in addition to the basolateral Na^+/K^+ exchange pump.

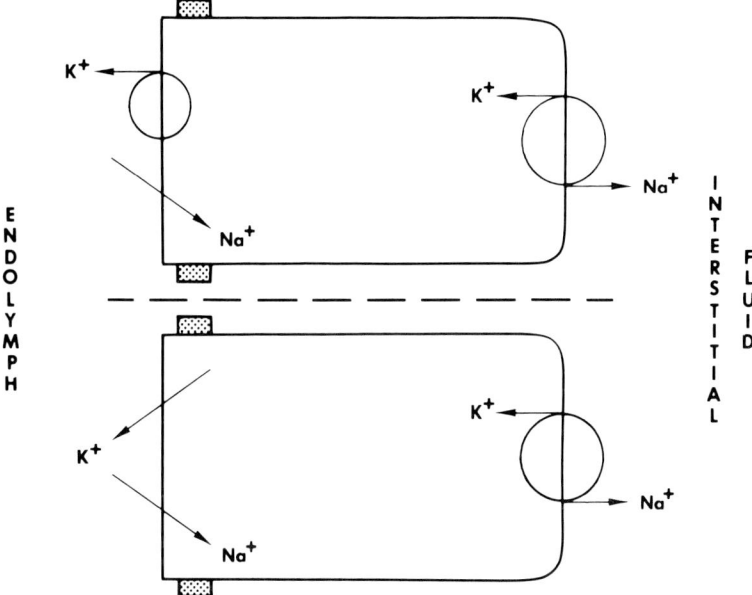

Figure 24-1. Models of Na^+ and K^+ movements across luminal and basolateral membranes of marginal cells of the stria vascularis. The lower (singlepump) model depends upon Na^+, K^+-ATPase as the sole driving force for K^+-secretion. The movement of K^+ (and Na^+) through the luminal membrane is passive. The upper (dual-pump) model includes, in addition, a luminal rheogenic "K^+-transporter". Neither model explicitly considers movements of Cl^-, paracellular ionic fluxes, or the intracellular electric potential. (Reprinted with permission from Marcus et al., 1983.)

As reported previously, we have tested the validity of these models by means of investigative drugs known to enhance or inhibit K^+-permeability (Marcus et al., 1983). Our experiments with the ionophores valinomycin and nystatin, which both enhance K^+ permeability, are too complex to be described and discussed in this brief report; the reader is referred to the detailed publication (Marcus et al., 1983). However, at this point it should suffice to mention that both sets of experiments with modifiers of K^+ permeability (ionophores as well as the Ba^{++} in studies described later) support the concept of a dual-pump system, with no evidence favoring the single-pump model. The weight of evidence seems to indicate that the secreting cells exhibit some kind of electrogenic K^+-transport mechanism in the luminal cell membrane. Even so, the existence of such a transport process in this cell membrane has not yet been proven. In addition, we have no information concerning the properties of this process (e.g., the immediate source of energy and pharmacologic sensitivities).

V. IONIC CONDUCTANCES OF LABYRINTHINE MEMBRANES: RELATION TO CATION TRANSPORT

As mentioned earlier, it is known that the Na^+ must be actively removed from endolymph to maintain its low endolymphatic concentration. This process has traditionally been attributed to the marginal cells of the stria vascularis, as illustrated in Fig. 1. By analogy, it has been postulated that the dark cells perform this function in the vestibular labyrinth (Kimura, 1969). We have recently obtained evidence which suggests that other cells (most likely the sensory cells) may be responsible for this function (Marcus and Marcus, 1984). However, since these observations have only been made in the utricle, it would be unwarranted at this time to conclude that a similar situation prevails in the cochlea.

The experimental set-up used in these experiments is diagrammed in Fig. 2. In essence, one block of liquid Sylgard (a liquid plastic compound used to isolate, electrically and chemically, two regions of the utricle) was injected into the endolymphatic space near the common crus and posterior ampulla. After a 1-h equilibration period, a second block of liquid Sylgard was placed. In one group of utricles, this second block was merely added to the first block as a control (lower left diagram of Fig. 2). In this case, the endolymph was in contact with both sensory and non-sensory tissue. In the other group of utricles, the second block was placed over the macula, in order to isolate a portion of endolymph which was in contact with non-sensory tissue alone (lower right diagram of Fig. 2). Samples of endolymph were taken at the beginning of this second period or at the conclusion of a 1-h incubation period. In all of these experiments, the blood supply of the labyrinth was left intact.[1]

The results of these experiments are shown in Table I. When the endolymph was in contact with both sensory and non-sensory tissues, there was no significant increase in endolymphatic Na^+ over a period of 1 h; however, when the sensory portion was excluded and the endolymph exposed to the dark-cell region alone, a significant elevation of the Na^+ level occurred, accompanied by an approximately equimolar decrease in K^+.

These results are *not* consistent with the prevalent notion that the dark

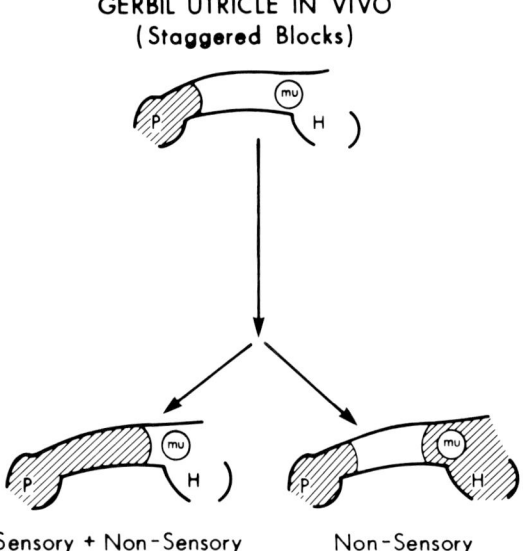

Figure 24-2. Diagram of the protocol for measurement of Na^+ absorption from the utricle (gerbil). Samples of endolymph were obtained from two groups of ears and analyzed for Na^+ concentration. The lumen of the utricle was in part occluded with two injections of liquid Sylgard 184. In one group (lower right), the Sylgard seals restrict the endolymph so that it contacts non-sensory tissue only. In the other group, endolymph remains in contact with both sensory and non-sensory cells. Further details of the protocol are given in the text. P, posterior ampulla; mu, utricular macula; H, horizontal ampulla.

TABLE 24-I
GERBIL UTRICLE IN VIVO
(Staggered Blocks)

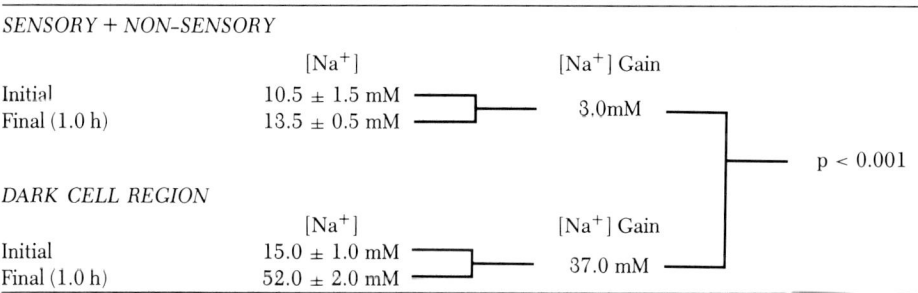

Table shows endolymphatic sodium levels in the gerbil utricle, which are the results of experiments diagrammed in Fig. 2.

cells are responsible for maintenance of the low Na^+ content of endolymph. Rather, it appears that the sensory hair cells or some other cells associated with them serve this function. Although it still remains to be demonstrated conclusively, we have preliminary evidence suggesting that K^+ secretion and Na^+ absorption may be regulated by more than one cell type in the cochlea, as well.

Another membrane property governing ion transport is the ion selectivity of the individual cell membranes and junctional complexes. We previously reported the results of perilymphatic perfusion of media containing Ba^{++} and of media with a high K^+-content (Marcus, 1984). When either of these media was introduced into scala tympani (in ears rendered anoxic to inhibit active K^+ secretion), the EP changed abruptly in the positive direction. No changes resulted when scala vestibuli was perfused with the same media. These changes of the EP in the positive direction are expected to occur if the cells of the organ of Corti are being depolarized and if their interior is electrically coupled to the scala media through a sufficiently low resistance of the endolymphatic membrane. We can conclude, therefore, that the organ of Corti contains a K^+-selective basolateral membrane, which, according to indirect evidence, can be localized to the basolateral membrane of the hair cells. Direct measurements by intracellular impalement with microelectrodes of amphibian vestibular hair cells lend support to this conclusion (Bracho and Budelli, 1978). This interpretation is further supported by the fact that Ba^{++} had no effect upon the EP when perfused through scala tympani or when injected into the scala media of waltzing guinea pigs (mutant animals with degenerated organs of Corti).

When we carried out analogous experiments in the utricle, we expected to obtain results comparable to those observed in the cochlea, namely, a minor effect of Ba^{++} or high-K^+ solutions on the isolated dark cell region, and more pronounced effects upon the macular region. This set of experiments was carried out by means of the *in vitro* preparation mentioned earlier. The effects of high-K^+ media were as predicted: the EP moved in the positive direction, and the increase was larger in the whole utricle than in the isolated dark-cell region. The application of Ba^{++}, on the other hand, led to a most unexpected response (Fig. 3). After a transient positive shift, the EP declined markedly, in a manner comparable to the situation obtained with quabain. The transient positive shift is probably due to partial blockage of a K^+-conductance in the sensory cells, as demonstrated by the earlier K^+-substitution experiments. However, the subsequent dramatic decline of the EP was unexplained, because this effect had *not* been observed in the cochlea and it was not consistent with the type of action attributed to Ba^{++} in other systems. Because the decline of the EP was not as sudden as typically seen when a conductance is blocked, but rather resembled in its time course the effect of ouabain and other agents which act upon the generator of the EP, we suspected that in this particular situation the decline of the EP was due to impairment of dark-cell function. Our supposition was borne out by subsequent experiments in which application of Ba^{++} was restricted to the isolated dark-cell region. This resulted in an almost identical pattern of decline of the EP, but, as expected, the initial shift in positive direction was absent.

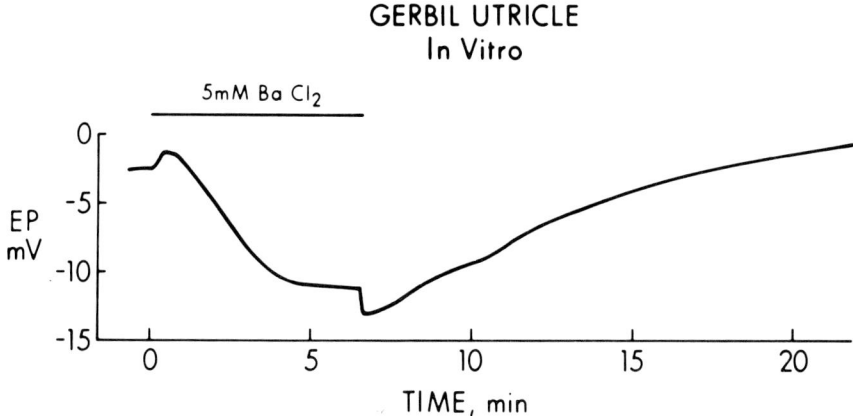

Figure 24-3. Effect of 5 mM barium in the bathing medium (perilymph) upon the endolymphatic potential of the utricle (gerbil). The temporal bone was removed from the animal and mounted *in vitro*. The exposed utricle was left *in situ* in the temporal bone and superfused with oxygenated artificial media.

Why, then, did we not see this type of response in the cochlea? It could be surmised that the divalent Ba^{++} ion is not capable of penetrating the junctional complexes of the basal-cell layer of the stria vascularis. To test this idea, we resorted once more to vascular perfusion, in an attempt to make Ba^{++} directly accessible to the basolateral membranes of the marginal cells. Fig. 4 shows a typical result of these experiments. When applied via the vasculature rather than via perilymph, Ba^{++} caused a marked decline of the EP, comparable to that observed in the utricle. Ba^{++} was effective in both cochlear and vestibular preparations at concentrations between 0.5 and 5.0 mM. This is the range of concentrations at which Ba^{++} has been found to block the K^+-permeability of numerous cell types in other systems.

What then can the Ba^{++} experiments tell us about the workings of the strial marginal cells? At this stage, we still have to speculate until more parameters can be measured, but two possible modes of action of Ba^{++} suggest themselves: 1. Ba^{++} may exert a pharmacologic action on the dark cells of the mammalian cochlea and vestibular labyrinth beyond that of

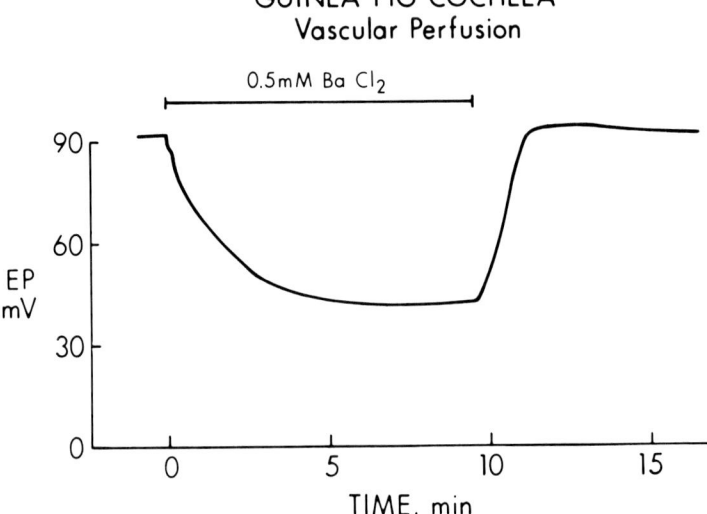

Figure 24-4. Effect of 0.5 mM barium in the vasculature on the endolymphatic potential of the cochlea (guinea pig). The vasculature of the ear was perfused with artificial media by techniques previously described (see text).

blocking a K^+-permeability. 2. Ba^{++} could be blocking a K^+-permeability in the basolateral membrane of the marginal cell, but this effect would not be reflected in a change of the EP. The latter could be the case if the luminal membrane had a relatively high resistance. In this situation, the marginal cell would be depolarized by the Ba^{++}, and it could then be postulated that such a depolarization of the cell is the cause for the inhibition of the electrogenic K^+-secretion process.

In order to throw some light upon these issues, we have carried out additional experiments in which we have demonstrated a distinct interaction between Ba^{++} and K^+. These interactions are similar to those observed in other tissues in which Ba^{++} acts by blocking a K^+ permeability. We, therefore, currently favor the second of the above expressed hypotheses.

On the basis of the above-described results and the results of other ion substitutions introduced by means of vascular perfusion (unpublished findings), we propose a new working hypothesis of the membrane characteristics of the marginal cells (Fig. 5). This model is an extension of the dual-pump model shown in Fig. 1. In addition to the features indicated in Fig. 1, the model postulates prominent conductive permeabilities of the basolateral cell membrane to K^+ and Cl^-. It is further proposed that the apical membrane has a much higher electrical resistance than the basolateral wall and that the tight-junction complex is more permeable to Na^+ than to Cl^-. This model needs to be validated using more sophisticated techniques and extended to include non-conductive pathways. In addition, the presumed "K^+-transporter" in the luminal membrane must be rigorously validated and characterized.

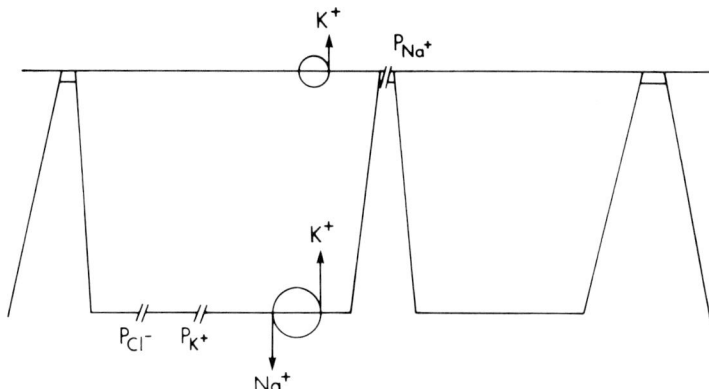

Figure 24-5. Working model for ion transport processes of the marginal cells, based upon the "dual-pump" hypothesis of Fig. 1 and upon results of vascular ionic-substitution experiments. The double-slash lines indicate prominent conductive permeabilities to the ions indicated. Because no ionic permeabilities are postulated to exist in the apical membrane, it is implicit that this membrane has a very high electrical resistance compared to the basolateral membrane. The tight-junction complex is thought to be more permeable to sodium than to chloride.

In summary, we believe that this brief review serves to illustrate that at the present stage much of our information concerning the mechanisms of formation of endolymph and generation of its electrical potentials is suggestive of certain features, but that it does not yet provide a satisfying, overall picture of the processes involved.

ACKNOWLEDGMENTS

These studies were supported by NIH grants NS 19490 and NS 06575.

NOTE

[1]While this particular experimental approach (with vasculature intact) has helped us to answer several important questions, an *in vitro* preparation of the utricle represents a more powerful approach for certain questions, such as the study of the effects of Ba^{++}, to be discussed later. As previously reported, we have developed an *in vitro* preparation of the utricle both with the structure *in toto* and with the dark cell region in isolation (Rokugo et al., 1983). As part of the validation procedures, the EP of the isolated dark cell region *in vitro* was shown to exhibit features similar to those characteristically seen in the utricle and the cochlea *in vivo*, including susceptibility to ouabain and bumetamide. In this preparation, we have used

the Mongolian gerbil, because of the favorable anatomical features of its vestibular labyrinth. All other experimental work from our laboratories that is discussed in this report has been performed upon the guinea pig.

REFERENCES

Bekesy, G. von: D-C resting potentials inside the cochlear partition. *J. Acoust. Soc. Am.* 24: 72-76, 1952.

Bosher, S. K., and Warren, R. L.: Observations on the electrochemistry of the cochlear endolymph of the rat. *Proc. Roy. Soc.* 171: 227-247, 1968.

Bracho, H., and Budelli, R.: The generation of resting membrane potentials in an inner ear hair cell system. *J. Physiol.* 281: 445-465, 1978.

Chou, J. T. Y., and Rogers, K.: Respiration of tissue lining the mammalian membranous labyrinth. *J. Larnygol. Otol.* 76: 341-351, 1962.

Greger, R; Oberleithner, H., Schlatter, E., Cassola, A. C., and Weidtke, C.: Chloride activity in cells of isolated perfused cortical thick ascending limbs of rabbit kidney. *Pflugers Arch.* 399: 29-34, 1983.

Kambayashi, J.; Kobayashi, T., DeMott, J. E., Marcus, N. Y., Thalmann, I., and Thalmann, R.: Effect of substrate-free vascular perfusion upon cochlear potentials and glycogen of the stria vascularis. *Hearing Res.* 6: 223-240, 1982.

Kimura, R. S.: Distribution, structure, and function of dark cells in the vestibular labyrinth. *Ann. Otol.* 78: 542-561, 1969.

Kobayashi, T.; Rokugo, M., Marcus, D. C., Comegys, T. H., and Thalmann, R.: Prolonged maintenance of endocochlear potential by vascular perfusion with media devoid of oxygen carriers. *Arch. Otorhinolaryngol.* 239: 243-247, 1984.

Konishi, T.: Effect of potassium deficiency on cochlear potentials and cation contents of the endolymph. *Acta Otolaryng.* 76: 410-418, 1973.

Konishi, T.; Butler, R. A., and Fernandez, C.: Effect of anoxia on cochlear potentials. *J. Acoust. Soc. Am.* 33: 349, 1961.

Konishi, T.; Hamrick, P. E., and Walsh, P. J.: Ion transport in guinea pig cochlea. I. Potassium and sodium transport. *Acta Otolaryng.* 86: 22-34, 1978.

Kusakari, J.; Ise, I., Comegys, T. H., Thalmann, I., and Thalmann, R.: Effect of ethacrynic acid, furosemide, and ouabain upon the endolymphatic potential and upon high energy phosphates of the stria vascularis. *Laryngoscope* 88: 12-37, 1978.

Marcus, D. C.: Characterization of the potassium permeability of the cochlear duct by perilymphatic perfusion of barium. *Am. J. Physiol.*, 1984. In press.

Marcus, D. C.; Marcus, N. Y., and Thalmann, R.: Changes in cation content of stria vascularis with ouabain and K-free perfusion. *Hearing Res.* 4: 149-160, 1981.

Marcus, D. C.; Rokugo, M., Ge, X.-X., and Thalmann, R.: Response of cochlear potentials to presumed alterations of ionic conductance: endolymphatic perfusion of barium, valinomycin and nystatin. *Hearing Res.* 12: 17-30, 1983.

Marcus, D. C.; Thalmann, R., and Marcus, N. Y.: Respiratory rate and ATP content of stria vascularis of guinea pig in vitro. *Laryngoscope* 88: 1825-1935, 1978.

Marcus, N. Y., and Marcus, D. C.: Utricular dark cells (gerbil) do not absorb sodium. *Assoc. Res. Otolaryngol. Abstr.* 7: 45, 1984.

Rodriguez-Echandia, E. L., and Burgos, M. H.: The fine structure of the stria vascularis of the guinea pig inner ear. *Z. Zellforsch.* 67: 600-619, 1965.

Rokugo, M.; Marcus, N. Y., and Marcus, D. C.: Maintenance of ion transport by cochlear and vestibular dark cells in vitro. *Assoc. Res. Otolaryngol. Abstr.* 6: 16-17, 1983.

Sellick, P. M., and Bock, G. R.: Evidence for an electrogenic potassium pump as the origin of the positive component of the endocochlear potential. *Pflugers Arch.* 352: 351-362, 1974.

Sellick, P. M., and Johnstone, B. M.: Production and role of inner ear fluid. *Progr. Neurobiol.* 5: 337-362, 1975.

Smith, C. A.; Lowry, O. H., and Wu, M. L.: The electrolytes of the labyrinthine fluids. *Larynogoscope* 64: 141-153, 1954.

Sterkers, O.; Saumon, G., Tran Ba Huy, P., and Amiel, C.: K^+, Cl^-, and H_2O entry in endolymph, perilymph, and cerebrospinal fluid of the rat. *Am. J. Physiol.* 243: F173-F180, 1982.

Tasaki, I., and Spyropoulos, C. S.: Stria vascularis as source of endocochlear potential. *J. Neurophysiol.* 22: 149-155, 1959.

Thalmann, I.; Kobayashi, T., and Thalmann, R.: Arguments against a mediating role of the adenylate cyclase-cyclic AMP system in the ototoxic action of loop diuretics. *Laryngoscope* 92: 589-593, 1982.

Thalmann, R.; Kusakari, J., and Miyoshi, T.: Dysfunction of energy releasing and consuming processes of the cochlea. *Laryngoscope* 83: 1690-1712, 1973.

Wada, J.; Kambayashi, J., Marcus, D. C., and Thalmann, R.: Vascular perfusion of the cochlea: effect of potassium-free and rubidium-substituted media. *Arch. Otorhinolaryngol.* 225: 79-81, 1979a.

Wada, J.; Paloheimo, S., Thalmann, I., Bohne, B. A., and Thalmann, R.: Maintenance of cochlear function with artificial oxygen carriers. *Laryngoscope* 89: 1456-1473, 1979b.

Chapter 25

NA$^+$, K$^+$-ACTIVATED ADENOSINE TRIPHOSPHATASE AND CARBONIC ANHYDRASE: INNER-EAR ENZYMES OF ION TRANSPORT

Dennis G. Drescher and Thomas P. Kerr

I. Introduction
II. Na$^+$, K$^+$-ATPase
   A. Properties of Na$^+$, K$^+$-ATPase
   B. Properties of Labyrinthine Na$^+$, K$^+$-ATPase
   C. Distribution and Localization of Labyrinthine Na$^+$, K$^+$-ATPase
   D. Function of Labyrinthine Na$^+$, K$^+$-ATPase
III. Carbonic Anhydrase
   A. Background
   B. Molecular Characteristics of Carbonic Anhydrases
   C. Properties of Cochlear Carbonic Anhydrase
   D. Localization of Carbonic Anhydrase in the Inner Ear
   E. Function of Carbonic Anhydrase
References

I. INTRODUCTION

In this chapter, we will discuss two enzymes found in the inner ear which are implicated in ion and fluid transport, the Na$^+$, K$^+$-stimulated adenosine triphosphatase (Na$^+$, K$^+$-ATPase) and carbonic anhydrase. Each has been localized in inner-ear tissues and characterized in some detail chemically. Another enzyme, adenylate cyclase, may eventually prove to be important in the regulation of labyrinthine transport processes, but will not be discussed here because it has not yet been well characterized chemically in the membranous labyrinth. The interested reader is referred to references on inner-ear adenylate cyclase (Ahlström *et al.*, 1975; Feldman and Brusilow, 1976; Paloheimo and Thalmann, 1977; Feldman, 1981).

II. NA$^+$, K$^+$-ATPASE

A. PROPERTIES OF Na$^+$, K$^+$-ATPase

An enzymatic basis for active cellular transport of sodium and potassium was initially identified within the context of neuronal excitability. Physiological experiments had proven that both Na$^+$ efflux and K$^+$ uptake in nerve fibers were reduced by cooling and by metabolic inhibitors. These results suggested the presence of "a coupled system which ejects Na from the axon on one limb of a cycle, and absorbs K on the other" (Hodgkin and Keynes, 1955). Skou (1957) first demonstrated, in particulate fractions from crab nerve, a Mg^{++}-dependent ATPase activity, stimulated by inclusion of Na$^+$ and K$^+$ in the assay medium. Skou (1960) subsequently found that the increment of activity obtained in the presence of Na$^+$ and K$^+$ was abolished when *ouabain*, a digitalis glycoside inhibitor of active cation transport (Schatzman, 1953), was simultaneously present. He suggested that this enzyme might participate in the process of axonal cation transport previously described by Hodgkin and Keynes (1955).

The observations of Skou (1957, 1960) illustrate several salient features of the enzyme. With regard to the recovery of enzymatic activity from a *particulate* (vs. soluble) fraction, it is now known that Na$^+$, K$^+$-ATPase is localized predominantly, if not exclusively, in the cell plasma membrane (Schwartz *et al.*, 1972; Deguchi *et al.*, 1977; Maunsbach *et al.*, 1980), at the interface between the high potassium concentration of the cell interior and the high sodium concentration of the interstitial spaces. Although Na$^+$, K$^+$-ATPase requires divalent magnesium ion for activity, it is not the only ATP-hydrolyzing enzyme with this requirement. The quantitative determination of Na$^+$, K$^+$-ATPase activity, therefore, necessitates a differential assay, in which nonspecific ATPase activity (usually, Mg^{++}-ATPase) is determined separately, under conditions unsuitable for activity of Na$^+$, K$^+$-ATPase (i.e., omission of K$^+$ and/or Na$^+$, or alternatively, inclusion of digitalis glycosides in the assay medium). Digitalis glycosides display an almost-absolute specificity for Na$^+$, K$^+$-ATPase; ouabain, the most water-soluble of these inhibitors, does not inhibit other ATPases, including Ca^{++}, Mg^{++}-ATPase, anion-sensitive ATPase, and K$^+$, H$^+$-ATPase (Schuurmans Stekhoven and Bonting, 1981). It is well established that ouabain exerts its effects by binding to the enzyme (Akera, 1977; Hansen, 1984).

Although Na$^+$, K$^+$-ATPase activity was first assayed by Skou (1957) in a broken-cell preparation, studies utilizing intact, or quasi-intact, membrane systems have provided much useful information. Of particular interest is the asymmetric behavior, or "sidedness", exhibited by the various reactants, cofactors, and inhibitors which influence catalytic and ion-transporting activity. This sort of information has been obtained from relatively simple systems, such as the resealed erythrocyte ghost (Hoffman, 1958; Hoffman et al., 1960; Whittam, 1962), or the internally-perfused squid giant axon (Caldwell et al., 1960). In these preparations, Na^+, Mg^{++}, and ATP exert their effects, and inorganic phosphate is released from ATP (Schatzmann, 1964), only at the inner, cytoplasmic membrane surface, while potassium ion and digitalis glycosides are effective only at the external surface (Schwartz et al., 1972). Systems like these lend themselves also to studies of reaction stoichiometry, since only two fluid compartments, cytoplasmic and extracellular, can participate in ion transport. Results from several studies suggest that the transmembrane Na^+/K^+ exchange, energized by the transport ATPase, adheres to the stoichiometry 3 Na^+ out: 2 K^+ in: 1 ATP hydrolyzed (e.g., Gardos, 1964; Sen and Post, 1964; Whittam and Ager, 1965; Garrahan and Glynn, 1967). Moreover, enzyme purified from diverse sources (eel electroplax, avian salt gland, renal medulla; see Jørgensen, 1980) exhibits the same transport stoichiometry when reconstituted into lipid vesicle membranes. Since this stoichiometric relation results in net loss of positive charge from the cell, the Na^+/K^+ "pump" is considered to be inherently electrogenic at the cellular level, and to participate in maintenance of the intracellular negative resting potential (Siegel et al., 1981).

While enzyme from several sources has been purified to some extent (e.g., Kyte, 1971; Hobbs and Albers, 1980) the highest purity yet reported (specific activity of 32–37 μmol phosphate/min/mg protein; Jørgensen, 1974) was associated with a preparation from the outer medulla of rabbit kidney. Studies of the renal enzyme have been reviewed by Jørgensen (1980). On the basis of maximum binding capacity for ouabain or ATP, or for phosphorylation from ATP, minimum molecular weight per binding site is approximately 270,000 to 280,000. From these data, the turnover rate (molecular activity) under optimal conditions is calculated to approach 10,000 per min.

Under physiological conditions, the enzymatic activity of Na^+, K^+-ATPase is completely dependent upon the presence of K^+ at high-affinity extracellular sites, together with Na^+ at high-affinity cytoplasmic sites (Robinson and Flashner, 1979; Siegel et al., 1981). The rate of enzyme-mediated ion transport, sustained by an intact cell, is consequently regulated by concentrations of the respective cations in the intra- and extracellular compartments. In kidney tubule cells, enzyme turnover rate is independent of extracellular K^+ within the physiological concentration range, and instead depends primarily upon the intracellular Na^+/K^+ ratio. From determinations of in vitro enzymatic activity over a wide range of cation ratios, and from measurements of cytoplasmic Na^+/K^+ concentrations, turnover rate of the renal enzyme *in vivo* is estimated to be only 6–18% of the maximum rate attainable at optimum cation ratio. However, the rate increases rapidly with increasing intracellular sodium activity: a change in Na^+/K^+ ratio from 20/130 to 30/120 results in a twofold increase of enzymatic activity (Jørgensen, 1980). With broken cell preparations, a "cation activation curve" can be constructed both for K^+ and for Na^+, by the measurement of enzymatic activity as a function of sodium or potassium concentration. From these curves, one may determine a K_m (the ion concentration giving half-maximal enzyme activity) for each cation. The K_m values, in turn, are assumed to provide an indication of the cation concentrations required at the respective high-affinity sites. The K_m for K^+ is near 1 mM in a number of enzyme preparations from various tissues (Kuijpers, 1969). This explains how enzymatic activity can be independent of extracellular K^+ *in vivo:* the extracellular potassium concentration normally exceeds the K_m by a wide margin.

The overall enzymatic reaction of Na^+, K^+-ATPase is thought to include a sequence of allosteric events associated with conformational changes in the enzyme molecule (Schwartz *et al.*, 1972; 1975). It is known, for example, that the enzyme can assume two distinct phosphorylated conformations. Of these, the form designated $E_1 \sim P$ is of higher energy, while the E_2-P form is lower in energy and sensitive to potassium. There is also an unphosphorylated conformation, E (or perhaps two such forms, E_1 and E_2: see Robinson and Flashner, 1979; Glynn and Karlish, 1975). Phosphorylation of the enzyme is Na^+-dependent; dephosphorylation is K^+-dependent (Glynn and Karlish, 1975; Hobbs and Albers, 1980). The transitions from one conformational state to the next result in the translocation of sodium and potassium across the plasma membrane. They may be denoted as a series of "partial reactions," according to the proposals of Albers (1967) and Post *et al.* (1969, 1975). The reaction sequence is written:

A. $E_1 + MgATP \xrightleftharpoons{Na^+} E_1 \sim P + ADP$

B. $E_1 \cdot P \xrightleftharpoons{Mg^{++}} E_2\text{-}P$

C. $E_2\text{-}P \xrightleftharpoons{K^+} E_2 + P_i$

D. $E_2 \rightleftharpoons E_1$

Certain nonphysiological organic phosphate substrates, such as paranitrophenyl phosphate, are able to promote direct formation of the E_2-P enzyme configuration, followed by dephosphorylation (reaction C, above). This potassium-dependent, ouabain-inhibitable phosphatase component of the overall reaction sequence provides an alternative method for the assay of enzymatic activity (Robinson and Flashner, 1979). In addition, ouabain binds preferentially to the E_2-P conformation of the enzyme (Post et al., 1969; Yoda and Yoda, 1982). Since the "life-time" of the E_2-P form is probably brief (Jørgensen, 1980), the rate of ouabain binding observed under particular circumstances is closely correlated with the rate of enzyme catalytic activity observed under similar conditions in the absence of ouabain. Enzymatic activity is, paradoxically, necessary for the binding of digitalis glycoside inhibitors, since it promotes formation of the enzyme configuration favorable for binding (Schwartz et al., 1975).

B. PROPERTIES OF LABYRINTHINE Na^+, K^+-ATPase

The enzyme associated with the stria vascularis of guinea pig cochlea seems generally similar in kinetic behavior to enzyme from other sources. In homogenates of lyophilized stria vascularis (Kuijpers, 1969; Kuijpers and Bonting, 1969) enzymatic activity is optimal at pH 7.3, and at a Mg^{++} concentration of 1–2 mM. The K_m for activation by K^+ is 0.9 mM; for Na^+, 4.5 mM. Only the latter parameter (K_m for sodium) is unusual, being somewhat lower than in other tissues (Kuijpers, 1969).

Kuijpers and Bonting (1969) reported that half-maximal inhibition of enzymatic activity in homogenates of guinea pig stria vascularis occurs at a ouabain concentration of 3.16 μM. In reasonable agreement with this observation, it was then found (Kerr et al., 1982; Kerr, 1983) that half-maximal occupancy of the ouabain binding sites, in a preparation of cochlear lateral wall assayed in vitro, occurs at a ouabain concentration of approximately 2 μM. The latter result was obtained by means of Scatchard analysis (cf. Bennett, 1978), which also showed that the lateral wall, a non-innervated tissue, contains only one class of ouabain binding site (and probably, therefore, a single form of the enzyme). This stands in contrast to results obtained in brain, where two forms of the enzyme have been identified, differing in affinity for ouabain (Sweadner, 1979), or for another inhibitor, erythrosin B (Silbergeld, 1981). It is possible that innervated regions of the labyrinth may, like brain, be provided with an additional form of the enzyme.

C. DISTRIBUTION AND LOCALIZATION OF LABYRINTHINE Na^+, K^+-ATPase

Na^+, K^+-ATPase is ubiquitous among animal cells (Schuurmans Stekhoven and Bonting, 1981), which share the requirement for a high-potassium intracellular environment. The enzyme concentration may, however, be markedly elevated in specialized cells whose distinctive function requires a high rate of cation transport. Evidence favoring a correlation between cation flux and Na^+, K^+-ATPase activity began to accumulate within a few years of the enzyme's discovery (see, for example, Bonting and Caravaggio, 1963). In epithelial cell types, which separate extracellular fluids of dissimilar composition, the enzyme may mediate transepithelial ion transport from one extracellular fluid compartment to another (as occurs, for example, in the nephron: see Jørgensen, 1980). In non-epithelial cell types, a high level of enzymatic activity signifies an augmented cation flux between the interstitial space and cytoplasm. This latter phenomenon is usually associated with cellular bioelectrical activity, as in nerve (Siegel et al., 1981) and muscle (Sjodin, 1982).

Biochemical investigations (Kuijpers, 1969; Kuijpers and Bonting, 1969; Matschinsky and Thalmann, 1970) furnished the first reliable information concerning the regional distribution of Na^+, K^+-ATPase in structures of the inner ear. Kuijpers found highest activity (8.0 mol phosphate released/kg dry tissue/h) in the stria vascularis, followed by that portion of the spiral ligament comprising the spiral prominence and external sulcus (1.6 mol/kg/h). Activity in the organ of Corti was considerably less (0.5 mol/kg/h), while lowest activities occurred in Reissner's membrane, and in that portion of the spiral ligament situated behind the stria vascularis (0.4 mol/kg/h in each of the two tissues). Of possible functional significance (see below) is the observation that enzymatic activity in stria vascularis (as well as in spiral ligament) decreases from the basal cochlear coil to the apex (Kuijpers, 1969; Kuijpers and Bonting, 1969). A similar base-to-apex gradient was noted in the quantitative distribution of ^3H-ouabain binding sites, when assayed in a preparation of cochlear lateral wall including both the stria and the spiral ligament (Kerr, 1983).

Biochemical studies of Na^+, K^+-ATPase in tissues of the vestibular apparatus were reported by Thalmann (1971), who assayed freeze-dried, microdissected tissue samples. Highest activity of Na^+, K^+-ATPase (0.9 mol phosphate/kg/h) was found in ampullar wall; intermediate levels in the macula of the saccule (0.7 mol/kg/h), macula of the utricle (0.5 mol/kg/h), and crista ampullaris (0.7 mol/kg/h); and lowest activity (0.3 mol/kg/h) in semicircular canal. Kusakari and Thalmann (1976) subsequently utilized improved methods of microdissection to separate nonspecialized from specialized epithelial regions in ampullar wall. The "specialized" epithelium consisted of "dark cells," with morphological features similar to the marginal cells of the stria vascularis (Kimura, 1969). Activity in the specialized preparation proved rather high (2.9 mol/kg/h).

A second research strategy to delineate the regional distribution of Na^+, K^+-ATPase within the membranous labyrinth involves the use of microscopical methods. Investigations of cochlear "ATPase" at the light-microscopic (Ishii and Nomura, 1968) and electron microscopic levels (Nakai and Hilding, 1966; 1967) preceded the biochemical studies cited above.

The earliest available methods (Wachstein and Meisel, 1957) for microscopical localization of the enzyme were intended to detect inorganic phosphate released by enzymatic hydrolysis of the physiological substrate, ATP. This was accomplished by the incubation of tissue in a medium containing both ATP and lead ion. Phosphate, released by enzymatic activity, formed an insoluble precipitate with Pb^{++}; the latter material could be visualized in the electron microscope, since its high atomic weight renders it opaque to the electron beam. It eventually became apparent, however, that this approach is unworkable. Lead ion inhibits Na^+, K^+-ATPase potently and preferentially, with much less inhibitory effect on non-specific ATPases (Jacobsen and Jørgensen, 1969; Ernst and Hootman, 1981). Conscientious researchers using lead-ion procedures (such as Nakai and Hilding, 1966, 1967) cautioned that the specificity of these techniques for Na^+, K^+-ATPase is uncertain. In retrospect, there is little question that much of the reaction product visualized by lead-capture methods reflects nonspecific enzymatic activity.

Two procedures with high specificity for Na^+, K^+-ATPase were subsequently introduced. One of these, developed by Ernst (1972a, b), employs strontium ion, rather than lead, to precipitate the inorganic phosphate released by enzymatic substrate hydrolysis. Strontium is much less inhibitory to Na^+, K^+-ATPase than is lead (Ernst, 1972a), and the specificity of the cytochemical reaction can be further enhanced by use of alkaline phosphatase inhibitors (Firth, 1974; Ernst, 1975) to suppress interfering enzymatic activity. Since this technique utilizes para-nitrophenyl phosphate as substrate, Na^+, K^+-ATPase is demonstrated as the K^+-dependent, ouabain-inhibitable phosphatase component of the overall Na^+, K^+-ATPase enzymatic reaction sequence (see above). A second technique, that of Stirling (1972), uses light-microscopic autoradiography to visualize enzyme sites labeled by their binding of tritiated ouabain. Immunocytochemical methods are also available (Kyte, 1976a, b; Wood et al., 1977), but have not yet been employed for investigation of the inner ear.

In a recent study of the regional distribution of Na^+, K^+-ATPase in the guinea pig cochlea (Kerr et al., 1981, 1982; Kerr, 1983), the cytochemical technique of Ernst (1972b) and the autoradiographic method of Stirling (1972) were used concurrently, so that results obtained with either method served as a check upon the other. The highest cytochemical activity and highest density of ouabain binding sites both were found in the marginal cell epithelial layer of the stria vascularis (Figs. 1–3). At the ultrastructural level (Fig. 3), it was observed that reaction product was absent from the luminal cell surface facing the endolymph, but present in copious deposits at the basolateral marginal cell extensions. This "polarized" cellular distribution of Na^+, K^+-ATPase activity, in epithelial cells characterized morphologically by elaborate mitochondria-filled basolateral membrane extensions (Hinojosa and Rodríguez-Echandía, 1966), has come to be recognized as a hallmark of ion-transporting epithelia (Ernst et al., 1980), whether the net ionic flux is directed toward or away from the free epithelial surface, and whether the major transported cation is sodium (Ernst, 1972b) or potassium (Ernst, 1975).

Reissner's membrane has sometimes been suggested as a probable site of cochlear ion transport. However, this structure, which consists of two cell layers, showed no detectable cytochemical reaction product, and no concentration of ouabain-binding sites above background levels (Fig. 1), with the possible exception of the attachment zone adjacent to the spiral limbus.

Another cochlear tissue sometimes considered to function in endolymphatic ion transport is the spiral prominence. Both the cytochemical and the autoradiographic evidence of Kerr et al. (1981, 1982; see Figs. 1 and 4) indicate that the highest enzymatic activity found in this tissue (and in the spiral limbus, as well) is associated with subepithelial stromal cells. These results confirmed and extended those of Kuijpers (1969), whose biochemical investigation demonstrated that Na^+, K^+-ATPase activity in the prominence "can only be localized . . . in the stroma cells behind the epithelium". The function of these reactive cells remains obscure. It is not likely that they are involved, in any direct fashion, with endolymphatic ion transport, since their membrane surfaces do not communicate with the endolymphatic compartment. Moreover, cytochemical reaction product is distributed along the stromal cell membrane in a symmetrical, non-polarized manner (Fig. 4).

Within the *organ of Corti*, Kerr et al. (1982) reported that only neural elements showed appreciable activity. Elevated levels of Na^+, K^+-ATPase were localized to synaptic regions beneath inner and outer hair cells, to the vicinity of the inner and outer spiral bundles, to unmyelinated fibers crossing the tunnel, to terminal nodes of Ranvier in the habenula perforata, and to fibers coursing through the osseous spiral lamina. Little activity was said to be associated with the hair cells. However, it was subsequently noted (Kerr, 1983) that only the outer hair cells were amenable to inspection at the ultrastructural level. Mechanical disruption of inner hair cells during the dissection and cytochemical incubation procedures precluded localization of possible inner-hair-cell enzymatic activity at the electron microscopic level. At the level of resolution afforded by light-microscope autoradiography of ^{3}H-ouabain binding, numerous enzyme sites were visualized at the basal pole of both cochlear hair cell types. The ultrastructural distribution of cytochemical reaction in the vicinity of the outer hair cells showed that product was associated exclusively with nerve endings, and not with the outer hair cells. It remains possible that some of the ouabain binding resolved by light microscopy near the base of the inner hair cell is actually associated with the hair cell rather than with neural elements.

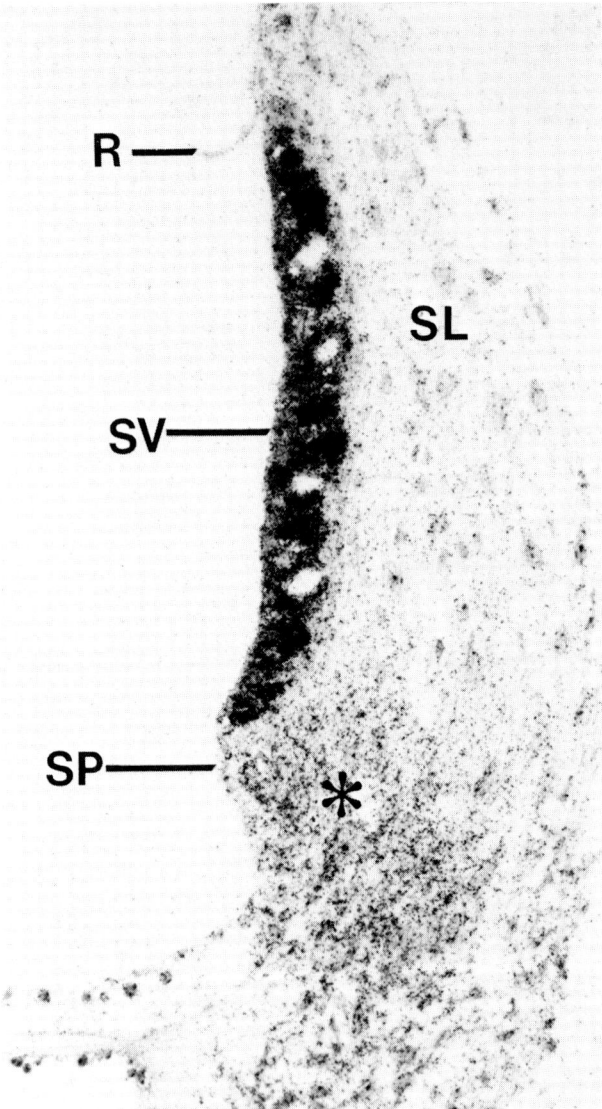

Figure 25-1. This survey light micrograph shows the distribution of autoradiographic silver grains in the lateral wall of guinea pig cochlea, after incubation with ^{3}H-ouabain. Dense black clusters of silver grains cover the basolateral compartments of marginal cells in the stria vascularis (SV), while the apical cytoplasm is relatively free of label. Neither the individual silver grains nor the cytoarchitectural detail of the marginal cells are resolved at this low magnification (see Figs. 2 and 3). A high density of ouabain binding sites is also associated with stromal tissue of the spiral prominence (SP), in the general region indicated with an asterisk (°). Note that the density of ouabain binding sites is higher in stromal tissue of the prominence (°) than in the prominence epithelium, facing the endolymphatic space (see also Fig. 4). The density of ouabain binding sites is much lower in the spiral ligament (SL) behind the stria vascularis, and in Reissner's membrane (R). Brightfield micrograph, methylene blue; second cochlear turn. (From Kerr *et al.*, 1982.)

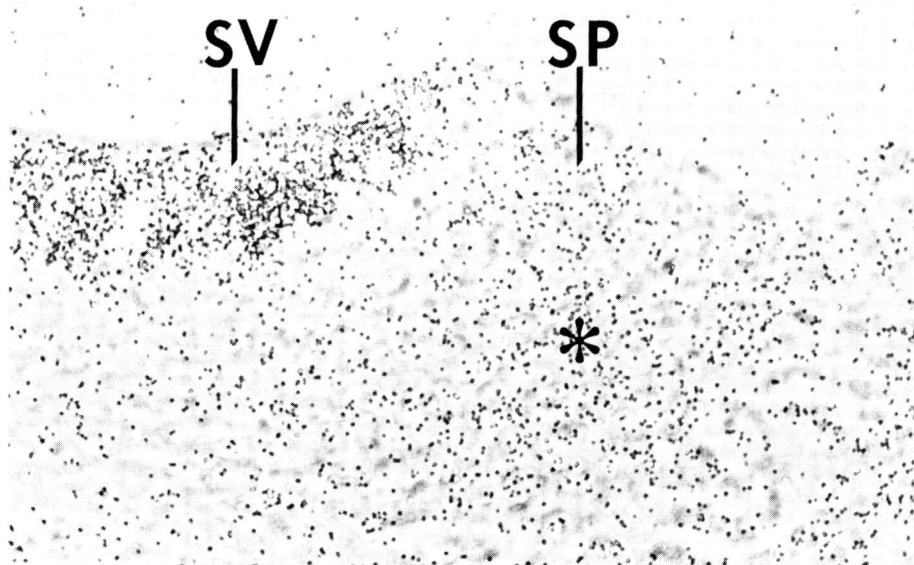

Figure 25-2. This light micrograph shows, at higher magnification, the distribution of ouabain-binding sites in the region of the spiral prominence (SP). Na^+/K^+ pump sites are associated with stromal cells of the prominence, in the general region indicated with an asterisk (*), as well as with marginal cells of the stria vascularis (SV). Phase-contrast micrograph; second cochlear turn. (From Kerr et al., 1982.)

No published account of the enzyme's cellular localization in mammalian vestibular tissues has appeared, excepting the report of Nakai and Hilding (1968). That investigation was completed prior to the introduction of techniques specific for Na^+, K^+-ATPase, and utilized instead the nonspecific lead ion capture method. Recently, however, the quantitative distribution of enzyme sites in frog (*Rana pipiens*) vestibule has been determined by Burnham and Stirling (1984a, b), using ^{3}H-ouabain autoradiography in combination with stereological techniques (i.e., methods for estimating membrane area and/or tissue volume). The highest "pump concentration" (42 μmol/L), and highest "membrane pump density" (4,500 sites/μm^2 of plasma membrane) both were found in vestibular dark cells. As in the marginal cell of the stria vascularis (Kerr et al., 1982), the highest pump concentration within the dark cell was found in the vicinity of the basolateral membrane extensions, with lower concentrations over the apical region. Comparison of *membrane pump density* (sites/μm^2) in dark cells with that found in various other vestibular cell types showed that the high *pump concentration* of the dark cell is mainly due to the augmented membrane surface area provided by the dark cell basolateral membrane extensions. Thus, dark-cell pump concentration is at least an order of magnitude higher than found in any

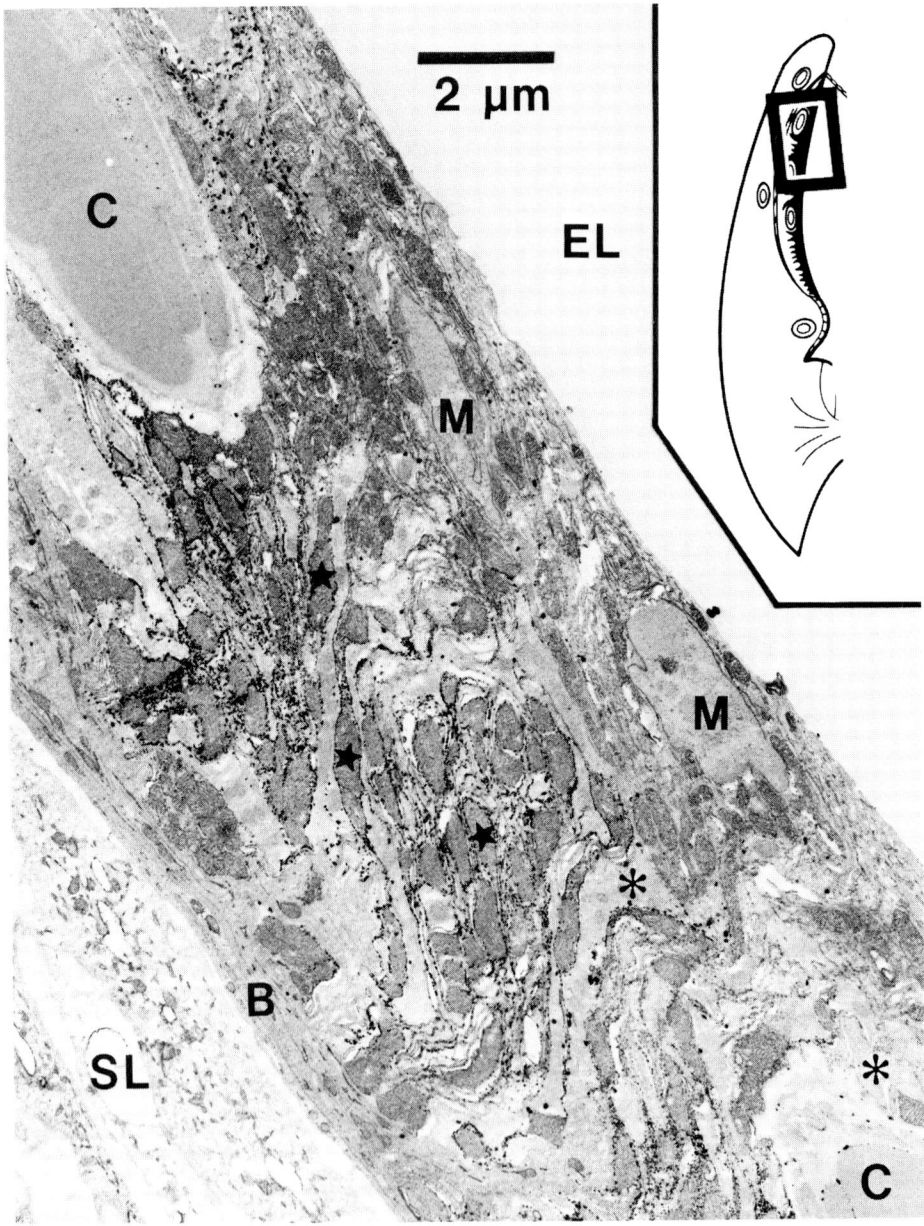

Figure 25-3. This electron micrograph shows the distribution of K^+-dependent, ouabain-inhibitable phosphatase activity in the stria vascularis. Electron-dense deposits of reaction product outline the basolateral extensions of marginal cells (e.g., ★) near the center of the field. The luminal surface of the marginal cells, bordering the endolymph (EL), is devoid of reaction product. Little or no reaction can be seen in basal cells (B), intermediate cells (*), capillaries (C), or in the spiral ligament (SL). M: marginal cell (nucleus). Second turn of guinea pig cochlea. (From Kerr et al., 1982.)

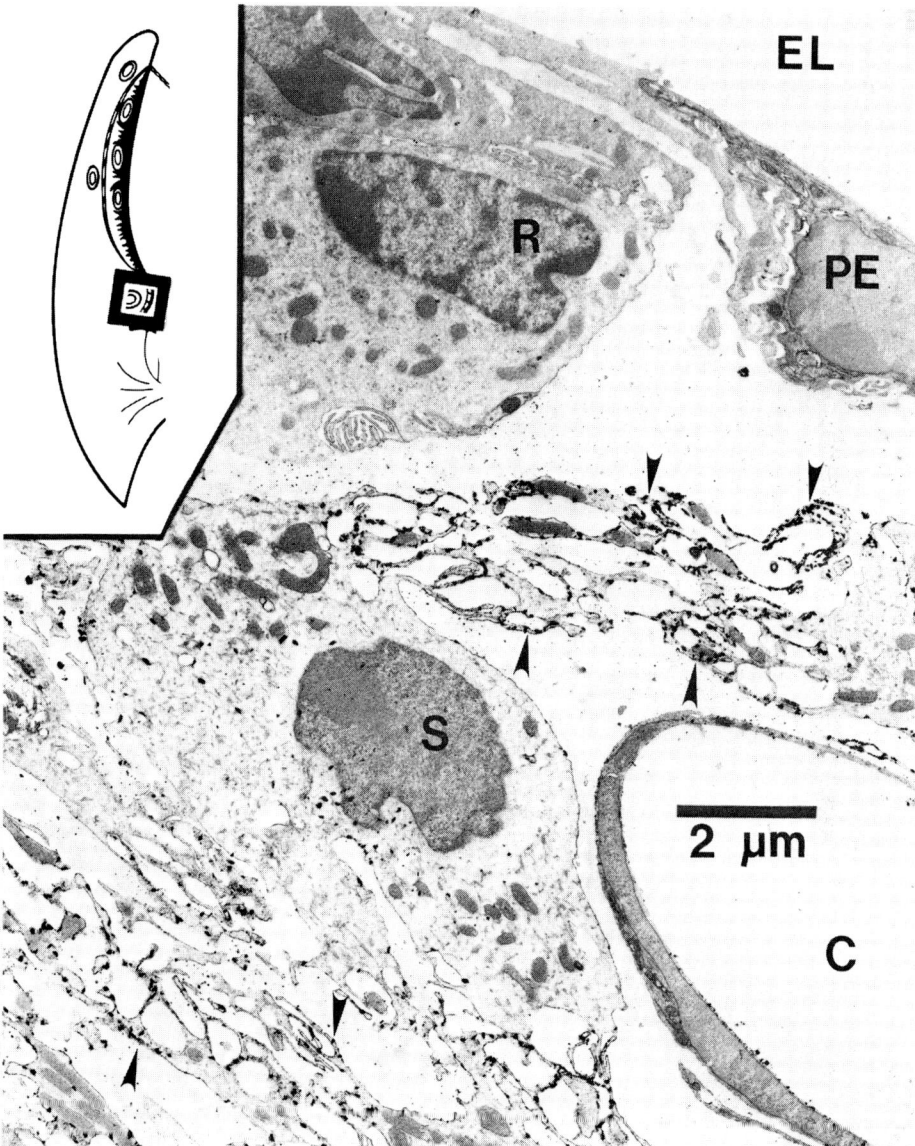

Figure 25-4. This electron micrograph shows the distribution of K^+-dependent, ouabain-inhibitable phosphatase activity in the spiral prominence. Near the center of the field, and at lower left, reaction product outlines the membrane extensions (arrowheads) of a stromal cell (S). Notice that the reaction product is distributed along the cytoplasmic membrane surface. Little or no activity can be seen in the prominence epithelial cell (PE), in the root cell (R), or in the capillary (C). EL: endolymph. Second turn of guinea pig cochlea. (From Kerr et al., 1982.)

other vestibular epithelium, while dark-cell membrane pump density exceeds that of other vestibular cells by a factor of only about two-fold. In comparison to the high pump concentration of the dark cell epithelium (42 μmol/L), pump concentration in other regions was much lower: in the macula of the saccule, 3.1 μmol/L; in the wall of the saccule, 3.8 μmol/L.

Burnham and Stirling (1984b) extended their analysis of pump site distribution in frog saccule to the individual cell types comprising the sensory epithelium. Denervation of the saccule was followed by a 45% decrease of ouabain-binding sites within the macula. Those pump sites which were lost were considered to have been associated with nerve terminals on the hair cells. Pump sites persisting after denervation were taken to represent the combined contribution of hair cells and supporting cells. Grain counts in the basal portion of the macular epithelium, where hair cell membranes are absent, yielded a value of 1,500 sites/μm^2 for the pump site membrane density of supporting cells. On the apparent assumption that this value remains constant in the supporting-cell lateral membrane along the basal-to-apical axis, it was estimated that in the region of the macular epithelium, where supporting-cell membranes are closely apposed to hair-cell membranes, pump site membrane density of the hair cells amounted to 3,000 sites/μm^2. This value was equivalent to that calculated for macular nerve terminals. The investigators noted that the presence of enzyme sites in hair cells of the frog saccule contrasts with the absence of detectable activity in hair cells of the guinea pig cochlea (Kerr et al., 1982).

D. FUNCTION OF LABYRINTHINE Na^+, K^+-ATPase

Interest in the Na^+, K^+-ATPase activity of the membranous labyrinth has been stimulated from the outset by the likelihood that this enzyme energizes the transduction mechanisms of acoustico-lateralis organs. The earliest evidence implicating an energy-dependent mechanism in hair-cell transduction was provided by Békésy (1951). He found that the AC cochlear potential, the cochlear microphonic, associated with hair-cell activity, draws upon some reservoir of external energy beyond that derived from the acoustic stimulus. Following the discovery that cochlear endolymph is maintained at a high positive DC potential (Békésy, 1952), and that the endolymph has a high potassium concentration unique among extracellular fluids (Smith et al., 1954), it was recognized (Davis, 1965) that these properties of the endolymph form the basis for a substantial gradient of DC potential and ionic concentration across the hair cells themselves. With respect to the gradient of DC potential, this drops from approximately +80 mV in cochlear endolymph to a negative value in the hair cell cytoplasm (cf. Konishi and Salt, 1983). Potassium concentration decreases from about 150 mM in endolymph (Bosher and Warren, 1968) to <10 mM in the extracellular fluid at the basal surface of the hair cell (Konishi and Salt, 1983). According to the proposal of Davis, these electrochemical gradients constitute a driving force which supports a flow of ionic current through the hair cell. Entry of ions at some region of the apical membrane surface, facing toward the endolymph, is considered to be modulated by membrane conductance changes induced by mechanical deformation. Calculations described by Sellick and Johnstone (1975) demonstrate that the driving force for ionic current flux through the hair cell is much higher for K^+ than for Na^+ or Cl^-, suggesting that current flow through the hair cell should be carried primarily by potassium ion. Speculating on the possible advantage conferred by such an arrangement, these authors noted that a hair cell ought to be able to sustain a large resting current carried by potassium ion, but not a large resting sodium current. This follows from the fact that potassium can exit passively across the basal membrane of the hair cell, owing to a favorable electrochemical gradient (Konishi and Salt, 1983). Sodium, however, having once entered the hair cell, would have to be removed by active transport (Matsuura et al., 1971). "The labyrinthine hair cell therefore gains sensitivity by reason of the large resting current but does not have to have the metabolic machinery to sustain it. This is located in the stria vascularis" (Sellick and Johnstone, 1975).

Certain versions of the foregoing analysis (e.g., Sellick and Johnstone, 1975) incorporate an additional assumption, viz., that the hair-bearing membrane of the sensory cell is selectively permeable to K^+. However, available evidence (Corey and Hudspeth, 1979) now suggests that the apical membrane of the hair cell (at least, that of the anuran saccule) may be permeable to many cations, including sodium. This latter result is not incompatible with the main features of the interpretation provided above (cf. Hudspeth, 1983), but does imply that endolymphatic sodium may gain entry to the hair cell in proportion to the endolymphatic driving force and sodium concentration.

It should also be noted that the electrochemical properties of endolymph show significant differences, in each of the endolymphatic chambers of a particular animal, as well as in different species. In mammals, the positive DC resting potential of the endolymph is $+80$ mV in the cochlea and less than $+10$ mV in the saccule and utricle (Smith et al., 1958). At the same time, endolymphatic Na^+ increases from <1 mM in the mammalian cochlea (Bosher and Warren, 1968) to 3 mM in the saccule (Sellick and Johnstone, 1972), and 12.6 mM in the utricle. With regard to interspecies differences, endolymphatic Na^+ concentration decreases, and K^+ increases, with phylogenetic advancement (Fernandez, 1967; Matsuura et al., 1971). In the frog, for example, utricular endolymph has a sodium concentration of 17.6 mM (Simon et al., 1973), as compared to 12.6 mM in guinea pig (Sellick et al., 1972).

Burnham and Stirling (1984b) erroneously stated that Kerr et al., (1982) "found some enzyme activity in the organ of Corti, mostly associated with nerve fibers, but little activity was found elsewhere in the cochlea." These investigators suggested that mammalian cochlear hair cells may have as many Na^+/K^+ pump sites as claimed by them for hair cells of the frog saccule. However, the considerations set forth above favor the possibility that hair cells of the frog saccule, exposed to a relatively high-sodium endolymph, may require a higher activity of Na^+, K^+-ATPase to compensate for sodium influx associated with the transduction current (see also Matsuura et al., 1971).

While it might be inferred that both the endolymphatic cation composition and the cochlear DC potential represent alternative manifestations of a single underlying transport mechanism, this has not been proven. Smith et al., (1958) concluded that the endolymphatic potential is not correlated with the endolymphatic potassium content, since potassium concentrations are similar in the cochlea and vestibular organs while the DC potential is quite different. In addition, it has been reported (Brusilow and Gordes, 1973) that ethacrynic acid intoxication may lead to a reversal of the normal sodium and potassium content in cochlear endolymph without concomitant change in amplitude of the DC potential. The latter result was interpreted to mean that the cochlear DC potential and cation concentrations are mutually independent (Brusilow and Gordes, 1973). However, it has subsequently been suggested that an apparent dissociation of cochlear ionic concentrations and DC potential may occur when active ion transport and passive membrane ion conductances both are altered concurrently, even though the DC potential and ion concentrations of cochlear endolymph may indeed depend conjointly upon active transport of potassium (Bosher, 1981; Juhn et al., 1981). There is, in fact, indirect evidence to suggest that the positive cochlear DC potential is closely correlated with potassium transport into scala media (Sellick and Bock, 1974).

In the paragraphs below, we summarize evidence bearing upon the possible function of Na^+, K^+-ATPase, first in maintaining endolymphatic cation gradients, and secondly, in generation of the cochlear DC potential. It should, however, be apparent at this point that one must maintain certain reservations in ascribing these electrochemical properties of the endolymph to any single underlying mechanism.

Evidence favoring a role for the enzyme in endolymphatic cation transport includes, first of all, the finding of extremely high enzymatic activity in the stria vascularis (Kuijpers, 1969; Kuijpers and Bonting, 1969; Matschinsky and Thalmann, 1970), and, to a lesser extent, in the dark cell epithelial regions of the vestibular apparatus (Kusakari and Thalmann, 1976). Because measured levels of Na^+, K^+-ATPase activity correlate closely with cation flux (Bonting and Caravaggio, 1963), the impressive activity found in the membranous labyrinth offers strong presumptive evidence that the enzyme participates in active transport of endolymphatic potassium and/or sodium. Consonant with this possibility, Sterkers et al., (1984) (also Sterkers, Chapter 26, this volume) found that the endolymphatic potassium concentration decreases between the basal and the middle turns of the rat cochlea. The gradient of endolymphatic potassium concentration, therefore, appears to coincide with the quantitative gradient of Na^+, K^+-enzymatic activity (Kuijpers, 1969; Kuijpers and Bonting, 1969), and of ouabain binding sites (Kerr et al., 1982; Kerr, 1983), in the guinea-pig cochlea. The similar gradient of cochlear oxygen consumption (Meyer zum Gottesberge et al., 1965; Mizukoshi and Daly, 1967) might indicate that oxidative formation of the substrate ATP proceeds at a rate determined largely by the local activity of Na^+, K^+-ATPase (cf. Ross et al., 1982).

A more direct form of evidence has been provided by studies which monitored endolymphatic ion concentrations during application of the specific enzyme inhibitor, *ouabain*. Konishi and Mendelsohn (1970) perfused the inhibitor through scala tympani of the guinea-pig cochlea, observing a dose-dependent increase in endolymphatic sodium, accompanied by a decrease in potassium. This result has been confirmed by Bosher (1980a), who used ion-sensitive electrodes to monitor Na^+ and K^+ concentrations in cochlear endolymph. Qualitatively similar findings have been reported for utricular endolymph of the frog, when the isolated membranous labyrinth was incubated with ouabain *in vitro* (Simon et al., 1973). Konishi et al., (1978) subsequently perfused the perilymphatic space of the guinea pig cochlea with $^{42}K^+$ or $^{22}Na^+$, demonstrating that the endolymph concentrates potassium, and extrudes sodium, against the concentration gradients for the two cations. These active transport processes showed a dose-dependent inhibition by ouabain, when the inhibitor was present in the perilymphatic perfusate. It is conceivable that the action of ouabain on endolymphatic cation content may, to some extent, reflect indirect effects (induced by enzyme inhibition) on general cellular metabolism (see Bosher, 1980a; Ross et al., 1982). On the balance, however, all reported studies concerned with the effect of ouabain on endolymphatic ion content are consistent with the hypothesis that Na^+, K^+-ATPase participates in endolymphatic ion transport.

The third, and most recent, category of evidence relating to endolymphatic ion transport is that concerning the cellular localization of Na^+, K^+-ATPase in presumptive ion-transporting epithelia of the membranous labyrinth. The asymmetrical, or "polarized" cellular distribution of enzyme sites in strial marginal cells of the mammalian cochlea (Kerr et al, 1981, 1982), and in dark cells of the anuran vestibular apparatus (Burnham and Stirling, 1984a, b) was described above. This pattern of enzyme distribution has been previously observed only in epithelia engaged in vectorial transport of salts and water, where the participation of the ATPase in transepithelial ion flux is well established (Ernst et al., 1980). The diversity of transport processes energized in different organs by basolateral Na^+/K^+ pump sites is impressive. The cation transported may be either Na^+ or K^+; the transport process may be either secretory or resorptive; and the fluid originating from salt-coupled transepithelial movements of water may be isotonic or hypertonic. From this, it is evident that different transporting epithelia must possess, in addition to the basolateral ATPase, various cellular specializations suited to their individual transport functions. Current hypotheses regarding the additional mechanisms involved in labyrinthine cation transport will be identified briefly in a section to follow.

For the most part, experimental approaches used to investigate the possible role of Na^+, K^+-ATPase in generation of the *cochlear DC potential* are analogous to those which indicate a relation between enzymatic activity and cation transport. There is, however, one additional item of available information concerning the DC potential: it emanates from the luminal surface of the strial marginal cells (Tasaki and Spyropoulos, 1959), the same cells distinguished by the intense enzymatic activity of their basolateral membranes. The magnitude of the endolymphatic DC potential decreases from the base to the apex of the cochlea (Misrahy et al., 1958; Kuijpers and Bonting, 1970), as does the Na^+, K^+-ATPase activity of the stria vascularis, and the endolymphatic potassium content (see above). Kuijpers (1974) determined that the post-natal development of the DC potential in rat cochlea proceeds in tandem with an increase of strial Na^+, K^+-ATPase activity to the adult level. Prazma (1969) first reported that perfusion of the cochlear perilymphatic channels with ouabain resulted in a decline of the positive DC potential. Konishi and Mendelsohn (1970) and Bosher (1980a) confirmed this observation, also demonstrating a concomitant tendency toward reversal of the normal endolymphatic Na^+/K^+ concentrations. Kuijpers and Bonting (1970) noted a strong resemblance between the dose-response curves for ouabain inhibition of Na^+, K^+-ATPase activity *in vitro* and the cochlear DC potential *in vivo*. Sellick and Johnstone (1974) found that the DC endolymphatic potential of the utricle, although normally of much lower magnitude than the cochlear potential, undergoes a qualitatively similar decline in response to perilymphatic ouabain perfusion. On the other hand, it is difficult to reconcile the great difference in magnitude of the cochlear and utricular endolymphatic DC potentials with the somewhat similar levels of enzymatic activity (Kusakari and Thalmann, 1976) and potassium concentration (Sellick and Johnstone, 1975), and the identical basolateral distributions of Na^+, K^+-ATPase in the respective ion-transporting epithelia (Kerr et al., 1982; Burnham and Stirling, 1984a). Until these discrepancies are resolved, it may be best to consider that Na^+, K^+-ATPase plays a permissive role in the generation of endolymphatic DC potentials, leaving open the possible involvement of other mechanisms.

To further clarify the function of Na^+, K^+-ATPase in the membranous labyrinth, it will be necessary to gain an understanding of *sodium* movements in the various labyrinthine fluid compartments. Recent studies of cation flux in the cochlea (Sellick and Bock, 1974; Sellick and Johnstone, 1975; Bosher, 1980b) point toward the conclusion that transport of potassium into endolymph greatly exceeds the outwardly-directed transport of sodium. Bosher (1980b), for example, estimated an endolymphatic cation transport ratio of at least 8 K^+ : 1 Na^+ in the mammalian cochlea. This transport stoichiometry is much different than the 3 Na^+ : 2 K^+ coupling of Na^+, K^+-ATPase in preparations such as resealed erythrocyte ghosts, and it was thought for a time that these findings might preclude the enzyme's participation in endolymphatic potassium transport. Sellick and his colleagues (Sellick and Bock, 1974; Sellick and Johnstone, 1975) recognized that the sodium concentration of cochlear endolymph ($<1mM$) is considerably lower than the K_m value for sodium determined by Kuijpers (1969) for the strial ATPase; they discussed the possibility that the strial Na^+, K^+-ATPase may normally remain inactive. Subsequent studies, however, showed that ouabain inhibition leads to a profound decline of strial energy utilization *in vivo* (Thalmann et al., 1977) and a drop in oxygen consumption amounting to nearly 50% *in vitro* (Marcus et al., 1978). The latter evidence almost certainly excludes the hypothesis of a non-functioning strial enzyme. It is, nevertheless, possible that only a fraction of the available strial enzyme sites are normally active, as seems to be the case in kidney (cf. Jørgensen, 1980, cited above).

Still unexplained, however, is the unusual Na^+ : K^+ transport stoichiometry of cochlear endolymph. If all potassium extruded from the marginal cell at the luminal membrane *enters* the cell through the basolateral Na^+/K^+ pump, and if the molecular transport stoichiometry of the strial enzyme conforms to the 3 Na^+ : 2 K^+ coupling found in many other systems, then it is clear that the pump can derive only a fraction of its cytoplasmic sodium requirement from endolymph. It has been suggested (Kerr et al., 1982; Kerr, 1983) that sodium may enter the marginal cell at the *basolateral* membrane, by passive diffusion from the adjacent strial interstitial spaces. Sodium, having once entered the cell, would be entrained by the Na^+/K^+ pump at the cytoplasmic surface of the basolateral membrane, and would then be returned by active transport to the interstitial spaces. At the same time, K^+ would be pumped from the interstitial spaces to the cell interior. Thus, Na^+, K^+-ATPase would furnish at least a portion of the energy required for the observed transepithelial flux of K^+ into endolymph, with minimal transepithelial sodium flux. Sodium movements, therefore, might be largely restricted to a "recycling" process at the basolateral membrane of the marginal cell. Passive entry of sodium at this membrane surface could be linked to a sodium-coupled uptake of glucose (Ullrich, 1979), or to uptake of other substances required for cellular metabolism.

Since estimates of transepithelial transport stoichiometry are not yet available for the vestibular endolymphatic compartments, it is not certain whether the same considerations will apply to potassium transport by the dark-cell epithelial regions. Sellick *et al.* (1972) noted that the Na^+ concentration of vestibular endolymph would be ample to maintain the activity of Na^+, K^+-ATPase. However, Thalmann and Marcus (Chapter 24, this volume) discuss evidence suggesting that vestibular dark cells do not accomplish transepithelial transport of sodium. If so, a similar mechanism for the recycling of sodium ion may be operative in the dark cells.

Two theories of cochlear potassium transport are currently considered viable. Both postulate a role for Na^+, K^+-ATPase in the active uptake of K^+ into strial marginal cells. The enzyme's basolateral distribution in marginal cells (Kerr *et al.*, 1982) is compatible with either scheme. These alternative hypotheses have been termed the "single-pump" model and the "two-pump" model (cf. Marcus *et al.*, 1983; Thalmann and Marcus, Chapter 24, this volume). According to the "single-pump" model (Johnstone and Sellick, 1972; Kusakari and Thalmann, 1976), the basolateral Na^+, K^+-ATPase concentrates K^+ in the marginal cell by active uptake; subsequent entry into endolymph is by passive diffusion across a potassium-permeable apical membrane. The "two-pump" model (Sellick and Johnstone, 1975) postulates an additional active transport step at the luminal membrane, mediated by a "potassium pump" (yet to be identified). A basolateral sodium-recycling mechanism (discussed above) could be integrated with either model, although Marcus *et al.*, (1983) have considered such a mechanism only in the context of the single-pump formulation. These investigators summarized indirect evidence which tends to favor the dual-pump model.

Other functional roles of labyrinthine Na^+, K^+-ATPase are addressed in the reviews of Kuijpers (1969), and Ross *et al.* (1982).

III. CARBONIC ANHYDRASE

A. BACKGROUND

In 1928, Henriques discovered that the escape of carbon dioxide from hemolyzed blood was significantly faster than could be accounted for by the rate of the uncatalyzed conversion of bicarbonate to carbon dioxide. Subsequently, Meldrum and Roughton (1933) found that hydration of carbon dioxide and dehydration of carbonic acid were catalyzed by an enzyme present in hemoglobin-free supernatants of chloroform-treated, hemolyzed blood. They named the enzyme carbonic anhydrase (Brinkman et al., 1932). Keilen and Mann (1939) demonstrated that carbonic anhydrase is a zinc-containing enzyme, thus establishing a physiological function for zinc. Mann and Keilen (1940) discovered that sulfonamides inhibit carbonic anhydrase, a finding which was to become important to functional studies on the enzyme. It was Davenport and colleagues who first described carbonic anhydrase in tissues other than blood, namely, those of the stomach (Davenport, 1939) and kidney (Davenport and Wilhelmi, 1941). In the latter organs, carbonic anhydrase functions in the transport of protons. The presence and action of carbonic anhydrase in a variety of other tissues has since been established (Maren, 1967).

In 1961, Erulkar and Maren reported the presence of carbonic anhydrase activity in the inner ear of the cat. Later, Drescher (1977) described the distribution of carbonic anhydrase activity in cochlear tissues of the guinea pig and purified a major soluble form of the enzyme from the membranous lateral wall. Because carbonic anhydrase has been implicated in endolymphatic ion and fluid transport (Maren, 1974; Sterkers et al., 1984; Sterkers, Chapter 26, this volume), the molecular properties and possible functions of this enzyme will be discussed here.

B. MOLECULAR CHARACTERISTICS OF CARBONIC ANHYDRASE

Carbonic anhydrases (carbonate hydro-lyases, EC 4.2.1.1.) catalyze the reaction

$$CO_2 + H_2O \rightleftharpoons HCO_3^- + H^+$$

This reaction proceeds slowly in the absence of enzyme (rate constant = 0.0375 s^{-1} for the hydration at 25°C; Gibbons and Edsall, 1963). The reason for the slowness of the uncatalyzed reaction is that linear carbon dioxide (O=C=O) must undergo an extensive electronic rearrangement to become bicarbonate ($^{HO}_{HO}$C-O$^-$), a planar triangular molecule, which is a time-consuming process. Carbonic anhydrase is a remarkably efficient catalyst, with a turnover number of 38 million (that is, 38 million carbon dioxide molecules are converted per min per molecule of enzyme), the fastest turnover number of any known enzyme. Erythrocytes, which contain carbonic anhydrase, increase the rate of the uncatalyzed reaction some 100 to 1,000-fold (Maren, 1967). In addition to its carbon-dioxide-hydrating ability, carbonic anhydrase also catalyzes the hydrolysis of aldehydes (Pocker and Meany, 1965) and esters (Tashian et al., 1963), although at a much slower rate than it hydrates carbon dioxide.

Four isoenzyme forms of carbonic anhydrase are presently known: carbonic anhydrases I, II, III, and IV (Holmes, 1977; Wistrand, 1984a). Each isoenzyme is probably the product of a separate gene (Whitney and Briggle, 1982; Hewett-Emmett *et al.*, 1984) and enzymes I, II, and III appear to be single polypeptide chains with molecular weights of about 30,000 (for enzyme IV, see below). Vertebrate carbonic anhydrases contain one atom of zinc per molecule, which forms part of the active catalytic site. Carbonic anhydrases I and II are found in erythrocytes (Edsall, 1968) and in other tissues (Carter, 1972). Of the isoenzyme forms, carbonic anhydrase II has the highest specific activity of the group. For example, enzyme II from chicken is 8 and 47 times more active, respectively, than carbonic anhydrases I and III from the same species (Holmes, 1977). Carbonic anhydrase II is also thought to be closely related to a common ancestral carbonic anhydrase (Hewett-Emmett *et al.*, 1984). Carbonic anhydrases I and II are strongly inhibited by sulfonamides such as sulfanilamide and acetazolamide (Mann and Keilen, 1940; Holmes, 1977; Maren, 1984). Carbonic anhydrases are also inhibited by monovalent anions such as chloride, with the most marked effects exerted on enzyme I (Roughton and Booth, 1946; Carter, 1972). The third form of carbonic anhydrase, carbonic anhydrase III, has been identified in red skeletal muscle (Holmes, 1976) and is now known to be the same as a sulfonamide-resistant carbonic anhydrase previously reported for male rat liver (King *et al.*, 1974). The most recently characterized carbonic anhydrase, carbonic anhydrase IV (Wistrand, 1984a), is tightly bound to membranes and exhibits a molecular weight of 52,000–68,000 in solutions of sodium dodecyl sulfate. Carbonic anhydrase IV is somewhat less sensitive to sulfonamide inhibition (five to ten-fold) than carbonic anhydrases I and II (Wistrand, 1984a). Carbonic anhydrase IV is found in the brush border of the proximal tubule of the kidney (McKinley and Whitney, 1976) and is probably also present in capillary endothelium of the lung (Whitney and Briggle, 1982).

The tertiary structures of carbonic anhydrases I and II have been determined in detail by X-ray crystallography (Liljas *et al.*, 1972; Kannan *et al.*, 1975). The molecules are ellipsoids, with dimensions of $41 \times 41 \times 47$ Å. In the protein, ten polypeptide-chain segments form a twisted, β-sheet structure (Edsall and Wyman, 1958), which roughly divides each molecule in half. The carbonic anhydrase polypeptide chain contains 17–20% helical segments, and two of the seven helical segments are nearly true α-helix (Edsall and Wyman, 1958). The enzyme's active site is located in a 12-Å-deep crevice at the center of the molecule where the zinc atom is situated. Zinc is bound by four coordinate bonds, forming a slightly misshapen tetrahedron; three of the bonds are to histidine residues and the fourth is to a water hydroxyl (Liljas *et al.*, 1972). During the catalytic hydration of carbon dioxide, the bound hydroxyl combines with carbon dioxide to form bicarbonate (Lindskog *et al.*,

1984). Sulfonamides and anionic inhibitors replace the bound water hydroxyl, thus preventing one substrate, water, from participating in the catalyzed reaction.

C. PROPERTIES OF COCHLEAR CARBONIC ANHYDRASE

In 1977, a soluble carbonic anhydrase was purified from the membranous lateral wall of the cochlea of the guinea pig (Drescher, 1977). This enzyme could not be accounted for by contribution from cochlear blood, and was similar to carbonic anhydrase II, the high-specific-activity or "C" form present in erythrocytes (Fig. 5). The purified cochlear enzyme has a molecular weight of about 30,000, a specific activity 60–80% that of carbonic anhydrase II from blood, and an electrophoretic mobility similar to that of the blood enzyme (Figs. 5B, 5C). In addition, about 3% of the total cochlear carbonic anhydrase appeared to be membrane bound. In Fig. 5A, the elution diagram shows that some cochlear carbonic anhydrase activity is associated with material of high molecular weight (small dashed peak at 40 ml), eluted under conditions in which nonspecific binding was minimized (Drescher, 1977). Thus, carbonic anhydrase in the cochlea may be present in soluble and particulate forms similar to those found in the kidney (McKinley and Whitney, 1976). Cochlear carbonic anhydrase is half-maximally inhibited by 4×10^{-9} M acetazolamide, is completely inhibited above 10^{-5} M acetazolamide, and forms a fluorescent complex with 5-dimethylaminonaphthalene-1-sulfonamide (DNSA), by which it can be distinguished on polyacrylamide gels (Fig. 6) (Chen and Kernohan, 1967; Drescher, 1978).

D. LOCALIZATION OF CARBONIC ANHYDRASE IN THE INNER EAR

Carbonic anhydrase has been localized in tissues of the inner ear by microdissection and direct biochemical assay (Erulkar and Maren, 1961; Drescher, 1977) and by cytochemical procedures (Lim et al., 1983; Hsu and Nomura, 1984; Watanabe and Ogawa, 1984).

Erulkar and Maren (1961), using a biochemical approach, showed that the concentration of carbonic anhydrase increases from base to apex in the whole tissues of the cat cochlea[1] and that carbonic anhydrase is also present in vestibular tissue (combined utricle, saccule, ampullae, and anterior semicircular canal) and in endolymphatic sac. Carbonic anhydrase was found to be present in the inner ear at one of the highest specific activities known in any organ (reviewed by Maren, 1967). Drescher (1977) localized carbonic anhydrase by direct assay (Wilbur and Anderson, 1948) in various fractions of the guinea-pig cochlea (Table I). Table I indicates that the highest specific activity of carbonic anhydrase (enzyme activity units/mg protein) is

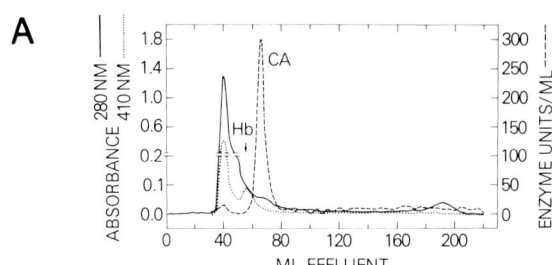

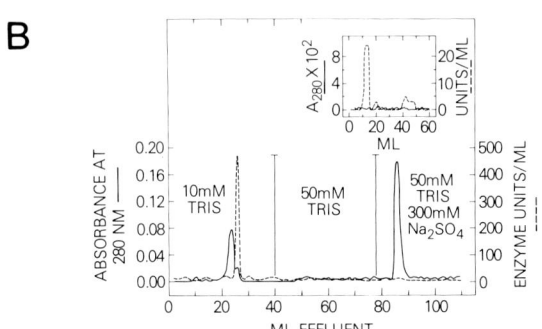

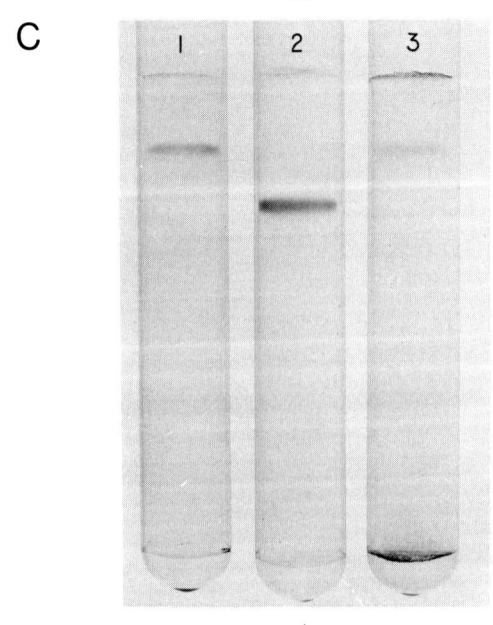

Figure 25-5. A. Isolation of crude carbonic anhydrase from a homogenate of membranous lateral wall (60 cochleas; 1 ml) by gel filtration on Bio-Gel P-100, 100–200 mesh (Bio-Rad, Richmond, CA). Bed dimensions were 2.6 × 32 cm and flow rate was 2.3 ml/(cm²h). Eluting buffers were 10 mM Tris sulfate, 100 mM sodium sulfate, pH 7.4. One unit of enzyme activity is the amount of carbonic anhydrase that increases the reaction rate 10% above that of the uncatalyzed reaction, as measured at 4°C by the method of Wilbur and Anderson (1948). The positions of elution of hemoglobin (Hb, small dotted peak) and carbonic anhydrase (CA, dashed peak) are shown. B. Further purification of cochlear carbonic anhydrase by anion exchange and gel filtration. Active fractions from the Bio-Gel P-100 column (60–74 ml, Fig. 5A) were pooled, dialyzed against 10 mM Tris sulfate, pH 8.9, and concentrated to 1.5 ml with a 25-mm Diaflo DM5 membrane (Amicon, Lexington, MA). The sample was applied to a DEAE Bio-Gel A column initially equilibrated with 10 mM Tris sulfate, pH 8.9; the elution pattern is shown in the lower half of the illustration. Column bed dimensions were 0.9 × 13 cm; flow rate, 3.9 ml/(cm²h). (Although not illustrated, carbonic anhydrase I from guinea pig blood is eluted in the region marked "50 mM Tris.") Cochlear material with carbonic anhydrase activity from fraction 26 was concentrated to 0.5 ml, applied to a 0.9 × 13-cm column of Bio-Gel P-60, 50–100 mesh, and eluted with 10 mM Tris sulfate, pH 8.9; flow rate, 3.5 ml/(cm²h) (inset). The active cochlear enzyme, eluted at 10–12 ml, was electrophoretically homogeneous. C. Polyacrylamide gel electrophoresis of purified carbonic anhydrases (1) type II and (2) type I from blood, and purified carbonic anhydrase from cochlear membranous lateral wall (3). Samples contained 8, 9, and 7 μg of protein, respectively. Electrophoretic techniques are described in Drescher (1977). Blood enzymes were isolated from guinea-pig erythrocyte lysates by methods similar to those described for the cochlear enzyme.

found in the spiral ligament fraction of the cochlear lateral wall, which contains about twice as much enzyme/mg protein as whole blood. Specific activity of carbonic anhydrase was also determined in a fraction of spiral ligament containing the spiral prominence and a fraction containing the remainder of the spiral ligament; the spiral prominence fraction showed slightly higher activity (about 10%) than the remainder of the spiral ligament (Drescher, unpublished results). Table I shows that the concentration of carbonic anhydrase in the stria vascularis is also high, about the same as that of whole blood. Little carbonic anhydrase activity was found in the organ-of-Corti fraction. The auditory nerve and cochlear nucleus also contain relatively low concentrations of carbonic anhydrase, in agreement with published studies on the enzyme in neural tissue (Giacobini, 1962; Maren, 1967; Wong et al. 1983; Sapirstein et al., 1984). In brain, carbonic anhydrase is located in glial cells (Ghandour et al., 1980), and glia probably account for activity in the cochlear nucleus (Table I).

The biochemical results of Erulkar and Maren (1961) and of Drescher (1977) regarding the location of carbonic anhydrase have been confirmed and extended by cytochemical studies. Lim et al. (1983), with the chinchilla, and Hsu and Nomura (1984) and Watanabe and Ogawa (1984), with the guinea pig, used the method of Hansson (1967) to localize carbonic anhydrase activity in inner-ear tissues. Hansson's (1967) method depends upon the

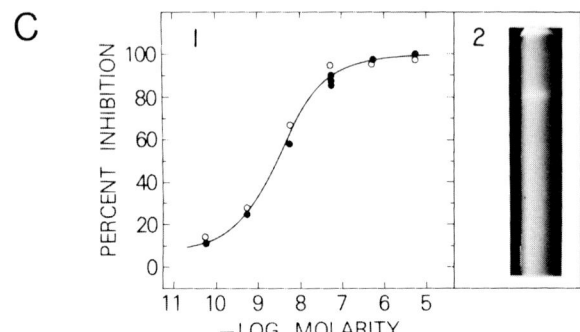

Figure 25-6. A, B. Chemical structures of aromatic sulfonamides (DNSA: 5-dimethylaminonaphthalene-1-sulfonamide). C. Interaction of cochlear carbonic anhydrase with the aromatic sulfonamides shown in A and B. 1. Inhibition by acetazolamide, where percent inhibition of carbonic anhydrase is plotted against the negative logarithm of the concentration of acetazolamide in the assay medium. (o) Homogenate of membranous lateral wall. (•) Purified enzyme. Concentration of crude and purified enzyme was about 10^{-9} M. 2. Polyacrylamide gel, showing single fluorescent band (at the position of carbonic anhydrase II) after electrophoresis of a homogenate of membranous lateral wall containing 10 μM DNSA.

nonenzymatic formation of OH^- and H_2CO_3 from H_2O and HCO_3^- present in the medium containing the tissue. The local production of OH^- causes formation of a cobalt-phosphate complex from $CoSO_4$ and KH_2PO_4 which are present in the medium. Carbonic anhydrase enhances this nonenzymatic reaction by enzymatically converting H_2CO_3 to H_2O and CO_2, the latter escaping into the atmosphere and driving the nonenzymatic reaction to form more H_2CO_3 and OH^-. The cobalt-phosphate complex is subsequently reacted with $(NH_4)_2S$ to form a black CoS precipitate at sites of enzyme activity. Tissues treated with acetazolamide serve as controls. Hansson's method has been critically examined and found to be specific for carbonic anhydrase (Lönnerholm, 1974; Maren, 1980).

TABLE 25-I
CARBONIC ANHYDRASE ACTIVITY OF INNER-EAR FRACTIONS
AND OTHER TISSUES OF THE GUINEA PIG

	Fresh tissue						Lyophilized tissue		
	Cochlear bone	Cochlear nucleus	Auditory nerve	Organ of Corti fraction	Membranous lateral wall	Whole blood	Membranous lateral wall	Spiral ligament fraction	Stria vascularis
Specific activity (units/mg protein)	0	45	24	47	545	301	488	611	286
Relative specific activity	0	0.2	0.1	0.2	1.8	1.0	1.6	2.0	0.9

The membranous lateral wall consisted of stria vascularis, spiral ligament, spiral prominence, and part of the outer sulcus. Relative specific activity refers to specific activity relative to that of whole blood. Enzyme activity units are the same as described for Fig. 5. (From Drescher, 1977.)

The results of Lim et al. (1983) on the localization of labyrinthine carbonic anhydrase are depicted in diagrammatic form in Fig. 7. Concentrated reaction product for carbonic anhydrase was observed in the cochlear spiral ligament cells, with somewhat less intense reaction in cells identified as intermediate and/or basal cells of the stria vascularis (Fig. 7A). In addition, reaction product was seen in external sulcus cells and Boettcher's cells. Weak reaction was observed in the organ of Corti and in Reissner's membrane. Thus, the results shown in Fig. 7A confirm previous biochemical findings of Drescher (1977) (Table 1). Figures 7B and 7C indicate that in the utricle and ampullae, intense carbonic anhydrase reaction is found in the dark cells, and in supporting cells of the sensory epithelia. Sensory (hair) cells in these vestibular organs show little cytochemical reaction. Hsu and Nomura (1984) also reported that vestibular supporting cells, but not vestibular hair cells, showed reaction for carbonic anhydrase. In a study employing light and electron microscopy, Watanabe and Ogawa (1984) confirmed the findings of Lim et al. (1983) (Fig. 7) for fibrocytes of the spiral ligament, strial intermediate cells, and dark cells of the utricle and cristae. Watanabe and Ogawa (1984) also observed reaction deposit in apical vesicles of strial marginal cells, in the spaces between marginal cells, and in endothelial cells of the strial capillaries.

Besides the biochemical and cytochemical methods discussed above, immunocytochemistry (Spicer et al., 1984) offers an approach for localizing carbonic anhydrase. Immunocytochemical methods have the advantage that the isoenzyme forms of carbonic anhydrase can be distinguished in tissues, a feat not possible with classical cytochemistry. We would, therefore, expect to see immunocytochemical techniques employed in-

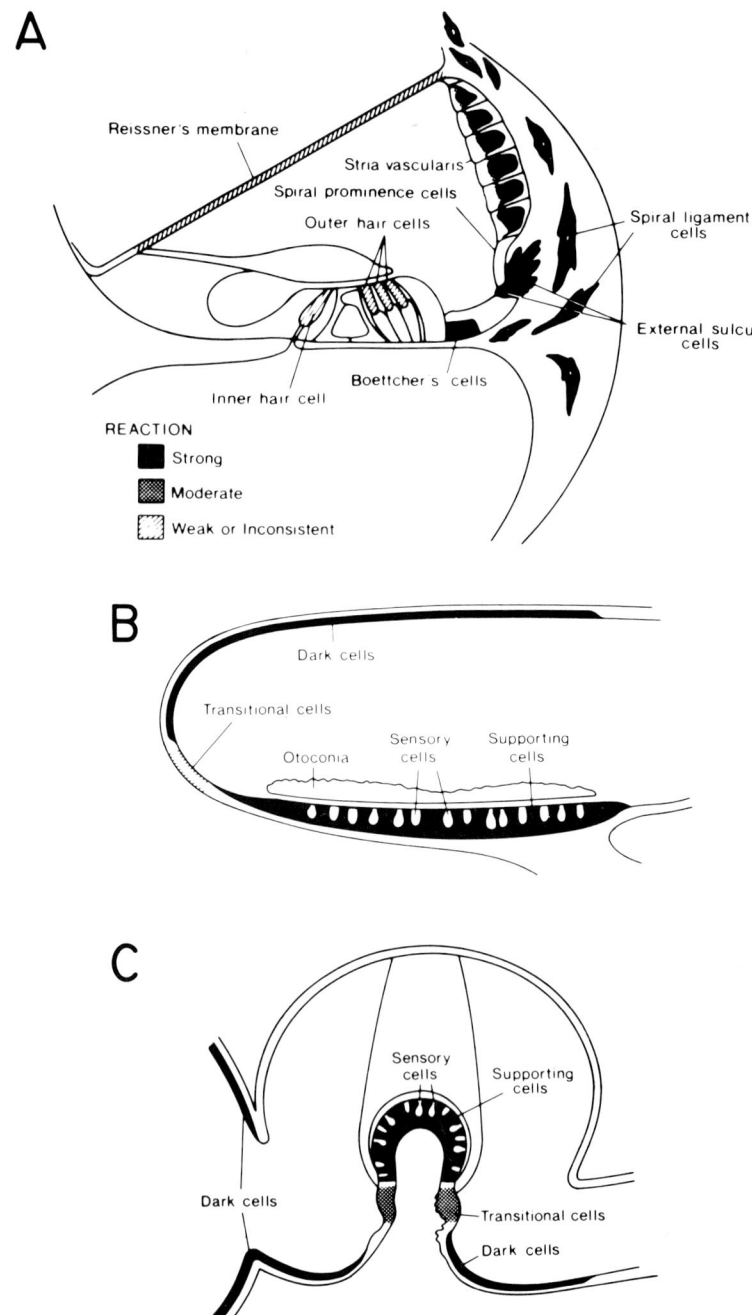

Figure 25-7. Diagrammatic representations showing cytochemical localization of carbonic anhydrase (Hansson, 1967) in the labyrinth of the chinchilla. A. Cochlea. B. Utricle. C. Ampulla. Key in Fig. 7A indicates intensity of the cytochemical reaction. (From Lim et al., 1983.)

creasingly in the future for localization of carbonic anhydrases in the inner ear.

E. FUNCTION OF CARBONIC ANHYDRASE

Carbonic anhydrase functions in two main ways in higher animals. The first is to carry carbon dioxide. In this connection, the enzyme speeds the hydration of metabolic carbon dioxide to bicarbonate (carbonic acid) which is carried by the blood to the lungs. There, the dehydration of bicarbonate (carbonic acid) is catalyzed, and carbon dioxide is exhaled, thus maintaining blood pH (see Maren, 1967, and Swenson, 1984, for reviews). The second function of carbonic anhydrase is to furnish bicarbonate or protons in ion-transporting, secretory systems.

The role of carbonic anhydrase in secretory processes can be described within the framework of a generalized secretory cell, where water is ionized to yield OH^- at one membrane surface and H^+ at the opposite surface (Maren, 1967). In such a cell, carbonic anhydrase allows CO_2 to combine rapidly with the OH^- to form HCO_3^- (OH^- is actually the form bound to the enzyme; see Lindskog et al., 1984). If the purpose of the cell is to secrete an alkaline, or mildly alkaline, fluid, then HCO_3^- is secreted at the "lumen" side of the cell, along with a counter ion such as sodium, and protons are buffered and carried away at the "blood" side of the cell. This is the case for the pancreas (Rawls et al., 1963), the choroid plexus of the brain (Vogh and Maren, 1975), the ciliary process of the eye (Zimmerman et al., 1976), and for other organs (Maren, 1967). In addition, the movement of bicarbonate in such organs often influences fluid movement in some yet unknown way (Maren, 1984). If, on the other hand, the secretory cell elaborates an acidic fluid, then H^+ is secreted at the lumen side and HCO_3^- is removed at the blood side of the cell. This is the case for the parietal cells of the stomach (Emas, 1962; Maren, 1984) and the cells of the proximal tubule of the kidney (Maren, 1967). Carbonic anhydrase is of functional importance in ion- and fluid-transporting systems only when there is a requirement for the formation of bicarbonate or protons at a rate so rapid as to need the enzyme for normal secretory function; thus, the functionality of carbonic anhydrase in an organ can be assessed by specifically inhibiting carbonic anhydrase with sulfonamides and noting whether or not there is a decrease in fluid flow or a change in fluid composition.

What is the function of the high concentration of carbonic anhydrase in the inner ear described in this chapter? Erulkar and Maren (1961) suggested that in the inner ear "carbonic anhydrase may play an important role in the mechanism of active secretion of potassium in the endolymph, and possibly in the overall formation of this fluid." This suggestion was based on the investigators' early data indicating that acetazolamide lowered endolymphatic potassium in the cat (Erulkar and Maren, 1961). However, that finding was not confirmed by Silverstein and Yules (1971) or by Sterkers et al. (1984) (see also Chapter 26, this volume). Acetazolamide has not proven to be effective in treating Meniere's disease, a condition which, presumably, is due to endolymphatic hydrops (Brookes et al., 1982). In contrast, acetazolamide has been impressively effective in the treatment of glaucoma (Wistrand, 1984b). The hypothesized involvement of carbonic anhydrase in the control of flow and pressure of endolymphatic fluid is open to question.

Maren et al. (1975) thoroughly studied the effect of acetazolamide on rates of movement of major ions into saccular endolymph of the dogfish. These workers found that large doses of acetazolamide decreased the rate of entry of radioactive bicarbonate into endolymph about three-fold, but had no effect on the entry of labeled sodium, rubidium (a marker for potassium), or chloride. Their data support the idea that carbonic anhydrase is important within the secretory epithelium of the saccule for forming endolymphatic bicarbonate (endolymphatic bicarbonate concentration in the fish saccule is about 11 mM). It is likely that carbonic anhydrase is also important in the mammalian stria vascularis for rapidly forming endolymphatic bicarbonate (about 20 mM concentration in the rat; Sterkers et al., 1984). Carbonic anhydrase does not appear to be important in controlling the flow of endolymphatic fluid in the fish saccule (Maren et al., 1975), and it appears that there is an analogous situation in the mammalian cochlea. In the rat, acetazolamide also lowers the endolymphatic potential, an interesting finding that is discussed by Sterkers in Chapter 26 of this volume.

The possible function of carbonic anhydrase in the spiral ligament (Table I) is considered by the present authors to be a problem apart from that of the enzyme's postulated role in the stria vascularis. It is possible that carbonic anhydrase in the spiral ligament functions in a manner similar to that of carbonic anhydrase in the choroid plexus (Vogh and Maren, 1975), namely, to furnish bicarbonate rapidly for the formation of perilymph. The perilymph would then transfer protons to blood in the capillaries of the membranous lateral wall (Smith, 1951, 1954). This possibility provides an interesting subject for future research.

ACKNOWLEDGMENTS

Part of the work presented in this chapter was supported by a NASA Space Biology Research Associate Award to T. P. Kerr. Figures 1-4 are reproduced with the permission of W. B. Saunders Company. Dr. David Lim kindly supplied the illustrations for Fig. 7. The authors thank Dr. Marian Drescher for critical comments.

NOTE

[1]The gradient of carbonic anhydrase from base to apex of the cochlea is opposite to that for Na^+, K^+-ATPase (see first section, this chapter). However, carbonic anhydrase is usually present in great excess of that needed physiologically (Maren, 1967), so the cochlear gradient of carbonic anhydrase may have no physiological significance.

REFERENCES

Ahlström, P.; Thalmann, I., Thalmann, R., and Ise, I.: Cyclic AMP and adenylate cyclase in the inner ear. *Larynogoscope 85:* 1241–1258, 1975.

Akera, T.: Membrane adenosinetriphosphatase: a digitalis receptor? *Science 198:* 569–574, 1977.

Albers, R. W.: Biochemical aspects of active transport. *Annu. Rev. Biochem. 36:* 727–756, 1967.

Békésy, G. V.: Microphonics produced by touching the cochlear partition with a vibrating electrode. *J. Acoust. Soc. Am. 23:* 29–35, 1951.

Békésy, G. V.: D-C resting potentials inside the cochlear partition. *J. Acoust. Soc. Amer. 24:* 72–76, 1952.

Bennett, J. P., Jr.: Methods in binding studies. In Yamamura, H.; Enna, S., and Kuhar, M. (Eds.): *Neurotransmitter Receptor Binding.* New York, Raven Press, 1978, pp. 57–90.

Bonting, S. L., and Caravaggio, L. L.: Studies on Na-K-activated ATPase. V. Correlation of enzyme activity with cation flux in six tissues. *Arch. Biochem. Biophys. 101:* 37–46, 1963.

Bosher, S. K.: The effects of inhibition of the strial Na^+, K^+-activated ATPase by perilymphatic ouabain in the guinea pig. *Acta Otolaryg. 90:* 219–229, 1980a.

Bosher, S. K.: The nature of the ototoxic actions of ethacrynic acid upon the mammalian endolymph system I. Functional aspects. *Acta Otolaryng. 89:* 407–418, 1980b.

Bosher, S. K.: Interpretation problems in identifying the nature of changes produced in the cochlear endolymph system. In Klinke, R.; Lahn, W., Querfurth H., and Scholtholt, J. (Eds.): *Ototoxic Side Effects of Diuretics.* Scand. Audiol. Suppl. 14: 51–60, 1981.

Bosher, S. K., and Warren, R. L.: Observations on the electrochemistry of cochlear endolymph of the rat: a quantitative study of its electrical potential and ionic composition as determined by means of flame spectrometry. *Proc. Roy. Soc. Lond. B 171:* 227–247, 1968.

Brinkman, R.; Margaria, R., Meldrum, N. U., and Roughton, F. J. W.: The CO_2 catalyst present in blood. *J. Physiol. (Lond.) 75:* 3P-4P, 1932.

Brookes, G. B.; Hodge, R. A., Booth, J. B., and Morrison, A. W.: The immediate effects of acetazolamide in Meniere's disease. *J. Laryngol. Otol. 96:* 57–72, 1982.

Brusilow, S. W., and Gordes, E.: The mutual independence of the endolymphatic potential and the concentrations of sodium and potassium in endolymph. *J. Clin. Invest. 52:* 2517–2521, 1973.

Burnham, J. A., and Stirling, C. E.: Quantitative localization of Na-K pump site in frog inner ear dark cells. *Hearing Res. 13:* 261–268, 1984a.

Burnham, J. A., and Stirling, C. E.: Quantitative localization of Na-K pump sites in the frog sacculus. *J. Neurocytol. 13:* 617–638, 1984b.

Caldwell, P. C.; Hodgkin, A. L., Keynes, R. D., and Shaw, T. I.: The effects of injecting energy rich phosphate compounds on the active transport of ions in the giant axons of loligo. *J. Physiol. (Lond.) 152:* 561–590, 1960.

Carter, M. J.: Carbonic anhydrase: isoenzymes, properties, distribution, and functional significance. *Biol. Rev. 47:* 465–513, 1972.

Chen, R. F., and Kernohan, J. C.: Combination of bovine carbonic anhydrase with a fluorescent sulfonamide. *J. Biol. Chem. 242:* 5813–5823, 1967.

Corey, D. P., and Hudspeth, A. J.: Ionic basis of the receptor potential in a vertebrate hair cell. *Nature 281:* 675–677, 1979.

Davenport, H. W., and Wilhelmi, A. E.: Renal carbonic anhydrase. *Proc. Soc. Exp. Biol. Med. 48:* 53–56, 1941.

Davenport, H. W.: Gastric carbonic anhydrase. *J. Physiol. (Lond.) 97:* 32–43, 1939.

Davis, H.: A model for transducer action in the cochlea. *Cold Spring Harbor Symp. Quant. Biol. 30:* 181–190, 1965.

Deguchi, N.; Jørgensen, P. L., and Maunsbach, A. B.: Ultrastructure of the sodium pump. Comparison of thin sectioning, negative staining, and freeze-fracture of purified, membrane bound (Na^+, K^+)-ATPase. *J. Cell Biol. 75:* 619–634, 1977.

Drescher, D. G.: Purification of a carbonic anhydrase from the inner ear of the guinea pig. *Proc. Natl. Acad. Sci. U.S.A. 74:* 892–896, 1977.

Drescher, D. G.: Purification of blood carbonic anhydrases and specific detection of carbonic anhydrase isoenzymes on polyacrylamide gels with 5-dimethylaminonaphthalene-1-sulfonamide (DNSA). *Anal. Biochem. 90:* 349–358, 1978.

Edsall, J. T., and Wyman, J.: *Biophysical Chemistry. Vol. I.* New York, Academic Press, 1958, pp. 47–135.

Edsall, J. T.: The carbonic anhydrases of erythrocytes. *Harvey Lectures 62:* 191–230 1968.

Emas, S.: Effect of acetazolamide on histamine-stimulated and gastrin-stimulated gastric secretion. *Gastroenterol. 43:* 557–563, 1962.

Ernst, S. A.: Transport adenosine triphosphatase cytochemistry. I. Biochemical characterization of a cytochemical medium for the ultrastructural localization of ouabain-sensitive, potassium-dependent phosphatase activity in the avian salt gland. *J. Histochem. Cytochem. 20:* 13–22, 1972a.

Ernst, S. A.: Transport adenosine triphosphatase cytochemistry. II. Cytochemical localization of ouabain-sensitive, potassium-dependent phosphatase activity in the secretory epithelium of the avian salt gland. *J. Histochem. Cytochem. 20:* 23–38, 1972b.

Ernst, S. A.: Transport ATPase cytochemistry: Ultrastructural localization of potassium-dependent and potassium-independent phosphatase activities in rat kidney cortex. *J. Cell Biol. 66:* 585–608, 1975.

Ernst, S. A., and Hootman, S. R.: Microscopical methods for the localization of Na^+, K^+-ATPase. *Histochem. J. 13.* 397–418, 1981.

Ernst, S. A.; Riddle, C. V, and Karnaky, K. J., Jr.: Relationship between localization of Na^+-K -ATPase, cellular fine structure, and reabsorptive and secretory electrolyte transport. In Bronner, F., and Kleinzeller, A. (Eds.): *Current Topics in Membranes and Transport, Vol. 13.* New York, Academic Press, 1980, pp. 355–385.

Erulkar, S. D., and Maren, T. H.: Carbonic anhydrase and the inner ear. *Nature 189:* 459–460, 1961.

Feldman, A. M., and Brusilow, S. W.: Effects of cholera toxin on cochlear endolymph production: Model for endolymphatic hydrops. *Proc. Natl. Acad. Sci. U.S.A. 73:* 1761–1764, 1976.

Feldman, A. M.: Cochlear biochemistry. In Brown, R. D., and Diagneault, E. A. (Eds.): *Pharmacology of Hearing.* New York, Wiley, 1981, pp. 51–80.

Fernandez, C.: Biochemistry of labyrinthine fluids. *Arch. Otolaryngol. 86:* 116–127, 1967.

Firth, J. A.: Problems of specificity in the use of a strontium capture technique for the cytochemical localization of ouabain-sensitive, potassium-dependent phosphatase in mammalian renal tubules. *J. Histochem. Cytochem. 22:* 1163–1168, 1974.

Gardos, G.: Connection between membrane adenosine triphosphatase activity and potassium transport in erythrocyte ghosts. *Experientia (Basel) 20:* 387, 1964.

Garrahan, P. J., and Glynn, I. M.: The stoichiometry of the sodium pump. *J. Physiol. (Lond.) 192:* 217–237, 1967.

Ghandour, M. S.; Langley, O. K., Vincendon, G., Gombos, G., Fillipi, D., Limozin, N., Dalmasso, C., and Laurent, G.: Immunochemical and immunohistochemical study of carbonic anhydrase II in adult rat cerebellum: a marker for oligodendrocytes. *Neuroscience 5:* 559–571, 1980.

Giacobini, E.: A cytochemical study of the localization of carbonic anhydrase in the nervous system. *J. Neurochem. 9:* 169–177, 1962.

Gibbons, B. H., and Edsall, J. T.: Rate of hydration of carbon dioxide and dehydration of carbonic acid at 25°. *J. Biol. Chem. 238:* 3502–3507, 1963.

Glynn, I. M., and Karlish, S. J. D.: The sodium pump. *Annu. Rev. Physiol. 37:* 13–55, 1975.

Hansen, O.: Interaction of cardiac glycosides with (Na^+ + K^+)-activated ATPase. A biochemical link to digitalis-induced inotropy. *Pharmacol. Rev. 36:* 143–163, 1984.

Hansson, H. P. J.: Histochemical demonstration of carbonic anhydrase activity. *Histochemie 11:* 112–128, 1967.

Henriques, O. M.: Die Bindungweise des Kohlendioxyds im Blute. *Biochem. Z. 200:* 1–4, 1928.

Hewett-Emmett, D.; Hopkins, P. J., Tashian, R. E., and Czelusniak, J.: Origins and molecular evolution of the carbonic anhydrase isozymes. *Ann. N. Y. Acad. Sci. 429:* 338–358, 1984.

Hinojosa, R., and Rodríguez-Echandía, E. L.: The fine structure of the stria vascularis of the cat inner ear. *Am. J. Anat. 118:* 631–664, 1966.

Hobbs, A. S., and Albers, R. W.: The structure of proteins involved in active membrane transport. *Annu. Rev. Biophys. Bioeng. 9:* 259–291, 1980.

Hodgkin, A. L., and Keynes, R. D.: Active transport of cations in giant axons from *Sepia* and *Loligo*. *J. Physiol. (Lond.) 128:* 28–60, 1955.

Hoffman, J. F.: Physiological characteristics of human red blood cell ghosts. *J. Gen. Physiol. 42:* 9–28, 1958.

Hoffman, J. F.; Tosteson, D. C., and Whittam, R.: Retention of potassium by human erythrocyte ghosts. *Nature (Lond.) 185:* 186–187, 1960.

Holmes, R. S.: Mammalian carbonic anhydrase isozymes: Evidence for a third locus. *J. Exp. Zool. 197:* 289–295, 1976.

Holmes, R. S.: Purification, molecular properties and ontogeny of carbonic anhydrase isozymes. *Eur. J. Biochem. 78:* 511–520, 1977.

Hsu, C.-J., and Nomura, Y.: Carbonic anhydrase activity in the inner ear. *Ear Res. Jpn. 15:* 213–215, 1984.

Hudspeth, A. J.: Mechanoelectrical transduction by hair cells in the acousticolateralis sensory system. *Ann. Rev. Neurosci. 6:* 187–215, 1983.

Ishii, T., and Nomura, Y.: Histochemical study on adenosine triphosphatase activity in the inner ear. *Pract. Oto-Rhino-Laryng. 30:* 237–244, 1968.

Jacobsen, N. O., and Jørgensen, P. L.: A quantitative biochemical and histochemical study of the lead method for localization of adenosine triphosphate-hydrolyzing enzymes. *J. Histochem. Cytochem. 17:* 443–453, 1969.

Johnstone, B. M., and Sellick, P. M.: The peripheral auditory apparatus. *Quart. Rev. Biophys. 5:* 1–57, 1972.

Jørgensen, P. L.: Purification and characterization of (Na^+ + K^+)-ATPase. III. Purification from the outer medulla of mammalian kidney after selective removal of membrane components by sodium dodecyl sulphate. *Biochim. Biophys. Acta 356:* 36–52, 1974.

Jørgensen, P. L.: Sodium and potassium ion pump in kidney tubules. *Physiol. Rev. 60:* 864–917, 1980.

Juhn, S. K.; Rybak, L. P., Morizono, T., and Green, L. P.: Pharmacokinetics of furosemide in relation to the alteration of endocochlear potential. In Klinke, R.; Lahn, W., Querfurth, H., and Scholtholt, J. (Eds.): *Ototoxic Side Effects of Diuretics. Scand. Audiol. Suppl. 14:* 39–47, 1981.

Kannan, K. K.; Notstrand, B., Fridborg, K., Lövgren, S., Ohlsson, A., and Petef, M.: Crystal structure of human erythrocyte carbonic anhydrase B. Three-dimensional structure at a nominal 2.2-Å resolution. *Proc. Natl. Acad. Sci. U.S.A. 72:* 51–55, 1975.

Keilen, D., and Mann, T.: Carbonic anhydrase. *Nature 144:* 442–443, 1939.

Kerr, T. P.: *Cytochemical and Autoradiographic Localization of Sodium- and Potassium-activated ATPase in Tissues of the Mammalian Cochlear Duct.* Ph.D. Thesis, University of Michigan, Ann Arbor, 1983.

Kerr, T. P.; Ross, M. D., and Ernst, S. A.: Cellular localization of transport ATPase in the lateral wall of the cochlear duct: a reassessment. *Anat. Rec. 199:* 138A, 1981.

Kerr, T. P.; Ross, M. D., and Ernst, S. A.: Cellular localization of Na^+, K^+-ATPase in the mammalian cochlear duct: significance for cochlear fluid balance. *Am. J. Otolaryngol. 3:* 332–338, 1982.

Kimura, R. S.: Distribution, structure, and function of dark cells in the vestibular labyrinth. *Ann. Otol. 78:* 542–561, 1969.

King, R. W.; Garg, L. C., Huckson, J., and Maren, T. H.: The isolation and partial characterization of sulfonamide-resistant carbonic anhydrase from the liver of the male rat. *Molec. Pharmacol. 10:* 335–343, 1974.

Konishi, T., and Mendelsohn, M.: Effect of ouabain on cochlear potentials and endolymph composition in guinea pigs. *Acta Otolaryng. 69:* 192–199, 1970.

Konishi, T.; Hamrick, P. E., and Walsh, P. J.: Ion transport in guinea pig cochlea. *Acta Otolaryng. 86:* 22–34, 1978.

Konishi, T., and Salt, A. N.: Electrochemical profile for potassium ions across the cochlear hair cell membranes of normal and noise-exposed guinea pigs. *Hearing Res. 11:* 219–233, 1983.

Kuijpers, W.: *Cation Transport and Cochlear Function.* Ph.D. Thesis, University of Nijmegen, The Netherlands, 1969.

Kuijpers, W.: Na-K-ATPase activity in the cochlea of the rat during development. *Acta Otolaryng. 78:* 341–344, 1974.

Kuijpers, W., and Bonting, S. L.: Studies on (Na^+, K^+)-activated ATPase. XXIV. Localization and properties of ATPase in the inner ear of the guinea pig. *Biochim. Biophys. Acta. 173:* 477–485, 1969.

Kuijpers, W., and Bonting, S. L.: The cochlear potentials I. The effect of ouabain on the cochlear potentials of the guinea pig. *Pflügers Arch. 320:* 348–358, 1970.

Kusakari, J., and Thalmann, R.: Effects of anoxia and ethacrynic acid upon ampullar endolymphatic potential and upon high energy phosphates in ampullar wall. *Laryngoscope 86:* 132–147, 1976.

Kyte, J.: Purification of the sodium- and potassium-dependent adenosine triphosphatase from canine renal medulla. *J. Biol. Chem. 246:* 4157–4165, 1971.

Kyte, J.: Immunoferritin determination of distribution of (Na^+ + K^+) ATPase over the plasma membranes of renal convoluted tubules. I. Distal segment. *J. Cell Biol. 68:* 287–303, 1976a.

Kyte, J.: Immunoferritin determination of the distribution of (Na^+ + K^+) ATPase over the plasma membranes of renal convoluted tubules. II. Proximal segment. *J. Cell Biol.* 68: 304-318, 1976b.

Liljas, A.; Kannan, K. K., Bergsten, P.-C., Waara, I., Fridborg, K., Strnadberg, B., Carlbom, U., Järup, L., Lövgren, S., and Petef, M.: Crystal structure of human carbonic anhydrase C. *Nature New Biol.* 235: 131-137, 1972.

Lim, D. J.; Karabinas, C., and Trune, D. R.: Histochemical localization of carbonic anhydrase in the inner ear. *Am. J. Otolaryngol.* 4: 33-42, 1983.

Lindskog, S.; Engberg, P., Forsman, C., Ibrahim, S. A., Jonsson, B.-H., Simonsson, I., and Tibell, L.: Kinetics and mechanisms of carbonic anhydrase isoenzymes. *Ann. N. Y. Acad. Sci.* 429: 61-75, 1984.

Lönnerholm, G.: Carbonic anhydrase histochemistry. A critical study of Hansson's cobalt-phosphate method. *Acta Physiol. Scand. Suppl.* 418: 1-43, 1974.

Mann, T., and Keilen, D.: Sulphanilamide as a specific inhibitor of carbonic anhydrase. *Nature* 146: 164-165, 1940.

Marcus, D. C.; Thalmann, R., and Marcus, N. Y.: Respiratory rate and ATP content of stria vascularis of guinea pig *in vitro. Laryngoscope* 88: 1825-1835, 1978.

Marcus, D. C.; Rokugo, M., Ge, X.-X., and Thalmann, R.: Response of cochlear potentials to presumed alterations of ionic conductance: Endolymphatic perfusion of barium, valinomycin and nystatin. *Hearing Res.* 12: 17-30, 1983.

Maren, T. H.: Carbonic anhydrase: chemistry, physiology, and inhibition. *Physiol. Rev.* 47: 595-781, 1967.

Maren, T. H.: Kinetics, equilibrium and inhibition in the Hansson histochemical procedure for carbonic anhydrase: a validation of the method. *Histochem. J.* 12: 183-190, 1980.

Maren, T. H.; Swenson, E. R., and Addink, A. D. F.: Rates of ion transport from plasma to endolymph in the dogfish. *Ann. Otol. Rhinol. Laryngol.* 84: 847-858, 1975.

Maren, T. H.: The cerebrospinal fluid, with notes on aqueous humor and endolymph. In Mountcastle, V. B. (Ed.): *Medical Physiology.* St. Louis, Mosby, 1974, pp. 1116-1141.

Maren, T. H.: The general physiology of reactions catalyzed by carbonic anhydrase and their inhibition by sulfonamides. *Ann. N. Y. Acad. Sci.* 429: 568-579, 1984.

Matschinsky, F. M. and Thalmann, R.: Energy metabolism of the cochlear duct. In Paparella, M. (Ed.): *Biochemical Mechanisms in Hearing and Deafness.* Springfield, IL, Charles C Thomas, 1970, pp. 265-288.

Matsuura, S.; Ikeda, K., and Furukawa, T.: Effects of Na^+, K^+, and ouabain on microphonic potentials of the goldfish inner ear. *Jap. J. Physiol.* 21: 563-578, 1971.

Maunsbach, A. B.; Skriver, E., Deguchi, N., and Jørgensen, P. L.: Ultrastructure of Na, K-ATPase. *Acta Histochem. Cytochem.* 13: 103-112, 1980.

McKinley, D. N., and Whitney, P. L.: Particulate carbonic anhydrase in homogenates of human kidney. *Biochim. Biophys. Acta* 445: 780-790, 1976.

Meldrum, N. U., and Roughton, F. J. W.: Carbonic anhydrase. Its preparation and properties. *J. Physiol. (Lond.)* 80: 113-142, 1933.

Meyer zum Gottesberge, A.; Rauch, S., and Koburg, E.: Unterschiede in Metabolismus der einzelnen Schneckenwindungen. *Acta Otolaryng.* 59: 116-123, 1965.

Misrahy, G. A.; Hildreth, K. M., Shinabarger, E. W., and Gannon, W. J.: Electrical properties of wall of endolymphatic space of the cochlea (guinea pig). *Am. J. Physiol.* 194: 396-402, 1958.

Mizukoshi, O., and Daly, F. J.: Oxygen consumption in normal and kanamycin damaged cochleae. *Acta Otolaryng.* 64: 45-50, 1967.

Nakai, Y., and Hilding, D.: Electron microscopic studies of adenosine triphosphatase activity in the stria vascularis and spiral ligament. *Acta Otolaryng.* 62: 411-428, 1966.

Nakai, Y., and Hilding, D.: Adenosine triphosphatase distribution in the organ of Corti. *Acta Otolaryng. 64:* 477–491, 1967.

Nakai, Y., and Hilding, D.: Vestibular endolymph-producing epithelium. Electron microscopic study of the development and histochemistry of the dark cells of the crista ampullaris. *Acta Otolaryng. 66:* 120–128, 1968.

Paloheimo, S., and Thalmann, R.: Influence of "loop" diuretics upon Na^+ K^+-ATPase and adenylate cyclase of the stria vascularis. *Arch. Otorhinolaryngol. 217:* 347–359, 1977.

Pocker, Y., and Meany, J. E.: The catalytic versatility of carbonic anhydrase from erythrocytes. The enzyme-catalyzed hydration of acetaldehyde. *J. Am. Chem. Soc. 87:* 1809–1811, 1965.

Post, R. L.; Kume, S., Tobin, T., Orcutt, B., and Sen, A. K.: Flexibility of an active center in sodium-plus-potassium adenosine triphosphatase. *J. Gen. Physiol. 54:* 306s–326s, 1969.

Post, R. L., Toda, G., and Rogers, F. N.: Phosphorylation by inorganic phosphate of sodium plus potassium ion transport adenosine triphosphatase. *J. Biol. Chem. 250:* 691–701, 1975.

Prazma, J.: Active ion transport from the scala vestibuli into the scala media. *Acta Otolaryng. 67:* 631–638, 1969.

Rawls, J. A.; Wistrand, P. J., and Maren, T. H.: Effects of acid-base changes and carbonic anhydrase inhibition on pancreatic secretion. *Am. J. Physiol. 205:* 651–657, 1963.

Robinson, J. D., and Flashner, M. S.: The (Na^+ + K^+)-activated ATPase. Enzymatic and transport properties. *Biochim. Biophys. Acta 549:* 145–176, 1979.

Ross, M. D.; Ernst, S. A., and Kerr, T. P.: Possible functional roles of Na^+, K^+-ATPase in the inner ear and their relevance to Ménière's disease. *Am. J. Otolaryngol. 3:* 353–360, 1982.

Roughton, F. J. W., and Booth, V. H.: The effect of substrate concentration, pH, and other factors upon the activity of carbonic anhydrase. *Biochem. J. 40:* 319–320, 1946.

Sapirstein, V. S.; Strocchi, P., and Gilbert, J. M.: Properties and function of brain carbonic anhydrase. *Ann. N. Y. Acad. Sci. 429:* 481–493, 1984.

Schatzmann, H. J.: Herzglykoside als Hemmstoffe fur den aktiven Kalium und Natrium Transport durch die Erythrocytenmembran. *Helv. Physiol. Pharmac. Acta II:* 346–354, 1953.

Schatzmann, H. J.: Intracellular phosphate release by the Na^+, K^+ activated membrane ATPase. *Experientia (Basel) 20:* 551–552, 1964.

Schuurmans Stekhoven, F., and Bonting, S. L.: Transport adenosine triphosphatases: properties and functions. *Physiol. Rev. 61:* 1–76, 1981.

Schwartz, A.; Lindenmayer, G. E., and Allen, J. C.: The Na^+, K^+-ATPase membrane transport system: importance in cellular function. In Bronner, F., and Kleinzeller, A. (Eds.): *Current Topics in Membranes and Transport, Vol. 3.* New York, Academic Press, 1972, pp. 1–82.

Schwartz, A.; Lindenmayer, G. E., and Allen, J. C.: The sodium-potassium adenosine triphosphatase: pharmacological, physiological and biochemical aspects. *Pharmacol. Rev. 27:* 3–134, 1975.

Sellick, P. M., and Bock, G. R.: Evidence for an electrogenic potassium pump as the origin of the positive component of the endocochlear potential. *Pflügers Arch. 352:* 351–361, 1974.

Sellick, P. M., and Johnstone, B. M.: The electrophysiology of the saccule. *Pflügers Arch. 336:* 28–34, 1972.

Sellick, P. M., and Johnstone, B. M.: Differential effects of ouabain and ethacrynic acid on the labyrinthine potentials. *Pflügers Arch. 352:* 339–350, 1974.

Sellick, P. M., and Johnstone, B. M.: Production and role of inner ear fluid. *Prog. Neurobiol. 5:* 337–362, 1975.

Sellick, P. M.; Johnstone, J. R., and Johnstone, B. M.: The electrophysiology of the utricle. *Pflügers Arch. 336:* 21-27, 1972.
Sen, A. K., and Post, R. L.: Stoichiometry and localization of adenosine triphosphate-dependent sodium and potassium transport in the erythrocyte. *J. Biol. Chem. 239:* 345-352, 1964.
Siegel, G. J.; Stahl, W. L., and Swanson, P. D.: Ion transport. In Siegel, G. J.; Albers, R. W., Agranoff, B. W., and Katzman, R. (Eds.): *Basic Neurochemistry.* Boston, Little, Brown, 1981, pp. 107-143.
Silbergeld, E. K.: Erythrosin B is a specific inhibitor of high affinity ^{3}H-ouabain binding and ion transport in rat brain. *Neuropharmacol. 20:* 87-90, 1981.
Silverstein, H., and Yules, R. B.: The effect of diuretics on cochlear potentials and inner ear fluids. *Laryngoscope 81:* 873-888, 1971.
Simon, E. J.; Hilding, D. A., and Kashgarian, M: Micropuncture study of the mechanism of endolymph production in the frog. *Am. J. Physiol. 225:* 114-118, 1973.
Sjodin, R. A.: Transport of electrolytes in muscle. *J. Membr. Biol. 68:* 161-178, 1982.
Skou, J. C.: The influence of some cations on an adenosine triphosphatase from peripheral nerves. *Biochim. Biophys. Acta 23:* 394-401, 1957.
Skou, J. C.: Further investigations on a $Mg^{++}+Na^+$-activated adenosinetriphosphatase, possibly related to the active, linked transport of Na^+ and K^+ across the nerve membrane. *Biochim. Biophys. Acta 42:* 6-23, 1960.
Smith, C. A.; Davis, H., Deatherage, B. H., and Gessert, C. F.: DC potentials of the membranous labyrinth. *Am. J. Physiol. 193:* 203-206, 1958.
Smith, C. A.; Lowry, O. H., and Wu, M. L.: The electrolytes of the labyrinthine fluids. *Laryngoscope 64:* 141-153, 1954.
Smith, C.: Capillary areas of the cochlea in the guinea pig. *Laryngoscope 61:* 1073-1095, 1951.
Smith, C.: Capillary areas of the membranous labyrinth. *Ann. Otol. Rhinol. Laryngol. 63:* 435-447, 1954.
Spicer, S. S.; Sens, M. A., Hennigar, R. A., and Stoward, P. J.: Implications of the immunohistochemical localization of the carbonic anhydrase isozymes for their function in normal and pathologic cells. *Ann. N. Y. Acad. Sci. 429:* 382-397, 1984.
Sterkers, O.; Saumon, G., Tran Ba Huy, P., Ferrary, E., and Amiel, C.: Electrochemical heterogeneity of the cochlear endolymph: effect of acetazolamide. *Am. J. Physiol. 246: (Renal Fluid Electrolyte Physiol. 15):* F47-F53, 1984.
Stirling, C. E.: Radioautographic localization of sodium pump sites in rabbit intestine. *J. Cell Biol. 53:* 704-714, 1972.
Sweadner, K. J.: Two molecular forms of (Na^+, K^+)-stimulated ATPase in brain. Separation and difference in affinity for strophanthidin. *J. Biol. Chem. 254:* 6060-6067, 1979.
Swenson, E. R.: The respiratory aspects of carbonic anhydrase. *Ann. N. Y. Acad. Sci. 429:* 547-560, 1984.
Tasaki, I., and Spyropoulos, C. S.: Stria vascularis as source of endocochlear potential. *J. Neurophysiol. 22:* 149-155, 1959.
Tashian, R. E.; Plato, C. C., and Shows, T. B.: Inherited variant of erythrocyte carbonic anhydrase in Micronesians from Guam and Saipan. *Science 140:* 53-54, 1963.
Thalmann, R.: Metabolic features of auditory and vestibular systems. *Laryngoscope 81:* 1245-1260, 1971.
Thalmann, R.; Thalmann, I., and Palaheimo, S.: Noxious effects upon cochlear metabolism. *Laryngoscope 87:* 699-721, 1977.

Ullrich, K. J.: Sugar, amino acid, and Na$^+$ cotransport in the proximal tubule. *Annu. Rev. Physiol. 41:* 181-195, 1979.

Vogh, B. P., and Maren, T. H.: Sodium, chloride, and bicarbonate movement from plasma to cerebrospinal fluid in cats. *Am. J. Physiol. 228:* 673-683, 1975.

Wachstein, M., and Meisel, E.: Histochemistry of hepatic phosphatases at a physiologic pH with special reference to the demonstration of bile canaliculi. *Am. J. Clin. Path. 27:* 13-23, 1957.

Watanabe, K., and Ogawa, A.: Carbonic anhydrase activity in stria vascularis and dark cells in vestibular labyrinth. *Ann. Otol. Rhinol. Laryngol. 93:* 262-266, 1984.

Whitney, P. L., and Briggle, T. V.: Membrane-associated carbonic anhydrase purified from bovine lung. *J. Biol. Chem. 257:* 12056-12059, 1982.

Whittam, R., and Ager, M. E.: The connection between active cation transport and metabolism in erythrocytes. *Biochem. J. 97:* 214-227, 1965.

Whittam, R.: The asymmetrical stimulation of a membrane adenosinetriphosphatase in relation to active cation transport. *Biochem. J. 84:* 110-118, 1962.

Wilbur, K. M., and Anderson, N. G.: Electrometric and colorometric determination of carbonic anhydrase. *J. Biol. Chem. 176:* 147-154, 1948.

Wistrand, P. J.: Properties of membrane-bound carbonic anhydrase. *Ann. N. Y. Acad. Sci. 429:* 195-206, 1984a.

Wistrand, P. J.: The use of carbonic anhydrase inhibitors in opthalmology and clinical medicine. *Ann. N. Y. Acad. Sci. 429:* 609-619, 1984b.

Wong, V.; Barrett, C. P., Donati, E. J., Eng, L. F., and Guth, L.: Carbonic anhydrase activity in first-order sensory neurons of the rat. *J. Histochem. Cytochem. 31:* 293-300, 1983.

Wood, J. G.; Jean, D. H., Whitaker, J. N., McLaughlin, B. J., and Albers, R. W.: Immunocytochemical localization of sodium, potassium activated ATPase in knifefish brain. *J. Neurocytol. 6:* 571-581, 1977.

Yoda, A., and Yoda, S.: Interaction between ouabain and the phosphorylated intermediate of Na, K-ATPase. *Molec. Pharmacol. 22:* 700-705, 1982.

Zimmerman, T. J.; Garg, J. C., Vogh, B. P., and Maren, T. H.: The effect of acetazolamide on the movements of anions into the posterior chamber of the dog eye. *J. Pharmacol. Exp. Ther. 196:* 510-516, 1976.

Chapter 26

ORIGIN AND ELECTROCHEMICAL COMPOSITION OF ENDOLYMPH IN THE COCHLEA

OLIVIER STERKERS

I. Introduction
II. Origin of the Endolymph
  A. Review of the Literature
  B. Analysis of Experiments Involving Perilymphatic Perfusion of Tracers
  C. Analysis of Experiments Involving Intravenous Infusion of Tracers
III. Electrochemical Heterogeneity of Endolymph
  A. Longitudinal Electrochemical Gradient in Endolymph
  B. Role of K^+ Transport in the Generation of the Longitudinal Gradient
  C. Na^+, Cl^-, H^+, HCO_3^-, and the Longitudinal Gradient
IV. Active H^+ Transport and the Endolymphatic Potential
V. Summary and Conclusions
  References

I. INTRODUCTION

The membranous labyrinth, a "tight" sensory epithelium (Jahnke, 1975), separates fluids of widely differing chemical composition. The lumen is filled with potassium-rich endolymph, while the surrounding spaces are filled with perilymph, whose composition resembles that of a plasma ultrafiltrate (Bosher and Warren, 1968). In the mammalian cochlea, an electrogenic K^+ transport system, presumably located in the stria vascularis, is believed to maintain the intracellular-like composition of endolymph and to produce the +80 mV (with respect to blood) endocochlear resting potential (EP), which can be recorded from the scala media (Tasaki and Spyropoulos, 1959; Bosher and Warren, 1968; Kuijpers and Bonting, 1970b; Sellick and Bock, 1974; Sellick and Johnstone, 1975; Bosher, 1980). The Na^+, K^+-ATPase in the stria vascularis probably energizes the electrogenic K^+ pump, because the magnitude of the EP and the Na^+, K^+-ATPase activity have been found to decrease in parallel from the base to the apex of the cochlea (Kuijpers and Bonting, 1969, 1970a). These functions of the stria vascularis led to the concept that endolymph originates from plasma. However, electrolyte transport has been demonstrated recently between perilymph and endolymph by means of local perfusion of radioactive tracers into perilymph (Konishi and Hamrick, 1978; Konishi et al., 1978), thereby suggesting that perilymph should be examined as a potential precursor of endolymph. Since a well-defined model of the epithelial transport involved in the homeostasis of the endolymphatic system cannot be given at present, investigations were designed in an attempt to answer the following basic questions:

1. Does endolymph originate from perilymph or from plasma?
2. Is the longitudinal electrical gradient within the cochlear epithelium accompanied by a concentration gradient?
3. Is the EP entirely produced by K^+ transport?

II. ORIGIN OF THE ENDOLYMPH

To elucidate the origin of the endolymph in the cochlea, kinetic studies of electrolyte entry into cochlear endolymph have recently been made, using radioactive material that was either perfused into the perilymphatic spaces (Konishi and Hamrick, 1978; Konishi et al., 1978) or administered intravenously (Sterkers et al., 1982b). "Compartmental" analysis of the data provided evidence for a perilymphatic source of endolymph. However, with such analysis, the assumptions made to reach a solution need to be clearly defined. It is the purpose of this chapter to clarify the assumptions that were used in the analysis of the data by "compartmental" calculations, to be discussed shortly.

A. REVIEW OF THE LITERATURE

Previous studies aimed at the assessment of the origin of the endolymph have not been conclusive. Observations regarding the permeability of both the cochlear epithelium and the capillaries of the stria vascularis to electron-opaque macromolecular tracers, administered either locally (Duvall and Quick, 1969; Ilberg and Vosteen, 1969; Hinojosa, 1972) or intravenously (Duvall et al., 1971; Osako and Hilding, 1971; Winther, 1971; Gorgas and Jahnke, 1974), have been conflicting. Indeed, tracers were reported to be transported by cells of Reissner's membrane in both directions, suggesting exchanges between endolymph and perilymph, while strial capillaries were permeable to horseradish peroxidase, indicating a possible role of plasma in endolymph production. However, tracers other than horseradish peroxidase were found not to permeate strial capillaries (Santos-Sacchi and Marovitz, 1980).

Opposite conclusions were drawn from studies using radioactive electrolytes, administered either perilymphatically (Rauch et al., 1963), or intraperitoneally (Choo and Tabowitz, 1964, 1965). Thus, Rauch et al. (1963) proposed a perilymphatic origin of endolymph, while Choo and Tabowitz (1964, 1965) postulated a plasma source. It should be emphasized that in these studies the sodium content in endolymph samples was either reported to be abnormally high, or not even determined (Choo and Tabowitz, 1964, 1965). However, it is now well accepted that the Na^+ concentration of endolymph samples must be monitored in order to exclude samples contaminated by perilymph (Peterson et al., 1978). Moreover, these previous investigations (Rauch et al., 1963; Choo and Tabowitz, 1964, 1965) did not fulfill the conditions for a compartmental analysis of the experimental data.

B. ANALYSIS OF EXPERIMENTS INVOLVING PERILYMPHATIC PERFUSION OF TRACERS

Konishi and his colleagues (Konishi and Hamrick, 1978; Konishi et al., 1978) provided evidence that electrolytes can exchange between perilymph and endolymph. Kinetics of electrolyte (Na^+, K^+, Cl^-) entry into endolymph, using perilymphatic perfusion of radioactive tracers, indicated that perilymph may be considered to be a potential precursor of endolymph. A two-compartment model, depicted in Fig. 1a, was applied to the analysis of the experimental data ("compartmental analysis"). A simple equation allowed the calculation of the transfer rate constant values between endolymph and perilymph, as given in Table I, since

$$\ln (C_e/C_p - {^*C_e}/{^*C_p}) = \ln C_e/C_p - (J_{ep}/V_e C_e)t, \quad (1)$$

where J indicates the electrolyte flow between the two compartments and e and p denote endolymph and perilymph, respectively. The data obtained in different animals are plotted against time (Fig. 2), and the slope of the regression line (line 1 of Fig. 2) provides the transfer rate constant k_{ep}, since

$$J_{ep} = k_{ep} V_e C_e, \quad (2)$$

where V_e is the volume of the compartment studied, i.e., the endolymph. The intercept with the zero time axis supplies the term $\ln (C_e/C_p)$, i.e., the ratio of electrolyte concentrations in the two compartments under stationary conditions.

A difficulty arises, however, in Konishi's analysis from the duration of the observation, which is rather short as compared to the slowness of electrolyte exchanges between endolymph and perilymph (for discussion, see Sterkers et al., 1982a). Solutions other than that given by equation 1, including a plasma-endolymph exchange, can fit the data as well, exemplified by curve 2

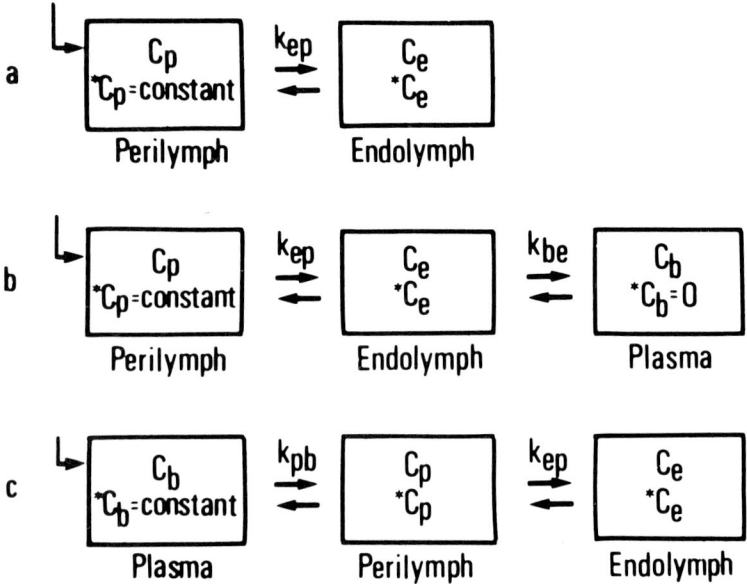

Figure 26-1. Compartmental models for exchange between endolymph, perilymph, and plasma. C is the concentration of non-radioactive material, *C is the concentration of radioactive material, and k_{ij} is the transfer rate constant from compartment j to compartment i (e, endolymph; p, perilymph; b, plasma). The right-angled arrow indicates the compartment into which the radioactive material was infused. (From Sterkers et al., 1982a.)

of Fig. 2, which follows the three-compartment model shown in Fig. 1b. The latter model is described according to the mass-balance principle by a differential equation,

$$d^*C_e/dt = k_{ep}(C_e/C_p)^*C_p - k_{pe}{}^*C_e + k_{eb}(C_e/C_b)^*C_b - k_{be}{}^*C_e. \quad (3)$$

Briefly, equation 3 indicates that the quantity of tracer in the endolymph at any given time is a function of the algebric sum of the tracer flow entering the perilymph and the tracer flow exiting toward both the perilymph and the plasma. According to Konishi's experimental conditions, $^*C_b = 0$ and $^*C_p =$ constant; thus, if we consider chloride kinetics, for which $C_e/C_p \simeq 1$, equation 3 becomes:

$$\ln(1 - {}^*C_e/{}^*C_p) = \ln\left[1 - \frac{k_{ep}}{k_{pe}+k_{be}}(1 - e^{(-k_{pe}+k_{be})t})\right] \quad (4)$$

This allowed a solution from the data of Konishi and Hamrick (1978), for example, that depicted by curve 2 of Fig. 2, where the entry of electrolyte into endolymph from plasma is estimated to be as high as one third that from perilymph.

Origin and Composition of Cochlear Endolymph

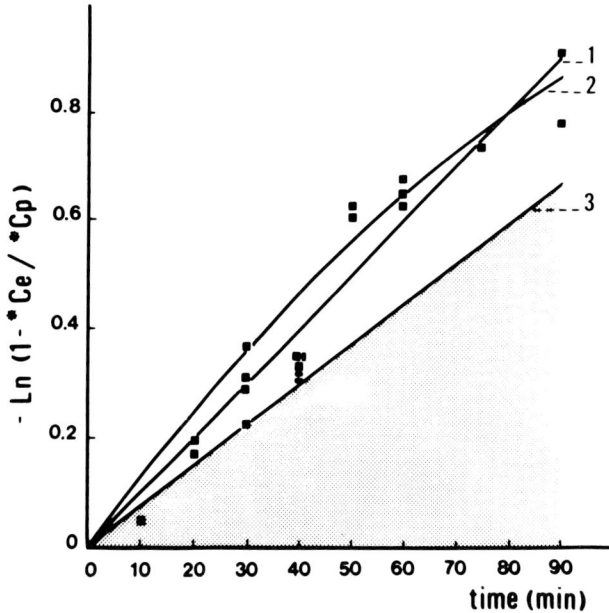

Figure 26-2. Comparative analysis of ^{36}Cl intravenous infusion (Sterkers et al., 1982b) and ^{36}Cl perilymphatic perfusion (Konishi and Hamrick, 1978). The experimental points are from the study of Konishi and Hamrick (1978) and are represented by ln (1 − *C_e/*C_p) as a function of time (*C_e = ^{36}Cl concentration in endolymph; *C_p = ^{36}Cl concentration in perilymph). Regression line 1 was calculated by the latter authors, assuming no exchange between endolymph and plasma (equation 1). Curve 2 indicates a possible fit of the Konishi and Hamrick (1978) data, assuming that endolymph chloride exchanges with both perilymph and plasma (see equation 4, text; k_{ep} = 0.013 min^{-1} and k_{be} = 0.005 min^{-1}). Line 3 is the upper, well-defined limit of the geometrical loci of all of the possible solutions drawn from our data, represented by the shaded area. The slope of line 3 is provided by our compartmental model (Fig. 26-1c), indicating the absence of chloride exchange between endolymph and plasma. Since all of the experimental data of Konishi and Hamrick (1978) lie just above the shaded area, the model tested (i.e., perilymph $\rightleftharpoons$ endolymph) is the best solution compatible with the two series of experiments. (From Sterkers et al., 1982b.)

TABLE 26-1
TRANSFER RATE CONSTANTS BETWEEN ENDOLYMPH AND PERILYMPH (k_1)
AND BETWEEN PERILYMPH AND PLASMA (k_2)

Tracer	k_1 (min^{-1})	k_2 (min^{-1})	Reference
^{42}K	0.0126	—	Konishi et al. (1978)
	0.0114	0.0201	Sterkers et al. (1982b)
^{36}Cl	0.010	—	Konishi and Hamrick (1978)
	0.0074	0.0239	Sterkers et al. (1982b)
	0.0047	0.0235	Sterkers et al. (1984)

C. ANALYSIS OF EXPERIMENTS INVOLVING INTRAVENOUS INFUSION OF TRACERS

An experimental protocol was designed to study the kinetics of K^+, Cl^-, and water entry into endolymph, perilymph, and cerebrospinal fluid (CSF) (Sterkers et al., 1982b). A plateau of plasma radioactivity was sustained for at least 6 h using a minimal amount of radioactive material, by intravenous injection of tracers (^3H_2O, ^{42}K, ^{36}Cl) into nephrectomized animals. For each inner-ear fluid sample, radioactivities were adequately counted, Na^+ and K^+ concentrations determined in 1-nl aliquots by ultramicroemission spectrometry (Morel and Lucarain, 1967), and Cl^- concentration by microelectrometric titration (Ramsay et al., 1955). The mean Na^+ concentration in the endolymph samples was 2.9 ± 0.44 (SE) mM, n = 28. Water entry into all of the fluids occurred very rapidly (Fig. 3), while K^+ and Cl^- entry were slower into perilymph and CSF, and much slower into endolymph, as shown for ^{36}Cl in Fig. 4.

A compartmental analysis was made to test the hypothesis that electro-

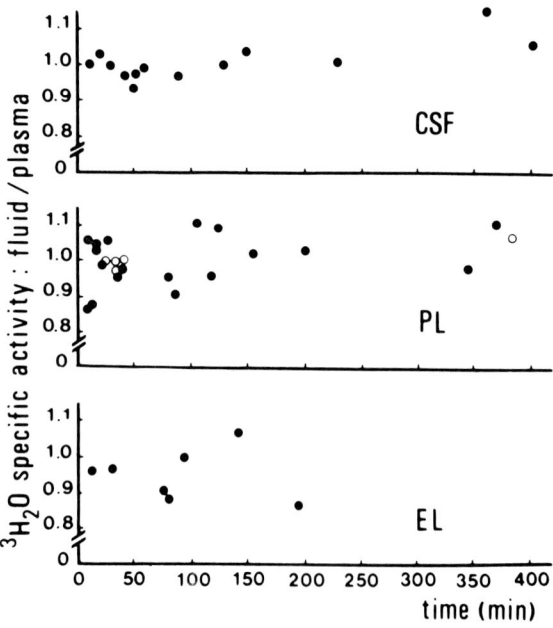

Figure 26-3. Penetration of tritiated water into cerebrospinal fluid (CSF), perilymph (PL) from scala vestibuli (○) and from scala tympani (●), and endolymph (EL), as indicated. Radioactive water entry into EL, PL, and CSF was such that in the earliest samples (EL at 12 min, PL at 10 min, and CSF at 13 min) the equilibrium with plasma was already achieved. (From Sterkers et al., 1982b.)

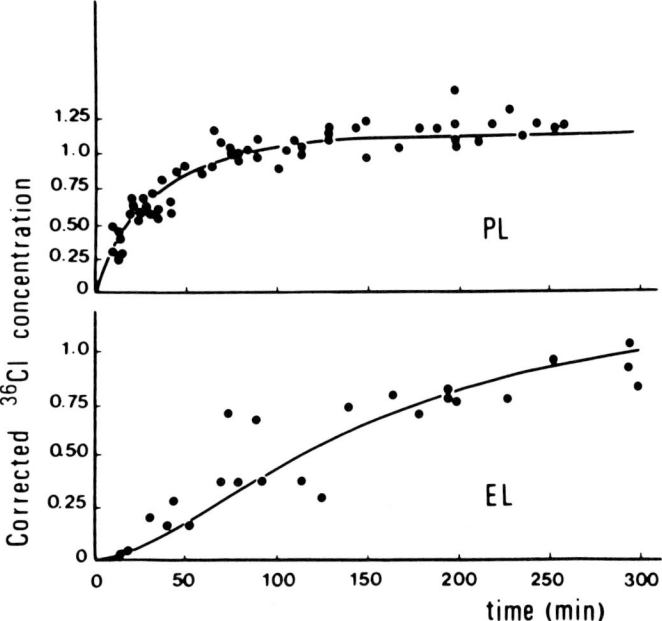

Figure 26-4. Kinetics of ^{36}Cl entry in perilymph (PL) and endolymph (EL). Best fit between curves obtained from the compartmental model (equations 5 and 6) and the experimental points for PL and EL. Transfer rate constants k_1 and k_2 were calculated from this best fit. (From Sterkers *et al.*, 1982b.)

lytes of endolymph exchange with those of perilymph and not with those of plasma. The model, depicted in Fig. 1c, is described by a two-differential-equation system. Assuming that endolymph originates only from perilymph, then the inward electrolyte flow is equal, at steady state, to the outward flow, and equation 3 simplifies to

$$d^*C_e/dt = k_1(C_e/C_p)^*C_p - k_1^*C_e, \quad (5)$$

because $J_{ep} = J_{pe}$ and $k_{ep} = k_{pe} = k_1$.

On the other hand, assuming that perilymph originates only from plasma and that no net flow occurs, the tracer kinetics into perilymph are described by

$$d^*C_p/dt = k_2(C_p/C_e)^*C_e - k_2^*C_p + k_3(C_p/C_b)^*C_b - k_3^*C_p. \quad (6)$$

The fit between the model and the experimental data, as shown for ^{36}Cl in Fig. 4, provides the transfer-rate-constant values given in Table I. The identification of the constants k_1 and k_2 between endolymph and perilymph associated with the determination of electrolyte concentrations in these two compartments yields the ratio of perilymph to endolymph volumes, because

J_{ep} is equal to J_{pe}, and according to equation 2,

$$k_1 V_e C_e = k_2 V_p C_p. \tag{7}$$

A volume ratio of 6.4 was calculated from ^{36}Cl kinetics, which is in good agreement with previous estimates for mammals (Fernandez, 1952; Maggio, 1966) and for humans (Morrison, 1981).

The conclusion that endolymph originates for perilymph rather than from plasma was reached by a comparison of the results obtained from the two series of experiments using perilymphatic (Konishi and Hamrick, 1978; Konishi et al., 1978) and intravenous (Sterkers et al., 1982b) administration of tracers. This conclusion could not have been obtained by one experiment alone. As shown in Fig. 2, the coherence between the two experiments (Konishi and Hamrick, 1978; Sterkers et al., 1982b) allows us to conclude that no substantial electrolyte exchange occurs between plasma and endolymph. This is in good agreement with the findings of Thalmann and coworkers (Wada et al., 1979; Kambayashi et al., 1982), that vascular perfusion of the cochlea with either K^+-free or metabolic-substrate-free, blood-like solutions does not alter the EP.

III. ELECTROCHEMICAL HETEROGENEITY OF ENDOLYMPH

A. LONGITUDINAL ELECTROCHEMICAL GRADIENT IN ENDOLYMPH

Since the initial report of the chemical composition of the endolymph (Smith et al., 1954), it has been well established that in the mammalian endolymphatic system the electrochemical composition of endolymph varies from one part of the labyrinth to another (Sellick and Johnstone, 1975). Furthermore, it has been observed that in the cochlea, the magnitude of the EP decreases from the base to the apex in a variety of species, including the rat (Bosher and Warren, 1971), chinchilla (Benitez et al., 1972), and guinea-pig (Misrahy et al., 1958b; Kuijpers and Bonting, 1970a; Syka et al., 1981). No investigation, however, on the ionic composition of the cochlear endolymph and of the resting potential recorded from the scala media has been made in different cochlear turns, except for the study of Syka et al. (1981). These authors found, using double-barreled micropipettes, that the K^+ activity in endolymph has a tendency to decline from the base to the apex of the guinea-pig cochlea in parallel with the EP, suggesting the presence of a longitudinal electrochemical gradient within the cochlear endolymphatic space.

This hypothesis was confirmed recently by the observation that rat cochlear endolymph is heterogeneous in electrical and electrolyte (K^+ and Cl^-) composition (Sterkers et al., 1984). The electrochemical composition of endolymph was studied at two adjacent cochlear turns (basal and middle). The resting potential measurement and the subsequent fluid sampling were begun randomly in the basal and middle turns, and it was verified that the sampling of endolymph from one turn did not alter the electrical and ionic composition in the other. As indicated in Table II, the EP and the K^+ and Cl^- concentrations in endolymph were found to be higher in the basal turn than in the middle turn, while no differences between the two turns was observed for Na^+ concentration and pH, as determined by the ^{14}C–DMO (5, 5-dimethyloxazolidine-2,4-dione) distribution method (Waddel and Butler, 1959). These results indicated the presence of a longitudinal gradient within the endolymph, involving the resting potential and K^+ and Cl^- concentrations, the values of which decline from the base to the middle turn, and presumably to the apex of the cochlea (although no attempt was made to study the composition of endolymph in the apical turn).

B. ROLE OF K^+ TRANSPORT IN THE GENERATION OF THE LONGITUDINAL GRADIENT

The longitudinal electrochemical gradient observed within the cochlear endolymph (Sterkers et al., 1984) may be largely accounted for by the variation of the transepithelial K^+ concentration along the cochlea. The EP is thought to result from transepithelial K^+ transport across the membranous labyrinth (for review, see Sellick and Johnstone, 1975). The EP appears to be composed of both a large positive potential, due to a postulated electrogenic K^+ transport system dependent on the Na^+, K^+-ATPase in the stria vascularis (Kuijpers and Bonting, 1970a, b; Sellick and Bock, 1974; Sellick and Johnstone, 1975; Bosher, 1980), and of a negative diffusion potential of about 40 mV, resulting mainly from the K^+ concentration gradient between the endolymph and perilymph (Kuijpers and Bonting, 1970b; Sellick and Johnstone, 1975; Bosher, 1979; Konishi et al., 1979).

Variation along the length of the cochlea of the activity of electrogenic K^+ transport into endolymph could account for the differences in EP and K^+ concentration that were observed between the basal and middle turns of the rat cochlea (Sterkers et al., 1984). This hypothesis is supported by the finding of Kuijpers and Bonting (1969) that the activity of the Na^+, K^+-ATPase in the stria vascularis decreased along the length of the cochlea toward the apex. Alternatively, should the electrochemical gradient within endolymph be dependent upon the transepithelial K^+ gradient, the permeability to K^+ of the cochlear epithelium would increase from apex to base. This is unlikely, because the K^+ diffusion potential is found to be more negative at the base than at the apex (Syka et al., 1981). Furthermore, the electrical resistance of the cochlear duct decreases toward the apex (Misrahy et al., 1958b).

C. NA^+, Cl^-, H^+, HCO_3^-, AND THE LONGITUDINAL GRADIENT

Since all the major ions are out of electrochemical equilibrium in the cochlear endolymph, transport systems for ions other than K^+ could be involved in the longitudinal electrochemical gradient within the endolymph.

A major contribution of the transepithelial Cl^- transport to potential generation has been ruled out by local perfusion studies (Prazma, 1969; Kuijpers and Bonting, 1970b; Sellick and Johnstone, 1975); instead, the voltage may affect Cl^- concentration in endolymph, thus accounting for the Cl^- concentration differences between the basal and middle cochlear turns (Sterkers et al., 1984). Assuming that Cl^- distribution in the endolymph is voltage-dependent (Fig. 5), an increase in EP produced by exposure to noise (Konishi et al., 1979) or a decrease of EP induced by either anoxia (Konishi and Hamrick, 1978), ouabain (Konishi and Hamrick, 1978), ethacrynic acid (Bosher et al., 1973), or acetazolamide (Sterkers et al., 1984) should be associated with a corresponding alteration of Cl^- concentration in cochlear endolymph. The absence of a Na^+ concentration difference in endolymph between two adjacent cochlear turns must be interpreted with caution, because very low Na^+ concentrations are involved. As shown in Table II, the proton (H^+) and bicarbonate (HCO_3^-) concentrations are not statistically different between the basal and the middle turns (Sterkers et al., 1984), and these ionic species do not seem to participate in the longitudinal gradient.

IV. ACTIVE H^+ TRANSPORT AND THE ENDOCOCHLEAR POTENTIAL

The pH of inner ear fluids has been evaluated in the cochlea by means of microelectrodes *in situ* (Misrahy et al., 1958a; Bosher, 1979, 1980) and the introduction of radioactive DMO (Sterkers et al., 1984). No difference was observed between the pH of endolymph and blood (Misrahy et al., 1958a; Bosher, 1979, 1980) or between the pH of endolymph, perilymph, and blood (Sterkers et al., 1984), which implies that H^+ is far from its electrochemical equilibrium in cochlear endolymph. To account for the absence of a pH gradient between the endolymph and the perilymph or blood, H^+ must be actively transported into endolymph. Indeed, a passive distribution of protons would yield a pH value higher than 8.0 in endolymph, because of the positive polarization of this fluid. However, little is known about the mechanisms involved in the acid-base balance in the cochlea.

The fact that the distributions of protons and bicarbonate ions do not

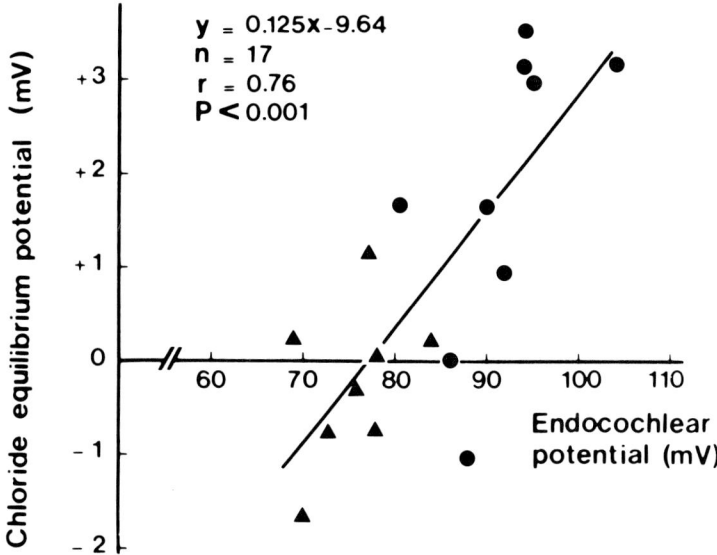

Figure 26-5. Relation between the chloride equilibrium potential, predicted by the Nernst equation, and the endocochlear potential. Each point corresponds to the chloride equilibrium potential predicted by the Nernst equation, using the Cl^- concentrations determined in EL and PL samples obtained from the same cochlea, as compared to endocochlear potential measured just before the EL sampling. Data were obtained from control rats (circles) or acetazolamide-administered rats (triangles). No change of ^{36}Cl rate of entry into endolymph was observed in the presence of acetazolamide, suggesting no variation of Cl^- permeability. (From Sterkers et al., 1984.)

TABLE 26-II
ELECTROCHEMICAL COMPOSITION OF ENDOLYMPH IN THE COCHLEA

	$EP(mV)$	$K^+(mM)$	$Na^+(mM)$	$Cl^-(mM)$	pH	$HCO_3^-(mM)$
Basal turn	96.6 ± 1.9 (5)	165.6 ± 3.0 (14)	1.74 ± 0.24 (14)	144.6 ± 2.1 (14)	7.32 ± 0.063 (4)	21.4 ± 3.3 (4)
Middle turn	87.0 ± 1.6 (6)	155.7 ± 2.5 (15)	2.05 ± 0.33 (15)	133.2 ± 1.5 (15)	7.31 ± 0.056 (6)	20.4 ± 2.2 (6)
Significance	$p < 0.005$	$p < 0.05$	ns	$p < 0.001$	ns	ns

Values are given as mean ± SEM, with the number of samples in parenthesis; ns = not significant. The above study was performed with nephrectomized animals. (From Sterkers et al., 1984.)

follow the longitudinal electrochemical gradient in the cochlear endolymph (Sterkers et al., 1984) argues against a main role of Na^+, K^+-ATPase in the transport of H^+ and HCO_3^- into endolymph. High activities of carbonic anhydrase are found in the cochlear tissues (Erulkar and Maren, 1961; Drescher, 1977; Drescher and Kerr, Chapter 25, this volume), especially in the stria vascularis and the spiral ligament. The inhibition of this enzyme by acetazolamide, administered by perilymphatic perfusion (Kuijpers and Bonting, 1970b) or by intravenous infusion (Sterkers et al., 1984), induces a 20% decrement of the EP. No alteration of the Na^+ and K^+ concentrations in perilymph or endolymph occur during the inhibition of carbonic anhydrase by acetazolamide (Mendelsohn and Mittelman, 1971; Silverstein and Yules, 1971; Sterkers et al., 1984). This finding does not support the hypothesis of Kuijpers and Bonting (1970b) that the EP decrement produced by acetazolamide administration is due to the inhibition of the strial Na^+, K^+-ATPase by acidosis. Furthermore, a participation of an active H^+ transport into endolymph for the EP generation is suggested by the fact that an increase in EP induced by hypercapnia, which is known to stimulate H^+ transport, is prevented by carbonic anhydrase inhibitors (Prazma, 1978). In addition, according to Maren et al. (1975), most of the bicarbonate may not be transported into endolymph by transepithelial transport, but generated in the stria vascularis from dissolved carbon dioxide. Native bicarbonate could then enter endolymph in exchange for chloride, since HCO_3^- and Cl^- concentrations vary inversely in the presence of acetazolamide (Sterkers et al., 1984).

V. SUMMARY AND CONCLUSIONS

The present chapter deals with the origin and the electrochemical composition of the endolymph in the cochlea. The three following conclusions can be drawn:

1. Perilymph is the precusor of the endolymph. This implies, as far as K^+ and Cl^- are concerned, that exchanges of substances occur between the two fluids across the cochlear epithelium. Electrolyte movements are rather slow between perilymph and endolymph, while water movement is found to be unrestricted.

2. Cochlear endolymph is heterogeneous in composition. Evidence for a longitudinal electrochemical gradient, involving the endocochlear resting potential and endolymphatic K^+ and Cl^- concentrations, is supported by differences in composition found between the basal and middle turns. This longitudinal gradient may be generated mainly by the transepithelial, active transport of K^+.

3. Protons are actively transported into endolymph. The inhibition of cochlear carbonic anhydrase by acetazolamide produces a partial inhibition of the endocochlear potential, which can be accounted for by a decrease of active H^+ transport. There are also grounds for a HCO_3^--Cl^- exchange process at the luminal surface of the transporting cells of the cochlear epithelium.

In conclusion, the cochlear epithelium, as other tight epithelia, allows selective movements of ions and water without obligatory coupling of the two processes. The location, as well as the mechanisms, of the transepithelial ion-transport systems in the cochlea remain to be established.

ACKNOWLEDGMENTS

This work originated from observations performed in collaboration with Prof. Claude Amiel, and Drs. Evelyne Ferrary, Georges Saumon, and Patrice Tran Ba Huy. The author acknowledges the secretarial assistance of Francoise Carlier. The research was supported in part by grants from INSERM (CRL826035), CNRS (Greco 24), and Université Paris 7 (Faculté Xavier Bichat).

REFERENCES

Benitez, L. D.; Eldredge, D. H., and Templer, J. W.: Temporary threshold shifts in chinchilla: electrophysiological correlates. *J. Acoust. Soc. Am. 52:* 1115-1123, 1972.

Bosher, S. K.: The nature of the negative endocochlear potential produced by anoxia and ethacrynic acid in the rat and guinea-pig. *J. Physiol. (Lond.) 293:* 329-345, 1979.

Bosher, S. K.: The nature of the ototoxic actions of ethacrynic acid upon the mammalian endolymph system. I. Functional aspects. *Acta Otolaryng. 89:* 407-418, 1980.

Bosher, S. K.; Smith, C., and Warren, R. L.: The effects of ethacrynic acid upon the cochlear endolymph and stria vascularis. A preliminary report. *Acta Otolaryng. 75:* 184-191, 1973.

Bosher, S. K., and Warren, R. L.: A study of the electrochemistry and osmotic relationship of the cochlear fluids in the neonatal rat at the time of the development of the endocochlear potential. *J. Physiol. (Lond.) 212:* 739-761, 1971.

Bosher, S. K., and Warren, R. L.: Observations on the electrochemistry of the cochlear endolymph of the rat: a quantitative study of its electrical potential and ionic composition as determined by means of flame spectrophotometry. *Proc. Roy. Soc. Lond. Ser. B. 171:*227-242, 1968.

Choo, Y. B., and Tabowitz, D.: The formation and flow of the cochlear fluids. I. Studies with radioactive sodium (^{22}Na). *Ann. Otol. Rhinol. Laryngol. 73:* 92-100, 1964.

Choo, Y. B., and Tabowitz, D.: The formation and flow of the cochlear fluids. II. Studies with radioactive potassium (^{42}K). *Ann. Otol. Rhinol. Laryngol. 74:* 140-145, 1965.

Drescher, D. G.: Purification of a carbonic anhydrase from the inner ear of the guinea pig. *Proc. Natl. Acad. Sci. U.S.A. 74:* 892-896, 1977.

Duvall, A. J., and Quick, C. A.: Tracers and endogeneous debris in delineating cochlear barriers and pathways. *Ann. Otol. Rhinol. Laryngol. 78:* 1041-1057, 1969.

Duvall, A. J.; Quick, C. A., and Sutherland, C. R.: Horseradish peroxidase in the lateral cochlear wall. *Arch. Otolaryngol. 93:* 304-316, 1971.

Erulkar, S. D., and Maren, T. H.: Carbonic anhydrase and the inner ear. *Nature (Lond.) 189:* 459-460, 1961.

Fernandez, C.: Dimensions of the cochlea (guinea-pig). *J. Acoust. Soc. Am. 24:* 519-523, 1952.

Gorgas, K., and Jahnke, K.: The permeability of blood vessels in the guinea pig cochlea. II. Vessels in the spiral ligament and the stria vascularis. *Anat. Embryol. 146:* 33-42, 1974.

Hinojosa, R.: Electron microscope studies of the stria vascularis and spiral ligament after ferritin injection. *Acta Otolaryng. 74:* 1-14, 1972.

Ilberg, C., and Vosteen, K. H.: Permeability of the inner ear membranes. *Acta Otolaryng. 67:* 165-170, 1969.

Jahnke, K.: The fine structure of freeze-fractured intercellular junctions in the guinea-pig inner ear. *Acta Otolaryng. Suppl. 336:* 1-40, 1975.

Kambayashi, J.; Kobayashi, T., De Mott, J. E., Marcus, N. Y., Thalmann, I., and Thalmann, R.: Effect of substrate-free vascular perfusion upon cochlear potentials and glycogen of the stria vascularis. *Hearing Res. 6:* 223-240, 1982.

Konishi, T., and Hamrick, P. E.: Ion transport in the cochlea of guinea-pig. II. Chloride transport. *Acta Otolaryng. 86:* 176-184, 1978.

Konishi, T.; Hamrick, P. E., and Walsh, P. J.: Ion transport in guinea-pig cochlea: I. Potassium and sodium transport. *Acta Otolaryng. 86:* 22-34, 1978.

Konishi, T.; Salt, A. N., and Hamrick, P. E.: Effects of exposure to noise on ion movement in guinea-pig cochlea. *Hearing Res. 1:* 325-342, 1979.

Kuijpers, W., and Bonting, S. L.: Studies on (Na^+-K^+) activated ATPase. XXIV. Localization and properties of ATPase in inner ear of the guinea pig. *Biochim. Biophys. Acta 173:* 477-485, 1969.

Kuijpers, W., and Bonting, S. L.: The cochlear potentials. I. The cochlear potentials. I. The effect of ouabain on the cochlear potentials of the guinea pig. *Pflügers Arch. 320:* 348-358, 1970a.

Kuijpers, W., and Bonting, S. L.: The cochlear potentials. II. The nature of the cochlear endolymphatic resting potential. *Pflügers Arch. 320:* 359-372, 1970b.

Maggio, E.: The humoral system of the labyrinth. *Acta Otolaryng. Suppl. 218:* 1-135, 1966.

Maren, T. H.; Swensor, E. R., and Addink, A. D.: Rates of ion movement from plasma to endolymph in the dogfish. *Ann. Otol. Rhinol. Laryngol. 84:* 847-858, 1975.

Mendelsohn, M., and Mittelman, J.: Diuretics and the cation content of guinea-pig endolymph. *Ann. Otol. Rhinol. Laryngol. 80:* 186-191, 1971.

Misrahy, G. A.; Hildreth, K. M., Clark, L. C., and Shinabarger, E. W.: Measurement of the pH of the endolymph in the cochlea of guinea-pigs. *Am. J. Physiol. 194:* 393-395, 1958a.

Misrahy, G. A.; Hildreth, K. M, Shinabarger, E. W., and Gannon, W. J.: Electrical properties of wall of endolymphatic space of the cochlea (guinea pig). *Am. J. Physiol. 154:* 396-402, 1958b.

Morel, F., and Lucarain, C.: Un spectrophotomètre à flamme pour le dosage du sodium et du potassium dans des échantillons biologiques de l'ordre de nanolitre. *J. Physiol. (Paris) 59:* 460-461, 1967.

Morrison, A. W.: Ménière's disease. *J. Roy. Soc. Med. 74:* 183-189, 1981.

Osako, S., and Hilding, D. A.: Electron microscopic studies of capillary permeability in normal and ames waltzer deaf mice. *Acta Otolaryng. 71:* 365-376, 1971.

Peterson, S. K.; Frishkopf, L. S., Lechene, C., Oman, C. M., and Weiss, T. F.: Element composition of inner ear lymphs in cats, lizards, and skates determined by electron probe microanalysis of liquid samples. *J. Comp. Physiol. 126:* 1-14, 1978.

Prazma, J.: Carbonic anhydrase in the generation of cochlear potentials. *Am. J. Physiol. 235 (Renal Fluid Electrolyte Physiol. 4):* F317-F320, 1978.

Prazma, J.: Passive ion transport through the Reissner membrane. *Acta Otolaryng. 68:* 53-61, 1969.

Ramsay, J. A.; Brown, R. H. J., and Crogham, P. C.: Electrometric titration of chloride in small volumes. *J. Exp. Biol. 32:* 822-829, 1955.

Rauch, S.; Kostlin, A., Schnieder, E. A., and Schindler, K.: Arguments for the permeability of Reissner's membrane. *Laryngoscope 73:* 135-147, 1963.

Santos-Sacchi, J., and Marovitz, W. F.: An evaluation of normal strial capillary transport using the electronopaque tracers ferritin and dextran. *Acta Otolaryng. 89:* 12-26, 1980.

Sellick, P. M., and Bock, G. R.: Evidence for an electrogenic potassium pump as the origin of the positive component of the endocochlear potential. *Pflügers Arch. 352:* 351-361, 1974.

Sellick, P. M., and Johnstone, B. M.: Production and role of inner ear fluid. *Prog. Neurobiol. 5:* 337-362, 1975.

Silverstein, H., and Yules, R. B.: The effect of diuretics on cochlear potentials and inner ear fluids. *Laryngoscope 81:* 873-888, 1971.

Smith, C. A.; Lowry, O. H., and Wu, M.: The electrolytes of the labyrinthine fluids. *Laryngoscope 64:* 141-153, 1954.

Sterkers, O.; Saumon, G., Tran Ba Huy, P., and Amiel, C.: Evidence for a perilymphatic origin of endolymph; Application to the pathophysiology of Meniere's disease. *Am. J. Otolaryngol. 3:* 367-375, 1982a.

Sterkers, O.; Saumon, G., Tran Ba Huy, P., and Amiel, C.: K, Cl, and H_2O entry in endolymph, perilymph, and cerebrospinal fluid of the rat. *Am. J. Physiol. 243: (Renal Fluid Electrolyte Physiol. 12):* F173-F180, 1982b.

Sterkers, O.; Saumon, G., Tran Bay Huy, P., Ferrary, E., and Amiel, C.: Electrochemical heterogeneity of the cochlear endolymph: effect of acetazolamide. *Am. J. Physiol. 246: (Renal Fluid Electrolyte Physiol. 15):* F47-F53, 1984.

Syka, J.; Melichar, J., and Ulehlova, L.: Longitudinal distribution of cochlear potentials and the K^+ concentration in the endolymph after acoustic trauma. *Hearing Res. 4:* 287-298, 1981.

Tasaki, I., and Spyropoulos, C. S.: Stria vascularis as source of endocochlear potential. *J. Neurophysiol. 22:* 149-155, 1959.

Wada, J.; Kambayashi, J., Marcus, D. C., and Thalmann, R.: Vascular perfusion of the cochlea: effect of potassium free and rubidium substituted media. *Arch. Otorhinolaryngol. 225:* 79-81, 1979.

Waddell, W. J., and Butler, T. C.: Calculation of intracellular pH from the distribution of 5,5-dimethyl-2,4-oxazolidine dione (DMO). Application to skeletal muscle of the dog. *J. Clin. Invest. 38:* 720-729, 1959.

Winther, F. O.: The permeability of the guinea-pig cochlear capillaries to horseradish peroxidase. *Z. Zellforsch. 114:* 193-202, 1971.

NOTE

Since this chapter was written, the cochlear endolymph was found to be hyperosmolar with respect to perilymph and plasma. Furthermore, the osmolality in endolymph decreases from the base to the apex of the cochlea in parallel with the EP and the potassium and chloride concentrations in endolymph. The active potassium transport into endolymph may account for both the internal and external osmotic gradients. [See Sterkers et al., *Am. J. Physiol.* 247 (Renal Fluid Electrolyte Physiology): F602-F606, 1984.]

Chapter 27

TRANSPORT CHARACTERISTICS OF THE BLOOD-LABYRINTH BARRIER

STEVEN K. JUHN, LEONARD P. RYBAK, AND TIMOTHY T. K. JUNG

I. Introduction
II. Transport of Glucose
III. Osmotic Relation Between Blood and Perilymph
IV. Entry of Ototoxic Drugs into the Inner-Ear Fluids
V. Biochemical Factors Involved in Inner-Ear Homeostasis
References

I. INTRODUCTION

The blood-labyrinth barrier (BLB) is thought to be an important homeostatic mechanism which protects most of the intricate biological structures contained within the labyrinth. It is well established that the maintenance of constant composition of the inner-ear fluids is essential for the functional integrity of the inner ear (Sellick and Johnstone, 1975). It is also known that the two types of inner-ear fluids, endolymph and perilymph, are distinct from each other in chemical composition and that they also differ from serum or cerebrospinal fluid (Makimoto *et al.*, 1978; Juhn *et al.*, 1981). The mechanisms which maintain these differences are complex and it is necessary to postulate barrier systems which regulate the transport of various substances into and out of the fluid-containing systems, (Juhn *et al.*, 1981).

The blood-brain barrier (BBB) offers resistance to entry of blood components into the fluids of the central nervous system. In this way it protects the brain from potentially toxic substances in the blood. It also protects the brain against wide fluctuations in systemic concentrations of neurotransmitters. The BBB modulates the entry of metabolic substrates into the brain. For example, glucose entry into this organ is almost entirely carrier-mediated (Fishman, 1964; Hochwald *et al.*, 1983). It is quite conceivable that similar barrier systems exist between the blood and the fluid-containing compartments of the inner ear. One way to characterize such barrier systems is to inject test substances by a physiological route and recover them from the inner-ear fluids. In this chapter, we will discuss some of the characteristics of the entry of various test substances into the perilymphatic space from the blood. We will also discuss the effect of antidiuretic hormone on inner-ear fluid composition, and present preliminary studies on prostaglandins ("local hormones") in inner-ear tissues and fluid (perilymph).

II. TRANSPORT OF GLUCOSE

Glucose appears to be the most important substrate for energy metabolism in many organs, including the inner ear (Kambayashi et al., 1982). The importance of glucose for normal cochlear function has been evaluated by Koide et al. (1960), who demonstrated a decrease of the cochlear microphonics in hypoglycemic cats. These authors also found that injections of glucose caused recovery of the cochlear microphonics. Juhn and Youngs (1976) demonstrated that glucose concentration in perilymph parallels glucose concentration in blood. Takeda et al. (1976) reported that the transport of ^{14}C-glucose from blood into perilymph and endolymph was reduced after injection of phlorizin, a competitive inhibitor of glucose transport. Since phlorizin is known to have a high affinity for the glucose-transport carrier, the latter authors suggested that glucose transport into inner-ear fluids may be partially carrier-mediated. Fig. 1 shows the effects of changes in serum glucose levels, after a continuous infusion, on the ratio of perilymph/serum glucose concentrations. It is interesting to observe that the ratio of perilymph/serum glucose concentration was constantly around 40%, and a slight decline was noticed when the serum glucose level exceeded 500 mg%. This may indicate that the blood-labyrinth barrier possesses rate-limiting saturation mechanisms. Similarly, Hochwald et al. (1983) demonstrated that glucose concentration in CSF was about 20% of serum glucose concentration when serum glucose in cats exceeded 11.1 mM (250 mg%).

III. OSMOTIC RELATION BETWEEN BLOOD AND PERILYMPH

Osmosis is a phenomenon of paramount importance in the movement of water and solutes through biological membranes. It accounts for fluid flux out of kidney tubules and gastrointestinal tract, into capillaries, and across cell membranes. When there is a major change in the osmolality of body fluids, this effect will be reflected in the central nervous system. For example, hyperosmolality draws water out of the brain, while hyposmolality causes an increase in brain water. Glycerol and urea have a well-documented osmotic effect of reducing cerebrospinal fluid (CSF) pressure. Angelborg and Agerup (1975) reported the reduction of intracochlear pressure, as well as of CSF pressure, after intravenous injection of glycerol. The effects of glycerol and

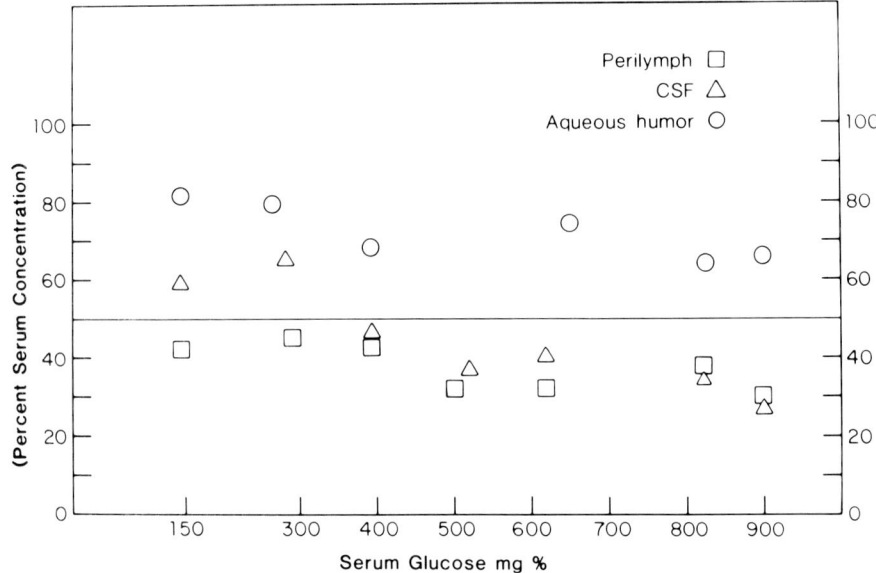

Figure 27-1. Effects of changes in serum glucose levels on the ratios of glucose concentrations between body fluids (perilymph, CSF, and aqueous humor) and serum.

urea on serum and perilymph osmolality have also been studied (Juhn *et al.* 1976, 1981), and the results indicate that perilymph osmolality parallels serum osmolality, with time lag of about 1 h (Fig. 2).

Recently, Carlborg and Farmer (1983) reported a reduction of CSF and perilymphatic hydrostatic pressure with and without blockage of the cochlear aqueduct. (In effect, hydrostatic and osmotic pressure are similar here.) It is clear from the previously-cited studies that inner-ear fluids respond to alterations in hydrostatic or osmotic pressure of the blood or CSF. However, the time lag caused by the presence of the BLB can induce a transient exit of water from the labyrinth. This might explain the transient improvement in hearing and tinnitus of Meniere's patients with endolymphatic hydrops after glycerol administration. On the other hand, drastic and sudden osmotic shifts after hemodialysis may cause transient or permanent loss of auditory function (Rizvi and Holmes, 1980).

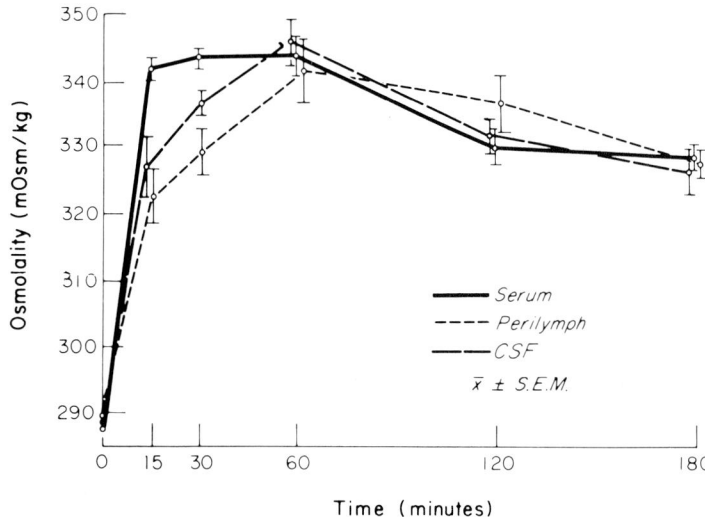

Osmolality of Body Fluids after Glycerol Injection (3.0g/kg)

Figure 27-2. Changes in osmolality of fluids from guinea pigs with time after intravenous administration of glycerol (3.0 gm/kg).

IV. ENTRY OF OTOTOXIC DRUGS INTO THE INNER-EAR FLUIDS

Ototoxic drugs are apparently transported into the inner-ear fluids after systemic injection. In studies of aminoglycoside antibiotics, it is generally observed that less than 10% of serum levels are reached in the inner-ear fluids at the peak fluid concentration of the antibiotic, and a distinct time delay exists before the peak concentration in perilymph is achieved (Juhn et al., 1981). The half-lives of aminoglycosides in guinea-pig perilymph are around 10–12 h, whereas in serum they are considerably shorter (Federspil, 1973, 1976). The pharmacokinetic behavior of aminoglycosides in endolymph appears to be similar to that in perilymph; for example, the concentrations of gentamicin in endolymph are found to be 80–90% of those in perilymph (Federspil, 1973, 1976). Based upon gentamicin infusion studies, it has been suggested that tissues of the inner ear bind the aminoglycosides and then slowly release the drugs into the innerear fluids (Tran Bay Huy et al., 1983). It thus appears that the concentrations of aminoglycosides in the inner-ear fluids reflect the levels in the tissues. It has also been reported that ethacrynic acid facilitates gentamicin entry into the endolymph of the rat, although the mechanisms for this facilitation are not clear (Tran Bay Huy et al., 1983).

In contrast to the potentiating effect of ethacrynic acid on the entry of gentamicin into endolymph, pretreatment with probenecid causes a marked reduction of furosemide entry into perilymph (Rybak et al., 1984). It is interesting to note that transport of ototoxic drugs into the innerear fluids through the blood-labyrinth barrier can be modified by simultaneous treatment with other substances which may alter the membrane permeability or alter the uptake of ototoxic drugs in the cochlear tissues.

In a furosemide (FSM) kinetics study (Juhn and Rybak, 1981), the concept of critical or threshold concentration of ototoxic drug in perilymph was suggested. By monitoring the endocochlear potential (EP) after intravenous injection of furosemide, it was observed that when the concentration of FSM in perilymph declines to a certain level, the recovery of EP takes place. Even though the time necessary to have full recovery of EP differs with the different doses administered, the level of FSM in perilymph at which recovery occurs was fairly constant, indicating that if the concentration of FSM increases above this threshold level, EP will remain depressed. It was also observed that the ratio of FSM concentration in serum to that in perilymph can be calculated, based on the kinetic study (Rybak et al., 1979; Green et al., 1981) (Fig. 3).

Salicylates are known to cause temporary hearing loss as well as tinnitus and vertigo. We recently studied the mode of entry of aspirin into perilymph, aqueous humor, and CSF after intraperitoneal injection of the drug. Salicylate concentration was measured by a colorimetric method (Trinder, 1954). It was interesting to observe that the time to reach the peak concentration of aspirin was different between aqueous humor, CSF, and perilymph (Fig. 4). The peak level was achieved in CSF 1 h after injection, wherease the peak levels in aqueous humor and perilymph were reached 2 h after injection. On the other hand, the peak level in serum occurred 0.5 h after injection. Slow clearance was observed in each of the above fluid compartments (Fig. 4). It was also observed that the amount of salicylate entering into perilymph at the peak perilymphatic concentration was about 20% of the serum level, whereas for CSF it was 25%, and for aqueous humor about 50%. The rates of transport of salicylate into perilymph and CSF were slower than into aqueous humor. In parallel studies, compound action potential thresholds were measured after aspirin injections. A threshold increase was observed for tone-burst stimuli of 8 kHz and above, but not for lower frequencies, starting from 0.5 h after injection and peaking 1 to 2 h after injection, sustained for the duration of experiment (3 h). There was no threshold change before 0.5 h (data not shown). The concept of threshold concentration can be realized by the fact that elevation of high frequency thresholds appeared when salicylate concentrations reached a certain level in perilymph (about 4 mg%).

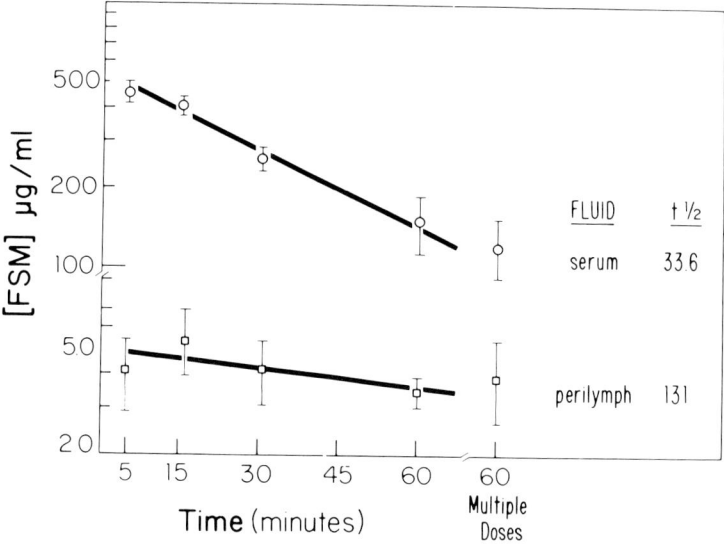

Figure 27-3. Furosemide kinetics in serum and perilymph after intravenous injection (100 mg/kg).

It should be possible to perform similar studies with aminoglycoside antibiotics so that dose-related changes of drug concentrations in inner-ear fluids can be correlated with morphological as well as physiological studies. This line of research will open the new possibility of preventing ototoxic side effects by controlling the dose. The threshold concentration can be estimated by knowing the blood concentration and kinetic behavior of the various ototoxic drugs.

V. BIOCHEMICAL FACTORS INVOLVED IN INNER-EAR HOMEOSTASIS

Several mechanisms can be postulated for the maintenance of the remarkable homeostasis of the inner ear-fluids. One of the most important mechanisms appears to be the permeability characteristics of the capillaries and membranes of the inner ear. It is generally believed that the unique structure of brain capillaries contributes to the characteristics of the blood-brain barrier (Oldendorf, 1977). The major structural differences between general and brain capillaries are that brain capillaries lack an intercellular cleft, pinocytosis, and fenestrae. The brain capillaries are

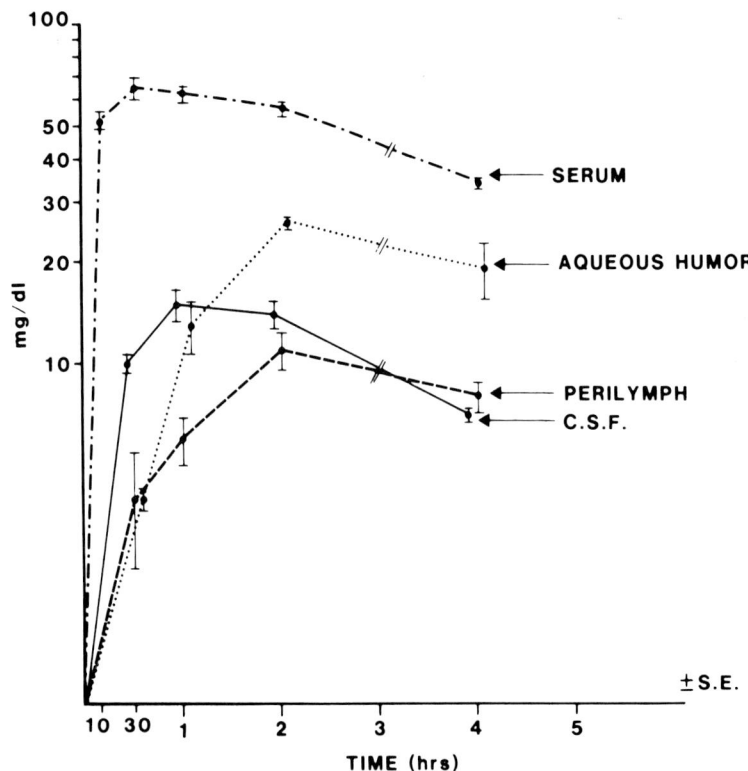

Figure 27-4. Aspirin levels in serum, perilymph, CSF, and aqueous humor of the chinchilla after intraperitoneal injection of aspirin (300 mg/kg).

composed of a single layer of endothelial cells with tight junctions.

The characteristics of capillaries in the stria vascularis and the spiral ligament have been studied by various investigators. In general, the morphology of the capillaries in the spiral ligament and stria vascularis is characterized by close endothelial cells with tight junctions (Kimura and Ota, 1974; Sakagami et al., 1982).

The entry characteristics of various substances into perilymph after systemic injections are shown in Fig. 5. When the molecular weight of the test substance is plotted against the percent of blood activity or concentration in perilymph 1 h after injection, the relative concentration or activity of the test substance in perilymph is found to be related to the inverse of its molecular weight.

An interesting question remains in how the inner ear responds to changes of internal environmental factors to maintain homeostasis. The effects of hormones on the inner-ear fluid composition have not been extensively studied. Antidiuretic hormone (ADH) is known to increase the water permea-

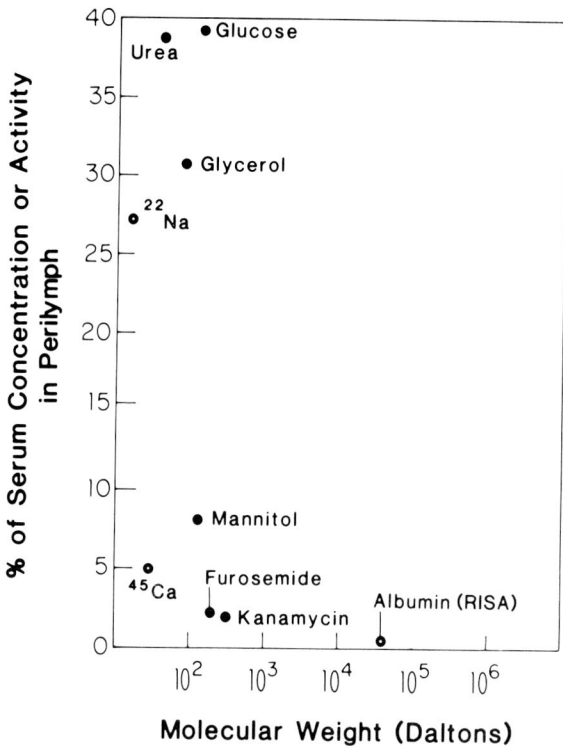

Figure 27-5. Relation between concentration of various test substances in perilymph and molecular weight of the test substances, in daltons. Substances shown were injected systemically.

bility of various epithelial cells by activating adenylate cyclase and increasing the cellular content of cyclic AMP. For example, at the luminal membrane of the renal collecting tubule, ADH decreases water excretion by changing the water permeability.

We have studied the effect of ADH on inner-ear fluid composition after intravenous infusion (Fig. 6). It was interesting to observe that the electrolyte concentration in serum and perilymph decreased after infusion of ADH. Membranes of the inner ear contain adenylate cyclase, which is coupled to membrane receptors for vasopressin and epinephrine; thus both of the latter hormones can stimulate the production of cyclic AMP (Zenner and Zenner, 1979). The results shown in Fig. 6 may be due to simply an increase of extracellular fluid volume stimulated by ADH, resulting in generalized dilution of extracellular fluid electrolytes or may be due to local effects of ADH on the inner-ear tissues (Fig. 7).

We have recently measured the concentrations of prostaglandins, "local hormones," in perilymph, and arachidonic-acid metabolites in the cochlear

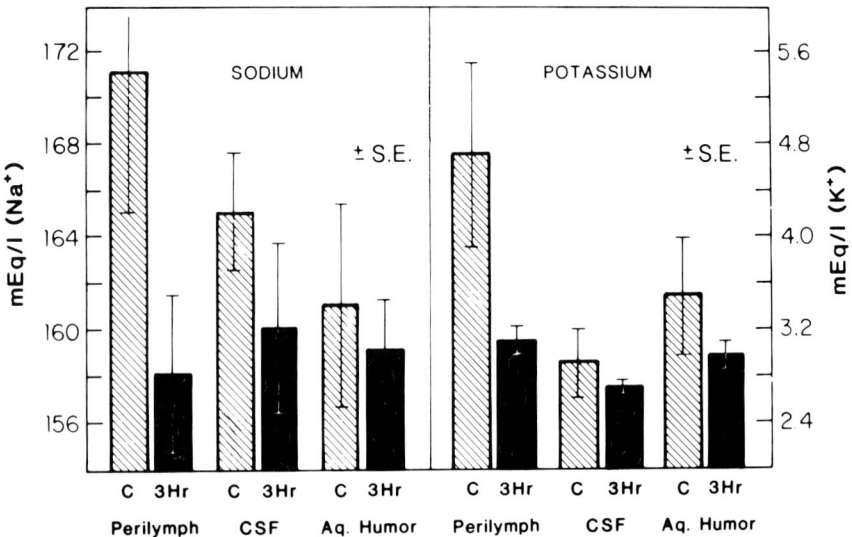

Figure 27-6. Electrolyte (Na^+, K^+) changes in perilymph, CSF, and aqueous humor after intravenous infusion of antidiuretic hormone (lysine vasopressin, 400 μU/min/kg for 3 h).

tissues. Both PGE_2 and 6-keto-PGF_1 were found in higher concentrations in perilymph than in CSF. It was also interesting to observe that prostaglandins in the perilymph decreased after treatment with indomethacin or aspirin (drugs known to inhibit prostaglandin synthesis). Arachidonic-acid metabolites found in chinchilla cochlear tissues included PGE_2, D_2, 6-keto-PGF_1, TxB_2, and HETEs (Jung and Juhn, 1984). Although the role of prostaglandins in arachidonic-acid metabolism in the inner ear is not clarified, prostaglandins may be involved in the regulation of inner-ear fluid dynamics. There seems to be a close relation between ADH and prostaglandins. ADH can stimulate adenylate cyclase activity and cyclic AMP accumulation, thereby resulting in increased water permeability, but ADH also stimulates arachidonic acid and PGE_2 synthesis, which results in decreased adenylate cyclase activity and an inhibition of water flow (Zusman, 1981). It thus appears that the ADH and prostaglandins are involved in some way in the regulation of water permeability and water flow and the maintenance of homeostasis of the inner ear (Fig. 8). Any changes that cause a shift in water balance may disrupt homeostasis and may cause inner-ear dysfunction. Further studies are necessary to clarify the role of prostaglandins in maintaining homeostasis in the inner ear.

Possible Mechanisms of ADH Induced Perilymph Alteration

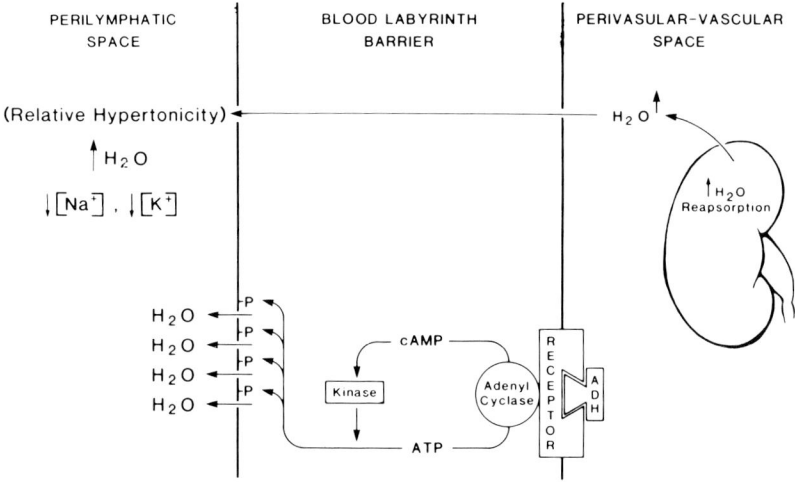

Figure 27-7. Possible mechanisms of antidiuretic-hormone-induced perilymph alteration.

POSSIBLE FACTORS INVOLVED IN WATER PERMEABILITY

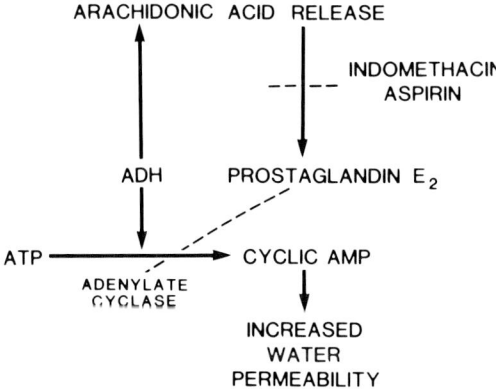

Figure 27-8. Possible factors involved in water permeability in inner-ear membranes. (Modified from Zusman, 1981.)

REFERENCES

Angelborg, C., and Agerup, B.: Glycerol effects on the perilymphatic and cerebro-spinal fluid pressure. *Acta Otolaryng.* 79: 81–82, 1975.

Carlborg, I. R., and Farmer, J. C.: Transmission of cerebrospinal fluid pressure via the cochlear aqueduct and endolymphatic sac. *Am. J. Otolaryngol.* 4: 273–282, 1983.

Federspil, P.: Pharmacokinetik des Gentamycins in Serum und Innenohrflussigkeiten. *Münch Med. Wochenschn.* 115: 82, 1973.

Federspil, P.; Schätzle, W., and Tiesler, E.: Pharmacokinetics and ototoxicity of gentamycin, tobramycin and amikacin. *J. Infect. Dis. Suppl.* 134: S200–S205, 1976.

Fishman, R. A.: Carrier transport of glucose between blood and cerebrospinal fluid. *Am. J. Physiol.* 206: 836–844, 1964.

Green, J. P.; Ryback, L. P., Mirkin, B. L., Juhn, S. K., and Morizono, T.: Pharmacologic determinants of furosemide ototoxicity in the chinchilla. *J. Pharmacol. Exp. Ther.* 216: 537–542, 1981.

Hochwald, B. M.; Gandhi, M., and Goldman, S.: Net transport of glucose from blood to cerebrospinal fluid in the cat. *Neuroscience* 10: 1035–1040, 1983.

Juhn, S. K.; Prado, S., and Pearce, J.: Osmolality changes in perilymph after systemic administration of glycerin. *Arch. Otolaryngol.* 102: 683–685, 1976.

Juhn, S. K., and Rybak, L. P.: Labyrinthine barriers and cochlear homeostasis. *Acta Otolaryng.* 91: 529–534, 1981.

Juhn, S. K.; Rybak, L. P., and Prado, S.: Nature of blood-labyrinth barrier in experimental conditions. *Ann. Otol. Rhinol. Laryngol.* 90: 135–141, 1981.

Juhn, S. K., and Youngs, J. N.: The effect on perilymph of the alteration of serum glucose on calcium concentration. *Laryngoscope* 86: 273–279, 1976.

Jung, T. T. K., and Juhn, S. K.: Prostaglandins in perilymph. *Assoc. Res. Otolaryngol. Abstr.* 7: 107, 1984.

Kambayashi, J.; Kobayashi, T., Marcus, N. Y., Demott, J. E., Thalmann, L., and Thalmann, R.: Minimal concentrations of metabolic substances capable of supporting cochlear potentials. *Hearing Res.* 7: 105–114, 1981.

Kimura, R. S., and Ota, C. Y.: Ultrastructure of the cochlear blood vessels. *Acta Otolaryng.* 77: 231–250, 1974.

Koide, Y.; Tajima, S.; Yoshida, M., and Konno, M.: Biochemical changes in the inner ear induced by insulin, in relation to the cochlear microphonics, *Ann. Otol. Rhinol. Laryngol.* 69: 1083–1097, 1960.

Makimoto, K; Takeda, T., and Silverstein, H.: Chemical composition in various compartments of inner ear fluid. *Arch. Otorhinolaryngol.* 220: 259–264, 1978.

Oldendorf, W. H.: The blood-brain barrier. *Exp. Eye Res. Suppl.* 25: 177–190, 1977.

Rizvi, S. S., and Holmes, R. A.: Hearing loss from hemodialysis. *Arch. Otolaryngol.* 106: 751–756, 1980.

Rybak, L. P.; Green, T. P., Juhn, S. K., Morizono, T., and Mirkin, B. L.: Elimination kinetics of furosemide in perilymph and serum of the chinchilla. *Acta Otolaryng.* 88: 382–387, 1979.

Rybak, L. P.; Green, T. P., Juhn, S. K., and Morizono, T.: Probenecid reduces cochlear effects and perilymph penetration of furosemide in chinchilla. *J. Pharmacol. Exp. Ther.* 230: 706–709, 1984.

Sakagami, M.; Matsunaga, T., and Hashimoto, P. H.: Fine structure and permeability of capillaries in the stria vascularis and spiral ligament on the inner ear of the guinea pig. *Cell Tiss. Res.* 226: 511–522, 1982.

Sellick, P. M., and Johnstone, B. M.: Production and role of inner ear fluid. *Prog. Neurobiol. 5:* 337–362, 1975.

Takeda, T.; Makimoto, K., Murata, K., Suchiro, S., and Iwai, H.: Effect of phlorizin on glucose transport in the inner ear. *Acta Otolaryng. 81:* 424–428, 1976.

Tran Bay Huy, P.; Manuel, C., Meulemans, A., Sterkers, O., Wassef, M., and Amiel, C.: Ethacrynic acid facilitates gentamicin entry into endolymph of the rat. *Hearing Res. 11:* 191–202, 1983.

Trinder, P.: Rapid determination of salicylate in biological fluids. *Biochem. J. 57:* 301–303, 1954.

Zenner, H. P., and Zenner, B.: Vasopressin and isoproterenol activate adenylate cyclase in the ginea pig inner ear. *Arch. Otorhinolaryngol. 222:* 275–283, 1979.

Zusman, R.: Prostaglandins and water excretion. *Ann. Rev. Med. 32:* 359–374, 1981.

Chapter 28

ANALYTICAL STUDIES OF THE ORGANIC MATERIAL OF OTOCONIAL COMPLEXES, INCLUDING ITS AMINO ACID AND CARBOHYDRATE COMPOSITION

Muriel D. Ross, Kenneth G. Pote, and Fulvio Perini

I. Introduction
II. Otoconia as Mosaics of Crystallites
III. The Organic Phase
IV. Amino Acid Composition
V. Carbohydrate Composition
VI. Glycoproteins as Potential Facilitators and Inhibitors of Otoconial Seeding and Growth
References

I. INTRODUCTION

Inner-ear sensory organs utilize an organic material, named as various "membranes" (otoconial/otolithic, cupula, tectorial), to help transmit signals to the receptor hair cells. In the case of the gravity receptors, the membranes are supplemented with deposits of inorganic mineral in the form of a single "stones," or otoliths, or as a multitude of minute particles called otoconia. The mineral adds mass to the organic membrane, increasing the sensitivity of the system to gravity and to other linear accelerations.

The inorganic deposit is typically a polymorph of calcium carbonate, usually aragonite in classes other than birds and mammals, and calcite in some other cases (Carlstrom, 1963). Vaterite is rare and normally confined to certain fishes of the *Actinopterygii* (Carlstrom, 1963). However, one of the most primitive of living vertebrates, *Lampetra*, has spherical otoconia that consist of amorphous calcium phosphate rather than calcium carbonate. Recently, we have described calcitic-type otoconial configurations in some of the gravity receptors of turtles and alligators (Ross and Pote, 1983). These modern reptiles are considered to have evolved from the same vertebrates which were the ancestors of mammals and birds. The mineral content of otoconia of ancestral vertebrates is unknown, but it is clear that the calcitic-type otoconial configuration has an evolutionary history that precedes the appearance of warm-blooded species.

The evolutionary trends in vertebrate inner-ear mineral deposits, then, were toward otoconia, mimicking the shapes of natural crystals of the mineral deposited (even though organic material was incorporated into the particles) and toward deposition of the most stable polymorph of calcium carbonate, calcite. With calcite deposition, the otoconia also lost their highly-faceted side faces, although the terminal faces retained planarity (Figs. 1 and 2) and collectively showed the three-fold symmetry characteristic of inorganically-produced calcite.

Figure 28-1. Scanning electron micrograph, illustrating highly faceted otoconia of the saccular gravity receptor of the newt (*Notophthalmus viridescens*). These otoconia contain aragonite. × 5,100.

II. OTOCONIA AS MOSAICS OF CRYSTALLITES

The mimicking of the single-crystal forms of calcium carbonate polymorphs led to the interpretation that the otoconia of numerous species were, indeed, single crystals (Carlstrom, 1963), even though they contained organic

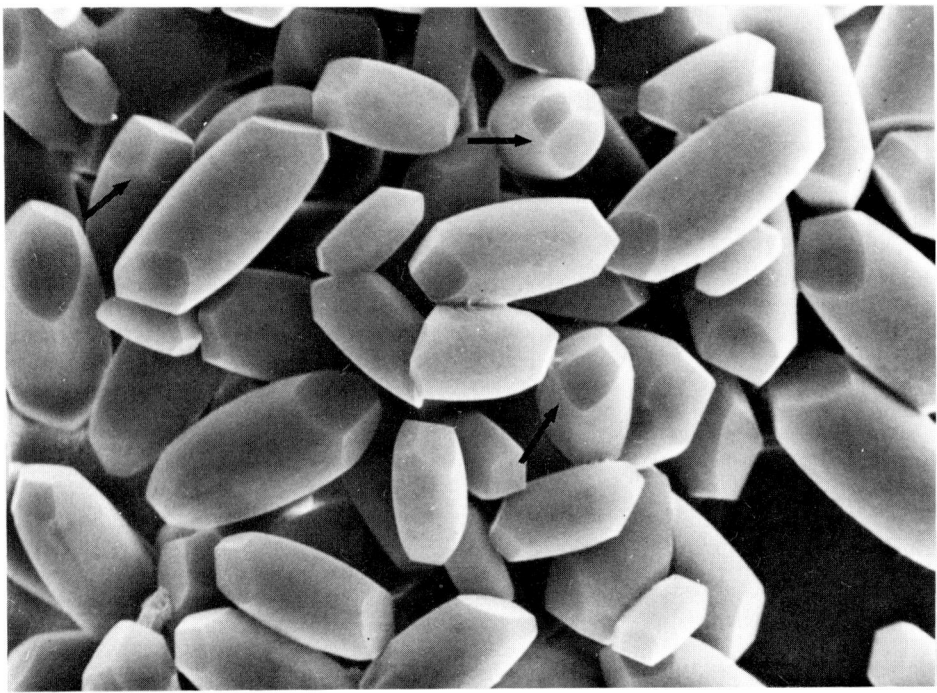

Figure 28-2. Otoconia of birds and mammals normally have rounded bodies and faceted terminals, which show the three-fold symmetry of calcite (arrows). Human saccular otoconia, × 1,400.

matter. Conclusive evidence to the contrary was obtained only recently, when ultra-high resolution, transmission electron-microscopic studies revealed that otoconia are composites of highly-ordered crystallites (Mann et al., 1983). The smallest particle observed, corresponding to a single crystal, was about 80 nm in diameter. It had some curved edges and some sharp edges, and had amorphous (assumedly organic) material in association with it (Fig. 3). Fragments of Iceland spar, in contrast, always have sharp edges, and the particles are not of a given size or shape.

The finding that rat otoconia are mosaics of crystallites (extremely small, single crystals), having a shape not duplicated in natural calcite crystals, has several implications. One of these is that the crystallites are not perfect and may become electrically charged when stressed, making the unit as a whole piezoelectric or, possibly, semiconducting. Alternatively, the organic material, or junctional regions between the crystalline and non-crystalline phases, might be piezoelectric, as has been suggested for bone (see discussion in Basset, 1968). This would have implications for how otoconia and otoliths (which have been shown to be piezoelectric by Morris and Kittleman, 1967)

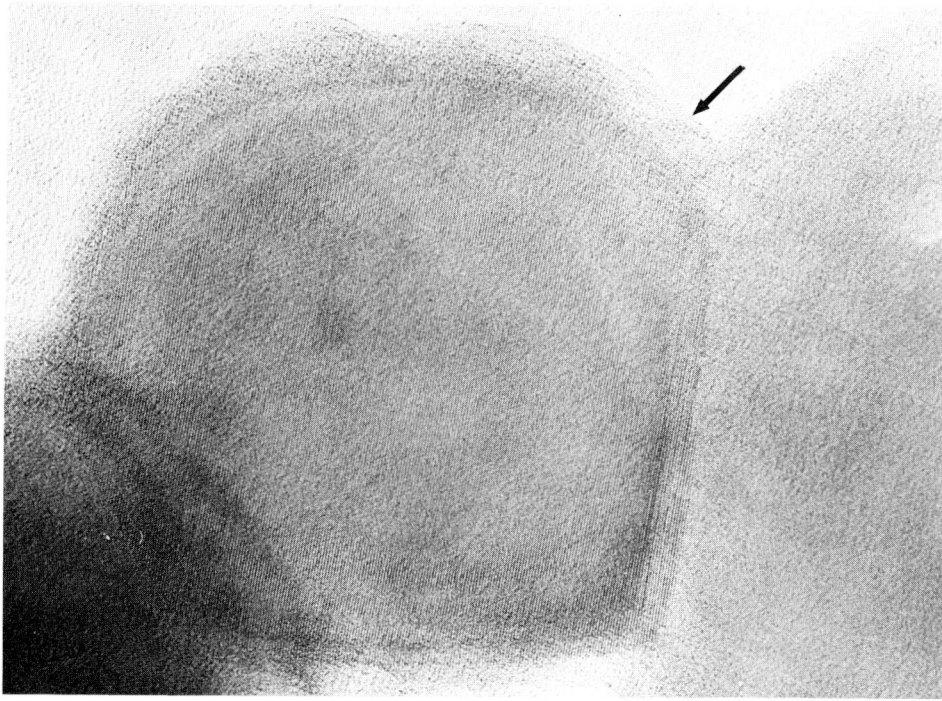

Figure 28-3. This ultra-high-resolution transmission electron micrograph features one of three crystallites of a rat saccular otoconial fragment. The straight, horizontal lines indicate the distribution of calcium atoms in the unit, which corresponds to a single crystal. The arrow points to some of the amorphous, assumedly organic, material in close association with the crystallites. × 702,500.

might function *in vivo*, namely, by altering the electric field around hair cells [see Ross (1983) for further discussion].

A further implication is that organic material might be responsible for the rapid uptake and release of Ca^{++} observed under both *in vivo* and *in vitro* conditions (Belanger, 1953, 1960; Preston *et al.*, 1975; Ross, 1979; Ross and Williams, 1979, among others). This uptake takes place within a time frame equal to that for bone mineral, while release is more rapid. The speed of calcium uptake is difficult to explain on the basis of purely inorganic processes of crystal growth.

Another, far-reaching conclusion is that organic material must be involved in the seeding and maturation of the crystallites, including the ordering of the subunits into the final otoconial shape. Earlier fetal studies demonstrated that the mineral and organic phases were arranged in identical fashion in growing rat otoconia (Salamat *et al.*, 1980). This suggested that organic material might be instrumental in directing mineral deposition to follow a predetermined pattern, which would ultimately result in an otoconium that mimicked a single crystal of calcite. The patterning was not interpreted in the sense of an organic envelope (or template) that formed first and then was filled in with mineral as suggested by others (Nakahara and Bevelander, 1979). Rather, Salamat *et al.* (1980) looked upon the depositions of organic and inorganic phases as nearly simultaneous, with organic factors both facilitating and inhibiting otoconial growth. Inhibitors are required to control otoconial size and mass in the system. The process would correspond to "matrix-mediated" mineralization, as proposed by Lowenstam (1981).

This hypothesis did not preclude a further period of maturation when the amount of organic substance would be reduced and mineral deposition increased. If this were true, then otoconial production could be a phenomenon of early development, never to be repeated if otoconial loss occurred in later life.

This issue has relevance clinically, because we know that age-related saccular otoconial loss occurs in humans and that utricular and/or saccular otoconial demineralization can be a result of disease or of drug administration (Ross *et al.*, 1976; Harada and Sugimoto, 1977; Harada *et al.*, 1978; Wright and Hubbard, 1978; Johnsson *et al.*, 1981, 1982). (Humans, like other mammals, have only utricular and saccular gravity receptors.) Sometimes, the otoconia can change configuration entirely to mimic the spherules seen in *Lampetra* (Fig. 4; see also Johnsson *et al.*, 1982) or the vaterite of *Actinopteyigii* (Wright *et al.*, 1982).

III. THE ORGANIC PHASE

The findings begin to raise certain questions. For example: What is the nature of the organic substance that predetermines a specific mineral deposition in a particular gravity receptor? What must happen to the organic material, or to the environment in which the crystals grow, to completely alter otoconial configuration in humans? What role does the organic material play in otoconial loss with age, drugs, and disease? Is the organic material constantly renewed, or does its nature change with age? What is the role of otoconial complexes in calcium regulation in the endolymph? What, if anything, is the clinical relevance of otoconial loss? To begin to answer these questions, we need to learn the composition of the organic material in otoconia of early-developmental stages as well as of adult stages, and in a variety of species.

Analytical work on otoconia is fraught with difficulty. However, the most formidable obstacle is the small amount of material available from a single animal. In the rat, for example, a pair of utricular otoconial complexes weighs approximately 14.5 μg and a pair of saccular complexes about 9.5

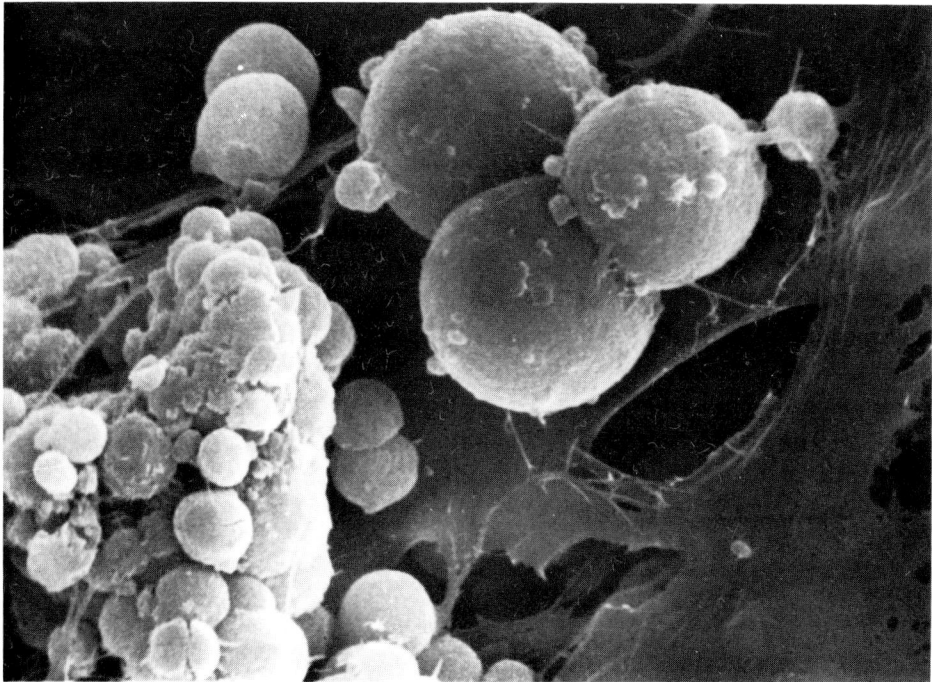

Figure 28-4. Human saccular otoconia are subject to age-, drug-, and disease-related demineralization and/or alteration in configuration. Conversion to a spherular form is one unusual example. Human saccular otoconia. × 3,900.

µg. Our current estimate is that, in the adult, only about 5–10% of this mass is represented by organic material. Thus, it is necessary to utilize very sensitive analytical methods, i.e., those sensitive in the picomole range.

The organic material is, or contains, glycoprotein (protein associated with carbohydrate), based upon its strong, positive reaction in the periodic-acid Schiff (PAS) stain (Ross, 1973). We had further indications, from work with enzymes (such as sialidase) and with PAS staining, that several proteins might be present and be variously distributed in and on the otoconia (unpublished observations). The first step then, appeared to be to separate the proteins of otoconial complexes (otoconia plus otoconial membrane) on the basis of their molecular weights. To do this, we employed microdisc gel electrophoresis, using SDS (sodium dodecyl sulfate) (Gainer, 1971). The results showed that prominent bands occurred at approximately 16,500, at 46,000, 56,000, 80,000, and >100,000 daltons (Ross et al., 1981). Minor bands were present between approximately 56,000 and 80,000 daltons. Saccular and utricular otoconial complexes yielded identical bands. Thus, the tendency for saccular otoconial demineralization with age and disease cannot be explained on the grounds of differences in organic material.

Comparison of our findings with those of Steel (1980) (Chapter 20, this volume) showed further that the organic material of otoconial complexes seems to be different in composition from that of another inner-ear membrane, the tectorial membrane. A major band occurred at approximately 140,000 daltons in the tectorial membrane, and the complete range was between about 62,000 and 180,000 molecular weight. Our material was more comparable to the organic substance present in other calcium carbonate-sequestering biosystems; for example, fish otolith organic material (otolin; Degens, 1976) and mollusc shells, which showed a range of approximately 70,000–80,000 and 20,000–80,000 daltons, respectively (Degens et al., 1969; Degens, 1976).

IV. AMINO ACID COMPOSITION

Our electrophoretic results were obtained by pooling of complexes from as few as four animals. The amount of protein actually present in any one band would be extremely small, bringing us to another degree of difficulty with respect to the analysis of these bands for amino acid and sugar content. In order to accomplish our new goals, we turned to high-performance liquid chromatography (HPLC). We began by analyzing entire otoconial patches rather than bands from gels, because our first objective was to prove that the method would work.

In our initial research, we pooled complexes from ten animals for analysis of seventeen amino acids. We used a method involving acid hydrolysis, cation-exchange chromatography, postcolumn reaction with ortho-phthalaldehyde (OPA) and a fluorescence detector theoretically sensitive in the ten-picomole range (See Ross and Pote, 1983, for details of methodology). It was necessary to wash the complexes several times in artificial endolymph (distilled, deionized water brought to K^+, Na^+, Mg^{++}, and Ca^{++} concentrations present in endolymph) to be certain that extraneous body fluids were not contaminating our samples. We learned empirically that ten washings, followed by centrifugation and removal of the supernatant, were necessary to bring the washing fluid (which was also analyzed) to near baseline levels. Scanning electron microscopy showed, to our surprise, that such relatively vigorous treatment did not break the binding material between the otoconia, although other organic material (such as that of the otoconial membrane) was not observed. The otoconia themselves had begun to demineralize slightly. The ultrastructural work led us to conclude that our analytical results truly reflect the amino-acid composition of the organic material within and between otoconia. They tell us little about the composition of the material on the outside of the otoconia, or of the inferior layers of the otoconial membrane, the so-called "honeycomb" and "meshwork" layers (Lim, 1979, 1980). These layers are represented in the washes. While analysis of supernatants from the washings did not reveal differences in amino-acid content, when compared to otoconial samples, further work is required to determine which washings included honeycomb and meshwork organic material.

Our findings were that otoconia have high quantities of acidic and much lesser amounts of basic amino acids (Ross and Pote, 1983), and that, once again, saccular and utricular otoconia were identical. The relative proportions of aspartate, glutamate, threonine and serine appear to be comparable to those found in neogastropod shells (Meenakshi et al., 1971), which are calcitic. High proportions of acidic amino acids are also present in otoliths (Degens, 1976), which are aragonitic. Thus, the organic phase of mammalian (rat) otoconia is generally comparable, with respect to its amino acid composition, to that of other biomineralized materials that contain a polymorph of calcium carbonate.

These results have been confirmed in our most recent analyses using hypochlorite prior to reaction with OPA to determine whether proline and hydroxyproline are present in organic otoconial material (Fig. 5). The initial preparative procedures were identical to those employed in our previous amino-acid analyses, except for the post-column reagent, which was composed of 0.2 M borate, pH 10.4, to which had been added 0.5 ml hypochlorite (Chlorox bleach)/L borate. The reagent was reacted with column effluent before mixing with OPA. The flow rate was 0.5 ml/min. with the reaction occuring at 50°C for 1 min. The excitation wavelength of the Kratos FS 970 fluorescence detector was 230 nm, and a 418-nm cutoff filter was used. The sensitivity of the detector was set at 450, range of 1 μA.

The chromatograms showed that a small amount of hydroxyproline and a trace of proline were present in the organic material of the otoconia. The peaks for the seventeen other, common amino acids were depressed because of the hypochlorite. Large quantities of acidic and low amounts of basic amino acids were present, as before (Fig. 5).

V. CARBOHYDRATE COMPOSITION

Chromatographic conditions were different for our carbohydrate analyses. For detection of amino sugars, the samples were hydrolyzed in 6 N HCl for 4 h at 100°C and then evaporated under nitrogen. Neutral sugars were hydrolyzed in 4 N trifluoroacetic acid for 2–3 h at 100°C. They were derivatized as follows: Up to 100 nmol total neutral sugars were placed in 50 μl of 0.2 M sodium cyanohydroboride in 1.0 M ammonium sulfate at pH 7.0. The sample was purged with nitrogen and heated at 100°C for 90 min. After cooling the sample, 50 μl 0.2 N HCl were added. After 5 min, the sample was evaporated at 40–50°C. Amino sugar and neutral sugar samples were

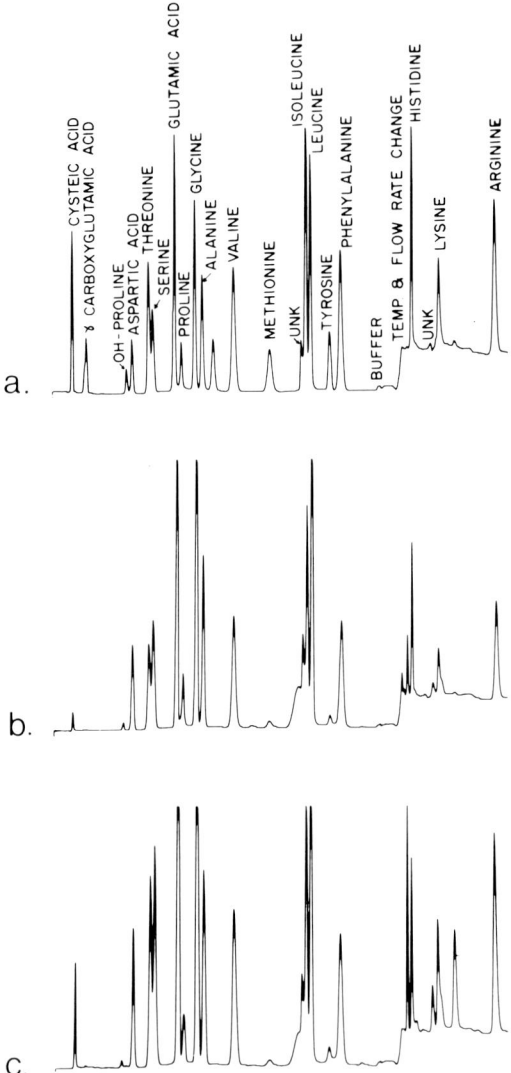

Figure 28-5. This series of chromatograms illustrates the amino acid composition of utricular (b) and saccular (c) otoconia compared to a standard (a), after hypochlorite treatment to demonstrate proline and hydroxyproline. Otoconia are high in acidic (to the left) and low in basic (to the right) amino acids, and contain small amounts of proline and hydroxyproline. All standards contained 0.5 nmol of each component.

then dissolved in 200 μl of 0.2 N sodium citrate buffer, pH 4.25. Chromatographic conditions were isocratic with 0.02 M sodium tetraborate, pH adjusted with HCl to pH 8.10. The flow rate was 0.5 ml/min at 70–75°C. A postcolumn reaction system using OPA, as used for amino acid analysis, was employed. A

Kratos FS 950 fluorescence detector was used, with a 365-nm excitation filter and a 426-mn cutoff filter. Detector sensitivity was set at 740, range 0.5 ×. The procedures outlined were modified for HPLC from Perini and Peters (1982).

Results are shown in Figs. 6 and 7. The presence of large quantities of galactose, and small quantities of mannose, along with glucosamine and galactosamine, could indicate the presence of glycoproteins containing N-linked oligosaccharides, possibly with a repeating galactose-N-acetylglucosamine sequence, as well as O-linked oligosaccharides of the type sialic acid-galactose-N-acetylgalactosamine. The presence of mucopolysaccharides cannot be ruled out, since our procedure does not detect uronic acids.

Differences in neutral sugar content were greater between utricular and saccular otoconia than were those for amino sugars (compare Figs. 6 and 7). This result was consistent and might have occurred due to some unknown experimental condition. Several other explanations are possible, however. For example, variability in the extent to which surface organic material was removed by washing may be reflected in carbohydrate content of the samples. It is also possible that saccular otoconia have been more completely mineralized than utricular otoconia. As discussed above, maturation of the crystals may result in increased mineralization at the expense of the organic substance. This is known to occur in a number of biomineralized materials (see Weiner *et al.*, 1983, for discussion). In the case of tooth enamel, where degradation and removal of organic matrix does occur during maturation, the remaining organic substances are mostly soluble acidic proteins (Glimcher *et al.*, 1977).

VI. GLYCOPROTEINS AS POTENTIAL FACILITATORS AND INHIBITORS OF OTOCONIAL SEEDING AND GROWTH

It is exciting to note that the soluble, organic fraction implicated in biomineralization in many other living organisms is heterogeneous but the major constituents are proteins rich in acidic amino acids and, in molluscs, include sulfated, high-molecularweight glycoproteins (reviewed by Krampitz *et al.*, 1983). Experimental results are consistent with the notion that the soluble protein occurs at the site(s) of initial crystal seeding and growth (Crenshaw, 1982). It has been postulated that the glycoprotein component might act as a template for crystal growth, with calcium bound to the aspartic acid residues. The residues would have to be ordered to correspond to some multiple of the calcium ion spacing present in the crystal, to serve as a nucleation site. Another line of evidence indicates that when soluble matrix material is in solution, it inhibits spontaneous calcium carbonate precipitation from supersaturated solutions (Wheeler *et al.*, 1981).

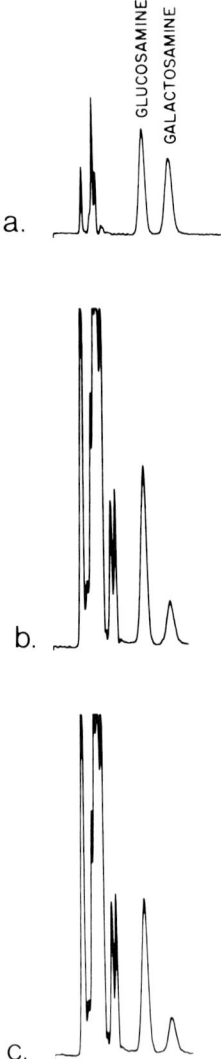

Figure 28-6. This series of chromatograms illustrates the amino sugar content of utricular (b) and saccular (c) otoconia, compared to a standard (a). Utricular and saccular otoconia contain relatively large amounts of galactosamine and glucosamine, and in about the same proportions.

An attractive hypothesis that has grown out of the research on soluble matrix is that it may act both as a nucleator and as an inhibitor. That is, it might be selectively absorbed to calcium sites on a growing crystal, blocking further crystal growth. However, its free acidic residues will attract calcium ions, then carbonate ions, to nucleate a new crystal by ionotropy (Crenshaw and Ristedt, 1976). At a specific point in the crystal growth, soluble matrix

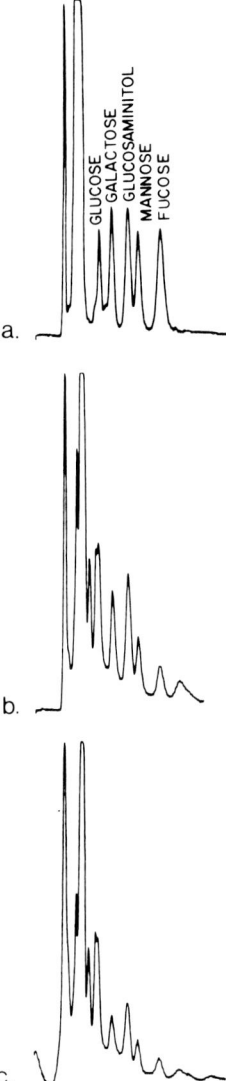

Figure 28 7. The neutral sugar content of otoconia is illustrated here. Saccular otoconia (c) consistently showed lesser amounts of these sugars than did utricular otoconia (b), when compared to the standard (a). The differences were slightly greater than those observed for amino sugars.

will again be adsorbed, blocking further growth of that crystal. Such a process could explain the nearly simultaneous deposition of organic and inorganic phases that seems to take place in developing otoconia (Salamat *et al.*, 1980), with crystallite size and orientations predetermined in part by configuration of the glycoprotein macromolecule. Limitation of total otoconial mass, as well as of the size of the inidividual otoconia, must depend upon complex local environmental factors and upon the presence of soluble matrix, adsorbed or in solution. It seems possible that our difficulty in reaching near-baseline levels of amino acids and sugars in our washes is due to the presence of soluble matrix, both present in endolymph and bound loosely to the otoconial "shell".

Our research into the composition of the organic material of otoconial complexes is still in its early stages. It seems possible, however, that the secrets of the composition of the organic material will yield to the powerfully sensitive analytical methods now at hand, and to others still emerging.

ACKNOWLEDGMENTS

This research was supported largely by NASA grant NSG 9047. Research on ultra-high-resolution microscopy of otoconia was carried out at the University of Oxford, England, with the aid of a Fogarty Senior International Fellowship (NIH 1 F06-TW00589) to Muriel Ross.

REFERENCES

Bassett, A. L.: Biologic significance of piezoelectricity. *Calc. Tiss. Res.* 1: 252–272, 1968.

Belanger, L. F.: Autoradiographic visualization of *in vitro* exchange in teeth, bones, and other tissues, under various conditions. *J. Dent. Res.* 32: 168–176, 1953.

Belanger, L. F.: Development, structure and composition of the otolithic organs of the rat. In Soggnaes, R. F. (Ed.): *Calcification in Biological Systems.* Washington, D.C., American Association for the Advancement of Science, 1960, pp. 151–162.

Carlstrom, D.: A crystallographic study of vertebrate otoliths. *Biol. Bull.* 125: 441–463, 1963.

Crenshaw, M. A.: Mechanisms of normal biological mineralization of calcium carbonates. In Nancollas, G. H. (Ed.): *Biological Mineralization and Demineralization.* Berlin-Heidelberg-New York, Springer, 1982, pp. 243–257.

Crenshaw, M. A., and Ristedt, H.: The histochemical localization of reactive groups in the septal nacre from Nautilus pompilius L. In Watabe, N., and Wilbur, K. M. (Eds.): *Mechanisms of Mineralization in the Invertebrates and Plants.* Columbia, University of South Carolina Press, 1976, pp. 335–367.

Degens, E. T.: Molecular mechanisms of carbonate, phosphate, and silica deposition in the living cell. *Topics Curr. Chem.* 64: 1–112, 1976.

Degens, E. T.; Deuser, W. G., and Haedrich, R. L.: Molecular structure and composition of fish otoliths. *Marine Biol.* 2: 105–113, 1969.

Gainer, H.: Microdisc electrophoresis in sodium dodecyl sulfate. An application to the study of protein synthesis in individual identified neurons. *Anal. Biochem.* 44: 515–589, 1971.

Glimcher, M. J.; Brickley-Parson, D., and Levine, P. T.: Studies of enamel proteins during maturation. *Calcif. Tiss. Res.* 24: 259–270, 1977.

Harada, Y., and Sugimoto, Y.: Metabolic disorder of otoconia after streptomycin intoxication. *Acta Otolaryngol. (Stockh.)* 84: 65–71, 1977.

Harada, Y.; Graham, M. D., Pulec, J. L., and House, W. F.: Human otoconia in surgical specimens. *Arch. Otolaryngol. 104:* 371-375, 1978.

Johnsson, L. -G.; Rouse, R. C., Hawkins, J. E., Jr., Kingsley, T. C., and Wright, C. G.: Hereditary deafness with hydrops and anomalous calcium phosphate deposits. *Am. J. Otolaryngol. 2:* 245-298, 1981.

Johnsson, L. -G.: Rouse, R. C., Wright, C. G., Henry, P. J., and Hawkins, J. E., Jr.: Pathology of neuroepithelial suprastructures of the human inner ear. *Am. J. Otolaryngol. 3:* 77-90, 1982.

Krampitz, G.; Drolshagen, H., Hausle, J., and Hof-Irmscher, K.: Organic matrices of mollusc shells. In Westbroek, P., and de Jong, E. W. (Eds.): *Biomineralization and Biological Metal Accumulation.* Dordrecht, Holland, D. Reidel Publishing Company, 1983, pp. 231-247.

Lim, D. J.: Fine morphology of the otoconial membrane and its relationship to the sensory epithelium. In Johari, O. (Ed.): *SEM/1979/III.* AMF O'Hare, IL, SEM, Inc., 1979, pp. 929-938.

Lim, D. J.: Morphogenesis and malformation of otoconia: A review. In *Birth Defects: Original Article Series, Vol. XVI, No. 4.* New York, March of Dimes Birth Defects Foundation, 1980, pp. 111-146.

Lowenstam, H. A.: Minerals formed by organisms. *Science. 211:* 1126-1131, 1981.

Mann, S.; Parker, S. B., Ross, M. D., Skarnulis, A. J., and Williams, R. J. P.: The ultrastructure of the calcium carbonate balance organs of the inner ear: an ultra-high resolution electron microscopy study. *Proc. R. Soc. (Lond.) B: 219:* 415-424, 1983.

Meenakshi, V. R.; Hare, P. E., and Wilbur, K. M.: Amino acids of the organic matrix of neogastropod shells. *Comp. Biochem. Physiol. 40:* 1937-1043, 1971.

Morris, R. W., and Kittleman, L. R.: Piezoelectric properties of otoliths. *Science, 158:* 368-370, 1967.

Nakahara, H., and Bevelander, G. An electron microscope study of crystal calcium carbonate formation in the mouse otolith. *Anat. Rec. 193:* 233-241, 1978.

Perini, F., and Peters, B.: Fluorometric analysis of amino sugars and derivatized neutral sugars. *Anal. Biochem. 123:* 357-363, 1982.

Preston, R. E.; Johnsson, L. G., Hill, J. H., and Schacht, J.: Incorporation of radioactive calcium into otolithic membranes and middle ear ossicles of the gerbil. *Acta Otolaryngol. (Stockh.) 80:* 269-275, 1975.

Ross, M. D.: Some histochemical and ultrastructural evidence that the otolithic membrane is a glycoprotein with ion exchange capacity. *Anat. Rec. 175:* 429-430, 1973.

Ross, M. D.: Calcium ion uptake and exchange in otoconia. *Adv. Otol. Rhinol. Laryngol. 25:* 26-33, 1979.

Ross, M. D.: Gravity and cells of gravity receptors in mammals. *Adv. Space Res. 3:* 179-190, 1983.

Ross, M. D.; Johnsson, L. -G., Peacor, D. R., and Allard, L.: Observations on normal and degenerating human otoconia. *Ann. Otol. Rhinol. Larnygol. 85:* 310-326, 1976.

Ross, M. D., and Pote, K. G.: Some properties of otoconia. *Phil. Trans. R. Soc. (Lond.) B. 304:* 445-452, 1983.

Ross, M. D.; Pote, K. G., Rarey, K. E., and Verma, L. M.: Microdisc gel electrophoresis in sodium dodecyl sulfate of organic material from rat otoconial complexes. *Ann. N. Y. Acad. Sci. 374:* 808-819, 1981.

Ross, M. D., and Williams, T.: Otoconial complexes as ion reservoirs in endolymph. *Physiologist (Wash.) 22:* S63-S64, 1979.

Salamat, M. S.; Ross, M. D., and Peacor, D. R.: Otoconial formation in the fetal rat. *Ann. Otol. Rhinol. Laryngol. 89:* 229-238, 1980.

Steel, K.: The proteins of normal and abnormal tectorial membranes. *Acta Otolaryngol. (Stockh.) 89:* 27-32, 1980.

Weiner, S., and Traub, W.: Organic matrix in calcified exoskeletons. In Webroek, P., and de Jong, E. W. (Eds.): *Biomineralization and Biological Metal Accumulation.* Dordrecht, Holland, D. Reidel, 1983, pp. 205-224.

Wheeler, A. P.; George, J. W., and Evans, C. A.: Control of calcium carbonate nucleation and crystal growth by soluble matrix of oyster shell. *Science 212:* 1397-1398, 1981.

Wright, C. G., and Hubbard, D. G.: Observations of otoconial membranes from human infants. *Acta Otolaryngol. (Stockh.) 86:* 184-194, 1978.

Wright, C. G.; Rouse, R. C., Johnsson, L. G., Weinberg, A. G., and Hubbard, D. G.: Vaterite otoconia in two classes of otoconial membrane dysplasia. *Ann. Otol. Rhinol. Laryngol. 91:* 193-199. 1982.

INDEX

A

Acetazolamide, 457, 460, 484
Acetylcholine, 42, 69, 104-109, 163-167, 187, 227-241, 245, 247
Acetylcholinesterase, 164-167, 169, 228-229
Acid-base balance, cochlear, 482-484
Acoustic emissions, 313
Acoustic information processing, 198, 218-221
Acousticolateralis systems, 46-48
Acoustic stria, 168
Actin, 281-303, 310-312, 317-322, 332
Actin, bend, 293-302
Actin filament bundles, crossbridges in, 283-291, 296-303
Actin, filament packing, 282-291
Actin, filaments, 281-303
Actin filaments, crossover points of, 282-291
Alanine, 262, 265, 268
Alpha-actinin, 324, 331-333
D-alpha-aminoadipate, 111-113, 188-191, 201, 205-208, 210, 213
D-alpha-aminosuberate, 188-189, 191
Amino acid analysis, 50-51
Amino acid neurotransmitters, 125-136, 198-218
Amino acids, 259-260
Amino acids, changes with age, 250-251
Amino acids, excitatory, 42, 108, 110, 187-191, 199-214
Amino acids, inhibitory, 108-109, 191-194, 214-221
Amino acids, label localization, 262-263
2-amino-5-phosphonovaleric acid, 188-189, 201, 205-206
Amino sugars, 507-510
Ampulla, 461-462
Anoxia, 31
Anteroventral cochlear nucleus, 167, 169, 171, 173-174, 193, 265-266

Antidiuretic hormone, 494-497
Arginine, 262, 265
Aspartate aminotransferase, 11, 131-133
D-aspartic acid, 129-130, 142-157
D-aspartic acid, high-affinity uptake, 148
D-aspartic acid, release, 142, 148-153
L-aspartic acid, 13, 22-23, 45, 47, 59, 64, 109-110, 126-138, 147, 187-191, 201-208, 245-247, 250-251, 260-262, 265-266, 272-273
ATP, 425, 439
Atropine, 111, 178, 230
Attention, 414-415
Audiogenic seizures, 249-254
Auditory development, 415-418
Auditory nerve, 134-136, 153-155, 167, 186, 336-349
Auditory nerve synapses, 199, 211
Auditory stimulation, 393-396
Autoradiography, 258-274, 442
Axonal transport, 337-340

B

Baclofen, 208-214
Barium, 423, 430-433
Basolateral membranes of strial marginal cells, 442-443, 445-446, 452-454
Beta-alanine, 58, 59, 61, 192
Bicarbonate, 454-455
Bicuculline, 111, 113, 192-193, 201, 210, 215
Blood-brain barrier, 488, 493-494
Blood-labyrinth barrier, 488-489, 494
Bone resorption, 366-368, 375, 377-379, 381-384
Brain slice, fresh, 259-260, 271-272
Brainstem auditory neurons, 198-221

C

Calcium, 106, 115, 363

Calcium-dependent neurotransmitter release, 129
Carbachol, 105–107
Carbonic anhydrase, cochlear, 455, 457–463
Carbonic anhydrase, cytochemistry, 459–461
Carbonic anhydrase, function, 463–464
Carbonic anhydrase, inner-ear localization, 461–462
Carbonic anhydrase, isoenzymes, 456
Carbonic anhydrase, tertiary structure, 456–457
Carbonic anhydrase, vestibular, 455, 457, 461–462
Catecholamines, 245, 247
Centrifugal pathways to cochlear nucleus, 165, 168, 170–171, 248
Cerebellum, 171–172, 250
Cerebrospinal fluid, primary-amine composition, 52, 54
Cholesteatoma, 374, 376–377, 380, 382
Choline acetyltransferase, 11, 12, 16–17, 164, 166–175
Choline acetyltransferase, vestibular, 90–94
Cholinergic agonists, 235–237
Cholinergic antagonists, 236–240
Cholinergic enzymes, 164–166
Cholinergic neurotransmission, 12, 17, 90–95, 163, 166
Cholinergic receptors, 176–178, 229–231
Cochlea, 6, 10, 15, 109–110, 118–119, 175, 178, 245–247, 281, 303, 423, 473
Cochlear ablation, 127
Cochlear nucleus, 131–134, 143–144, 164, 169, 193–194, 202, 247–248, 410
Cochlear pathology, 397
Co-containment, 12, 19
Collagenase, 368–369
Collagenase immunocytochemistry, 370, 372–374
Compartmental analysis of cochlear fluids, 475–476
Curare, 111

D

Dark-cell region, 428–431
Dark cells, 445, 448, 450, 461
Decremental response, 34–36
Deoxyglucose, 391–394, 401–418
Deoxyglucose autoradiography, 401–402
Depletion, transmitter, 34
Displacement potentials, 363
Donnan equilibrium, 361–363
Dopamine, 42, 109
Dorsal cochlear nucleus, 166–167, 173, 261–265

Dorsal cochlear nucleus, deep layer, 173
Dual pump model, 427

E

Efferent neurons, cochlear, 12–15, 17–23, 112
Efferent neurotransmitter, 104, 117
Eighth nerve, 393, 405–407
Electrochemistry of endolymph, 423–425
Electrolyte transport, 481
Electron microscopic autoradiography, 260
Endocochlear potential, 482–484
Endolymph, 423–425, 448–451, 473–484
Endolymphatic DC potentials, 423–424, 452
Endolymphatic perfusion, 424–425
Endolymph, chemical gradient, 480–483
Endolymph, electrical gradient, 480–483
Energy metabolism, 389–391
Enkephalin, 246
Enkephalin-like immunoreactivity, 11–15, 18–19
Epinephrine, 109
Epithelial debris, middle-ear, 377
Erythrocyte membrane proteins, 318, 327–329
Experimental epilepsy, 249

F

Facial motor root, 168
Fibroblasts, 377
Fimbrin, 312, 314, 326, 328
Fine tuning of hair cells, 302–307
Fixation of cochlea, 7–9
Forward masking, 34
Frequency and stereociliary length, 303–307
Furosemide, 492–493
Fusiform soma layer, 166–167, 173

G

GABA, 12, 42, 69–70, 72–77, 108–109, 192–193, 201, 208–212, 214, 218, 245, 248, 262–269
GABA-like high-performance liquid chromatographic component, 20, 59, 64
GABA receptors, 86–91
GABA, possible role as acousticolateralis transmitter, 63–64
Gamma-D-glutamylglycine, 188
Gas-liquid chromatography, 359
Gerbil, 429–431, 434

Glucose and endocochlear potential, 426
Glucose metabolism, 401–402
Glucose-6-phosphatase, 391–393
Glucose transport, 489–490
Glucose utilization, 392–393
L-glutamic acid, 13, 22, 46–47, 107–109, 115, 126–131, 147, 187, 189, 201–203, 245–247, 250, 262–266
Glutamic acid decarboxylase, 11–12, 20, 248
Glutamic acid decarboxylase, vestibular, 87–89
Glutamic acid diethyl ester, 111–112, 188–191, 201, 205
Glutamic acid, possible role as hair-cell transmitter, 62–63, 118–119
Glutaminase, 11–15, 131–133
Glutamine, 262, 265
Glycerol, 489–491
Glycine, 109, 191–193, 201, 214–218, 250, 252, 262–274
Glycoproteins, 336, 338, 341, 359, 509–512
Granular regions of cochlear nucleus, 166–167
Granulation tissue, middle-ear, 367, 370, 374, 378
Granule cells of cochlear nucleus, 144, 154–155
Guinea pig, 155, 432

H

HA-966, 188
Hair cell contraction, 24, 331–332
Hair cell environment, 363
Hair cells, type I, 81
Hair cells, type II, 81
Hexokinase, 391
High-performance liquid chromatography, 250, 506, 508, 510–512
High-performance liquid chromatography, methods, 50–52
Histamine, 109
Hydroxyproline, 507

I

Immunoblotting, 329–330
Immunocytochemistry, 8, 12–15
Immunofluorescence, 6, 13
Immunohistochemistry, 248
Immunoperoxidase, 6–7, 14
Incremental response, 34–36
Indomethacin, 378

Inferior colliculus, 168, 227, 248–249, 394, 410–411
Inhibition, neural, 415–416, 418
Inhibitory agents, 108–109, 191–193
Inner ear, chicken, 83–86
Inner-ear fluids, 473–484
Inner-ear fluids, homeostasis, 493–496
Inner-ear neurochemical ontogeny, 92–94, 96
Inner hair cells, 9, 12, 15, 394–395, 406–407
Inner spiral bundle, 14, 22
Interstitial nucleus, 169
Ionic conductances, inner-ear, 428–433
Ionic permeability, inner-ear, 426–432
Ion substitution, inner-ear, 430–432
Ion transport, inner-ear, 426–432, 449–454

K

Kainic acid, 110, 175, 188, 192
Keratin, 371

L

Labyrinthine spontaneous nystagmus, 76–77
Labyrinthine vertigo, 76–77
Lateral lemniscus, nuclei of, 415
Lateral line, 102–103, 117–119
Lateral superior olivary nucleus, 165–169, 215–218, 268–271
Lesions, 170
Leucine enkephalin, 60
Light microscopic autoradiography, 260
Limulus, 299
Lizard cochlea, 289–290, 303
Loop diuretics, 425
Low temperature and neurotransmission, 31
Lucifer Yellow, 37–38

M

Macrophages, 375–377
Magnesium, 106, 115, 188
Marginal cells, strial, 442, 446, 461
Mechanoreception, 81–82
Medial geniculate nucleus, 410
Medial nucleus of trapezoid body, 410
Medial superior olive, 266–268
Methionine enkephalin, 60

Microiontophoresis, 68, 72, 185, 192, 199, 202–203, 234–241
Molecular layer of cochlear nucleus, 166, 173
Monoclonal antibody, 23–25
Multiple release sites model, 36–38
Muscarinic receptors, 177–178
Muscimol, 193
Myosin, 311–312, 317–318, 322–323

N

Na^+, K^+-ATPase, 436–454
Nernst equation, 361
Neurotransmission, GABAergic, 85–90
Neurotransmitters, 13, 20, 43–44, 69, 76, 102–119, 164
Neurotransmitters, criteria for, 83, 85, 95
Neurotransmitter, release, 129, 134
Neurotransmitters, synthesis, 131–134
Neurotransmitters, uptake, 129
Neutral sugars, 507, 509
Nicotinic receptors, 177
N-methyl-D-aspartate, 110, 187, 189, 192, 201, 203, 206–210
Noise damage, 292–293, 389, 396
Non-prepotential units, 186, 189–190
Norepinephrine, 42, 109
Nucleus magnocellularis, 190–191

O

Off-supression, 34–35
Olivocochlear bundle, 165–168, 176, 246
Olivocochlear fibers, degenerated, 10
Organ of Corti, 9, 12, 17–22, 389, 393, 403, 406
Ornithine, 262, 265
Ortho-phthalaldehyde, 51, 250
Osmotic pressure, 362–363
Osmotic relation between blood and perilymph, 489–491
Osteoclasts, 375, 377
Otitis media, 367, 374–375
Otoconia, 500–513
Otoconia, amino acid composition, 506–507
Otoconia, carbohydrates in, 507–509
Otoconia, inorganic phase, 501–504
Otoconia, organic phase, 504–506
Ouabain, 425, 431, 443, 445, 451
Outer hair cells, 12–15, 406–408
Outer spiral bundle, 14

P

PAS reaction, 352
Peptides, 246–247
Perilymph, 473–475
Perilymphatic perfusion, 430–431
Perilymph, from sound-exposed animals, 56–62
Perilymph, primary amine composition, 52–56
Picrotoxin, 73–74, 76, 192
2,3-cis-piperidine dicarboxylate, 188–191
Polyacrylamide gel electrophoresis, 340, 345
Polyelectrolytes, 362–363
Positron emission tomography, 397
Posteroventral cochlear nucleus, 169
Potassium, 448–449, 453
Potassium-induced transmitter release, 117
Potassium secretion, 423–425
Prepotential units, 186, 190
Pressure, middle-ear, 377–378
Primary afferent transmitter of hair cells, 42–48, 117–119, 245–246
Primary vestibular afferents, 70–75
Probabilistic model of transmitter release, 36
Proline, 269, 272, 507
Prostaglandins, 379–380, 495–497
Protein synthesis, 338
Putative neurotransmitters, 259–275

Q

Quisqualic acid, 110, 187, 192

R

Radioactive tracers in cochlear fluids, 475–480
Release parameters, 37
Replenishment of transmitter, 34

S

Saccular macula, 70–72, 76
Saccule, 448, 464
Salicylates, 492, 494
Serotonin, 42, 46, 109, 245
S1 fiber, 31, 37–38
Single pump model, 427
Single unit responses, 234

Sodium, 449, 454
Sodium absorption, inner ear, 423, 426
Sodium-dodecyl-sulfate polyacrylamide gel electrophoresis, 356–357
Spectrin, 328–329, 333
Sperm, 299–301
Spinal trigeminal tract, 168, 171
Spiral ganglion cells, 13, 145, 154, 245, 344–347, 403, 406
Spiral ligament, 403, 405, 408, 417–418, 459, 461, 464
Spiral prominence, 459
Stain histochemistry, 164–166
Statoacoustic system, 80
Stereocilia, 281–307, 319, 322
Stereociliary stiffness, 310, 312–315
Stria vascularis, 393, 395, 403, 406, 425–426, 440–442, 461
Strychnine, 192, 201, 217–218
Substance P, 246
Subtectorial fluid, 363
Superior olivary complex, 165–175, 199, 215–221
Synaptic delay, 31
Synaptic proteins, 346
Synaptic terminal populations, 259
Synaptic vesicles, 36

T

Taurine, 17, 21, 117–119, 192–193, 250, 253–254, 262, 265, 269
Tectorial membrane, amino-acid composition, 354–355
Tectorial membrane, carbohydrates, 356–359, 361, 363
Tectorial membrane, charged matrix, 361–363
Tectorial membrane, elasticity, 355
Tectorial membrane, fibrils, 352–353
Tectorial membrane, histochemistry, 352–354

Tectorial membrane, physical properties, 359
Tectorial membrane, proteins, 354–357
Tetrodotoxin, 32–33
Tonotopic organization, 410–412, 418
Transepithelial transport, cochlear, 481–482
Transfer rate constant, cochlear electrolyte flow, 475
Transmitter-specific labeling of neurons, 154–155
Transport of ototoxic drugs, 491–494
Transport of substances into perilymph, 494–495
Trapezoid body, 165, 168
Tropomyosin, 312, 318, 324
Tubulin, 312–313
Tunnel crossing fibers, 14, 16, 18
Tunnel spiral bundle, 14, 23

U

Utricle, 428–431

V

Vascular perfusion, 431–432
Ventral nucleus of trapezoid body, 168, 170, 175
Vestibular labyrinthine organs, 80–82, 428–431
Vestibular sensory transmission, 82–83
Video-enhanced contrast microscopy, 20, 24–25
Villin, 327
Vinculin, 324

W

Waltzing guinea pig, 346–347, 430
Water permeability, regulation of, 494–497

X

Xenopus laevis, 102–105
X-ray microphobe analysis, 359

RAYMOND H. FOGLER LIBRARY
DATE DUE

BOOKS ARE SUBJECT TO
RECALL AFTER TWO W